VOLUME FORTY NINE

ADVANCES IN CHEMICAL ENGINEERING

Thermochemical Process Engineering

Edited by

KEVIN M. VAN GEEM
Laboratory for Chemical Technology
Ghent University

ELSEVIER

AMSTERDAM • BOSTON • HEIDELBERG • LONDON
NEW YORK • OXFORD • PARIS • SAN DIEGO
SAN FRANCISCO • SINGAPORE • SYDNEY • TOKYO
Academic Press is an imprint of Elsevier

Academic Press is an imprint of Elsevier
50 Hampshire Street, 5th Floor, Cambridge, MA 02139, United States
525 B Street, Suite 1800, San Diego, CA 92101-4495, United States
The Boulevard, Langford Lane, Kidlington, Oxford OX5 1GB, United Kingdom
125 London Wall, London, EC2Y 5AS, United Kingdom

First edition 2016

ISBN: 978-0-12-809777-9
ISSN: 0065-2377

Publisher: Zoe Kruze
Acquisition Editor: Poppy Garraway
Editorial Project Manager: Shellie Bryant
Production Project Manager: Vignesh Tamil
Cover Designer: Greg Harris

Typeset by SPi Global, India

CONTENTS

CONTRIBUTORS

L.J. Broadbelt
Northwestern University, Evanston, IL, United States

B. Cuenot
CFD Combustion Team, CERFACS, Toulouse cedex, France

W. Du
Key Laboratory of Advanced Control and Optimization for Chemical Processes of Ministry of Education, East China University of Science and Technology, Shanghai, China

T. Faravelli
Politecnico di Milano, Materiali e Ingegneria Chimica "Giulio Natta", Milano, Italy

G. Hu
Key Laboratory of Advanced Control and Optimization for Chemical Processes of Ministry of Education, East China University of Science and Technology, Shanghai, China

F. Manenti
Politecnico di Milano, Materiali e Ingegneria Chimica "Giulio Natta", Milano, Italy

F. Qian
Key Laboratory of Advanced Control and Optimization for Chemical Processes of Ministry of Education, East China University of Science and Technology, Shanghai, China

E. Ranzi
Politecnico di Milano, Materiali e Ingegneria Chimica "Giulio Natta", Milano, Italy

R. Vinu
Indian Institute of Technology Madras, Chennai, India

Y. Zhang
Key Laboratory of Advanced Control and Optimization for Chemical Processes of Ministry of Education, East China University of Science and Technology, Shanghai, China

X. Zhou
Northwestern University, Evanston, IL, United States

PREFACE

Thermochemical processes are and will be the most important chemical process affecting our daily lives in the coming decades. They are the key processes that provide us energy and the major base chemicals. This volume of *Advances in Chemical Engineering* provides a perspective on Best Practices, Recent Advances, and Future Challenges by the World experts in the respective fields of thermochemical reaction engineering. The focus hereby is not only on the classically applied fossil feedstocks, such as coal, natural gas, and oil but also on alternative feedstocks such as plastic waste and biomass. It is expected that the energy landscape will change substantially in the coming decades, with a gradual shift toward the use of renewable feedstocks and what is currently being considered waste. The transition to a circular economy, an industrial system in which products and materials are maintained at their highest value at all times, is more than ever needed. Waste and resource use should be minimized, and resources are to be kept within the economy to be reused. Chemical kinetic models are extremely powerful and valuable for this purpose. More and more public policy and business decisions are made on the basis of kinetic model predictions. For example, the Montreal Protocol, which imposed a worldwide ban on certain halocarbons, was based on a fundamental knowledge of the ozone layer problem established by kinetic modeling. In the chemical industry, kinetic models are widely applied, e.g., to simulate steam cracking, refining, or vinyl chloride production. However, for the majority of technologically important chemical processes, including combustion, pyrolysis, and oxidation of heteroatomic mixtures, complete detailed kinetic models are not yet available. This is because constructing a reliable model remains very difficult and time consuming. Moreover, these models typically contain thousands of reactions, involving hundreds of intermediates, while only a small fraction of the reaction rate coefficients have been determined experimentally. Moreover, it is usually impossible to measure the concentrations of all the kinetically significant chemical species. Numerically solving these large systems of differential equations in a reasonable time also remains a challenge, in particular when these models need to be implemented in computational fluid dynamics codes. All these challenges are discussed in the four chapters of this volume, and guidelines are provided to resolve even the most difficult ones.

In Chapter 1, Ranzi and coworkers discuss the newest developments in the field of the detailed kinetic modeling of pyrolysis, gasification, and combustion of solid fuels. One of the main challenges there lies in the characterization and representation of solid fuels. The latter determines the level of detail that can be accounted for in modeling, and the authors demonstrate that with proper understanding of the chemistry, it is possible to model even the most complex systems can be modeled accurately and fast if proper lumping procedures are applied during development and validation of the overall mathematical model. The authors demonstrate the power of this technique for a range of feedstocks starting from coal, biomass, and municipal solid waste. In Chapter 2, Vinu and Broadbelt go deeper in this topic focusing on pyrolysis of polyolefins. In the last century, we have rapidly moved to a society that consumes large amounts of disposable plastic products. Although plastic has a wonderful array of properties that have made it ideal for many of these applications, the fact that nature needs 100 or thousands of years to break it down is a huge problem. Pyrolysis is one of the most promising routes to partially recycle plastic waste. However, we need to improve chemical understanding of plastic waste conversion for the production of key chemicals (short and long olefins, aromatics, oxygenates) before this becomes a commercially viable technology. The presence of food and biomass in combination with plastic waste makes it even more challenging.

In Chapter 3, Du and Qian focus on a different pillar of sustainable chemical production, i.e., doing more with less. It is clear that fossil feedstocks will be the primary feedstock for the chemical industry and in particular for base chemical production. However, although steam cracking and ethylene dichloride production are mature technologies, still substantial improvements are possible. As demonstrated by the authors, multiscale modeling is the key tool for this purpose. A combination of computational fluid dynamics and detailed (kinetic) models is the key tool to make progress. This is further illustrated in Chapter 4 where Cuenot et al. illustrate the capabilities of Large Eddy Simulations (LES) for modeling turbulent combustion. These computational extremely demanding simulations are the current state of the art for designing internal combustion engines, gas turbines, and rocket engines. It is expected that LES methodology is now mature enough to be applied to any kind of turbulent reacting flow and in particular in the field of chemical processing.

Prof. Kevin M. Van Geem
Laboratory for Chemical Technology
Ghent University

CHAPTER ONE

Pyrolysis, Gasification, and Combustion of Solid Fuels

E. Ranzi[1], T. Faravelli, F. Manenti
Politecnico di Milano, Materiali e Ingegneria Chimica "Giulio Natta", Milano, Italy
[1]Corresponding author: e-mail address: eliseo.ranzi@polimi.it

Contents

Abstract

The aim of this chapter is to discuss and summarize the research activities done at Politecnico di Milano in the field of the detailed kinetic modeling of pyrolysis, gasification, and combustion of solid fuels. Different critical steps are involved in this multicomponent, multiphase, and multiscale problem. The first complexity relies in

Advances in Chemical Engineering, Volume 49
ISSN 0065-2377
http://dx.doi.org/10.1016/bs.ache.2016.09.001

the characterization of the solid fuels and their pyrolysis and devolatilization process. Detailed kinetic mechanisms, both in the solid and gas phase, involve a large number of species and reactions, which make the computations expensive and strongly reduce model applicability. For this reason, they need to be reduced and simplified, while still maintaining their description capability. Therefore, chemical lumping procedures are extensively applied to allow the development and validation of the overall mathematical model. Whereas the composition of plastics is usually well defined, coals, biomasses, and MSW (municipal solid waste) are typical fuels with a large composition variability and they require a characterization in terms of a few reference components. Multistep kinetic mechanisms with a lumped characterization of gas, tar, and residue are discussed, for the different solid fuels. Successive or secondary gas-phase reactions involve gas and tar components released during the devolatilization phase, while heterogeneous gasification or combustion reactions further modify the solid residue. Finally, the mathematical modeling of solid fuel gasification or combustion requires a comprehensive description of the coupled transport and kinetic processes, both at the particle and at the reactor scale. Several examples illustrate the capabilities and limitations of this model.

NOMENCLATURE

$\hat{C}$ specific heat
Da Darcy tensor
Da Darcy scalar in radial direction
F Forchheimer tensor
g gravitational acceleration
h heat exchange coefficient
$\hat{h}$ gas mass enthalpy
I identity matrix
j gas diffusive flux
k_c convective mass exchange coefficient
k_R rate constant
$\dot{m}$ mass flow rate
NC number of species
NL number of layers
N_p number of particles
NS number of shells
p pressure
Py pyrolysis number
$\dot{Q}_R$ reaction heat
q conductive fluxes according the Fourier's law
q_{rad} radiative heat
R radius
S surface
T temperature
t time
Th Thiele number

u velocity
u* relative velocity
V volume

GREEK SYMBOLS

ε solid porosity
μ dynamic viscosity
ρ gas
ω mass fraction
$\dot{\Omega}_k$ net formation rate of species *k*

SUPERSCRIPTS

(*I*) interface
bulk region outside the particle
G gas phase
S solid phase

SUBSCRIPTS

k species
j shell
p particle

1. INTRODUCTION

Pyrolysis is the thermal treatment of solid fuels in the absence of oxygen and producing a liquid fuel (Bridgwater, 2012) or a gas stream, mainly constituted by H_2 and CO, together with some CH_4 and CO_2. Syngas can be used either as raw material for the synthesis of methanol and liquid fuels (Olah, 2005) or as fuel for the generation of electric power. Gasification is the partial oxidation of solid fuels with steam and air and has several potential benefits over traditional combustion, mainly related to the possibility of combining temperature and equivalence ratio to obtain an appropriate syngas (Arena, 2012). BTL (biomass to liquids), CTL (coal to liquids), and IGCC (integrated gasification combined cycle) are emerging technologies based on solid fuel gasification (Leckner, 2015). Fig. 1 schematically shows how the solid fuel particles entering the hot region of a gasifier or a combustion device are affected by the chemistry at least at three different

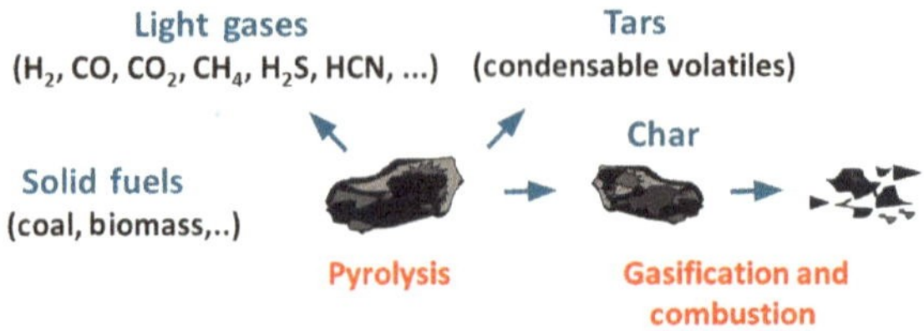

Fig. 1 Thermal treatment of solid fuels. Pyrolysis, gasification, and combustion.

levels: pyrolysis or devolatilization of the solid fuel, heterogeneous reactions of residual char, and successive gas-phase reactions of volatile products. Thus, pyrolysis is the common initial step also in the gasification and combustion processes and it accounts for the primary release of volatile products. Gas, condensable hydrocarbons and oxygenated species (tars), and residual char are always produced from the solid fuel pyrolysis, but their nature and relative amount significantly vary as a function both of the solid fuel and of the process operating conditions.

A comprehensive mathematical modeling of the thermal degradation of solid fuels is a very difficult and challenging problem, as its complexity occurs at several levels:

- Multicomponent problem: solid fuels are usually complex mixtures of organic and inorganic components and they require a proper characterization.
- Multiphase problem: the solid materials first react in a condensed (metaplast) phase and result in the formation of a solid (char or biochar), a liquid (tars), and a gas phase. Secondary reactions of released gas and tar species often take place in a multicomponent gas phase. Successive gasification and combustion processes of solid residue involves heterogeneous gas–solid reactions.
- Multiscale problem: the transport phenomena in the gas and solid phase and between the solid and gas phases need to be considered at both the particle and the reactor scale in order to characterize their relative role with respect of chemical kinetics.

Fig. 2 offers a schematic representation of these features, whose complexity is enhanced by the structure of this coupled and comprehensive approach. The development of these models is challenging because of the complexity of the solid fuel as well as the multiphase and multiscale nature of the conversion process (Mettler et al., 2012).

Modeling thermochemical processes of solid fuels is clearly a multiscale problem. Fig. 3 schematically shows the multiscale nature of a countercurrent gasifier, from the angstroms of the molecular scale to the meters of

VOLUME FORTY NINE

ADVANCES IN
CHEMICAL ENGINEERING
Thermochemical Process Engineering

ADVANCES IN CHEMICAL ENGINEERING

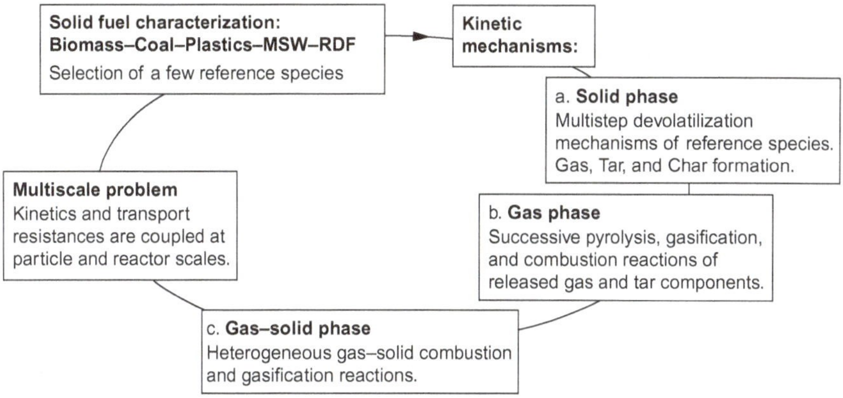

Fig. 2 Schematic representation of the thermal decomposition of solid fuels as a multicomponent, multiphase, and multiscale problem. *After Barker Hemings E: Detailed kinetic models for the thermal conversion of biomass, PhD Thesis, Politecnico di Milano, Italy, January 2012.*

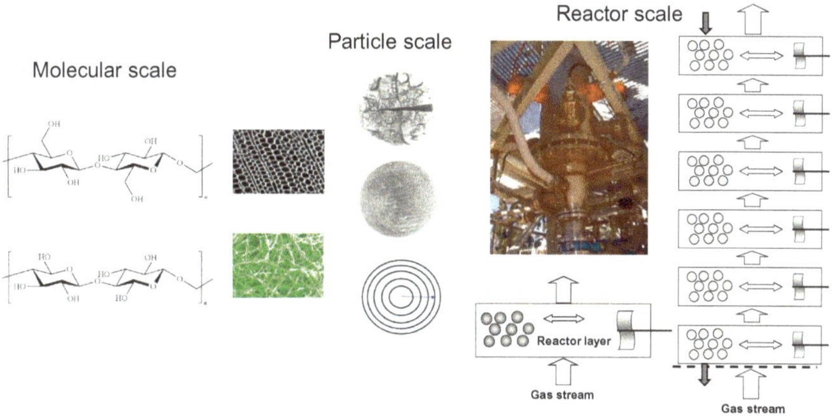

Fig. 3 Multiscale nature of a countercurrent gasifier of solid fuels.

the gasifier reactor. A similar variety is observed with respect to the timescale, which moves from the hours of the residence time of solid fuels in the combustor or gasifier reactors to the very short lifetimes of the propagating radicals involved in the pyrolysis and oxidation reactions. Finally, the multiscale mathematical modeling of thermochemical units of solid fuels requires to combine complex chemical mechanisms with transport phenomena, both at the particle and at the reactor scale. Fig. 3 also shows the complexity of the problem in terms of the nonideal and anisotropic nature of the solid particles, with possible fractures and comminutions along the decomposition process. Moreover, thermo and transport properties of the solid residue also vary with the conversion progress. This complexity strongly demands well-balanced efforts in the development of the mathematical model of combustor and gasifier units. Thus, strong simplifications need to be applied both to the kinetic mechanisms (lumping) and to the description level of mass, momentum, and energy balance equations.

This chapter updates and summarizes the research activities done at Politecnico di Milano in the field of the mathematical modeling of pyrolysis, gasification, and oxidation of solid fuels. The multistep kinetic mechanisms here discussed are an extension of the previous ones already presented by Marongiu et al. (2007) for plastics, by Ranzi et al. (2008) for biomass, by Sommariva et al. (2010) for coal, and by Cuoci et al. (2009) for waste and refuse-derived fuels. One of the peculiarities of these models lies in their ability to provide detailed information on the composition of gas and tar released as well as of solid residue. The kinetic models also involve the heterogeneous char gasification and combustion reactions, as well as the secondary gas-phase reactions of the plenty of species released during the fuel pyrolysis. This very large kinetic mechanism of pyrolysis and combustion of hydrocarbon and oxygenated species takes advantage of a well-consolidated experience, both in pyrolysis (Dente et al., 1979) and in combustion processes (Ranzi et al., 1994a). Meanwhile, by saving the previous agreement, all these kinetic models are progressively modified in order to continuously account for new available experimental data and theoretical findings.

After this general introduction, the chapter is structured as follows. Section 2 describes the characterization and the kinetic mechanisms of solid fuel pyrolysis. Namely, plastics, biomass, coal, and refuse-derived fuels are first characterized by means of a limited number of reference components. Then, their pyrolysis products are simply obtained by a linear combination of char, tar, and gas products released by the individual reference components. Attention is also devoted to the release of N and S components. Sections 3 and 4 discuss the heterogeneous reactions of char gasification and combustion and the secondary gas-phase reactions of volatile species released by solid fuel devolatilization. Section 5 presents mass and energy balances at the particle scale, together with three different application examples, in order to emphasize the effect of the coupling of reaction kinetics with mass and heat transfer resistances. The first and second examples relate to thick particles with the temperature overshooting of the center of particles (Corbetta et al., 2014) and the possibility to have combustion or gasification regimes depending on the residence times of the solid fuels. Attention is also given to the multiplicity of steady-state solutions and the importance of the start-up procedures. Finally, the third example refers to the fast pyrolysis of biomass and the crucial role of gas residence time to maximize bio-oil production (Calonaci et al., 2010). Section 6 presents the mass and energy balances at the reactor scale, with particular attention to the numerical methods and to the structure of the Jacobian matrix of resulting algebraic-differential system. Here two application examples are discussed and they refer to a traveling grate combustor (Ranzi et al., 2011) and to a countercurrent gasifier (Corbetta et al., 2015). The enhancing effect of sulfur components in syngas production is particularly discussed. Finally, some conclusions are drawn in Section 7. Again, it is important to highlight that the main goal of this chapter is to provide an overall view of our recent research activities on the modeling of solid fuels pyrolysis, gasification, and combustion. More than the direct comparisons with experimental data, the aim of the quoted application examples, partially discussed also in previous papers (Ranzi et al., 2014), is to show the possibilities as well as the limitations of the adopted lumped approach.

2. SOLID FUEL CHARACTERIZATION AND MULTISTEP PYROLYSIS MODEL

The characterization of plastics, biomass, coal, and waste is sequentially discussed in this section, along with the corresponding devolatilization models. Only a few reference components, together with their multistep pyrolysis models, allow to describe the solid-phase decomposition reactions of a very wide range of solid fuels. Aiming at a unifying approach, the van Krevelen diagram is extensively used to characterize all these different solid fuels (van Krevelen, 1950).

2.1 Plastics

Polyethylene (PE), polypropylene (PP), polystyrene (PS), and polyvinyl chloride (PVC) account for approximately 80 wt% of total plastic fraction, making them the most abundant compounds in the waste products. As already discussed by Marongiu et al. (2007), a unifying approach allows the description of the thermal degradation of the vinyl polymers PS, PP, and PE on the basis of the same classes of reactions, with only a few reference kinetic parameters. In fact, initiation, hydrogen abstractions, β-decompositions, and isomerizations through intramolecular abstractions and terminations are the controlling reactions in the pyrolysis process (Faravelli et al., 1999, 2001, 2003; Ranzi et al., 1997). The reference kinetic parameters of these reactions in the liquid phase are derived from the ones already adopted in the gas-phase pyrolysis of hydrocarbon feedstocks (Dente et al., 1979, 1992). The corrections account for the transposition in the liquid phase, and they become significant for reactions with high activation energies, such as initiation reactions, and are also applied to termination reactions to account for the diffusive limitations (Dente et al., 2007). This unified and mechanistic kinetic model was validated by comparison with experimental measurements in a wide range of operating conditions (Marongiu et al., 2007). The same reaction steps, or rather the same reaction classes, explain the pyrolysis process of the three different polymers. Fig. 4 schematically shows the main reaction steps, while the reference kinetic parameters are reported in Table 1. The three different polymers are represented through a simplified notation in which the side chain (G) can be either H, CH_3, or phenyl in the case of PE, PP, or PS, respectively. The kinetic scheme for the different polymers is then built by selecting only the proper reactions in the different classes.

Fig. 5 shows the typical behavior of the thermal degradation of the three polymers when heated at 10K/min. They behave quite similarly: one-step

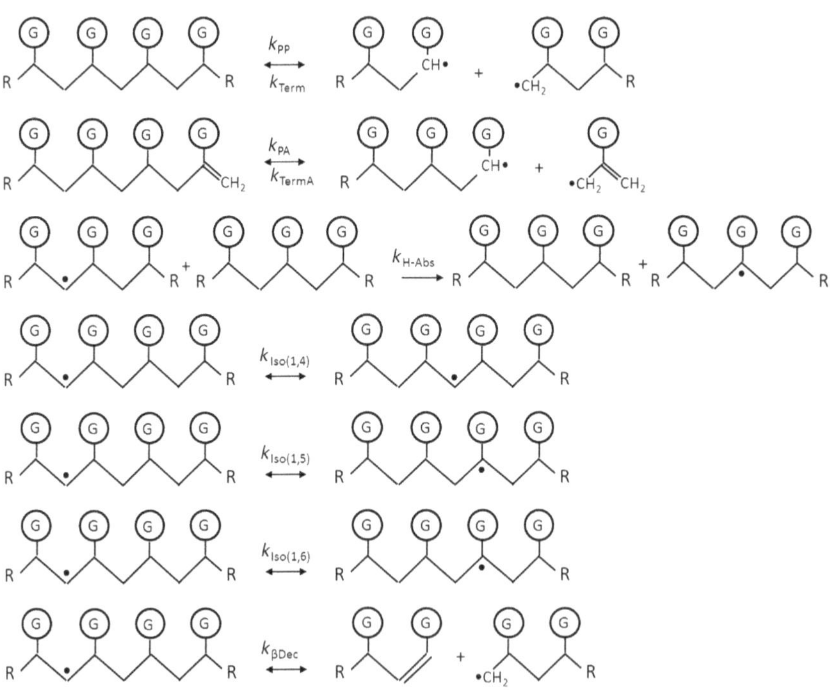

Fig. 4 Main reaction steps in the chain radical mechanism of PS, PP, and PE pyrolysis. *After Marongiu A, Faravelli T, Ranzi E: Detailed kinetic modeling of the thermal degradation of vinyl polymers,* J Anal Appl Pyrolysis *78:343–362, 2007.*

Table 1 Pyrolysis of Polystyrene, Polypropylene, and Polyethylene: Reference Rate Parameters

	Polystyrene			Polypropylene			Polyethylene		
	A	*n*	E_a	*A*	*n*	E_a	*A*	*n*	E_a
Initiation	5.0E + 13	0	63.7	6.0E + 14	0	73.8	8.0E + 14	0	77.9
Allyl initiation	5.0E + 12	0	58.5	6.0E + 14	0	67.4	8.0E + 14	0	73.1
H-abstraction (end chain)	5.0E + 07	0	13.5	2.0E + 08	0	12.2	3.0E + 08	0	11.9
H-abstraction (mid chain)	5.0E + 07	0	16.5	2.0E + 08	0	13.5	3.0E + 08	0	13.1
β-Scission (end chain)	1.6E + 13	0	25.8	1.0E + 14	0	30.0	3.5E + 14	0	30.1
β-Scission (mid chain)	1.2E + 13	0	27.0	1.0E + 14	0	30.0	1.5E + 14	0	30.1
Back-biting (1,4)	4.0E + 09	0	17.2	5.0E + 10	0	19.6	1.0E + 11	0	20.5
Back-biting (1,5)	5.0E + 09	0	15.8	1.4E + 10	0	13.9	1.6E + 10	0	14.3
Back-biting (1,6)	1.0E + 08	0	17.2	1.0E + 09	0	19.6	5.0E + 09	0	20.5
Termination	5.0E + 06	1	14.0	1.0E + 07	1	6.0	5.0E + 07	1	6.0

Rate constants are expressed as $AT^n \exp(-E_a/RT)$ (units: L, mol, s, kcal).

After Marongiu A, Faravelli T, Ranzi E: Detailed kinetic modeling of the thermal degradation of vinyl polymers, *J Anal Appl Pyrolysis* 78:343–362, 2007.

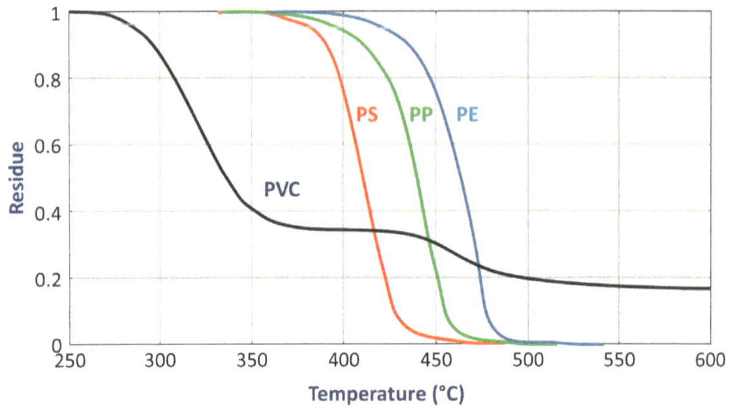

Fig. 5 Thermogravimetric analysis of PS, PP, PE, and PVC degradation (10K/min).

degradation without char formation. Because of the stability of the benzyl-like radicals formed during the pyrolysis, PS shows higher reactivity and higher propensity to unzip to styrene monomer.

The behavior of the three different degradation curves confirms the relative reactivity of the three polymers: PS > PP > PE, as expected as a consequence of the different bond dissociation energies (BDEs) and the corresponding rate parameters reported in Table 1.

Fig. 5 also shows the quite different behavior of PVC degradation under the same heating conditions (Marongiu et al., 2003). There is a two-step mechanism, where the first step corresponds to the very fast dehydrochlorination reaction with the formation of unsaturated poly-acetylene chains $(-CH{=}CH-)_n$. The successive degradation step releases tar species with a significant formation of a char residue. A lumped kinetic mechanism of 40 species (molecules and radicals) involved in about 250 reactions allows to reproduce the main characteristics of PVC degradation in a reliable way. From these detailed mechanisms of PS, PP, PE, and PVC, corresponding multistep pyrolysis models are easily derived and validated.

2.2 Biomass

2.2.1 Biomass Characterization and Reference Species

It is well known that cellulose (30–60 wt%), hemicellulose (15–35 wt%), and lignin (15–40 wt%) are the building blocks of woody biomass (Debiagi et al., 2015; Miller and Bellan, 1997; Vinu and Broadbelt, 2012). Biomass has a porous structure where cellulose microfibril represents the important element surrounded by other substances, which act as ligand (hemicellulose and pectin) and embed lignin materials. Moisture is also present and is found as hygroscopic water (hydrogen bonded to the hydroxyl groups of cellulose and hemicellulose), capillary water in the lumens, and water vapor in the gas phase (Grønli, 1996).

Cellulose is a long-chain polymer built by glucose, the monomeric unit of a six carbon sugar, bonded through β-1,4 glycosidic bonds. The chains are kept together by hydrogen bonds, which confer to the polymer an almost fully crystalline structure with few amorphous zones. Cellulose elementary microfibrils contain 36 chains. Hemicellulose is closely associated to the surface of the rigid cellulose crystallite forming the microfibril network. Pectins are cross-linked polysaccharides forming a hydrated gel that glues the cell-wall components together (Himmel et al., 2007). Cellulose is the most abundant structural component in biomasses, ranging from 30% to 60% of the total dry mass. Hemicellulose is a second structural polymer, and it is a mixture of hexose and pentose sugars (mainly xylose, mannose, galactose, and arabinose). Different from cellulose, it has a shorter chain and a much more amorphous structure, due to its irregular composition and the branches present on the chain. Holocellulose is commonly referred as the combination of cellulose and hemicellulose. Lignin is a racemic polymer composed by monomeric units of aromatic alcohols (coniferyl, sinapyl, and *p*-coumaryl), whose composition changes widely between gymnosperm and angiosperm biomasses.

Biomass offers important advantages as a solid fuel due to the high volatility and the high reactivity of the fuel and the residual char. In comparison with coal, biomass has a lower density and a lower heating value (LHV), because of the higher oxygen and moisture content. Biomasses are primarily composed of C, H, and O elements, with a smaller amount of N, S, Cl, and metal and metal oxides. Several correlations between the heating value and ultimate or elemental analysis are proposed in the literature (Demirbas, 2004). Table 2 reports typical examples of biomass compositions, both with proximate and with ultimate analysis (Williams et al., 2012).

A complete and structural analysis of biomass samples gives significant information on the relative content of carbohydrates (glucose, xylose, galactose, arabinose, and mannose), lignin, extractable materials, protein, and ash. Table 3 shows a sample of these structural or biochemical analyses of typical biomasses (Zhang, 2008).

Compared to the elemental analysis, these analytical methods are more complex and involve thermal, chemical, and/or enzymatic separations, which could also modify the original biomass structure. Despite several research efforts in this direction (Sluiter et al., 2010), data reporting both

Table 2 Typical Proximate and Ultimate Analysis of Different Biomass Samples

	Proximate Analysis				Ultimate Analysis					
Biomass	**Moisture**	**VM**	**FC**	**ASH**	**C**	**H**	**O**	**N**	**S**	**Cl**
Wood pine chips	4.00	81.30	14.60	0.10	52.00	6.20	41.59	0.12	0.08	0.01
Willow, SRC	6.96	75.70	16.31	1.03	51.62	5.54	42.42	0.38	0.03	0.01
Miscanthus giganteus	14.20	70.40	14.10	1.30	49.10	6.40	43.98	0.26	0.13	0.13
Switchgrass	7.17	73.05	15.16	4.62	49.40	5.70	44.25	0.45	0.10	0.10
Wheat straw	7.78	68.83	17.09	6.30	49.23	5.78	43.99	0.64	0.10	0.26
Rice husks	9.40	74.00	13.20	12.80	42.30	6.10	50.56	1.10	0.10	0.04
Palm PKE	7.60	72.12	16.18	4.10	51.12	7.37	38.21	2.80	0.30	0.20
Sugarcane bagasse	10.40	76.70	14.70	2.20	49.90	6.00	43.15	0.40	0.04	0.51
Olive residue	6.40	65.13	19.27	9.20	54.42	6.82	37.29	1.40	0.05	0.04
Cow dung	13.90	60.50	11.90	13.70	54.00	6.40	36.70	0.83	0.03	1.00

After Williams A, Jones JM, Ma L, Pourkashanian M: Pollutants from the combustion of solid biomass fuels, *Prog Energy Combust Sci* 38:113–137, 2012.

Table 3 Typical Structural or Biochemical Analysis of Different Biomass Samples

Wood Species	Cellulose	Hemicellulose	Lignin
Softwoods			
Picea glauca	41	31	27
Abies balsamea	42	27	29
Pinus strobus	41	27	29
Tsuga canadensis	41	23	33
Norway spruce	46	25	28
Loblolly pine	39	25	31
Thuja occidentalis	41	26	31
Hardwoods			
Eucalyptus globulus	45	35	19
Acer rubrum	45	29	24
Ulmus americana	51	23	24
Populus tremuloides	48	27	21
Betula papyrifera	42	38	19
Fagus grandifolia	45	29	22
Agricultural residues			
Corn stover	40	17	25
Wheat straw	30	20	50
Switchgrass	45	12	30

After Zhang YP: Reviving the carbohydrate economy via multi-product lignocellulose biorefineries, *J Ind Microbiol Biot* 35:367–375, 2008.

elementary and biochemical composition are not easily available in the open literature. This lack of information creates some difficulties to characterize biomasses for modeling purposes.

A method to characterize the biomass feedstock simply based on the elemental analysis has been proposed elsewhere (Ranzi et al., 2008). If only the elemental analysis in terms of C, H, and O content is available, then a suitable combination of a few reference species can be simply derived from the three atomic balances. As already mentioned, cellulose, hemicellulose, and lignin, together with extractives, constitute the largest portion of the biomass, and these are the reference species. Biomass pyrolysis products are then assumed as a linear combination of the pyrolysis products of these reference compounds. When direct information on biochemical composition is unavailable, cellulose, hemicellulose, lignin, and extractive content are derived through the elemental biomass composition in terms of H/C/O (Debiagi et al., 2015; Ranzi et al., 2008). As reference species, together with cellulose and hemicellulose, three different types of lignins, rich in carbon, hydrogen, and oxygen respectively (Faravelli et al., 2010), are considered. Finally, two lumped reference species account for the hydrophobic and the hydrophilic extractives. Fig. 6 reports the structure and formula of the seven species described earlier, while Fig. 7 shows the same reference species in the H% vs C% plot along with several biomass samples. Three reference mixtures (RM-1, RM-2, and RM-3) are first defined as different combinations of the seven reference species, in order to reduce the total number of degrees of freedom. RM-1 is representative of holocellulose, while RM-2 and RM-3 are mixtures of lignins with some content of extractives. The combinations of these mixtures are derived from experimental findings and can be easily modified. The relative amount of the seven reference components in the different biomass samples is then derived from the three reference mixtures and from the biomass elemental composition, respecting the C, H, and O mass balances. While further details on this characterization method are reported in Debiagi et al. (2015), a couple of examples can be useful to explain this approach. The hybrid poplar, whose elemental mass composition is H/C/O = 0.0565/0.5092/0.4343 (reported as A in Fig. 7), is characterized including 20% TANN in RM-3. The solution of the linear system of H/C/O balance equations gives the following mass composition of the reference mixtures:

$$\mathrm{RM}-1=0.5597 \quad \mathrm{RM}-2=0.0020 \quad \mathrm{RM}-3=0.4384$$

The amount of the reference mixture RM-2 is very low, because of the low hydrogen content, and RM-1 and RM-3 are major constituents of this biomass. From these values and the reference mixture composition, the following mass amounts of the seven reference species are obtained:

$$\begin{array}{lll} \mathrm{CELL}=0.3627 & \mathrm{HCELL}=0.1970 & \\ \mathrm{LIGH}=0.0017 & \mathrm{LIGO}=0.3181 & \mathrm{LIGC}=0.0489 \\ \mathrm{TGL}=0.0000 & \mathrm{TANN}=0.0716 & \end{array}$$

Similarly, the following mass compositions of the reference species characterize the olive husks with H/C/O = 0.0696/0.5489/0.3815 (reported as B in Fig. 7):

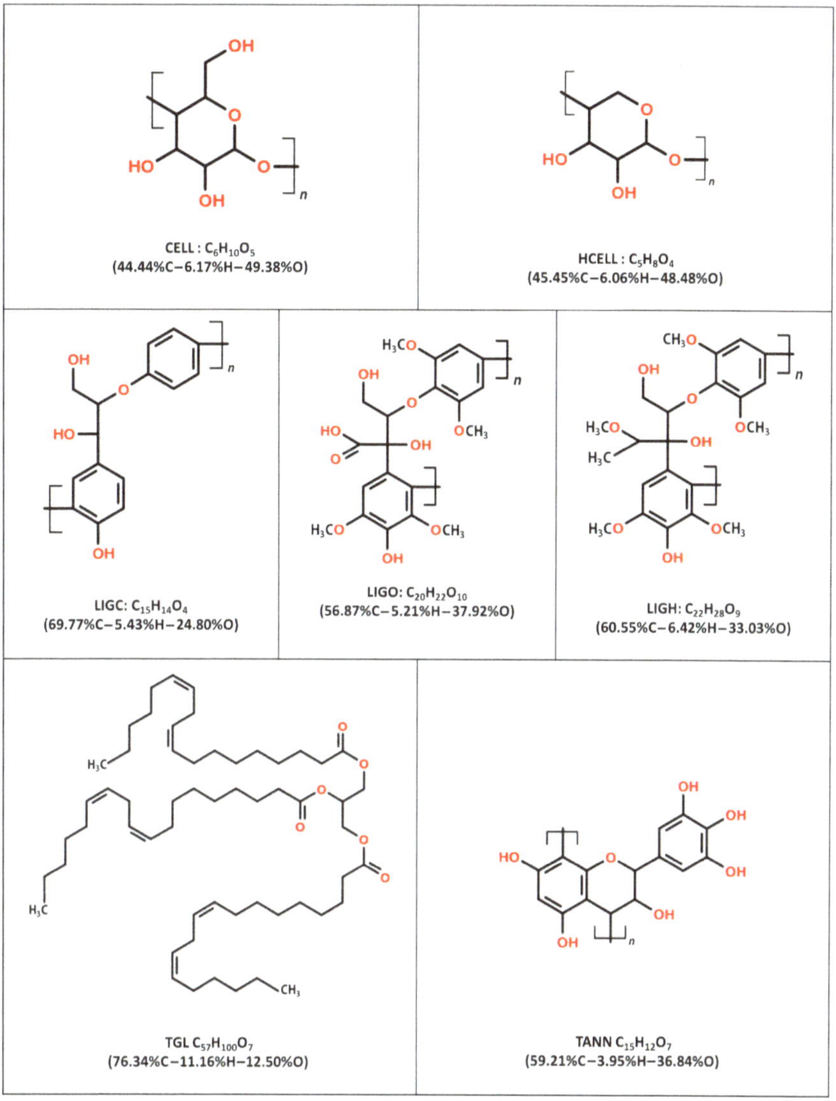

Fig. 6 Reference species for biomass characterization. *After Debiagi PEA, Pecchi C, Gentile G, et al.: Extractives extend the applicability of multistep kinetic scheme of biomass pyrolysis,* Energy Fuel *29(10):6544–6555, 2015.*

$$
\begin{aligned}
&\text{CELL} = 0.3484 \quad \text{HCELL} = 0.1892 \\
&\text{LIGH} = 0.2474 \quad \text{LIGO} = 0.0170 \quad \text{LIGC} = 0.0392 \\
&\text{TGL} = 0.1589 \quad \text{TANN} = 0.0000
\end{aligned}
$$

A linear combination of the seven reference components can describe all the biomasses contained in the shadow area of Fig. 7, and this

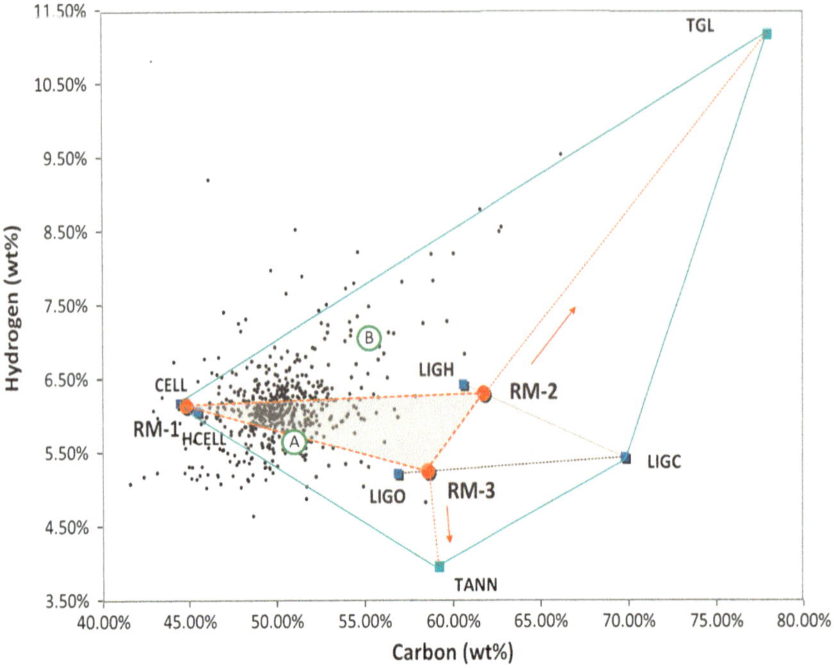

Fig. 7 Biomass characterization. Reference species in the H% vs C% plot along with several biomass samples. A and B refer to hybrid poplar and olive husks (see text). *After Debiagi PEA, Pecchi C, Gentile G, et al.: Extractives extend the applicability of multistep kinetic scheme of biomass pyrolysis,* Energy Fuel *29(10):6544–6555, 2015.*

characterization procedure is able to process more than the 90% of the wide range of lignocellulosic biomasses analyzed by Debiagi et al. (2015).

The chemical percolation devolatilization (bio-CPD) model uses a very similar approach, assuming that biomass pyrolysis occurs as a weighted average of its individual components (cellulose, hemicellulose, and lignin). The light gas and tar yields of a particular biomass are then calculated, together with the residual char, as a weighted average of the pyrolysis products of the reference components (Lewis and Fletcher, 2013).

2.2.2 Multistep Kinetic Model of Biomass Pyrolysis

The differences in the biomass composition as well as in the operating conditions of thermal treatments significantly change the decomposition product distribution, but a similar set of products are always obtained at least on a qualitative basis: water, sugars together with smaller quantities of aldehydes, ketones, alcohols, phenolics, along with light gases, and char residues.

Always referring to the previous seven reference components of Fig. 6, a multicomponent and multistep kinetic mechanism of primary biomass pyrolysis is reported in Table 4.

Each reference component decomposes independently through a multistep, branched mechanism of first-order reactions. These lumped reactions model the formation of char, intermediate solid and chemisorbed species, tars, and permanent gases. The lumped reactions, both in terms of rates and stoichiometries, were derived from experimental findings (Ranzi et al., 2008) and they are progressively and continuously extended and updated, based on new experimental data and comparisons across a wider range of experimental conditions. Recently, the experimental data on temperature profiles in thick particles with the overshooting of the center temperature allowed to better validate the endothermic release of tars and the exothermic charring process (Corbetta et al., 2014).

A peculiarity of this kinetic model is a detailed characterization of the pyrolysis products, including not only water vapor and permanent gases (H_2, CO, CO_2, CH_4, and C_2H_4), several alcohols, aldehydes, and carbonyl compounds but also different sugars together with heterocyclic and phenolic components. At high temperatures, several chemisorbed species contribute to describe the successive steps of char devolatilization with the progressive release of H_2, CO, and CO_2.

The cellulose pyrolysis mechanism (Antal and Varhegyi, 1995; Lédé, 2012), recently revised by Broadbelt's group (Burnham et al., 2015), is characterized by a first depolymerization step producing active cellulose with an apparent activation energy of 47 kcal/mol. This reaction reduces the polymerization degree without any volatile release. Active cellulose then decomposes with two competitive reactions: a slow reaction that produces char plus permanent gases, and a main reaction releasing levoglucosan. Only at high temperatures, decomposition reaction prevails over tar release. A side charring and exothermic reaction of cellulose is also considered. The multistep kinetic mechanism of hemicellulose pyrolysis resembles that of cellulose. Both cellulose and hemicellulose are polymeric sugar chains, releasing together with tar components (levoglucosan and anhydro-sugars), permanent gases, a wide number of oxygenated species, including methanol, acetic acid, hydroxyacetaldehyde, acetone, acetol, furfural, and 5-hydroxymethl-furfural (Shen et al., 2015). Fig. 8 schematically reports these lumped multistep kinetic mechanisms. Fig. 9 shows some comparisons between model predictions and experimental data of different thermogravimetric analyses (TGAs) relating to cellulose and hemicellulose pyrolysis at different heating rates.

Table 4 Multistep Kinetic Scheme of Biomass Pyrolysis

Pyrolysis Reactions			Kinetic Parameters A (s^{-1}), E_a (kcal/kmol)	ΔH_r (700K) (kcal/kmol)
Cellulose				
CELL	→	CELLA	$1.5 \times 10^{14} \times \exp(-47{,}000/RT)$	−1300
CELLA	→	0.45 HAA + 0.2 GLYOX + 0.1 MECHO + 0.25 HMFU + 0.3 ALD3 + 0.15 CH_3OH + 0.4 CH_2O + 0.31 CO + 0.41 CO_2 + 0.05 H_2 + 0.83 H_2O + 0.02 HCOOH + 0.2 G{CH_4} + 0.05 G{H_2} + 0.61 CHAR	$2 \times 10^{6} \times \exp(-19{,}100/RT)$	27,100
CELLA	→	LVG	$4 \times T \times \exp(-10{,}000/RT)$	23,200
CELL	→	5 H_2O + 6 CHAR	$6.5 \times 10^{7} \times \exp(-31{,}000/RT)$	−62,700
Hemicellulose				
HECELL	→	0.58 HCE1 + 0.42 HCE2	$1 \times 10^{10} \times \exp(-31{,}000/RT)$	−5000
HCE1	→	0.025 H_2O + 0.5 CO_2 + 0.025 HCOOH + 0.5 CO + 0.8 CH_2O + 0.125 ETOH + 0.1 CH_3OH + 0.25 C_2H_4 + 0.125 G{H_2} + 0.275 G{CO_2} + 0.4 G{COH_2} + 0.45 G{CH_3OH} + 0.325 G {CH_4} + 0.875 CHAR	$1.2 \times 10^{9} \times \exp(-30{,}000/RT)$	−500
HCE1	→	0.25 H_2O + 0.8 CO_2 + 0.05 HCOOH + 0.1 CO + 0.15 G{CO} + 0.15 G{CO_2} + 0.2 G{H_2} + 0.3 CH_2O + 1.2 G{COH_2} + 0.625 G{CH_4} + 0.375 G{C_2H_4} + 0.875 CHAR	$1.5 \times 10^{-1} \times T \times \exp(-8000/RT)$	−42,400
HCE1	→	XYLAN	$3 \times T \times \exp(-11{,}000/RT)$	17,900
HCE2	→	0.2 H_2O + 0.175 CO + 0.275 CO_2 + 0.5 CH_2O + 0.1 ETOH + 0.2 HAA + 0.025 HCOOH + 0.25 G{CH_4} + 0.3 G{CH_3OH} + 0.275 G{C_2H_4} + 0.4 G{CO_2} + 0.925 G{COH_2} + CHAR	$5 \times 10^{9} \times \exp(-33{,}000/RT)$	12,000

Lignins				
LIGC	→	0.35 LIGCC + 0.1 COUMARYL + 0.08 FENOL + 0.41 C_2H_4 + 1.0 H_2O + 0.7 G{COH_2} + 0.3 CH_2O + 0.32 CO + 0.495 G{CH_4} + 5.735 CHAR	$1.33 \times 10^{15} \times \exp(-48{,}500/RT)$	−10,300
LIGH	→	LIGOH + 0.5 ALD3 + 0.5 C_2H_4 + 0.25 HAA	$6.7 \times 10^{12} \times \exp(-37{,}500/RT)$	30,700
LIGO	→	LIGOH + CO_2	$3.3 \times 10^{8} \times \exp(-25{,}500/RT)$	26,000
LIGCC	→	0.3 COUMARYL + 0.2 FENOL + 0.35 HAA + 0.7 H_2O + 0.65 G{CH_4} + 0.6 G{C_2H_4} + G{COH_2} + 0.4 CO + 0.4 G{CO} + 6.75 CHAR	$1.67 \times 10^{6} \times \exp(-31{,}500/RT)$	−31,100
LIGOH	→	LIG + 0.9 H_2O + 0.1 CH_4 + 0.6 CH_3OH + 0.1 G{H_2} + 0.3 G{CH_3OH} + 0.05 CO_2 + 0.55 CO + 0.6 G{CO} + 0.05 HCOOH + 0.85 G{COH_2} + 0.35 G{CH_4} + 0.2 G{C_2H_4} + 4.15 CHAR	$1 \times 10^{8} \times \exp(-30{,}000/RT)$	−26,100
LIG	→	0.7 FE2MACR + 0.3 ANISOLE + 0.3 CO + 0.3 G{CO} + 0.3 MECHO	$4 \times T \times \exp(-12{,}000/RT)$	46,200
LIG	→	0.95 H_2O + 0.2 CH_2O + 0.4 CH_3OH + CO + 0.2 CH_4 + 0.05 HCOOH + 0.45 G{CO} + 0.5 G{COH_2} + 0.4 G{CH_4} + 0.65 G{C_2H_4} + 0.2 MECHO + 0.2 ALD3 + 5.5 CHAR	$4 \times 10^{8} \times \exp(-30{,}000/RT)$	−21,100
LIG	→	0.6 H_2O + 0.4 CO + 0.2 CH_4 + 0.4 CH_2O + 0.2 G{CO} + 0.4 G{CH_4} + 0.5 G{C_2H_4} + 0.4 G{CH_3OH} + 2 G{COH_2} + 6 CHAR	$8.3 \times 10^{-2} \times T \times \exp(-8000/RT)$	−83,600

Continued

Table 4 Multistep Kinetic Scheme of Biomass Pyrolysis—cont'd

Pyrolysis Reactions			Kinetic Parameters A (s^{-1}), E_a (kcal/kmol)	ΔH_r (700K) (kcal/kmol)
Extractives				
TGL	→	ACRO + 3 FFA	$7 \times 10^{12} \times \exp(-45{,}700/RT)$	1300
CTANN	→	FENOL + ITANN	$5 \times 10^{1} \times \exp(-11{,}000/RT)$	1300
ITANN	→	6 CHAR + 3 CO + 3 H_2O	$1.5 \times 10^{-2} \times \exp(-6100/RT)$	10,100
Metaplastic				
G{CO_2}	→	CO_2	$1 \times 10^{6} \times \exp(-24{,}000/RT)$	−29,100
G{CO}	→	CO	$5 \times 10^{12} \times \exp(-50{,}000/RT)$	−13,400
G{COH_2}	→	CO + H_2	$5 \times 10^{11} \times \exp(-71{,}000/RT)$	48,600
G{H_2}	→	H_2	$5 \times 10^{11} \times \exp(-75{,}000/RT)$	0
G{CH_4}	→	CH_4	$5 \times 10^{12} \times \exp(-71{,}667/RT)$	0
G{CH_3OH}	→	CH_3OH	$2 \times 10^{12} \times \exp(-50{,}000/RT)$	0
G{C_2H_4}	→	C_2H_4	$5 \times 10^{12} \times \exp(-71{,}667/RT)$	0

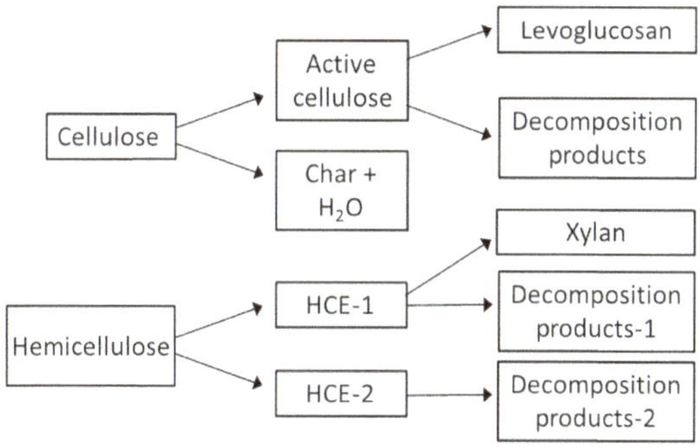

Fig. 8 Multistep kinetic mechanism of cellulose and hemicellulose pyrolysis.

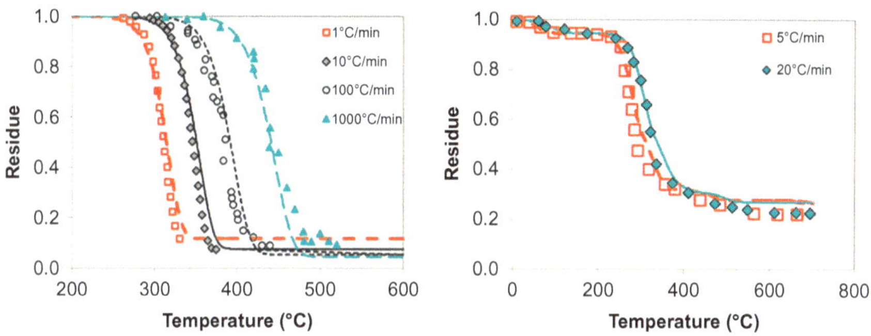

Fig. 9 Pyrolysis of cellulose and hemicellulose. *Left panel*: Cellulose. TGA at 1 and 10°C/min (Antal et al., 1998), 100 and 1000°C/min (Milosavljevic and Suuberg, 1995). *Right panel*: Hemicellulose. TGA at 5 and 20°C/min (Williams and Besler, 1996). Comparisons of model predictions (*lines*) and experimental data (*symbols*).

As far as the mechanism of cellulose is concerned, the heats of reaction agree well with the observation of Milosavljevic et al. (1996). While the tar release is an endothermic process and it absorbs ~500 kJ/kg, the char formation is an exothermic process releasing ~2000 kJ/kg of char formed. As it will be further discussed in the next section, this kinetic model and relating reaction heats were recently validated by Corbetta et al. (2014) by comparing model predictions with several experimental data of pyrolysis of thick biomass particles from different sources (Bennadji et al., 2013; Gauthier et al., 2013; Park et al., 2010).

Klein and Virk (2008) statistically described lignin structure as the juxtaposition of methoxy phenol (MP) and a propanoid side chain (PC) attribute on an aromatic ring. They developed a detailed pyrolysis model, which alters the state of the MP and PC attributes while leaving the aromatic ring conserved. In this way, it is possible to describe gases, aqueous liquid, tar species (the majority related to guaiacol, catechol, and phenol), and the char residue (multiple-ring aromatic products) (Hou et al., 2009). The multistep lignin decomposition scheme here considered is a simplification of the detailed mechanism of Faravelli et al. (2010) and it is schematically reported in Fig. 10. This multistep kinetic mechanism fairly fits the one more recently discussed by Zhou et al. (2014a). Fig. 11 compares model predictions and experimental data of a TGA of two different lignins, at heating rates of 20K/min (Jakab et al., 1995). The lignin pyrolysis reactions are active in a wide temperature range and release phenolic components. Phenol, anisole (methoxy-benzene), 2,6-dimethoxy-phenol, 4-(3-hydroxy-1-propenyl)

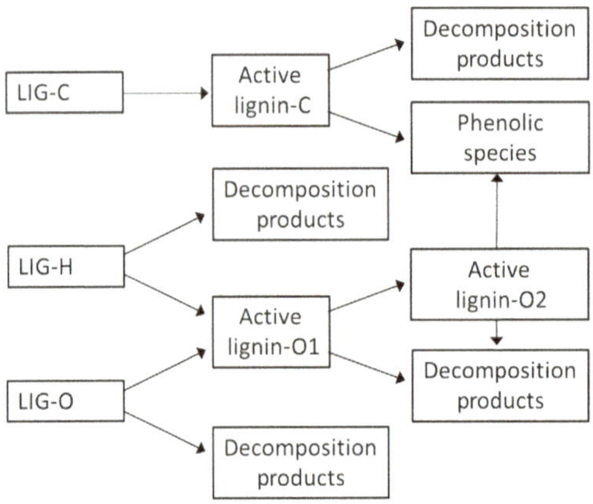

Fig. 10 Multistep kinetic mechanism of pyrolysis of the three reference lignins.

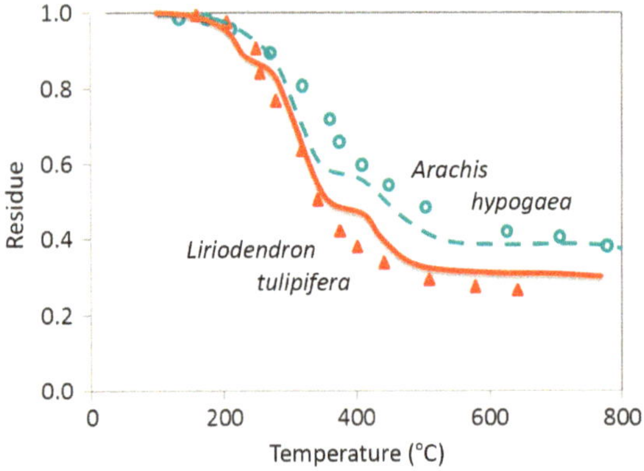

Fig. 11 Pyrolysis of two different lignins (heating rates 20K/min). Comparisons of model predictions (*lines*) and experimental data (*points*) (Jakab et al., 1995).

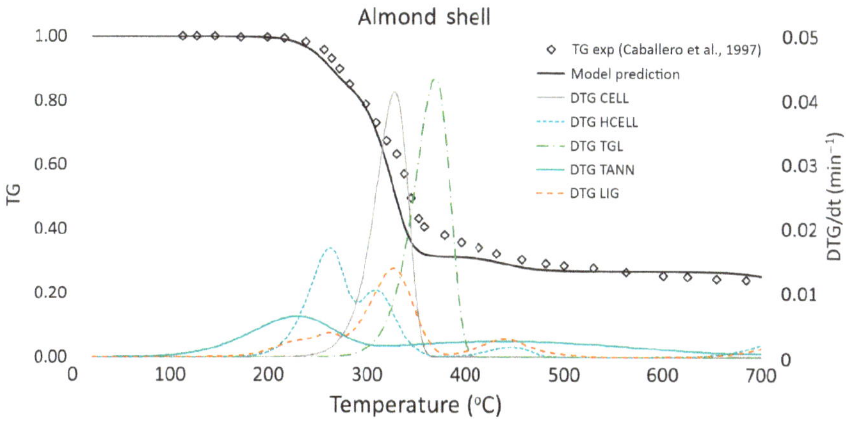

Fig. 12 Pyrolysis of almond shell (2K/min) (Caballero et al., 1997). Comparisons between experimental data (*points*) and model predictions (*lines*). DTG curves of individual reference components are also shown.

phenol, and 3-(4-hydroxy-3,5-dimethoxy-phenyl)-acrylaldehyde are the selected lumped species representatives of these compounds.

Phenol is also released by the first decomposition step of the hydrophilic extractives (TANN), while triglycerides (TGL) decompose with a fast single step to a lumped species representative of the free fatty acid (FFA).

As already mentioned, biomass is treated as a mixture of the seven reference components. Fig. 12 shows a comparison between the model predictions and the experimental data of the pyrolysis of almond shells (Caballero et al., 1997). Based on the elemental composition C/H/O = 0.509/0.061/0.430, the following mass composition of the reference species is obtained:

$$\text{CELL} = 0.446 \quad \text{HCELL} = 0.203 \quad \text{LIGH} = 0.077 \quad \text{LIGO} = 0.143$$
$$\text{LIGC} = 0.063 \quad \text{TGL} = 0.037 \quad \text{TANN} = 0.032$$

Fig. 12 also shows the differential contributions (DTG) of the individual reference components to the overall TG curve.

The lumped kinetic mechanism of Table 4 is very simplified, aiming at an effective use not only at the particle but also at the reactor scale. In fact, computational time limitations are certainly very severe when simulating a gasifier or a biomass combustor at the reactor scale (Ranzi et al., 2011, 2014, Stark et al., 2015). Anyway, it is evident that this multistep kinetic mechanism can be further improved in terms of new reaction steps, kinetic parameters, detail of reaction products (Anca-Couce, 2016; Anca-Couce et al., 2014). Shen et al. (2015) recently revised biomass fast pyrolysis discussing the yields of liquid and gas products, focusing on the primary and secondary formation pathways of oxygenated compounds. Moreover, gas chromatography with flame ionization detector and two-dimensional gas chromatography with time-of-flight mass spectrometry (Yildiz et al., 2013), as well as the application of tunable synchrotron vacuum ultraviolet photoionization mass spectrometry (SVUV-PIMS) (Weng et al., 2013), allow to identify and quantify hundreds of compounds, thus describing a large portion of bio-oil.

Moving from their preliminary work (Vinu and Broadbelt, 2012), Broadbelt's team extensively studied, both from a theoretical and experimental viewpoint, the fast pyrolysis of neat glucose-based carbohydrates (Mayes et al., 2014; Zhou et al., 2014b,c). They also developed a detailed mechanistic model for fast pyrolysis of glucose-based carbohydrates, involving about 100 species in more than 300 reactions. The mechanistic model describes the decomposition of cellulosic polymer chains, reactions of intermediates, and formation of several low molecular weight compounds. Similarly, Seshadri and Westmoreland (2012) investigated and highlighted the implications of concerted molecular reactions for cellulose and hemicellulose kinetics.

The large extent and continuous research efforts in the pyrolytic behavior of biomass easily allow the extensions and improvements of the lumped kinetic mechanism (Anca-Couce and Obernberger, 2016). It is worth underlining that the interactions among reference species and the ash effect on the biomass pyrolysis are not addressed in the present model, even though it is known that ashes can catalyze and significantly modify the overall biomass pyrolysis process (Trendewicz et al., 2015).

2.3 Coal

Coal was the first fossil fuel and still represents today more than 20% of our global primary energy source. In most industrialized countries, coal has been extensively replaced by gas and oil. Worldwide, almost 40% of electric power is generated using more than 60% of the whole coal production. Coal can be seen as the product of a very slow, low temperature, high pressure pyrolysis of biomass. The result of this process is a devolatilization, which increases the carbon content and reduces both the hydrogen and oxygen content with respect to the original biomass. Coal composition and structure are then the result of the age and the conditions of this very slow pyrolysis, strongly differing from coal to coal. Low-rank coals, i.e., young coals, still contain large amounts of oxygen, longer side chains, and smaller aromatic clusters. Increasing the rank (i.e. increasing the pyrolysis time) oxygen content decreases, side chains become unlikely and shorter, while the aromatic clusters gradually increase evolving toward graphite like structures. Peat is the first coal precursor, which is progressively converted into lignite or brown coal. As the process of maturation continues with exposure to high pressure and temperatures, lignite is further transformed into bituminous coals and finally anthracite (Olah et al., 2011). As in the case of biomass, coal characterization can be based on its elemental composition and a triangular definition of key reference components (Sommariva et al., 2010). The difference relies in the choice of reference components: instead of main constituents, a proper definition assumes coals of very different compositions, which reflect the coal ranks. Thus, neglecting in the initial phase sulfur and nitrogen compounds released and assuming ashes as inert, which do not catalytically affect the coal devolatilization process, the elemental analysis of the coal is corrected and simply normalized to the C, H, and O content, on dry, ash (and S, N)-free basis. Fig. 13 shows the van Krevelen diagram with the composition of several coals of practical interest, investigated by several researchers (Fletcher et al., 1990; Hercog and Tognotti, 2008;

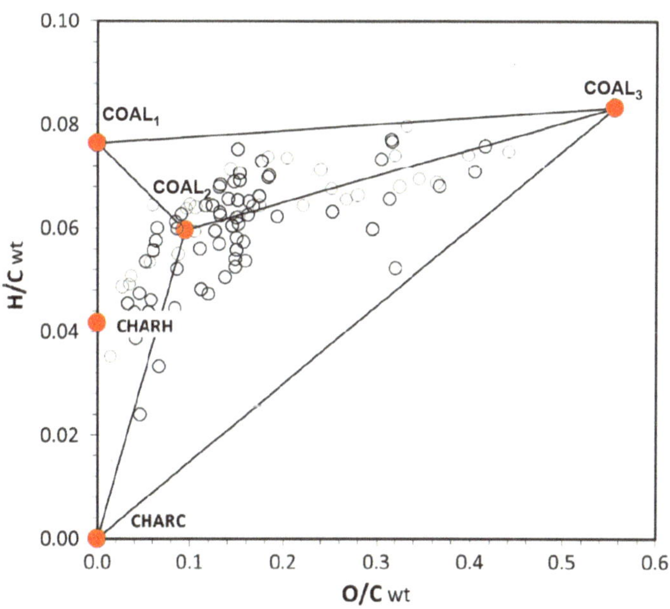

Fig. 13 Composition of some literature coals and reference coal components. *After Sommariva S, Maffei T, Migliavacca G, Faravelli T, Ranzi E: A predictive multi-step kinetic model of coal devolatilization,* Fuel *89:318–328, 2010.*

Jupudi et al., 2009; Matsuoka et al., 2003; Solomon et al., 1990; Sommariva et al., 2010). Carbon content is always higher than 60–65 wt%, while hydrogen content is usually lower than 7 wt%. The rank of the coal increases with the rising carbon content, moving from the low rank of lignite, to average values for bituminous coals, up to the high rank and high carbon content of anthracite. Fig. 13 also shows the assumed reference components: pure carbon (CHARC), and two other reference coals, a lignite-like with high oxygen content ($COAL_3$) and a reference coal rich in hydrogen ($COAL_1$), which does not contain oxygen. A further reference coal ($COAL_2$) lies in the middle of the triangle formed by the other three reference coals, close to a great number of bituminous coals. The structure of these reference coals is described by three lumped or equivalent monomer molecules, which stand for reference configurations, saving the elemental C/H/O composition.

The plot of Fig. 13 results thus divided into three triangles, and each coal lies inside one of them. As in the case of biomass, any coal is then considered as constituted by a linear combination of the three closest reference coals, and its volatilization is assumed a straightforward weighted combination of the volatilization of the reference coals. Fig. 14 sketches the average structures of the reference monomers. $COAL_1$ is considered as a 50/50 mol mixture of ($-C_{12}H_{10}-$) and ($-C_{12}H_{12}-$).

Zhao et al. (1994) proposed a similar approach to define the properties of any coals through an interpolation from the properties of reference coals. This approach was applied to define the preliminary structural parameters of unknown coals (Solomon et al., 1988) and the functional group parameters used to estimate the release of light gas from an unknown coal (Xu and Tomita, 1987a).

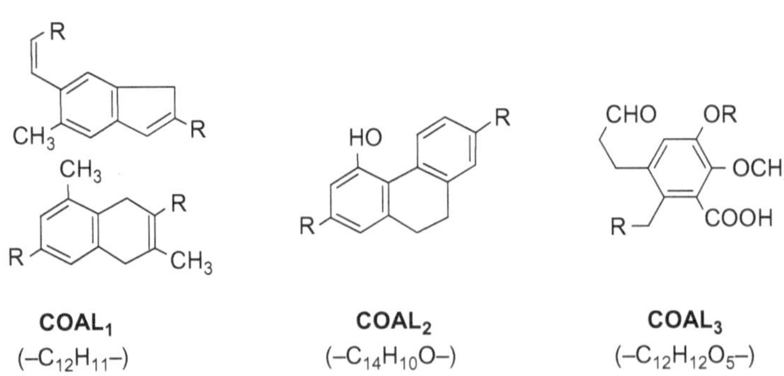

Fig. 14 Reference coals and reference monomer structures.

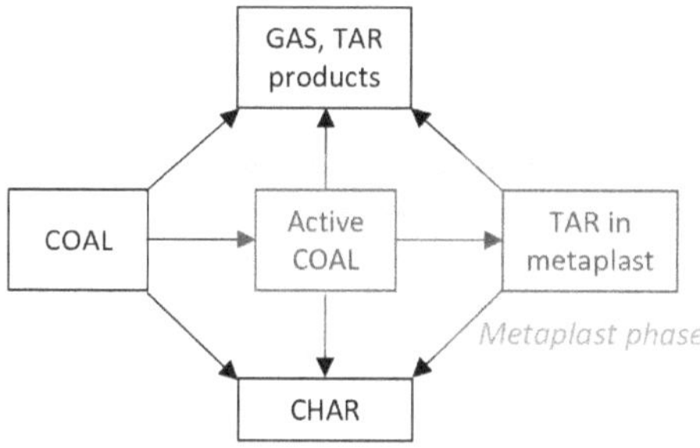

Fig. 15 Schematic representation of the multistep kinetic mechanism of coal pyrolysis.

A multistep pyrolysis mechanism is proposed for the three reference coals, with different product distributions and different kinetic parameters. Fig. 15 very schematically shows the main volatilization steps.

The mechanism considers that initially the coal forms a metaplastic phase. Successively, gas and tar species are released with different mechanisms at low and high temperatures. At low temperatures (or low heating rates), the reference coals initially form char and volatile species, which still are trapped in the condensed phase. The apparent activation energy of this thermal decomposition is about 33–40 kcal/mol. The tar in the metaplast can either be released with a proper kinetics or interact with CHAR through cross-linking and reticulation reactions. The tar release reactions account for gas–liquid equilibrium in a simplified form, which does not explicitly include the tar molecular weight. At high temperatures (or high heating rates) the reference coals more directly decompose to gas and tar, and always form more aromatic char structures. The activation energy of the high-temperature decomposition reactions is in the range of 61–75 kcal/mol.

The description of gas species products is simplified, but it can be easily extended if of interest. Light nonoxygenated gases are H_2, CH_4, and a lumped pseudo-component (CH_2–), which represents the C_2–C_5 hydrocarbons. Main oxygenated products are CO, CO_2, and H_2O. The model accounts for the primary production of minor oxygenated species considering only an equimolar formaldehyde and methanol mixture (Ox–C). Tar species from the different coals are grouped in pseudo-components, whose elemental composition reflects that of the corresponding reference coal. Monoaromatic components (mainly benzene, toluene, and xylene) are accounted in terms of a single lumped component (BTX) with the equivalent formula $C_{6.5}H_7$, which corresponds to an internal molar split B:T:X of 6:3:1. Table 5 reports the whole decomposition mechanism for the three reference coals.

The stoichiometric coefficients of the reaction products are evaluated saving the atomic (H/C/O) balances of the initial reference coal. The very high number of degrees of freedom of this procedure in respect of the few available experimental information could be better satisfied on the basis of new experimental information.

As in the case of biomass pyrolysis model, gas and tar species trapped in the condensed phase and/or in the solid matrix are in graph brackets. The model also accounts for possible annealing effect by considering the dehydrogenation reaction to form a completely carbonaceous structure (CHARC) from a partially hydrogenated char (CHARH, assumed as C_2H, representative of a coronene-like structure: $C_{24}H_{12}$). Successive annealing reactions progressively transform CHARC into a graphitic carbon (CHARG), less reactive and with a more ordered turbostratic graphitic structure (Maffei et al., 2011). An extensive validation of the model can be found in Sommariva et al. (2010). As an example, Fig. 16 shows the comparison between the experimental data (Xu and Tomita, 1987a,b) and the predictions of the solid residue of three different coals of different ranks obtained from a Curie point analyser, by varying the final temperature. Table 6 reports the elemental analysis of these coals and their corresponding characterization in terms of reference coals.

2.4 Municipal Solid Wastes and Refuse-Derived Fuels

There is nowadays a growing attention to the recovery of energy from municipal solid waste (MSW). The appropriate selection of MSW fractions, typically paper, biomass and plastics, and their recycle is a valid alternative to their direct combustion. Production lines of refuse-derived fuels (RDFs) consist of several major subprocesses, including bag ripping, magnetic sorting, shredding, size reduction, and trammel screening. Inerts, such as metal, glass, and ceramic materials, are separated from the MSW stream. Refuse-derived fuels are made by drying, crushing, compressing, and extruding the combustible fraction of MSW into pellets or briquettes. RDF constitutes a promising feed for pyrolysis, gasification, and combustion since it grants high heating value, easy transport, proper size, and a more constant and homogeneous composition. Nevertheless, the differences in composition and in the thermal degradation behavior make the modeling, design, and operation of energy

Table 5 Multistep Kinetic Model of Coal Devolatilization

		A^a	$E_a{}^a$
	$COAL_1$ ($-C_{12}H_{12}-$)		
1	$COAL_1 \rightarrow 5.\ CHAR_H + 0.1\ CHAR_C + 0.2\ H_2 + 0.9\ CH_4 + \{C_{2-5}\}$	2.0×10^8	40,000
2	$COAL_1 \rightarrow \{TAR_1\}$	1.0×10^8	40,000
3	$COAL_1 \rightarrow 5.\ CHAR_H + 0.25\ CHAR_C + 0.5\ H_2 + 0.75\ CH_4 + C_{2-5}$	1.0×10^{14}	75,000
4	$COAL_1 \rightarrow \{TAR_1\}$	1.0×10^{14}	75,000
5	$\{TAR_1\} \rightarrow TAR_1$	2.5×10^{12}	50,000
6	$\{TAR_1\} + CHAR_H \rightarrow 5.3\ CHAR_H + 3.\ CHAR_C + 2.55\ H_2 + 0.4\ CH_4$	2.5×10^7	32,500
7	$\{TAR_1\} + CHAR_C \rightarrow 4.3\ CHAR_H + 4.\ CHAR_C + 2.55\ H_2 + 0.4\ CH_4$	2.5×10^7	32,500
	$COAL_2$ ($-C_{14}H_{10}O-$)		
8	$COAL_2 \rightarrow 2.\ CHAR_C + 3.94\ CHAR_H + 0.25\ COAL_1 + 0.04\ \{BTX\} + 0.31\ \{CH_4\} + 0.11\ \{C_{2-5}\} + 0.11\ \{COH_2\} + 0.15\ \{CO_2\} + 0.41\ \{H_2O\} + 0.18\ \{CO\} + 0.265\ H_2$	6.0×10^{10}	36,000
9	$COAL_2 \rightarrow 0.61\ CHAR_C + 4.33\ CHAR_H + 0.21\ COAL_1 + 0.16\ \{BTX\} + 0.27\ CH_4 + 0.7\ CO + 0.1\ H_2O + 0.2\ \{COH_2\} + 0.28\ H_2$	4.0×10^{18}	63,000
10	$COAL_2 \rightarrow \{TAR_2\}$	5.0×10^{10}	36,000
11	$COAL_2 \rightarrow TAR_2$	4.0×10^{17}	63,000
12	$\{TAR_2\} \rightarrow TAR_2$	2.4×10^9	39,000
13	$\{TAR_2\} + CHAR_H \rightarrow 1.5\ CHAR_C + 7.CHAR_H + \{H_2O\} + 0.5\ CH_4$	4.5×10^9	30,000
	$COAL_3$ ($-C_{12}H_{12}O_5-$)		
14	$COAL_3 \rightarrow 2.73\ CHAR_C + 1.8\ CHAR_H + 0.22\ COAL_1 + 0.08\ \{BTX\} + 0.2\ Ox\text{-}C + 0.1\ \{CH_4\} + 0.11\ \{C_{2-5}\} + 0.2\ H_2 + 0.6\ \{COH_2\} + 2.2\ \{H_2O\} + 0.1\ CO_2 + 0.4\ \{CO_2\} + \{CO\}$	2.0×10^{10}	33,000
15	$COAL_3 \rightarrow \{TAR_3\}$	5.0×10^{18}	61,000
16	$\{TAR_3\} \rightarrow 1.5\ CHAR_H + 0.82\ CHAR_C + 2.08\ CO + 0.25\ Ox\text{-}C + 0.14\ CH_4 + 0.7\ C_{2-5} + 0.5\ CO_2 + 0.47\ \{COH_2\} + 0.16\ \{BTX\} + 0.25\ COAL1 + 1.2\ H_2O + 0.29\ H_2$	1.2×10^8	30,000
17	$COAL_3 \rightarrow \{TAR_3\} + \{CO_2\} + H_2O$	1.6×10^9	33,000
18	$COAL_3 \rightarrow TAR_3 + CO_2 + H_2O$	2.0×10^{18}	61,000
19	$\{TAR_3\} \rightarrow TAR_3$	5.0×10^9	32,500
20	$\{TAR_3\} + CHAR_H \rightarrow 4\ CHAR_H + 2.5\ CHAR_C + 0.2\ \{CH_4\} + 2\ \{COH_2\} + 0.8\ H_2 + 0.3\ C_{2-5}$	1.4×10^8	30,000

[a] $k = A\ \exp\ (-E_a/RT)$ (units are kcal, kmol, m, K, s).

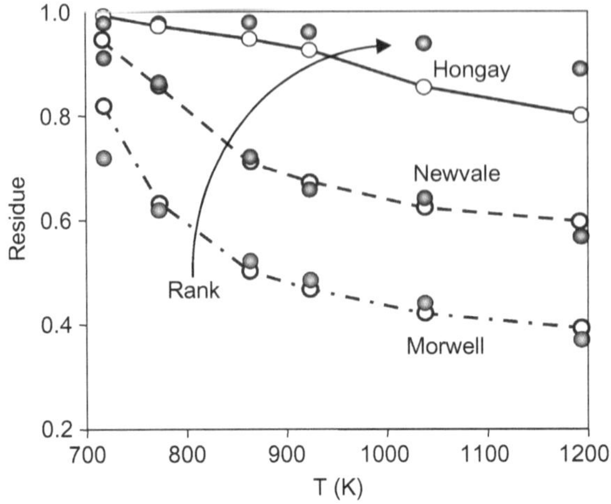

Fig. 16 Solid residues vs final temperature (K). *Symbols with lines*: model predictions. *Filled symbols*: experimental data (Xu and Tomita, 1987a,b).

Table 6 Elemental and Reference Composition of the Coals

	Morwell	Newvale	Hongay
C%	67.94	85.83	95.32
H%	5.04	5.10	3.36
O%	27.02	9.07	1.32
COAL1	0.0000	0.0000	0.3566
COAL2	0.2522	0.9537	0.1604
COAL3	0.7536	0.0356	0.0000
CHARC	0.0122	0.0107	0.4830

After Sommariva S, Maffei T, Migliavacca G, Faravelli T, Ranzi E: A predictive multi-step kinetic model of coal devolatilization, *Fuel* 89:318–328, 2010.

recovery units a challenge (Dou et al., 2007). The kinetic model of RDF and waste is simply based on a linear combination of the devolatilization models of its main constituents: lignocellulosic and plastic materials ($-CH_2-$), together with ash and moisture. Similar to the biomass and coal approach, also the characterization of RDFs is obtained as a suitable combination of three mixtures of reference species, based on the elemental H/C/O balances. The same approach was very recently applied by Younan et al. (2016). Fig. 17 gives a very simple example of a possible selection of reference species useful to

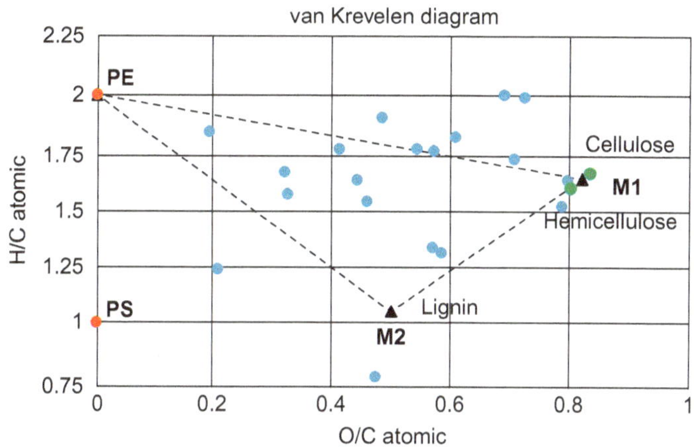

Fig. 17 RDF compositions in the van Krevelen diagram with selected reference mixtures.

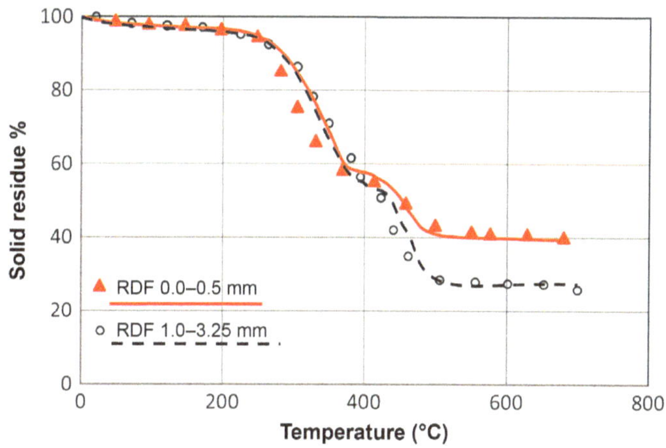

Fig. 18 Effect of RDF particle sizes on TGA at 10K/min (Buah et al., 2007).

characterize refuse-derived fuels. Here two mixtures of lignocellulosic materials are considered together with a more hydrogenated plastic material ($-CH_2-$), simply referred to as PE. Based on the devolatilization models of these reference species, it is then possible to evaluate the TG curves of different RDF samples (Sommariva et al., 2011).

Buah et al. (2007) reported interesting thermogravimetric data of RDF and they highlighted that the particle sizes significantly affect pyrolysis products and heating value of RDF. Fig. 18 shows the TG pyrolysis curves at 10K/min of RDF particles of two different sizes. Predicted curves are obtained by varying RDF composition for fine and coarse particles. Plastic content, responsible of the second devolatilization step at 400–500°C, is higher in coarse particles, while ashes or inert materials are more abundant in fine particles.

This dependence of RDF composition on the particle size was also observed in terms of different heating value by Skodras et al. (2008). They analyzed two RDF samples from different locations with different elemental compositions and heating values: RDF1 (HV = 28.5 MJ/kg) and RDF2 (HV = 21.3 MJ/kg). The higher content of plastic material in RDF1, richer in hydrogen, is evident from the extent of the second devolatilization step in the TG curves of Fig. 19. In a different way, Chouchene et al. (2010) analyzed the effect of the particle size on the thermal degradation of olive solid waste. They simply observed a higher reactivity of samples having sizes lower than 0.5 mm with respect to particles having diameter between 2 and 2.8 mm, possibly because of the transport resistances. The experimental data of the first devolatilization step, reported in Fig. 18, show a very similar behavior.

Fig. 20 summarizes on the van Krevelen diagram the hydrogen, carbon, and oxygen content of the different solid fuels (van Krevelen, 1950). Solid fuels (plastics, biomass, coal, and RDF) are always characterized based on the elemental H/C/O balances, by referring to a suitable combination of only three mixtures based on a very limited number of reference species. As a conclusion, it is possible to observe that the products from the thermochemical treatments of the different solid fuels can be simply obtained as a linear combination of the pyrolysis products of these reference species.

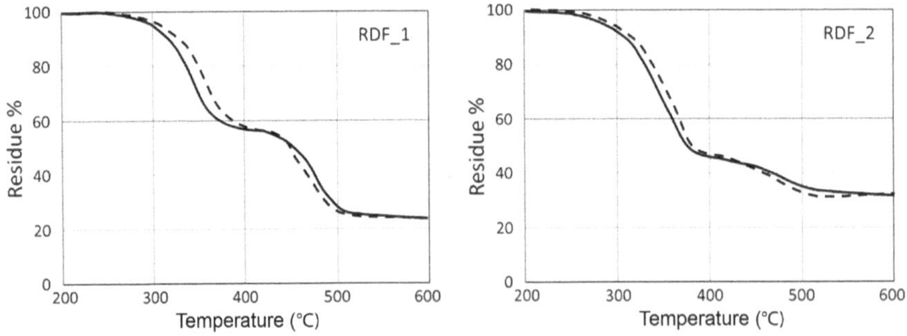

Fig. 19 Experimental (*solid lines*) and predicted (*dashed lines*) weight losses of RDF at 20K/min (Skodras et al., 2008).

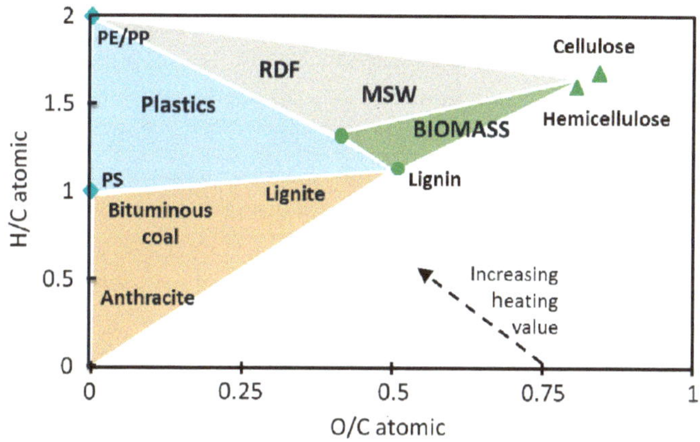

Fig. 20 van Krevelen diagram: atomic H/C and O/C ratios in the different solid fuels. *After van Krevelen DW: Graphical-statistical method for the study of structure and reaction processes of coal,* Fuel *29:269–284, 1950.*

The continuous and progressive reduction of the O/C atomic ratios from biomass to lignite, bituminous coal, and anthracite is clear on this figure. The trend lines in the van Krevelen diagram are indicative of structural relationships among families of compounds. Thus, wood and biomass are the precursors of coal through a coalification process largely characterized by dehydration reactions. Finally, this diagram shows that there is a clear increase of the heating value of the different solid fuels by increasing the H/C and decreasing the O/C ratio.

2.5 Nitrogen and Sulfur Emissions From Solid Fuel Volatilization

One of the main problem arising during solid fuel volatilization is the release of sulfur and nitrogen compounds, which contribute to the pollutant emissions. The characterization of the formation of these compounds is quite difficult, being S and N bond to both organic and inorganic species. Moreover, the number of experimental data about their release is quite limited, especially considering their large variability in terms of amount and molecular structure. Considering coal in particular, because of the quite high values of the content of both N and S, a simplified approach can be found in two specific papers (Maffei et al., 2012, 2013a). This approach allows giving a quantitative estimation of the formation of these compounds.

Less than 5 wt% of sulfur is usually present in coal. The kinetic model of sulfur release first requires the identification of the relative amounts of organic and inorganic sulfur species (Maffei et al., 2012). The inorganic sulfur, slightly more than a half of total sulfur, is not directly bonded but is simply enclosed in the carbon matrix and is made up mostly of pyrite, marcasite, and sulfates. The mass fraction of sulfate is about a 10th of the whole inorganic fraction. It was shown that the inorganic sulfur amount is linearly dependent on the total sulfur content of the coal. Organic sulfur consists of S-atoms inside the carbon structure. Three main families of organic sulfur compounds with different reactivities were identified: aliphatic sulfur (cyclic and aliphatic sulfides, thiols, disulfides, mercaptan); aromatic sulfur (aryl sulfides); and thiophenic sulfur. The distribution of the organic fraction is heavily dependent on the coal's rank or, in other words, on the carbon content. Despite the major uncertainties in these internal distributions, very simple linear relations have been proposed (Maffei et al., 2012).

Two different mechanisms (low and high temperature) compete during the release of the sulfur components. As in the case of coal pyrolysis, the apparent activation energy for the low-temperature mechanism is ~31–40 kcal/mol. At high temperatures, the sulfur species directly decompose to gas and tar components with an activation energy, which varies between 61 and 70 kcal/mol. Fig. 21 shows an example of the model results

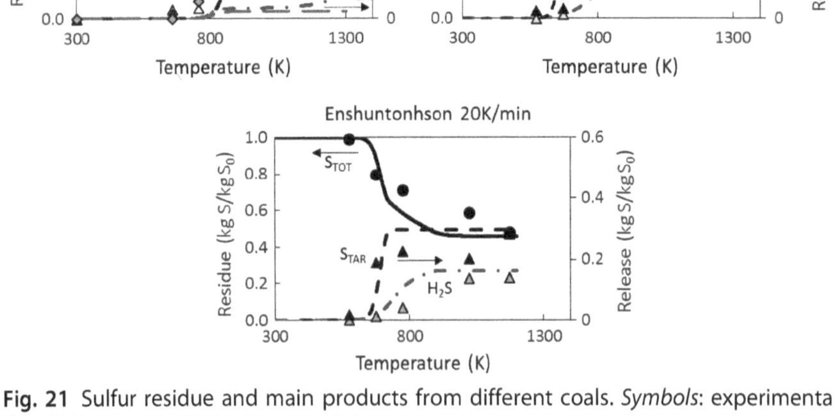

Fig. 21 Sulfur residue and main products from different coals. *Symbols*: experimental data (Miura et al., 2001); *lines*: model predictions.

in comparisons with experimental data (Miura et al., 2001) in terms of residue, tar, and gas formation.

The model predicts belated weight loss in the Illinois coal and a small release at higher temperatures. This corresponds to lower gas release, while the sulfur tar is in reasonable agreement. The initial estimated H_2S formation is in line with the experimental data, but in the temperature range between about 600°C and 800°C, the model indicates an intermediate plateau, which is the result of slower pyrite decomposition.

Quaternary, pyridinic, and pyrolytic compounds are typical forms of the nitrogen in coals (Maffei et al., 2013a). Differently from the sulfur case, these species do not show any evident correlations with the elemental composition or others physical properties of coals. Anyway, the few available experimental information (Chen and Niksa, 1992; Perry, 1999) showed similar behaviors between the nitrogen volatile compounds and the total volatile matter released from coal pyrolysis. On this basis, an analogy between the release of nitrogen components and the release of hydrocarbon species from coal can be presumed. Four nitrogen solid references compounds are assumed $COAL_1$-N, $COAL_2$-N, $COAL_3$-N, and CHAR-N. Thus, according to the elemental composition (in terms of C, H, and O), the nitrogen compounds of each coal are described as a combination of the three reference whose pyrolysis mechanism can be found in Maffei et al. (2013a) and is similar to the one already proposed for coal by Sommariva et al. (2010). The nitrogen model includes low- and high-temperature reactions, with the activation energy of 33–40 and 61–75 kcal/mol, respectively. This multistep kinetic mechanism contains 11 species involved in 17 reactions. Cross-linking and annealing reactions are also accounted for. Once again, the products are not directly released to the gas phase at low heating rate conditions, but they are first entrapped in the metaplastic phase. These metaplastic species are finally released to the gas phase with a kinetics, which represents the volatilization step. At high temperatures, the nitrogen species are directly released to the gas phase. Fig. 22 shows a sample of model comparisons with the experimental data carried out in a wide range of operating conditions (Fletcher and Hardesty, 1992; Genetti et al., 1999; Hambly, 1988).

The NH_3/HCN ratio depends on both the experimental conditions and the coal rank. At low heating rates (Bassilakis et al., 1993) NH_3 formation prevails. At high heating rates, comparable release of HCN and NH_3 is predicted for low-rank coals, while a higher release of HCN in respect of NH_3 is observed for medium- and high-rank coals.

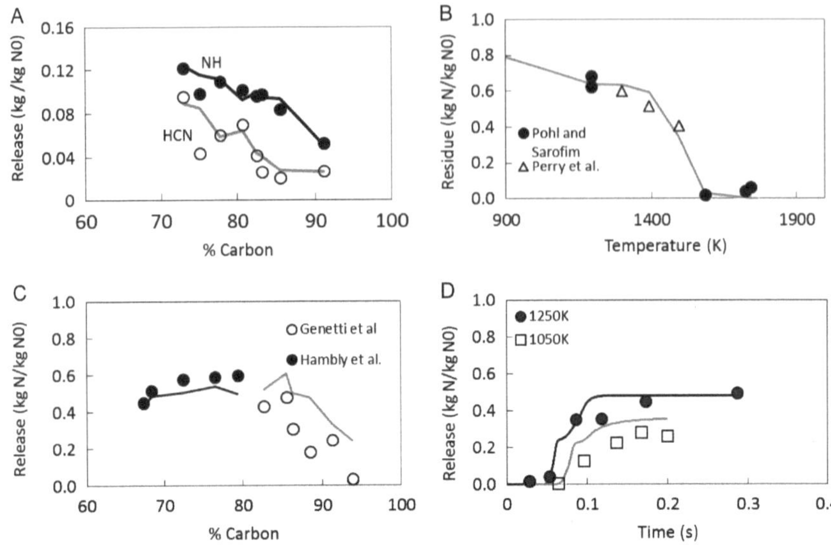

Fig. 22 (A) Release of N as NH_3 and HCN as a function of coal rank (wt% C, DAF) (Bassilakis et al., 1993). Pyrolysis condition: $h = 0.5$K/s, $T_{pyrolysis} = 1173$K, hold time = 3 min; (B) nitrogen residue (Pohl and Sarofim, 1977). Pyrolysis condition: $h = 1$K/s, $T_{pyrolysis} = 900$–2300K, hold time = 20 min; (C) release of total volatile nitrogen (Genetti et al., 1999; Hambly, 1988). Pyrolysis conditions: $h = 10^5$K/s, $T_{pyrolysis} = 1641$K, $t_{pyrolysis} = 18$ ms (Genetti et al., 1999; Hambly, 1988), 78 ms (Fletcher and Hardesty, 1992); (D) release of total volatile nitrogen of Pittsburgh #8 coal (Genetti et al., 1999). Pyrolysis conditions: $h = 10^4$K/s, $T_{pyrolysis} = 1050$–1250K. *Symbols*: experimental data; *line* and *line with symbols*: model predictions ($w_{N,species}/w_{N0,coal}$).

3. HETEROGENEOUS REACTIONS OF RESIDUAL CHAR

The behavior of the heterogeneous gas-solid reactions at the particle scale remains very similar, regardless of gasification or combustion process, fixed or fluid bed reactors. The dimensionless Biot number, that is the ratio between conduction and convection times, is a first useful information to define whether the particle is isothermal ($Bi < 1$: thin particle) or with internal temperature gradients ($Bi > 1$: thick particle) during the conversion process.

Two asymptotic regimes are usually defined to characterize the particle conversion (Wen, 1968):

- a "kinetic controlled regime" for thin particles at low temperatures, where mass and thermal diffusivity are faster than kinetics.
- a "transport controlled regime" for thick particles at high temperatures, where kinetics are very fast and transport resistances are the rate-determining step.

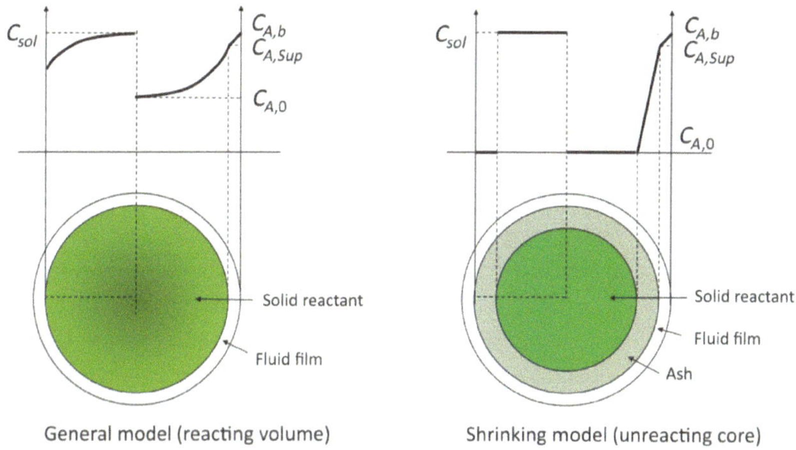

Fig. 23 Concentration profiles of residual solid (C_{sol}) and reactant gas (C_A) for the general model and for the unreacted-core shrinking model.

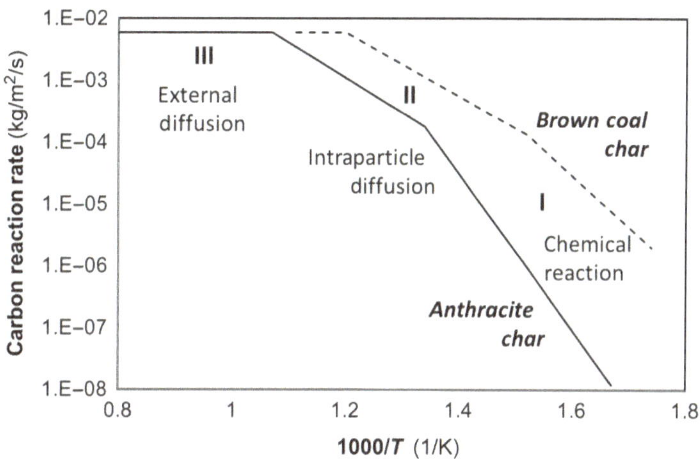

Fig. 24 Controlling regimes in the oxidation reactivity of char particles. *After Griffiths JF, Barnard JA: Flame and combustion, London, UK, 1995, Chapman & Hall.*

Fig. 23 schematically shows these two regimes, in which either the entire volume of the particle is progressively reacting (General Model), or the reaction mainly interests the external solid surface with an unreacted core (shrinking model) and a coherent external ash layer (Wen, 1968).

Coal and char combustion provides a good example of the way gas-solid heterogeneous reactions can take place as a function of different controlling processes. Fig. 24 illustrates the presence of three different regimes during the combustion of two different charcoal particles. These regimes depend on the competition between chemical reaction and reactant diffusion inside the coal particle. Kinetic control occurs when the surface reaction is slow compared with the diffusion process. Thus, regime I predominates typically at low temperatures, since the diffusion is weakly temperature dependent. The higher reactivity of the Brown coal with respect to the anthracite char is recognizable. At very high temperatures (Regime III), the rate of combustion is equal to the rate at which the reactant (normally oxygen) diffuses to the external particle surface. Due to the very high reaction rate, the shrinking core model is a good representation of the system. At intermediate temperatures, reaction rate increases with temperature and oxygen concentration decreases from the surface to the center of the particle. Thiele modulus, that is the ratio between reaction and diffusion times, gives a very useful index of the transition between the first two regimes. For isothermal particles and accounting for the reactant concentration profile, effective reaction rates inside the particle are easily derived (Froment and Bischoff, 1990). When intraparticle diffusion becomes dominant, the apparent activation energy of the process asymptotically becomes one-half of the activation energy of the chemical reaction. At even higher temperatures, the apparent activation energy of the process further decreases when entering the external diffusion regimes, where the reaction takes place only on the external particle surface. In these conditions, both the charcoal particles behave in a similar way, being the external diffusion the rate-limiting step.

During the thermal conversion of the particle, several structural changes interest the solid matrix, shrinking and swelling, ash deposition, pore size variation, and so on. All these morphological variations significantly modify the transport properties of the system and make the solid fuel and char combustion more complex, because of its intrinsic dynamic nature (Di Blasi, 1993).

In thermally thick particles, where the heating and reaction front move from the external surface to the center of the particle, the char heterogeneous reactions are initially inhibited by the diffusion of volatile pyrolysis products (Williams et al., 2012). Often, char gasification and combustion occur after the end of the biomass pyrolysis process, also in fine particles. The surface area and reactive properties of the residual char strongly depend on the solid fuel and on pyrolysis conditions. Despite the high porosity of the char, these heterogeneous reactions are usually the rate-determining step in the overall gasification or combustion process. As already reported in Ranzi et al. (2014) for biomass, Table 7 summarizes

Table 7 Biochar Gasification and Combustion Reactions (Units: kcal, kmol, m^3, K, s)

Reaction	*k*
$CHAR + O_2 \rightarrow CO_2$	$1.2 \times 10^{10} \exp(-32,300/RT)$ [CHAR] $[O_2]$
$CHAR + 0.5\,O_2 \rightarrow CO$	$2.5 \times 10^{11} \exp(-38,200/RT)$ [CHAR] $[O_2]^{0.78}$
$CHAR + H_2O \rightarrow CO + H_2$	$2.5 \times 10^{9} \exp(-52,000/RT)$ $[CHAR]^{0.5}$ $[H_2O]^{0.70}$

Note that [CHAR] is here considered as the ratio of actual char to initial char concentration.

the reference kinetic parameters of char combustion and gasification reactions (Groeneveld and Van Swaaij, 1980; Kashiwagi and Nambu, 1992; Tognotti et al., 1991).

Similar gasification and combustion reactions for the different char species considered in coal pyrolysis model are reported by Maffei et al. (2013b).

4. SECONDARY GAS-PHASE REACTIONS OF RELEASED PRODUCTS

As already mentioned, during the thermochemical conversion of solid fuels, primary volatile products are often exposed to high temperatures, where gas-phase decomposition and combustion reactions can play a significant role (Carstensen and Dean, 2010; Mettler et al., 2012). These secondary gas-phase reactions of released species (tars and gases) are tackled by using an extension of a general and detailed kinetic mechanism of pyrolysis and combustion of hydrocarbon and oxygenated fuels (Ranzi et al., 2012). Due to the modular structure, the extension of the kinetic mechanism to the new species released from solid fuels simply requires to include the primary reactions of these species. Typically, the reaction classes to be included are initiation, H-abstraction, and addition reactions, together with molecular and successive radical decompositions until the formation of intermediate products already accounted for in the kinetic mechanism. The complete kinetic model in CHEMKIN format together with thermodynamic and transport properties of all involved species is available at the website: http://creckmodeling.chem.polimi.it/. The overall dimensions of the kinetic scheme, in terms of species and reactions, need always to be a good compromise between model accuracy and computational efforts. For this reason, tars and heavy species are lumped into a limited number of equivalent components representative of analogous species and/or isomers with similar reactivity. Due to the limited availability of reliable thermochemical data for all these species, the biomass and coal pyrolysis products, such as substituted aromatic compounds (alkyl, hydroxy, methoxy, formyl-aromatics), received particular attention (Catoire et al., 2008; Ince et al., 2015).

As already discussed in a previous paper (Sommariva et al., 2010), the secondary gas-phase reactions during coal devolatilization can explain relevant temperature and pressure effects, not only related to the primary solid-phase reactions. The formation of soot particles due to successive gas-phase reactions of polycyclic aromatic hydrocarbons (PAHs) during coal devolatilization was also discussed by Mitra et al. (1987). Similarly, the secondary reactions of volatile species from biomass pyrolysis are responsible for the reduction of oil and the increase of gas yields, at high temperatures. Thus, it is possible to find the maximum in bio-oil formation and the corresponding optimal operating temperature and conditions (Bridgwater, 2003; Calonaci et al., 2010; Di Blasi, 2008). Recently, Norinaga et al. (2014) discussed the reaction pathways leading to benzene and naphthalene in cellulose vapor-phase cracking.

When discussing the secondary gas-phase pyrolysis of biomass, Carstensen and Dean (2010) clearly highlighted that it is not feasible to perform ab initio high-level calculations of the rate constants for all the reactions, because of the large dimension of the kinetic model. From first-principle calculations, they systematically derived kinetic laws on a series of small reactants for several reaction classes and used these data to generate rate estimation rules, to be extrapolated to all members of the same reaction class.

While coal and several plastics (PE, PP, PS) mainly release hydrocarbon species with a low or no oxygen content, tar components released by biomasses are typically carbohydrates, phenolics, alcohols, and aldehydes, together with species with two or more oxygenated groups. Table 8 reports a list of relevant oxygenated species released from biomasses and involved in the POLIMI kinetic mechanism, whose primary decomposition reactions are shortly discussed in this Section.

4.1 Generic Rate Rules for H-Abstraction Reactions

H, OH, and CH_3 are the dominant reactive radicals in pyrolysis and oxidation conditions. Since several years (Dente et al., 1979, 1992), the kinetic modeling of steam cracking reactions highlighted that the rate constant of H-abstraction reactions from pure hydrocarbons can be obtained, with reasonable accuracy, by generic rate rules. H-abstraction or metathesis reactions are written in the generic form:

Table 8 Formation Enthalpy $\Delta H_{f,298}$ (kcal/mol) and Formation Entropy $\Delta S_{f,298}$ (cal/mol/K) of Major Oxygenated Species Released From Biomasses

Chemical Name		ΔH_f	ΔS_f
Glyoxal	$C_2H_2O_2$	−50.6	65.4
Acetaldehyde	C_2H_4O	−39.5	63.0
Hydroxy-acetaldehyde	$C_2H_4O_2$	−73.5	73.5
Ethyleneglycol	$C_2H_6O_2$	−92.0	76.3
Hydroxyl-oxo-propanal	$C_2H_4O_3$	−102.7	88.4
Acrolein	C_3H_4O	−20.3	67.4
Propanedial	$C_3H_4O_2$	−62.4	73.7
Propanal	C_3H_6O	−45.3	72.8
1-Propanol	C_3H_8O	−60.9	76.4
2-Propanol	C_3H_8O	−65.5	74.5
Acetol	$C_3H_6O_2$	−87.4	80.6
3-Hydroxypropanal	$C_3H_6O_2$	−80.3	83.3
1,3-Propanediol	$C_3H_8O_2$	−45.5	86.0
Glycerol	$C_3H_8O_3$	−137.1	95.8
Furan	C_4H_4O	−10.2	60.2
Butanedione	$C_4H_6O_2$	−78.4	84.2
C4 O-heterocycles	C_4H_8O	−27.7	73.6
Furfural	$C_5H_4O_2$	−36.1	77.8
Xylosan	$C_5H_8O_4$	−151.6	104.8
Phenol	C_6H_6O	−23.0	75.3
Hydroxymethyl-furfural	$C_6H_6O_3$	−79.8	98.2
Levoglucosan	$C_6H_{10}O_5$	−200.9	113.5
Anisole	C_7H_8O	−17.1	84.0
Syringol	$C_8H_{10}O_3$	−95.3	111.0
Coumaryl alcohol	$C_9H_{10}O_2$	−49.2	109.0
Sinapyl aldehyde	$C_{11}H_{12}O_4$	−0.3	186.7

$$R^{\bullet} + R'H \leftrightarrow RH + R^{\bullet\prime}$$

where $R^{\bullet}$ is the H-abstracting radical. Rate constant of this reaction can be decomposed in the product of two terms:

$$k_f = k^0_{ref,R} \cdot C_{R'H}$$

where $k^0_{ref,R}$ represents the reference rate constant of the R radical to abstract an H-atom from a methyl group and $C_{R'H}$ is the reactivity of the specific H-atom with respect to the primary one (Ranzi et al., 1994b). Fig. 25 shows the rate constants of H-abstraction from primary, secondary, tertiary, allyl, and vinyl positions. These rate constants greatly correlate with the corresponding C–H BDEs. Rate constants of H-abstraction reactions from aromatics to form phenyl-like radicals are similar to those of H-abstraction of a vinyl H-atom. On the other side, the rate constant to form benzyl-like radicals is more similar to the one to form allyl radicals (Violi et al., 2004).

Similar generic rate rules can be formulated not only for abstraction reactions involving different H-sites in hydrocarbons but also in oxygenated species. Thus, in order to verify the reactivity of the H-atoms close to carbonyl of hydroxyl groups, Fig. 26 shows the BDEs (kcal/mol) for butane, 1-butanol, and butanal calculated at G4 level (0K). These values, very useful to understand and define the relative reactivity of the different H-atoms, well agree with the similar ones recently discussed by Oyeyemi et al. (2014a,b). The differences in the BDEs allow to explain the relative selectivities of the H-abstraction reactions from the different H-sites, as reported in Fig. 27 for *n*-butanol and butanal with respect to H-abstraction by OH radical.

The rate constant of the H-abstraction from the hydroxyl hydrogen by a generic radical (H, OH, and CH_3) is lower than the corresponding reference rate parameters of the abstraction of a single primary H-atom, because of the higher BDE. Moreover, the nature of the alkyl group does not significantly affect this value. Successive decomposition reactions of alkoxyl radicals are discussed by Curran (2006).

Fig. 27 shows that the H-abstractions from C–H sites in the α-position to the hydroxyl group are the dominant one in *n*-butanol. In fact, the corresponding BDE is lower than one of the secondary H-atoms. Generic rate rule for the rate parameters of the abstraction of a secondary H-atom in α-position to the hydroxyl group (such as those of ethanol, *n*-propanol, 1-butanol, and iso-butanol) are assumed ~1.5–2 times faster than the corresponding rate parameters of the secondary H-atoms (Fig. 25). Similarly,

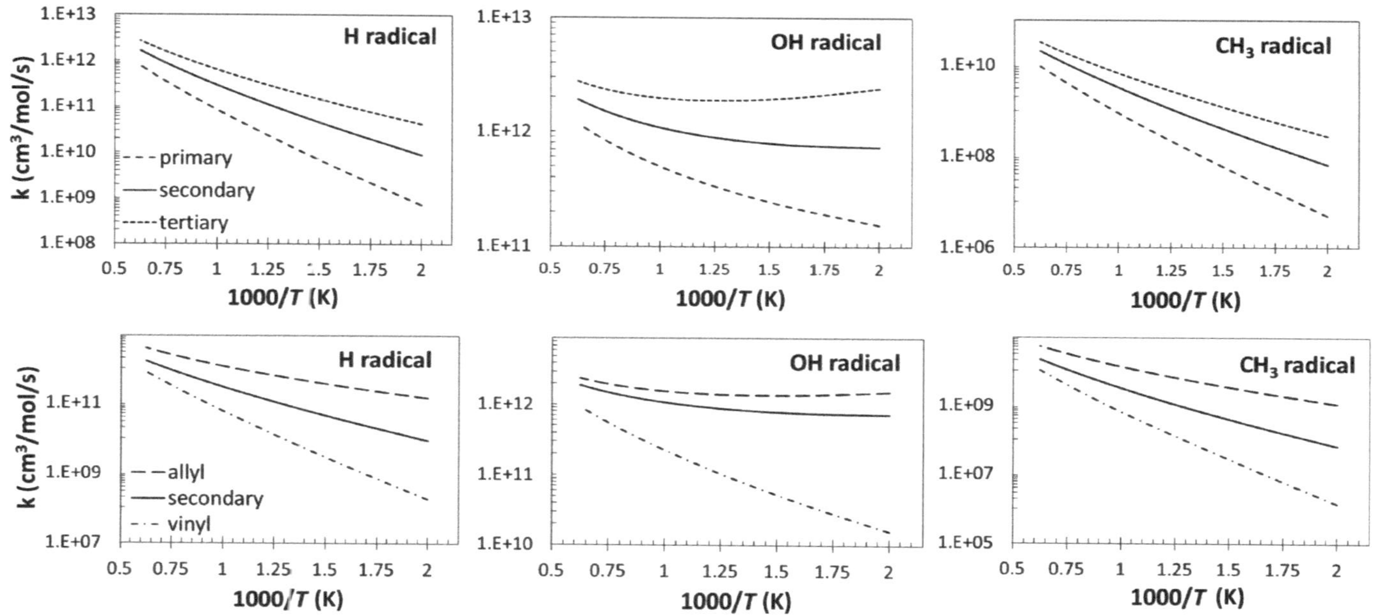

Fig. 25 H-abstraction reactions. Rate constants (per H-atom) for simple primary, secondary, tertiary H-atoms (*top*) and for secondary H-atoms in alkyl, vinyl, and allyl sites (*bottom*).

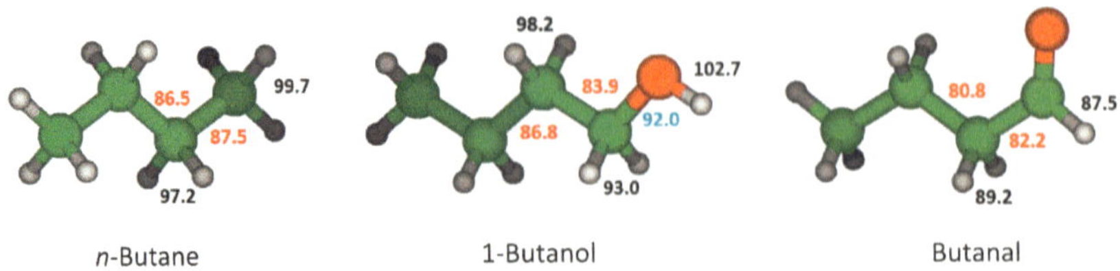

Fig. 26 Bond dissociation energies (kcal/mol) for butane, 1-butanol, and butanal calculated at G4 level (0K). *Adapted from Pelucchi M, et al: Relative reactivity of oxygenated fuels: alcohols, aldehydes, ketones and methyl esters,* Energy Fuel *2016, doi:10.1021/acs.energyfuels.6b01171.*

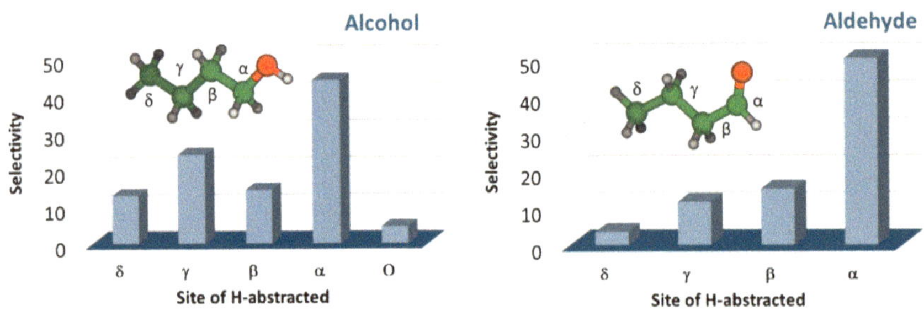

Fig. 27 Relative selectivities of the H-abstraction reactions by OH radical from the different H-sites in *n*-butanol (*left*) (Frassoldati et al., 2012) and butanal (*right*) (Pelucchi et al., 2015).

the abstraction parameters of tertiary H-atoms (such those of iso-propanol and 2-butanol) are ~1.5 times the corresponding ones of tertiary H-atoms in alkanes. Only short-range forces (i.e., in the order of magnitude of the bond length) can affect the reaction rates (Benson, 1976). Therefore, the influence of the OH group on the reactivity of C–H bonds practically vanishes at the β-position, also in agreement with Carstensen and Dean (2010).

The removal of the acyl H-atom is highly favored in aldehydes, due to the low BDE of the C–H bond in the carbonyl group (Gray et al., 1981). Based on kinetic studies on formaldehyde, acetaldehyde, and heavier aldehydes (Pelucchi et al., 2015), we assumed the following rate parameters for the abstraction of the acylic H-atom by H, OH, and CH_3 radicals:

$$k^{H} = 2.5 \times 10^{14} \exp(-6360/RT) \quad (cm^3/s/mol)$$
$$k^{OH} = 2.0 \times 10^{13} \exp(-630/RT) \quad (cm^3/s/mol)$$
$$k^{CH_3} = 7.0 \times 10^{11} \exp(-7235/RT) \quad (cm^3/s/mol)$$

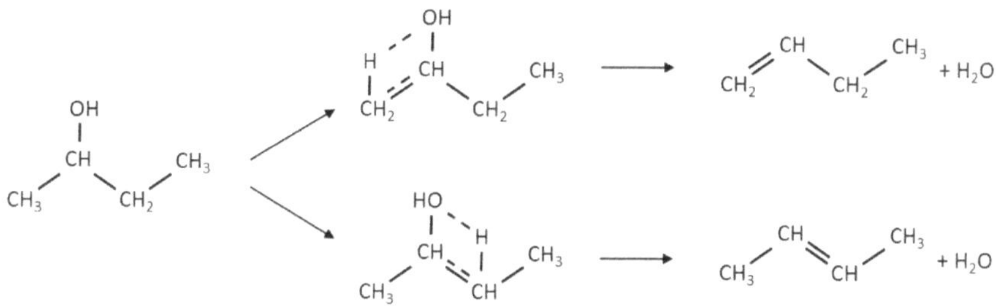

Fig. 28 Water elimination reactions of 2-butanol to form 1-butene and 2-butene, via four center molecular reactions.

All these generic rate rules for H-abstraction reactions are useful to create a first reasonable set of rate parameters for the secondary gas-phase reactions. Successive rate and sensitivity analysis are then able to identify sensitive reactions, requiring more accurate evaluations.

4.2 Alcohols, Carbohydrates, and Water Elimination Reactions

As already discussed by Carstensen and Dean (2010) and also highlighted in the kinetic studies of alcohol fuels (Frassoldati et al., 2010; Grana et al., 2010), water elimination reactions are an important class of molecular reactions. Thus, Fig. 28 shows the water elimination reactions of 2-butanol to form 1-butene and 2-butene, via four center molecular reactions.

According to Grana et al. (2010), the following rate constants are assumed for the different butanol isomers:

$$
\begin{array}{ll}
1-C_4H_9OH \rightarrow 1-C_4H_8+H_2O & k=10^{14}\exp(-67{,}600/RT)\ (s^{-1}) \\
2-C_4H_9OH \rightarrow 2-C_4H_8+H_2O & k=10^{14}\exp(-66{,}100/RT)\ (s^{-1}) \\
2-C_4H_9OH \rightarrow 1-C_4H_8+H_2O & k=1.5\times10^{14}\exp(-67{,}100/RT)\ (s^{-1}) \\
iso-C_4H_9OH \rightarrow iso-C_4H_8+H_2O & k=5\times10^{13}\exp(-65{,}600/RT)\ (s^{-1}) \\
tert-C_4H_9OH \rightarrow iso-C_4H_8+H_2O & k=4.5\times10^{14}\exp(-65{,}100/RT)\ (s^{-1})
\end{array}
$$

These kinetic values, well confirmed by the recent review of Sarathy et al. (2014), show that reference rate parameters for this reaction class are site specific, i.e., the position of the OH group inside the carbon skeleton affects the kinetic constants. While the different alcohol has little impact on the reference rate constant, large deviations are observed for substituted aldehydes, when water elimination reactions form unsaturated species with conjugated double bonds.

Fig. 29 depicts the two successive molecular dehydration reactions in glycerol pyrolysis (Barker-Hemings et al., 2012).

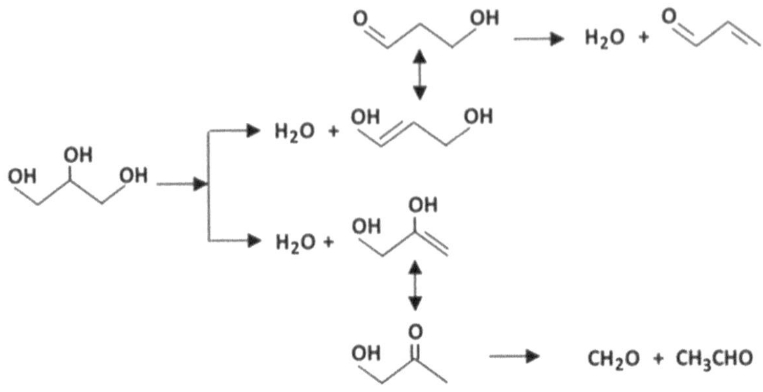

Fig. 29 Water elimination reactions in glycerol pyrolysis.

Prop-1-ene-1,3-diol or prop-1-ene-2,3-diol is formed through the first dehydration reaction. The isomerization, via keto-enol tautomerism, transforms prop-1-ene-1,3-diol into 3-hydroxypropanal, which rapidly forms acrolein through a second dehydration reaction. The aldehyde moiety strongly influences the reactivity and stabilizes the transition state and the products with a reduction of the activation energy of more than 10 kcal/mol (Carstensen and Dean, 2010).

These reactions govern the first molecular dehydration with ring opening of carbohydrates, specifically of levoglucosan and xylan. Together with the molecular dehydration and the chain initiation reactions, H-abstraction reactions for these species are always considered with the previously referred rules. Primary radicals progressively decompose forming the major intermediates, such as formaldehyde, hydroxyl-acetaldehyde, glyoxal, acetol, and other small-oxygenated components. A second dehydration reaction forms the 5-hydroxymethyl-furfural ($C_6H_6O_3$: HMFU), whose successive reactions form furfural ($C_5H_4O_2$) and furfuryl-alcohol ($C_5H_6O_2$) (Shen et al., 2015). According to Carstensen and Dean (2010), retro-Diels–Alder reactions are molecular reaction paths to form C_2–C_4 oxygenated species.

4.3 Secondary Gas-Phase Reactions of Aromatics. PAH and Soot Formation

As already observed in Table 5, tar components released by coal pyrolysis largely belong to alkenes and aromatic species, such as benzene, toluene, xylene, and phenol, together with naphthalene and phenanthrene. As already discussed by Mitra et al. (1987), soot formation from coal devolatilization is also due to the successive pyrolytic reactions of the released aromatic and PAH species. They also observed the largest release of PAH and soot formation from bituminous coals, while the anthracitic coals, releasing the lowest amount of PAH species, were also lower than lignitic coals in sooting tendency.

Saggese et al. (2013) revised and discussed the intermediate- and high-temperature reactions of benzene with particular attention to the successive reaction paths to form heavy PAHs and soot (Saggese et al., 2015). Together with the kinetics of benzene pyrolysis and oxidation, also phenol reactions were carefully discussed. In fact, the high-temperature reactions of phenol are also important in pyrolytic and combustion systems. Moreover, phenol and phenolic species merit special attention, for their presence not only as tar components released by lignins but also as precursors of dibenzofurans and possibly of dibenzodioxins. Fig. 30 shows a comparison of model predictions with experimental data of phenol thermolysis in H_2, in a flow reactor at atmospheric pressure (Manion and Louw, 1989).

Kinetic studies on phenol, cresol, and anisole chemistry highlight the importance of CO elimination from unsubstituted and substituted phenoxy radicals. The reference reaction rate refers to the phenoxyl radical:

$$\mathrm{C_6H_5O \rightarrow cyC_5H_5 + CO} \quad k = 5 \times 10^{11} \exp(-43{,}920/RT) \quad \left(\mathrm{s^{-1}}\right)$$

Similar reactions to form CO and cyclopentadienyl radicals from phenoxy-substituted species were discussed by Carstensen and Dean (2010). Successive reactions of cyclopentadienyl radicals are responsible for the formation of naphthalene and heavier PAHs (Djokic et al., 2014), as clearly evident from the species reported in Fig. 30. While phenol and cresol were extensively investigated for their interest in combustion systems (Brezinsky et al., 1998), anisole ($C_6H_5OCH_3$) was mainly studied as a very simple surrogate of tar from lignin pyrolysis (Barker-Hemings et al., 2011; Nowakowska et al., 2014).

Chain initiation reactions of aromatic species containing one or more methoxy groups ($-OCH_3$) involve the breaking of the weak O–CH_3 bond (BDE ~63.2 kcal/mol) (Brezinsky et al., 1998; Pecullan et al., 1997). Indicating with Ph the unsubstituted or substituted phenyl groups, we have the following reference reaction (Barker-Hemings et al., 2011):

$$\mathrm{Ph - OCH_3 \rightarrow Ph - O + CH_3} \quad k = 3 \times 10^{15} \exp(-63{,}200/RT) \quad \left(\mathrm{s^{-1}}\right)$$

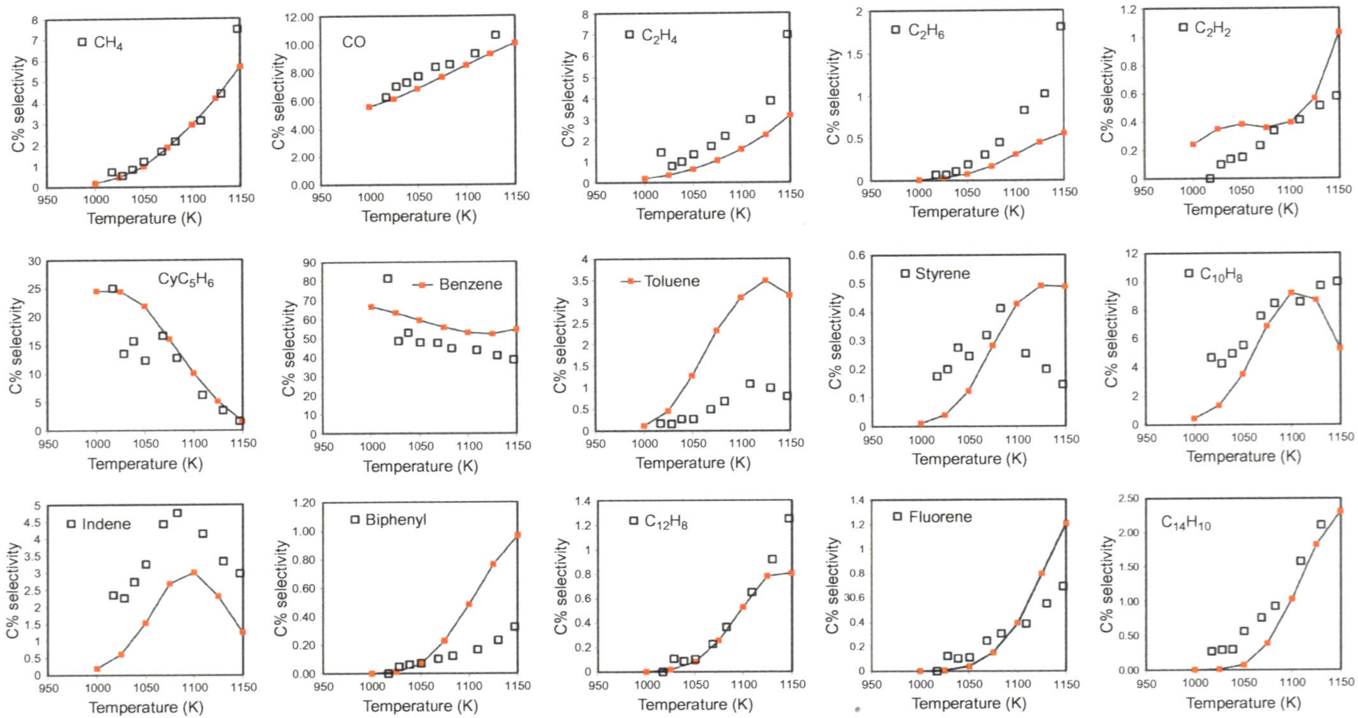

Fig. 30 Carbon selectivity of main and minor products in the thermolysis of phenol/H_2 mixtures. Experimental data (*symbols*) and model predictions (*lines with small symbols*) (Manion and Louw, 1989).

Table 9 Reference Reactions of the ipso-Addition Reaction Class

H + anisole → phenol + CH_3 $k = 1 \times 10^{13} \exp(-6000/RT)$ $(cm^3/s/mol)$
H + phenol → benzene + OH $k = 1.2 \times 10^{13} \exp(-6000/RT)$ $(cm^3/s/mol)$
H + anisole → benzene + OCH_3 $k = 1 \times 10^{13} \exp(-8000/RT)$ $(cm^3/s/mol)$
OH + toluene → cresol + H $k = 1.1 \times 10^{12} \exp(-11{,}000/RT)$ $(cm^3/s/mol)$
OH + toluene → phenol + CH_3 $k = 4.4 \times 10^{12} \exp(-6700/RT)$ $(cm^3/s/mol)$
CH_3 + phenol → cresol + H $k = 1.3 \times 10^{12} \exp(-16{,}200/RT)$ $(cm^3/s/mol)$
CH_3 + phenol → toluene + OH $k = 1 \times 10^{12} \exp(-15{,}000/RT)$ $(cm^3/s/mol)$

The same rate rules are also assumed when replacing CH_3 with a different alkyl radical.

The ipso-addition reactions constitute a further important reaction class. Table 9 gives some reference rate parameters for a sample of reference reactions, mainly derived from the kinetic studies of pyrolysis and oxidation of phenol and anisole. All these reactions progressively convert the aromatic and phenolic species.

4.4 Secondary Gas-Phase Reactions of Cellulose and Lignin Products

Recently, Norinaga et al. (2013, 2014) and Yang et al. (2015) developed a two-stage tubular reactor for evaluating first the rapid biomass pyrolysis and then the secondary reactions of the products from cellulose and lignin pyrolysis, while minimizing the interactions among char and volatile species. They investigated the pyrolysis system at residence times up to 6 s in a wide temperature range. These experimental data are interesting tests for the validation of the secondary gas-phase reactions of released species. Fig. 31 shows some comparisons of the time evolution of predicted and experimental yields of several products from cellulose (Norinaga et al., 2013). The model correctly predicts the time evolution of the most abundant products (CO, H_2O, CH_4, H_2, and methane) as well as the decomposition of heavy tar species and the formation of benzene.

Fig. 32 shows similar comparisons between experimental data of Yang et al. (2015) and model predictions for the pyrolysis products

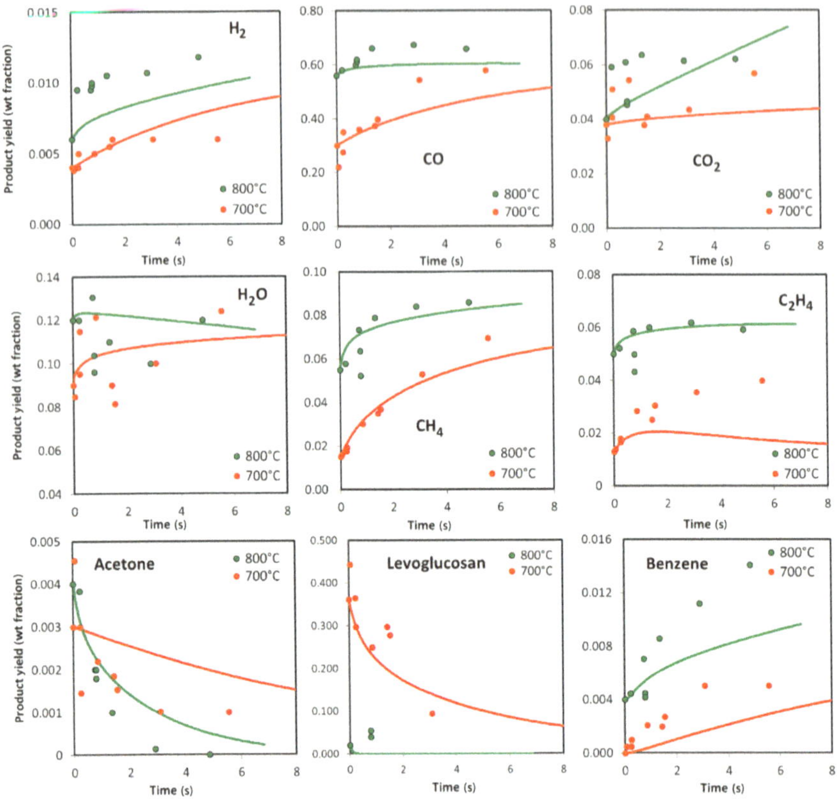

Fig. 31 Secondary pyrolysis of cellulose products at 700 and 800°C. Comparison of experimental (*symbols*) and predicted yields (*solid lines*) of CO, CO_2, methane, H_2O, H_2, ethylene, levoglucosan, acetone, and benzene (Norinaga et al., 2013).

of lignin samples prepared by enzymatic hydrolysis (EHL) of empty fruit bunches, in the temperature range of 773–1223K. The primary volatile products released by lignin include a large amount of heavy undetected phenolic species (>30% at 773K). In simulating the experimental data of Fig. 32, phenolic species, including the undetected ones, were distributed according to the prediction of the primary devolatilization model of the EHL sample. It is also relevant to observe that, because of the very high temperatures and the severe pyrolysis conditions, there is significant formation of heavy PAHs and soot. Model predictions indicate the formation of ~5% of species heavier than C_{20}, in agreement with the experimental observation of early soot deposition on reactor walls earlier 1023K.

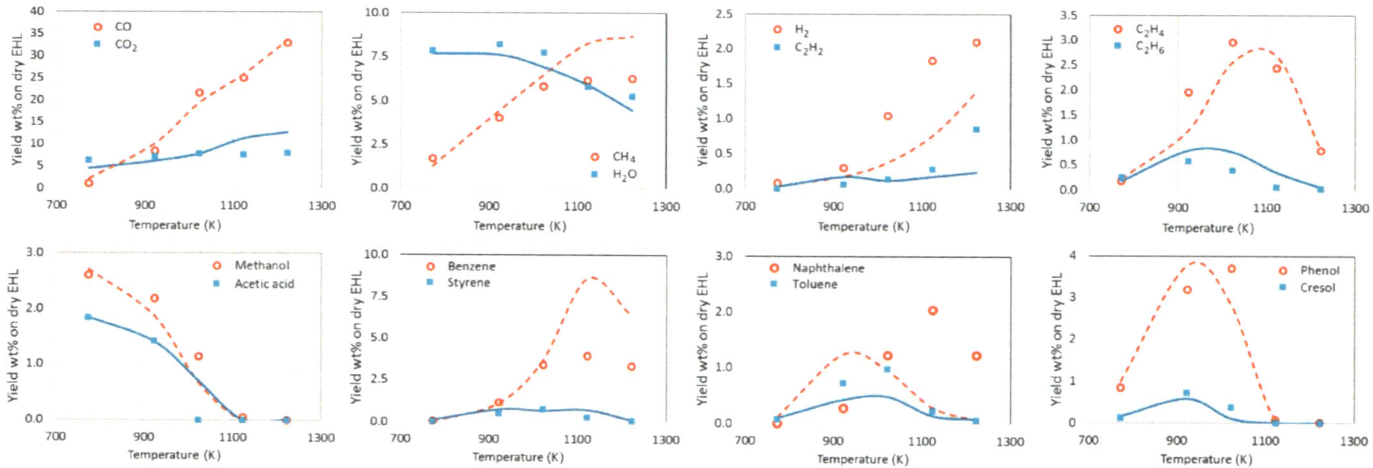

Fig. 32 Secondary gas-phase reactions of volatile products from lignin pyrolysis at 3.6 s. Comparison between model predictions (*lines*) and experimental data (*symbols*) (Yang et al., 2015).

5. BALANCE EQUATIONS AT THE PARTICLE SCALE (FROM GENERAL TO 1D-MODEL)

Intra- and interphase heat and mass transfer phenomena need to be considered and coupled with the kinetics when modeling reactors treating thick particles. According to previous works (Pierucci and Ranzi, 2008; Ranzi et al., 2011), a convenient way to present the mass and energy balance equations is to distinguish the particle and the reactor scale. The particle model should be able to predict temperature profiles and product distribution as a function of time. This model requires not only reaction kinetics but also reliable rules for estimating transport properties to account for morphological changes during the pyrolysis process. Biomass particles shrink by as much as 60–70% in the different directions, during the conversion process. Heat transfer should account for variable transport properties during the pyrolysis process: namely, in virgin biomass, dry and reacting biomass, and the residual char (Di Blasi, 1993, 2008).

The mathematical model of the particle solid fuel evolution can be described based on the fundamental governing equations of conservations of total mass, momentum, and energy for both fluid and solid phases. The particle is conveniently considered as a porous medium including both the solid volume and the fluid contained inside its pores. Heat transfer occurs by conduction, convection, and radiation. A local thermal equilibrium between the solid and the gas phase is assumed, being the Peclet number for heat transfer sufficiently large.

Continuity equation

$$\frac{\partial}{\partial t}\left(\rho^G\varepsilon\right)+\nabla\left(\rho^G\mathbf{u}\right)=\dot{\Omega} \tag{1}$$

Momentum equation

$$\frac{\partial}{\partial t}\left(\rho^G\varepsilon\mathbf{u}\right)+\nabla\left(\rho^G\mathbf{uu}\right)=-\nabla p+\nabla\cdot\left[\mu\left(\nabla\mathbf{u}+\nabla\mathbf{u}^T\right)-\frac{2}{3}\mu(\nabla\mathbf{u})\mathbf{I}\right]+\rho^G\mathbf{g}+\mathbf{S} \tag{2}$$

The source term **S** showed in Eq. (2) is computed according the Darcy–Forchheimer law:

$$\mathbf{S}=-\left(\mu\mathbf{Da}+\frac{1}{2}\rho^G|u_{kk}|\mathbf{F}\right)\mathbf{u} \tag{3}$$

S is composed of two terms, a viscous loss term and an inertial loss term, creating a pressure drop that is proportional to the velocity and velocity squared, respectively.

Gas-phase species equations

$$\frac{\partial}{\partial t}\left(\rho^G \varepsilon \omega_k^G\right) + \nabla\left(\rho^G \mathbf{u} \omega_k^G\right) = -\nabla j_k + \dot{\Omega}_k \quad k = 1, \dots, NC^G \tag{4}$$

Solid-phase species equations

$$\frac{\partial}{\partial t}\left(\rho^S (1-\varepsilon) \omega_k^S\right) + \nabla\left(\rho^S \mathbf{u} \omega_k^S\right) = \dot{\Omega}_k \quad k = 1, \dots, NC^S \tag{5}$$

Energy equation

$$\begin{aligned} &\hat{C}_p^G \frac{\partial(\rho^G \varepsilon T)}{\partial t} + \hat{C}^S \frac{\partial(\rho^S (1-\varepsilon) T)}{\partial t} + \hat{C}_p^G \nabla\left(\rho^G \mathbf{u} T\right) \\ &= -\nabla q - \left.\frac{\partial \ln \rho^G}{\partial \ln T}\right|_p \frac{Dp}{Dt} + \dot{Q}_R - \nabla j_k \sum_{k=1}^{NC^G} \hat{H}_k \end{aligned} \tag{6}$$

The solution of this system is quite complex, taking into account the relatively high number of components, the potential anisotropy of the particle, and the need of a moving mesh to describe the particle shrinking. A simplified condition refers to an isotropic, spherical particle. In this case, it is possible to discretize the system with an onion-like structure of concentric NS shells, as sketched in Fig. 33.

Some other assumptions can be adopted to simplify the solution. The characteristic times of the gas phase are much lower than those of the solid phase; thus, it is possible to assume steady-state conditions for the momentum conservation. Moreover, when the flow velocity is very low, as in this case, and the flow is steady, the inertial term and the quadratic drag force term can be neglected in the momentum equation. Thus, substituting the

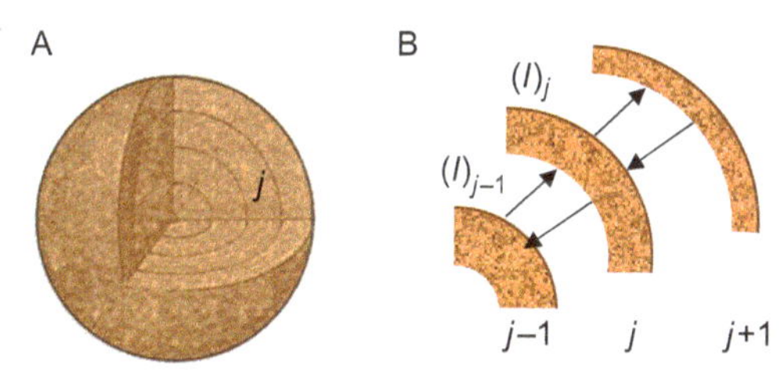

Fig. 33 Discretization of an isotropic, spherical particle.

source term (3), disregarding the very low Forchheimer contribution, in the momentum equation in radial direction, it is possible to obtain:

$$0 = -\frac{\partial p}{\partial r} - \mu \mathrm{Da} u_r \tag{7}$$

The radial velocity is then estimated

$$u_r = -\frac{1}{\mu \mathrm{Da}} \frac{\partial p}{\partial r} \tag{8}$$

The kth species balance in the gas phase of the shell j can be expressed as

$$\frac{dm_{k,j}^G}{dt} = j_{k,j-1} S_{j-1} + j_{k,j} S_j + \dot{\Omega}_{k,j}^G V_j^G \quad k = 1, \ldots, NC^G \tag{9}$$

where

$$j_{k,j} = +\mathcal{D}_{k,j}^{(I)} \rho_j^{(I)} \frac{\omega_{k,j+1}^G - \omega_{k,j}^G}{r_{j+1} - r_j} - \rho_j^{(I)} \omega_{k,j} \frac{1}{\mu_j^{(I)} \mathrm{Da}} \frac{p_{j+1} - p_j}{r_{j+1} - r_j} \tag{10}$$

are the mass fluxes induced by both the ordinary diffusion and the Darcy law.

Eq. (10) approximates the derivative as finite differences. The radial velocity is derived from Eq. (8). The superscript symbol (I) refers to the interface. Properties at the interface are evaluated by interpolating the values of the center of the adjacent shells. S_j is the outer surface of the shell j, while the radius r_j refers to the center of the shell. $\mathcal{D}_{k,j}$ is the effective diffusion coefficient of the kth component at the interface between shell j and $j+1$.

The balance equation of the solid species for each shell is:

$$\frac{\partial m_{k,j}^S}{\partial t} = \dot{\Omega}_{k,j}^S V_j^S \quad k = 1, \ldots, NC^S \tag{11}$$

Finally, the discretized energy balance of shell jth can be easily derived:

$$\begin{aligned} &\frac{dT_j}{dt} \left(\hat{C}_p^G \sum_{k=1}^{NC^G} m_{k,j}^G + \hat{C}^S \sum_{k=1}^{NC^S} m_{k,j}^S \right) + \sum_{k=1}^{NC^G} \hat{h}_{k,j}^G \frac{dm_{k,j}^G}{dt} + \sum_{k=1}^{NC^S} \hat{h}_{k,j}^S \frac{dm_{k,j}^S}{dt} \\ &= -q_{j-1} S_{j-1} + q_j S_j + V_j \dot{Q}_{Rj} \end{aligned} \tag{12}$$

where

$$q_j = \kappa_j^{(I)} \frac{T_{j+1} - T_j}{r_{j+1} - r_j} - \sum_{k=1}^{NC^G} \hat{h}_{k,j}^G j_{k,j} \tag{13}$$

The specific heats are evaluated neglecting mixing effects. The pressure work induced by the gas expansion is not considered, being its effect very low.

$NC^G + NC^S + 1$ equations (9), (11), (13) together with the ideal gas law have to be solved, assuming the following boundary conditions:

$$\begin{aligned} \text{Pressure (Dirichlet condition)}: p_{NL} &= \bar{p} \\ \text{Gas species (flux continuity)}: k_c\left(\rho_k^{bulk} - \rho_{k,NL}\right) &= j_{k,NL} \quad k = 1, \ldots, NC^G \\ \text{Solid species (Neumann condition)}: m_{k,NL}^S &= m_{k,NL-1}^S \quad k = 1, \ldots, NC^S \\ \text{Temperature (flux continuity)}: h\left(T^{bulk} - T_{NL}\right) + q_{rad} &= q_{NL} \end{aligned}$$

Hereinafter some application examples of solid fuel pyrolysis, gasification, and combustion are investigated at the particle scale, including also the effect of secondary gas-phase reactions. The model is first applied to investigate the temperature profiles during the pyrolysis of thick biomass particles. A second example discusses and analyzes the gasification of thick biomass particles, emphasizing the need of a comprehensive model to foresee the possible presence of a combustion regime, because of the competition between transport phenomena and kinetic processes. A third and final example refers to the fast pyrolysis of biomass and the optimal operating temperature for the maximum bio-oil yield.

5.1 Pyrolysis of Thick Biomass Particles and Overshooting of the Internal Temperature

Large biomass particles are often used when charcoal is the desired product or when rapid heating rates are not required. The slow pyrolysis of wood chips and centimeter-scale wood particles is useful to optimize the production of biochar or charcoal for soil amendment (Bennadji et al., 2013; Wang et al., 2013). From a different perspective, pyrolysis of centimeter-scale wood particles provides a very sensitive test of kinetic models of biomass pyrolysis, especially with respect to the thermochemistry. Corbetta et al. (2014) already discussed how the competition between the exothermic charification reactions and the heat resistances inside the wood particle could result in an overshooting of the internal temperature. Park et al. (2010) already observed this interesting feature when studying the pyrolysis of a spherical wood particle at moderate temperatures (638–879K). They measured the global mass losses, along with the temperature profiles at the surface and center of the particle. After an initial increase of the core temperature, the temperature profile exhibits a plateau followed by a sharp peak, which overtakes the surface temperature profile. A similar behavior was also observed by Milosavljevic et al. (1996) in the study of the thermochemistry of cellulose pyrolysis.

Following the approach of Paulsen et al. (2013) the pyrolysis and Biot numbers are conveniently used to compare heat transfer and reaction timescales for the fuel particles. Biot number is very useful to evaluate the relative importance of external convective and internal conductive heat transfer:

$$Bi = \frac{h \cdot R_p}{k_p}$$

where h is the external heat transfer coefficient, k_p is the thermal conductivity of the particle, and R_p is its equivalent spherical radius $R_p = \frac{3 \cdot V_p}{S_p}$, being V_p and S_p the volume and the surface. Large external heating rates and low thermal conductivity determine large Biot numbers for thick particles, causing the presence of temperature gradients within the particle. Fuel particles greater than 0.1–0.2 mm have Biot numbers usually greater than 1.

Pyrolysis (Py) numbers are the ratios of reaction timescale and conduction or convection timescales:

$$Py_1 = \frac{k_p}{\rho_p \hat{C} k_R R_p^2} = \frac{1}{Th^2} \quad Py_2 = \frac{h}{\rho_p \hat{C} k_R R_p} = Py_1 \cdot Bi$$

being k_R the rate constant of the fuel pyrolysis reaction. Py_1 is equivalent to the reverse of thermal Thiele modulus (Th).

It is possible to highlight at least two typical regimes and to distinguish between thin and thick particles by using the pyrolysis and Biot numbers and comparing pyrolysis reactions with conductive and convective heat transfer (Corbetta et al., 2014; Paulsen et al., 2013). At $\text{Bi} < 1$ and $\text{Py} > 1$, there is an isothermal and kinetically limited region, where the entire thin particle is at one uniform temperature. At $\text{Bi} > 1$ and $\text{Py} < 1$, there is a conduction-limited region where there are significant temperature gradients inside the thick particle.

Fig. 34 compares the measured and predicted profiles of biomass residue and of center and surface temperatures from pyrolysis experiment of a wood sphere of 2.54 cm at 688K (Park et al., 2010). This behavior clearly highlights the presence of two different thermal regions. The temperature first increases until achieving an inflexion point at 600–650K, where the temperature profile becomes flat because of the latent heat requirement for

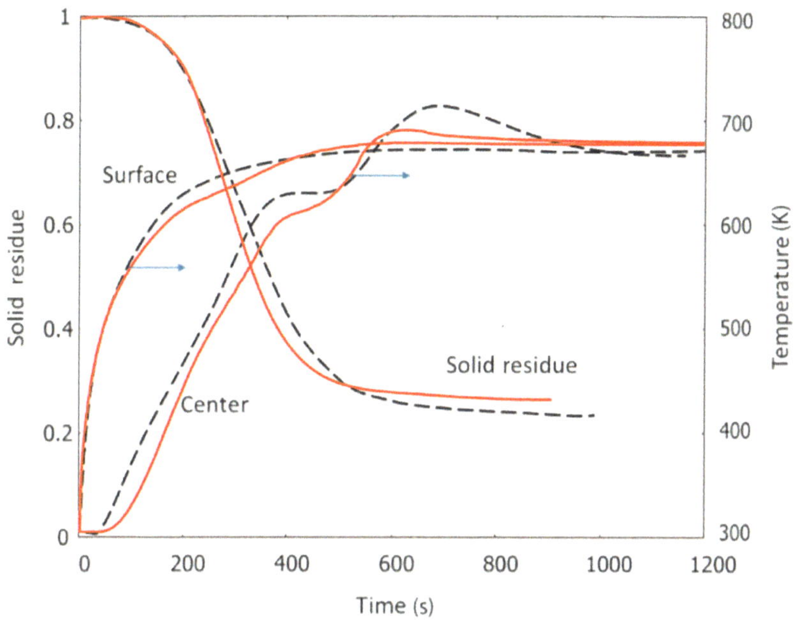

Fig. 34 Solid mass fraction residue and temperature profiles in a wood sphere at 688K. Comparison between experimental data (*dashed lines*) and model predictions (*solid lines*) (Park et al., 2010).

vaporizing the tar products. After the plateau, the temperature increases even exceeding the steady-state values of the nominal surface temperatures. According to Lédé (2012) and limiting our attention to cellulose, only levoglucosan (with a boiling temperature of 612K) would rapidly vaporize, while the dimer cellobiosan has a boiling temperature higher than 800K. The second region leads to the rising of the center temperature, which temporarily overcomes the surface temperature. This peak in the center temperature profile is due to the exothermic character of char formation. The behavior of these temperature profiles is highly sensitive to the thermochemical properties of the biomass pyrolysis as well as to biomass composition. Rather than an accurate fitting on specific operating conditions, the major interest of general model relies on a fair agreement with experimental data from different sources. Thus, Corbetta et al. (2014) deeply analyzed these temperature profiles also in comparison with other experimental data (Bennadji et al., 2013; Gauthier et al., 2013). They also investigated the released products and the effect of the secondary gas-phase reactions. In the Cornell experiments (Bennadji et al., 2013), pyrolysis reactions of the released products were practically negligible, because of the geometry of the pyrolysis reactor and to the relatively low temperatures. Fig. 35 reports

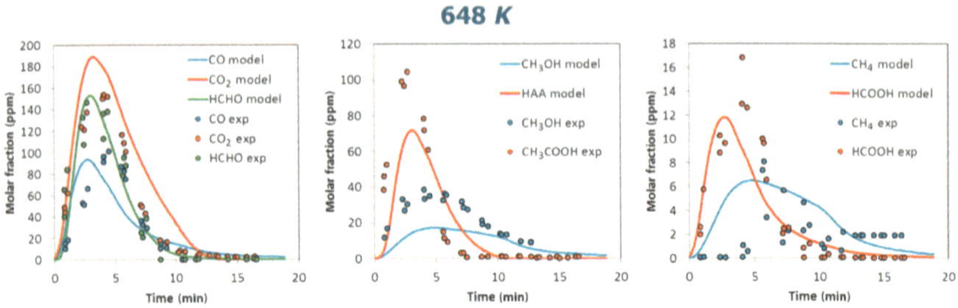

Fig. 35 Pyrolysis of thick biomass particles. Cornell experiments: gas and light oxygenated species released from 2.54 cm wood spheres at 648K (Bennadji et al., 2013). Comparisons of experimental data (*symbols*) and model predictions (*lines*).

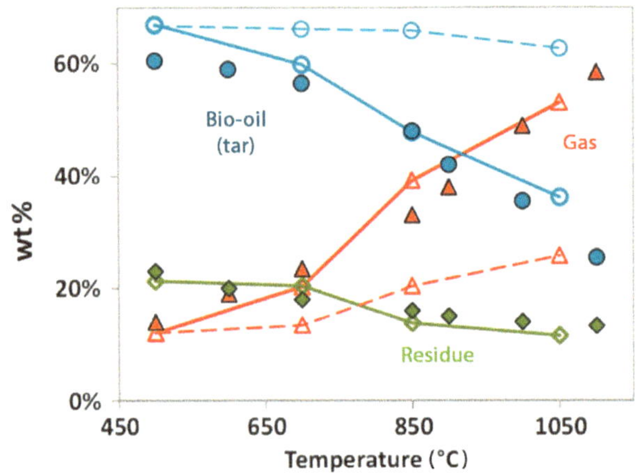

Fig. 36 Pyrolysis of thick biomass particles. Effect of secondary gas-phase reactions on released products. CEA experiments (Gauthier et al., 2013). Comparisons of experimental data and model predictions. Experimental data (*symbols*). Model predictions: pyrolysis only (*dashed lines*); pyrolysis + gas-phase reactions (*solid lines*).

detailed comparisons between experiments and predictions for production rates of released gases and light oxygenated species from Cornell experiments with 2.54 cm poplar spheres at 648K. The model predictions agree reasonably well with experiment except for discrepancies in the timing of CO_2 and CH_4 peaks.

On the contrary, the effect of the secondary gas-phase reactions in the PYRATES reactor used at CEA (Gauthier et al., 2013) was very important, at temperatures higher than 750K. The effective residence time in the PYRATES apparatus is ~100–200 ms, as evaluated on the basis of the pyrolysis products of a calibrated light hydrocarbon mixture injected into the heated reactor. Fig. 36 shows comparisons between measured and predicted yields of gas, tar, and solid residue in the CEA experiments. In order to highlight the contribution of secondary reactions, yield predictions are shown with and without the effect of homogeneous reactions. Secondary reactions clearly play a relevant role in gas production at temperatures higher than 750K. A large fraction of tar components decomposes, and the gas fraction increases significantly. Moreover, the model largely overestimates the low experimental yields of levoglucosan (2.5% of initial mass). The lumping of carbohydrate species in the model can only account for the minor part of this difference. A more reasonable hypothesis is that levoglucosan can undergo a catalytic dehydration reaction inside the thick biomass particle. In fact, large selectivities toward levoglucosan are experimentally observed in the fast pyrolysis of cellulose (even exceeding 50% of initial cellulose) in an isothermal thin film of micrometer scale (Mettler et al., 2012).

The relevant effect of the secondary gas-phase reactions will be further discussed in Section 5.3, where the biomass fast pyrolysis and the bio-oil formation are analyzed.

5.2 Gasification and Combustion Regimes of Thick Biomass Particles

Let us consider thick biomass (or coal) particles fed into a 10-cm single reactor layer countercurrent to a steam/air stream with a fuel equivalent ratio typical of a gasification process ($\Phi > 3$). Both the streams enter at ambient temperature in a gasifier layer. Depending on the start-up policy, the system could reach an ignited or a cold steady-state solution. This is a typical example of thermal feedback occurring in autothermal reactors. In order to reach the hot and desired solution, it is necessary to ignite the system. One possibility is to use an auxiliary fuel and heat up the inlet air stream at high temperatures ($T > 1000$K), until the ignition of residual char or released volatiles is observed in the gas phase. Only then, the steam/air stream is fed at ambient temperature and the system can evolve toward the hot steady solution. The system is able to maintain the hot solution only if the fuel particles have already stored enough energy; otherwise the dynamic evolution of the system reaches the cold steady solution. This is only a simple example of the complexity also related to the start-up policy and operations of countercurrent gasifiers, where multiple steady-state solutions could pertain to the different reactor layers.

In this application example, the effect of heat conduction as the controlling step for thermally thick particles is first analyzed through different

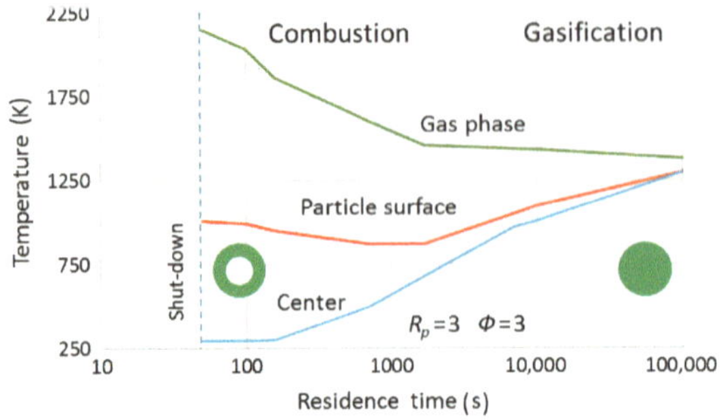

Fig. 37 Predicted gas and solid temperature profiles vs the residence time of the biomass inside the reactor.

simulations with spherical wood particles (R_p=30 mm) at Φ=3. Fig. 37 shows the predicted gas and solid temperature profiles vs the residence time of the biomass particles inside the reactor layer (inversely proportional to the solid flowrate). If the thermal penetration time is higher than this residence time, the biomass particles are not uniformly heated. As a consequence, only the external sectors can devolatilize, while the cold core of the particles remains unreacted. Fig. 37 also highlights the presence of two different regimes. At high residence times, the gas temperature is lower than 1500K and the particles are quasi isothermal. This is the behavior of the gasification regime. The biomass particles uniformly devolatilize, char gasification is completed, and released tar and gas react with oxygen in the rich conditions (Φ=3). The expected syngas is obtained. In contrast, the gas-phase temperature increases to more than 2000K when decreasing the residence time. Internal temperature gradients become significant, and the cold core of the particle remains unconverted. As a consequence, the biomass releases only partially the gas and tar species, leading to a fuel mixture approaching the stoichiometric conditions of a complete combustion. The partial devolatilization of biomass particles results in a large amount of CO_2 and H_2O as final products. After a further decrease of the residence time, the system is not any more able to sustain the combustion regime, leading to the complete shutdown of the system. It is thus clear the need of comprehensive models in order to analyze the behavior of these systems and to manage and/or optimize the operation of similar process units.

This behavior is also observed in Fig. 38, where the gas and particle temperature profiles are analyzed by varying the particle diameter, at fixed residence time. Again, very small and thin particles are near isothermal and

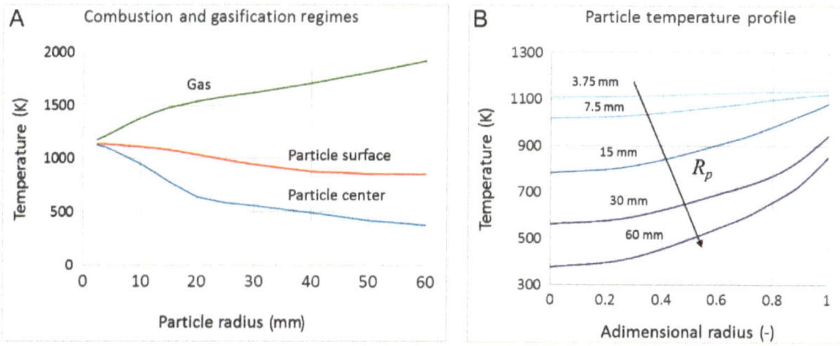

Fig. 38 Combustion and gasification regimes. (A) Predicted gas and solid temperature profiles vs the particle diameter. (B) Particle temperature profiles.

reach temperatures very close to the gas temperatures, while the large internal temperature gradients characterize thick particles. Small particles achieve ideal gasification conditions, while the incomplete pyrolysis of large particles drives the system toward combustion conditions with very high gas temperatures and a very poor gasification efficiency.

5.3 Fast Biomass Pyrolysis and Bio-Oil Formation

Gas, tar, and residual char are always produced from biomass pyrolysis, but their proportions significantly vary as a function of the operative conditions of the process. It is well established that low temperatures and long residence times of gas and tar released by biomass favor biochar production, while short residence times and moderate temperatures optimize bio-oil yields (Bridgwater, 2003). In fact, high severity conditions, i.e., high temperatures and residence times, favor the successive conversion of tar species to gas products.

Depending on the heating rate and solid residence time, three main modes of biomass pyrolysis including slow, fast, and flash pyrolysis can be highlighted. Slow pyrolysis is typically used for biochar production and is characterized by low temperatures (300–500°C), a wide range of sizes including very thick particles, and long residence times. The biomass pyrolysis proceeds under low heating rates to maximize the char yields, with enough time for tar condensation and cross reticulation reactions to take place.

Fast pyrolysis typically involves high heating rates and short residence times. Bio-oil yield can be as high as 50–70% on weight basis, while the flash pyrolysis process can produce even higher bio-oil yields (Kan et al., 2016). Table 10 summarizes the different modes of pyrolysis and gasification processes with relating operative conditions and product yields

Table 10 Different Modes of Biomass Pyrolysis and Gasification Processes

		Residence Time		Product Yields (wt)		
Mode	**Temperature (°C)**	**Vapor**	**Solid**	**Liquid**	**Solid**	**Gas**
Fast	500	1–2 s		75	12	13
Intermediate	500	5–30 s		50	25	25
Slow (carbonization)	400	Hours-days	Hours	30	35	35
Gasification	750–900	1–5 s		3	1	95
Torrefaction	280		10–60 min	0	80	20

Bridgwater AV: Review of fast pyrolysis of biomass and product upgrading, *Biomass Bioenergy* 38:68–69, 2012.

(Bridgwater, 2012). Relatively small or thin particles in fluidized-bed reactors with short vapor residence times are typical fuels for fast heating, while thick particles (3–6 cm) and biomass briquettes in packed-bed reactors are examples of solid fuels for slow heating process (Bridgwater, 2003; Di Blasi, 2009).

There are two reasons for the lower bio-oil yield at low temperatures: incomplete devolatilization of the solid particles and favored charcoal formation, due to condensation reactions of liquid and tar species. The gas-phase pyrolysis reactions of tar components cause the lower bio-oil yields at high temperatures. It is thus clear that both chemistry, and heat and mass transfer processes play a critical role in conditioning the optimal bio-oil yields from fast pyrolysis processes. The reacting biomass particles need to be rapidly heated to the optimum temperature in order to minimize their exposure to the lower temperatures that favor secondary char formation.

The pyrolysis of small biomass particles in fluidized-bed reactors, with or without recirculation, is a common way to realize fast biomass pyrolysis, where the contact times of bio-oil products at high temperatures are minimized.

Fig. 39 schematically shows a circulating fluidized-bed reactor for fast pyrolysis process. Biomass is dried and ground to minimize the water in the bio-oil and to improve the fast heating of small particles. Combustion of the char product warrants the required pyrolysis heat. Bio-oil is obtained after condensation of the tar products. It is typically a dark red-brown liquid, highly polar, with a density of ~1200 kg/m^3. As already mentioned, the residence time of the product gases must be short, while the endothermic

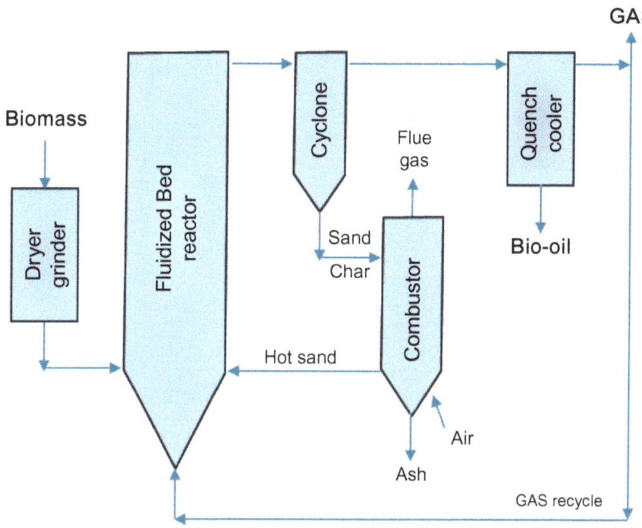

Fig. 39 Schematic of a circulating fluidized-bed fast pyrolysis process. *After Bridgwater AV: Review of fast pyrolysis of biomass and product upgrading,* Biomass Bioenergy *38:68–69, 2012.*

pyrolysis reaction requires a high rate of heat transfer. Currently, bubbling and circulating fluidized-bed processes produce bio-oil on a commercial scale, using wood or wood waste (Bridgwater, 2012). Circulating fluidized-bed reactors are potentially suitable for larger throughputs with respect to the bubbling ones even though the hydrodynamics is more complex.

Although large particles give slightly lower liquid yields, they are not as costly to grind. Torrefaction, that is a mild thermal pretreatment of biomass under anoxic conditions, is a useful way for improving the quality of the feed in terms of energy density and grindability properties. In fact, through decomposition of the hemicelluloses coupled with depolymerization of cellulose and thermal softening of lignin, the cell wall in the biomass sample is greatly weakened. Moreover, torrefaction ensures a more consistent quality of the biomass fuel for pyrolysis and combustion applications (Bridgeman et al., 2008; Shankar Tumuluru et al., 2011; Van der Stelt et al., 2011).

As already discussed by Calonaci et al. (2010), Table 11 compares experimental data and model predictions for the fast pyrolysis of three different biomass samples (Aguado et al., 2000; Ates et al., 2004; Westerhof et al., 2010). The first example of these comparisons refers to the fast pyrolysis of pine spruce sawdust. Aguado et al. (2000) studied the biomass pyrolysis in a conical spouted

Table 11 Optimal Conditions of Fast Biomass Pyrolysis

Biomass Composition	Aguado et al. (2000) Pine Spruce Sawdust		Ates et al. (2004) Sesame Stalk		Westerhof et al. (2010) Pine Wood	
Cellulose	48.7		26.1		35.0	
Hemicellulose	21.4		21.3		29.0	
Lignin	21.9		43.9		28.0	
Moisture	8.0		8.7		8.0	
Reactor type	Spouted bed		Fixed bed		Fluidized bed	
Temperature (K)	720		770		750	
Weight Fractions	**Model**	**Experiment**	**Model**	**Experiment**	**Model**	**Experiment**
Solid residue	0.14	0.12	0.23	0.22	0.27	0.17
Gases	0.17	0.17	0.23	0.26	0.18	0.23
Total liquids	0.69	0.70	0.54	0.52	0.55	0.58
H_2O	0.12	0.09	0.14	0.16	0.13	0.12
Organic liquids	0.57	0.61	0.40	0.36	0.42	0.46

Comparison of experimental data and model predictions (Aguado et al., 2000; Ates et al., 2004; Westerhof et al., 2010).

bed proving the reactor stability and flexibility with low operating costs. Moreover, this flash pyrolysis reactor is very convenient for reaching: short gas residence time, rapid feed heating, because of an effective heat and mass transfer. In agreement with the experimental information, a quite flat maximum of ~70 wt% is predicted in the temperature range of 690–740K. Two similar comparisons in the same table refer to the fast pyrolysis of sesame stalk in a fixed bed (Ates et al., 2004) and of pine wood in a fluidized-bed reactor (Westerhof et al., 2010). In line with the experimental data, the model predicts maxima of bio-oil yields of 52% and 58%, at 770K and 750K, respectively.

For small particles, the model predicts flat bio-oil yields and gas formations in a temperature range of ±50K around the maximum, where the biomass devolatilization is completed. These facts agree quite well with the experiments. Carbon oxides are the main gas species from primary devolatilization, together with small quantities of CH_4 and C_2 hydrocarbons. Primary H_2 yield is very limited and only occurs at high temperatures, where residual char is nearly constant. Chemical compositions of product liquids predicted by the model agree fairly well with the experimental data reported in the literature (Aguado et al., 2000; Lindfors et al., 2014; Oasmaa et al., 2015). Bio-oil mainly consists of carbohydrates and substituted phenols mostly derived from lignins. Alcohols, aldehydes, furans, and small-oxygenated species constitute up to 15–25%, while water yield ranges between 10% and 20% of dry biomass.

At 750–800K, maximum oil content usually ranges between 55% and 80%, due to the different cellulose content of biomass samples. Maximum oil yields are obtained from biomass with a large cellulose content. The maximum in bio-oil yield is due both to the primary biomass pyrolysis and to the decomposition reactions of tar species. Consequently, gas yield increases continuously with temperature. At moderate temperatures ($T<700$K), and mainly for large particles a relatively long residence time is necessary to complete the pyrolysis process. A comprehensive model of biomass devolatilization at the reactor and the particle scale is then required to describe bio-oil formation. Heat diffusivity inside the solid particle and heat transfer resistances explain the partial volatilization of the biomass at low and intermediate temperatures. Fig. 40 shows the spread of predicted yields of oil, char, and gas from a cellulosic and a lignin biomass. The upper and lower curves combine and show the effect of incomplete devolatilization and/or large particles at low temperatures, along with the role of secondary pyrolysis reactions, at high temperatures.

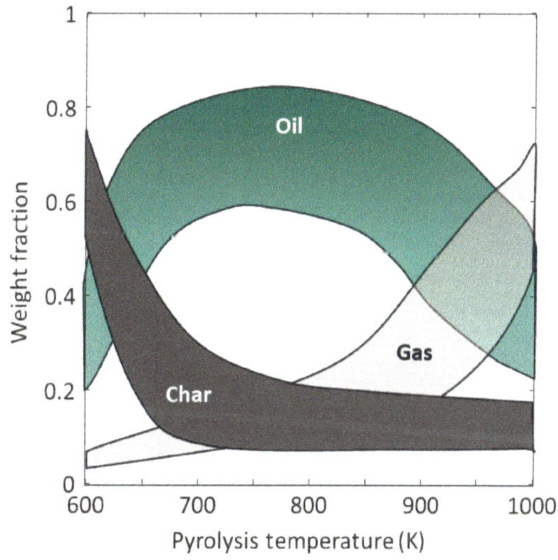

Fig. 40 Predicted typical yields of oil, char, and gas from fast pyrolysis of biomass.

6. BALANCE EQUATIONS AT THE REACTOR SCALE

While the description of entrained bed and fluid bed reactors can simply refer to the particle model, the modeling of an elemental reactor layer characterizing the gas–solid interactions is very useful to describe packed or fixed-bed reactors. In fact, the solid bed can be simulated as a series of *NR* elemental layers, as schematically shown in Fig. 41. The height of each layer is of the size order of the biomass particle, accounting in this way for the vertical dispersion phenomena. Gas and solid phases are assumed as perfectly mixed inside the layer, and the mixing of the gas phase is further improved due to the jets of volatile species released during solid fuel pyrolysis (Frigerio et al., 2008).

The mass balance equations for the gas-phase of each elemental reactor are:

$$\frac{dm_k}{dt} = \dot{m}_{k,in} - \dot{m}_{k,out} + J_k S N_p + V_R \dot{\Omega}_k^G$$

where m_k is the mass of the kth gas species within the reactor volume V_R; $\dot{m}_{k,in}$ and $\dot{m}_{k,out}$ are the inlet and outlet flowrate; $\dot{\Omega}_k^G$ is the net formation from gas-phase reactions; and the term J_k is the gas-solid mass exchange multiplied by the particle surface S and the number N_p of particles inside the layer.

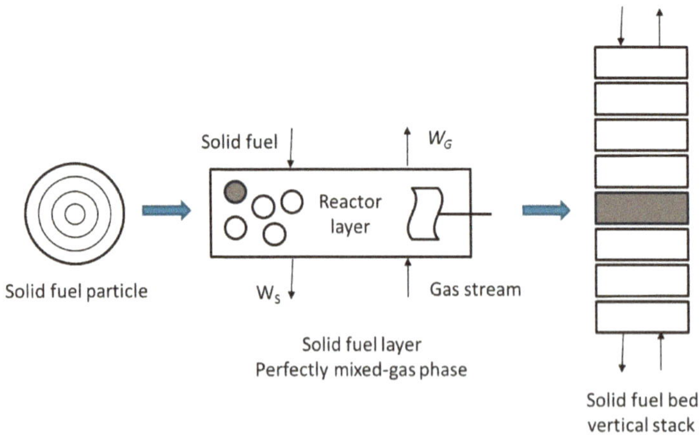

Fig. 41 Multiscale nature and structure of a countercurrent biomass gasifier.

The energy balance equation for the gas-phase of each elemental reactor is:

$$\frac{d\sum_{k=1}^{NC^G} m_k \hat{h}_k^G}{dt} = \sum_{k=1}^{NC^G} \dot{m}_{k,in} \hat{h}_{k,in}^G - \sum_{k=1}^{NC^G} \dot{m}_{k,out} \hat{h}_{k,out}^G + \sum_{k=1}^{NC^G} J_k \hat{h}_k^G SN_p + h\left(T - T^{bulk}\right) SN_p + V_R \dot{Q}_R^G$$

where T^{bulk} is the gas-phase temperature; the terms $\dot{m}_k \hat{h}_k^G$ are the species enthalpies of inlet and outlet flowrates; and $J_k \hat{h}_k^G$ is the flux of enthalpy relating to the mass transfer of each component of a single particle. Finally, $\dot{Q}_R^G$ is the overall heat of gas-phase reactions. As a matter of simplicity, the index of the reactor layer (from 1 to NL) is not reported in the previous balance equations.

In order to give an idea of the dimension of the overall problem, let us assume kinetic mechanisms involving 30 solid species, 100–200 gas species, and only 30 gas species really interacting with the solid matrix. The number of balance equations easily overcomes 10,000 and leads to numerical difficulties, by assuming 5–10 sectors to discretize the particles and again 5–10 reactor layers. Solvers of ordinary differential and differential-algebraic equations system belonging to BzzMath library (Buzzi-Ferraris, 2010; Buzzi-Ferraris and Manenti, 2010, 2012; Manenti et al., 2009) are adopted. Dsmoke and OpenSmoke codes are adapted and used for calculations of gas-phase ideal reactors (Cuoci et al., 2013, 2015).

It is evident from the analysis of Fig. 41 that the overall Jacobian matrix relating to the balance equations of a countercurrent gasifier has an embedded highly sparse and large-scale structure with diagonal blocks and upper and lower bands, as schematically reported in Fig. 42.

In order to reduce the numerical problem, a simplification can be adopted when simulating the whole gasifier. Gas species are assumed not reacting inside the particle; thus only a reduced number $NC^g\left(NC^g \leq NC^G\right)$ of gas species are formed in the particle. They can interact with the solid structure and they can move through the boundary and diffuse in the gas phase. Two different matrices are then adopted. A first matrix is used to characterize the biomass particle. It is a dense square matrix of dimension $\left(NC^S + NC^g + 1\right)^2$ accounting for NC^S solid species and only NC^g gas species. This matrix accounts for the intraparticle solid and gas–solid evolution within each sector

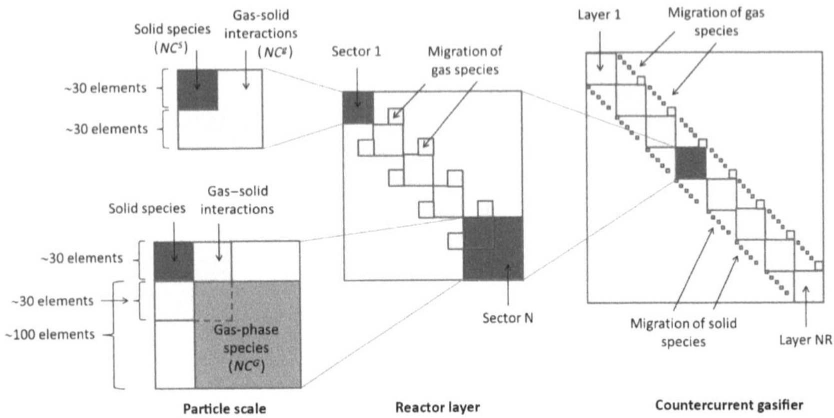

Fig. 42 Countercurrent gasifier. Structure of the Jacobian matrices at different scales.

of the solid particle. Only the external sector N interacts with all the NC^G gas species. The second matrix accounts for all the solid and gas species in the external sector and it is a $(NC^S + NC^G + 1)^2$ partially structured matrix. Biomass devolatilization, heterogeneous reactions, and secondary gas-phase reactions are accounted for.

At the scale of the reactor layer, the upper and lower bands involve only gas species and NC^g is the size of the band block, since the solid species are not diffusing. Both upper and lower bands are present due to the gas diffusion inside the particle. Finally, the external sector accounts for all the gas species. At this scale, more than 500 variables are easily involved.

At the reactor scale, there is a cascade of reactor layers, and each layer interacts with the gas stream coming from the upper or the lower layer, depending on the countercurrent or the concurrent configuration. Similarly, there is the migration of solid variables across the different layers. At the reactor scale, the Jacobian matrix assumes a diagonal-block structure with asymmetric bands. Referring to the countercurrent biomass gasifier, the lower band represents the solid particles that migrate toward the lower layers, whereas the upper band characterizes the gas species rising the biomass bed. The asymmetry of lower and upper bands comes from the large number of gas species ($NC^G > NC^S$). The sparsity and structure of the Jacobian matrix allow to reduce the numerical complexity of this large and heavy numerical problem. Stiffness remains the main responsible of the numerical difficulties, and computation time may vary from a few minutes to several hours, depending on the dimensions of the kinetic mechanisms and the adopted discretization.

In order to show some application example at the reactor scale, the following sections will analyze a traveling grate biomass combustor and countercurrent coal gasifiers. The enhancing effect in syngas production when gasifying coals containing large amounts of sulfur components is particularly discussed.

6.1 Traveling Grate Combustor

This application example deals with a traveling grate combustor where a bed of coal, biomass, or RDF particles is progressively dried, devolatilized, and burnt as schematically reported in Fig. 43. Volatile components released by the solid fuel traveling on the grate are involved in secondary gas-phase pyrolysis, gasification, and combustion reactions in the freeboard volume, over the particle bed. Finally, hot flue gases leave the freeboard and enter the boiler for steam generation. The complexity in the mathematical modeling of this combustor is first due to the morphological variations and the shrinking of the fuel particles, which need to be taken into account at particle and reactor scale. In fact, during the fuel conversion, the size, density, and porosity of the solid particles change, due to drying, devolatilization, and char gasification and combustion. Moreover, the effective gas-phase combustion in the freeboard involves and requires a complete mixing of primary and secondary air with the volatile released species in order to improve combustion and minimize

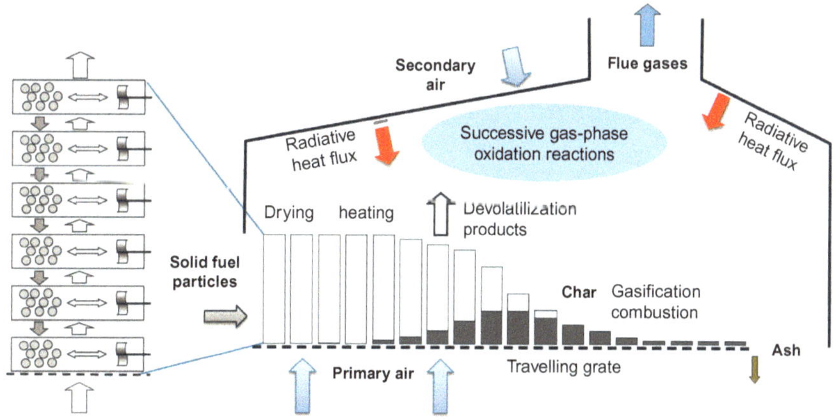

Fig. 43 Schematic representation of the traveling grate combustor. *After Ranzi E, Pierucci S, Aliprandi PC, Stringa S: Comprehensive and detailed kinetic model of a traveling grate combustor of biomass,* Energy Fuel *25:4195–4205, 2011.*

pollutant emissions. The fixed bed of solid fuel particles is assumed as successive stacks of several reactor layers, composed by spherical particles that exchange mass and heat with a completely mixed gas phase. The solid bed moves on the grate with fixed grate velocity and this velocity determines the effective residence time of the fuel particles inside the combustor unit. Finally, the combustion is completed in the freeboard volume over the grate. The energy balance equation on the overall reactor requires the proper closure conditions accounting for the radiative heat flux from the reactor walls. In this simplified model, the effective radiating temperature is an average of wall and flame temperature.

Due to the wall radiating heat and the cold primary air, the top reactor layers are the first to heat and pyrolyze, while the bottom layers take more time to heat up and decompose. On the contrary, the combustion reactions of the residual char follow the reverse order. Due to the limited availability of the oxygen in the primary air, there is initially the combustion of the char in the bottom layer and only then the combustion of the char in the top layer can be completed. This mathematical model, in principle suitable for all different solid fuels, was tested and validated in comparison with experimental data from an industrial biomass combustor of 12 MW designed by Garioni Naval and operating in Belgium (Ranzi et al., 2011). A more complete description of this application to the biomass combustor is reported elsewhere (Ranzi et al., 2011). Here, it seems useful to simply highlight the viability of this modeling approach to the control of industrial-scale combustors. Of course, the model is able to give all the detailed information relating to the progressive decomposition of the fuel particles and released species, together with the internal and external temperatures of the solid and the gas phase, along the grate. For control purposes, the position of the ignition and the combustion fronts are very useful performance indicators. Fig. 44

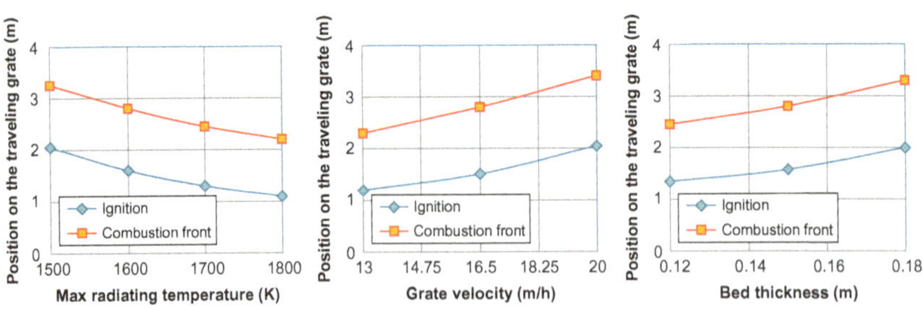

Fig. 44 Control parameters of the traveling grate biomass combustor. Ignition and combustion fronts vs radiating wall temperature, grate velocity, and bed thickness.

shows how some major control parameters such as the effective radiating temperature, the grate velocity, and the thickness of the fuel bed affect these dependent variables. These control variables ensure the appropriate fuel devolatilization, the char conversion, and the complete combustion in the freeboard, controlling in this way the residual carbon content in the ashes and the emissions in flue gases.

Higher radiating temperature leads to ignition and combustion fronts closer to the fuel inlet. The width of the front between ignition and combustion is ~1 m and it mainly depends on the primary air. The grate velocity is varied by maintaining the combustion stoichiometry, i.e., with corresponding variations of primary and secondary air. This corresponds to vary the whole boiler capacity. The increase in grate velocity shows the limits of combustor capacity and operability, and it moves the ignition and combustion fronts toward the end of the grate. The increase of the bed thickness on the grate, by maintaining the same grate velocity and combustion conditions, corresponds to an increase of combustor capacity. Again, the limits of safe operating conditions are easily found.

6.2 Countercurrent Gasifiers

Fig. 45 schematically shows the stack of elemental reactor layers of the countercurrent gasifier. The solid fuel is fed from the top, whereas steam and air inlet streams enter the bottom of the gasifier. The nominal operating conditions usually require a fuel equivalence ratio ~3–4, that is, a very rich

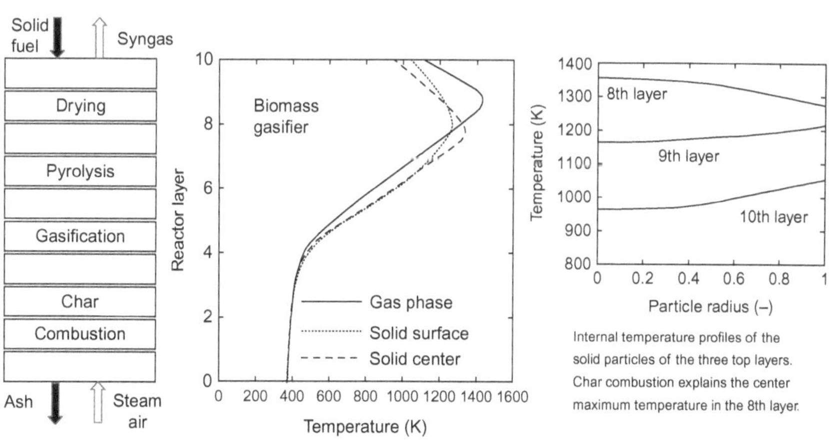

Fig. 45 Predicted temperature profiles in a countercurrent biomass gasifier.

mixture containing only a small amount of oxygen, useful to maintain autothermal reactor conditions. The weight steam to biomass ratio is around 0.3, the gas contact time is in the order of few seconds, whereas the solid residence time is significantly higher and in the order of 1 h. According to the multiscale modeling approach already analyzed in the previous example, the GASDS program is applied (Pierucci and Ranzi, 2008; Sommariva et al., 2011). Fig. 45 shows the cascade of 10 reactor layers of a biomass gasifier and also shows the gas and solid temperature profiles along the gasifier and inside the fuel particles. These profiles are reported according to the layer number, not to the real and steady shrink height. The effective volume of the first bottom layers, where also the residual char is completely burned out, only contains ash and is significantly reduced. Gas and ash temperatures are very similar. Rising on the bed, the gas is first heated up by the hot particles and then reaches temperatures higher than 1500K in the eighth to ninth layers. Here, the exothermic tar oxidation reactions provide the heat necessary to biomass pyrolysis. The maximum temperature of the center of the particle in the eighth layer is due to the combustion of the residual char. Finally, in the top reactor layer, the temperature of the gases leaving the gasifier decreases, due to cold biomass fed. The role of heterogeneous and secondary gas-phase reactions is well evident, in the definition of temperature profiles, and in the characterization of bio-syngas composition, including residual tars and organic volatile components (Dupont et al., 2009). The GASDS model can also describe the transient conditions to reach the final steady operation of the gasifier.

Lab (Grieco and Baldi, 2011) and pilot scale (Pettinau et al., 2011, 2013) gasifiers have been recently simulated and discussed (Corbetta et al., 2015). Only the first set of experimental data is here reported. The operating conditions and the main input parameters for model simulations are summarized in Table 12. These conditions refer to the lab-scale gasifier reported in Grieco and Baldi (2011). Before comparing model predictions with experimental data, it is convenient to analyze and assess some model features particularly with respect to the spatial discretization and the dynamic behavior to reach the steady-state conditions.

An accurate model sensitivity to particle and reactor discretization was performed by changing both reactor and particle radial discretization. Five to seven reactor layers and two particle discretizations are analyzed. This is a compromise between simulation accuracy and computational time. Due to the diameter of coal particles (2.54 cm), a thermally thick regime occurs within the particles, which requires to account for thermal resistances

Table 12 Input Parameters for the Simulation of the Lab-Scale Gasifier

	Grieco and Baldi (2011)
COAL: C/H/O	80.1/5.1/15.8
Ash/moisture	9/17.4
Particle diameter (cm)	2.54
Inlet gas temperature (K)	373
Peak temperature (K)	~1650
Equivalence ratio	0.176
Air to coal ratio	1.81
Steam to coal ratio (STC)	0.30
Specific gasification rate (kg/m^2/h)	342.7

(Corbetta et al., 2014). Increasing the number of particle sectors the combustion zone moves toward the top of the gasifier because of the thermal penetration time necessary to heat up the core of thick particles. The number of reactor layers mainly impacts on the position of the combustion front and the corresponding peak temperature.

A suitable start-up policy is required in order to reach the desired "hot" gasification conditions, avoiding the "cold" steady solution. This is true from both an operational and a numerical point of view. In the industrial practice, a duct burner is usually adopted to preheat inlet gas in order to start up the gasifier, providing the required heat to the endothermic pyrolysis and gasification reactions. Once the char gasification and combustion take place, the system self-maintains the hot conditions, it becomes autothermal, and the auxiliary burner is shut down. Simulation conditions mimic this start-up procedure. Inlet gas temperature is initially preheated up to 1300K until combustion occurs in the first layer ($T>1800$K), only then the preheating ends and the temperature is lowered to 300K. The steady-state condition inside the gasifier is reached only after several hours, with a progressive shift of the combustion front toward the upper layers. It is important to observe that the complete gasification reached at the bottom of the gasifier determines more than 90% mass and volume reduction, due to the presence of ~10% of ash.

It is necessary to account for significant changes of the effective dimensions of reactor layers, due to the shrinking and morphological modifications of fuel particles along the gasification process. Fig. 46 shows the height

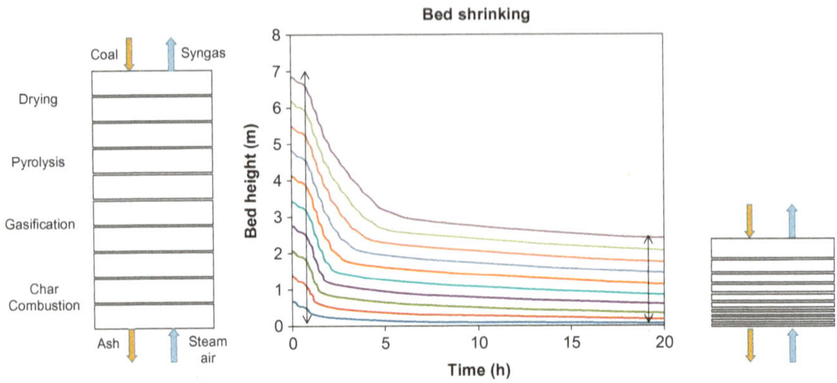

Fig. 46 Model predictions of the coal updraft gasifier (Grieco and Baldi, 2011). Evolution of reactor layers height during the start-up procedure.

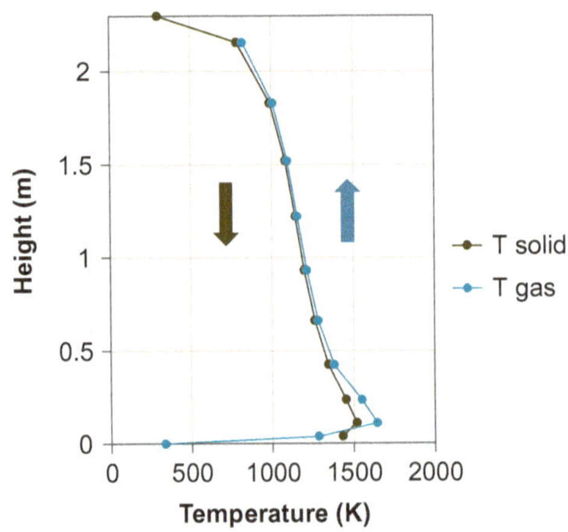

Fig. 47 Model predictions of the coal updraft gasifier (Grieco and Baldi, 2011). Gas and Coal temperature profiles along the gasifier height.

evolution of the solid bed. The system reaches the steady-state condition only after the completion of coal devolatilization and char gasification, which is the time-limiting step of the overall process. The mathematical model provides detailed information on chemical and physical phenomena occurring inside the gasifier. Fig. 47 shows the axial temperature and composition profiles of both the solid and the gas phases, giving an overall picture of the chemistry involved in the countercurrent gasification process. Moving from the top downward, solid particles are dried, pyrolyzed, gasified, and burnt. The temperatures at the bottom of the gasifier overcome

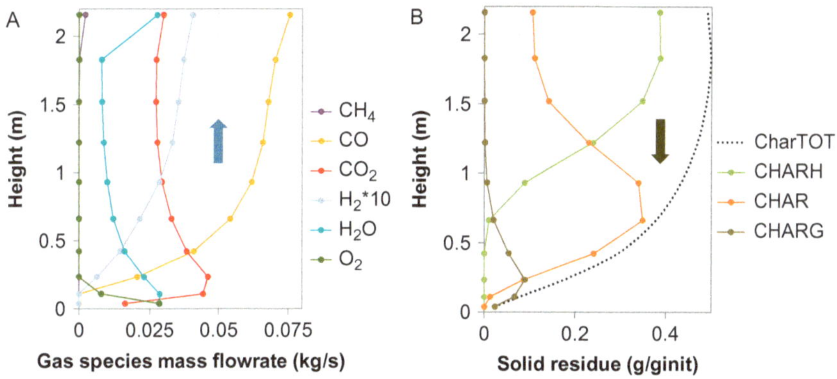

Fig. 48 Model predictions of the coal updraft gasifier (Grieco and Baldi, 2011). (A) Mass flowrate of key gas components. (B) Residual char transformation along the gasifier height.

1500–1600K. In the bottom layer the hot residual char and ash heat up the rising oxidizer stream completing the solid-phase combustion. Further oxidation reactions in the gas phase are responsible for the relevant production of CO_2 and for the temperature peak that overcomes the solid temperature. At these high temperatures, the endothermic gasification reactions contribute to CO and H_2 production, as clearly shown in Fig. 48.

Carbon dioxide is initially produced in the bottom reactor layers and it is then consumed by the gasification reactions, while it is slightly produced in the upper reactor zone due to secondary gas-phase oxidation reactions. A significant steam increase is observed at the top reactor layer, mainly due to the drying of wet coal. Methane is produced in the upper part of the gasifier due to coal devolatilization and secondary gas-phase pyrolysis reactions. These reactions are also responsible for the final inflection of the CO profile. Fig. 48 also shows the char formation and successive transformations along the reactor. The char presence in the first layer of the gasifier indicates the occurring of a relevant coal devolatilization. CHARG is the final product of the pyrolytic char transformations, and it is the rate-determining step for the overall gasification process and the major responsible for carbon content in residual ash.

Grieco and Baldi (2011) presented a gasification study of a leucite subbituminous coal (C/H/O/N/S=78.1/5/14.4/1/1.2). Fig. 49 shows comparisons between experimental data and model predictions, both in terms of bulk temperature profiles and syngas composition. Model predictions refer to 10 reactor layers and 1 and 2 particle sectors.

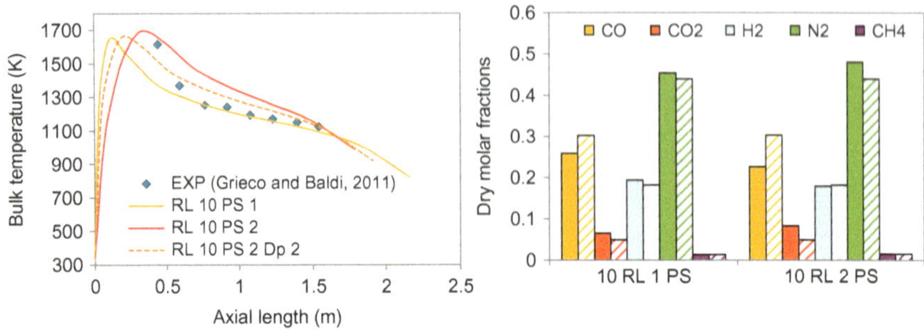

Fig. 49 Lab-scale coal updraft gasifier (Grieco and Baldi, 2011). Comparisons of experimental data (*dashed bars*) with model predictions (*solid bars*) with 10 reactor layers and 1 and 2 particle sectors.

As already observed, the hot spot location is more correctly predicted by the simulation performed with two particle sectors (PS 2), while the one particle sector simulation (PS 1) shows a slightly lower temperature peak, too close to the gas inlet zone, but better agrees with the experimental temperature profile in the upper zone of the gasifier. Both these simulations reasonably fit experimental data both in terms of temperature and product gas distribution. A better agreement, mainly in terms of temperature profiles, is obtained by using a slightly smaller equivalent spherical diameter (2 cm, instead of 2.54 cm), as shown by the dotted line in Fig. 49. Further details of the predicted syngas composition, also including tar components, are reported in Corbetta et al. (2015). Figs. 50 and 51 summarize the results of a sensitivity study of this lab-scale gasifier with respect to the following three operating parameters: effective air (λ), inlet gas temperature, and mass steam to coal ratio (STC = steam flow/coal flow).

λ is the ratio between actual and stoichiometric air, i.e., the reverse of the effective ratio. Air in the inlet gas (λ) strongly affects the temperature profile, bed height, and syngas composition. The oxygen increase improves coal combustion and gasifier temperature. A corresponding sharp reduction of solid residue and of bed height, along with a reduction of the H_2/CO ratio, is also observed. Due to this high sensitivity, this parameter only spans between 0.15 and 0.20. The lowest value of oxygen in the feed implies a limited coal gasification with a large carbon content in the residual ash. Thus, the gasifier efficiency initially increases with λ, due to the completion of the gasification process, and then decreases, due to the successive oxidation of H_2 and CO to form H_2O and CO_2. The λ values adopted here are relatively low when compared with those usually applied in gasifier units ($\sim$0.3). For

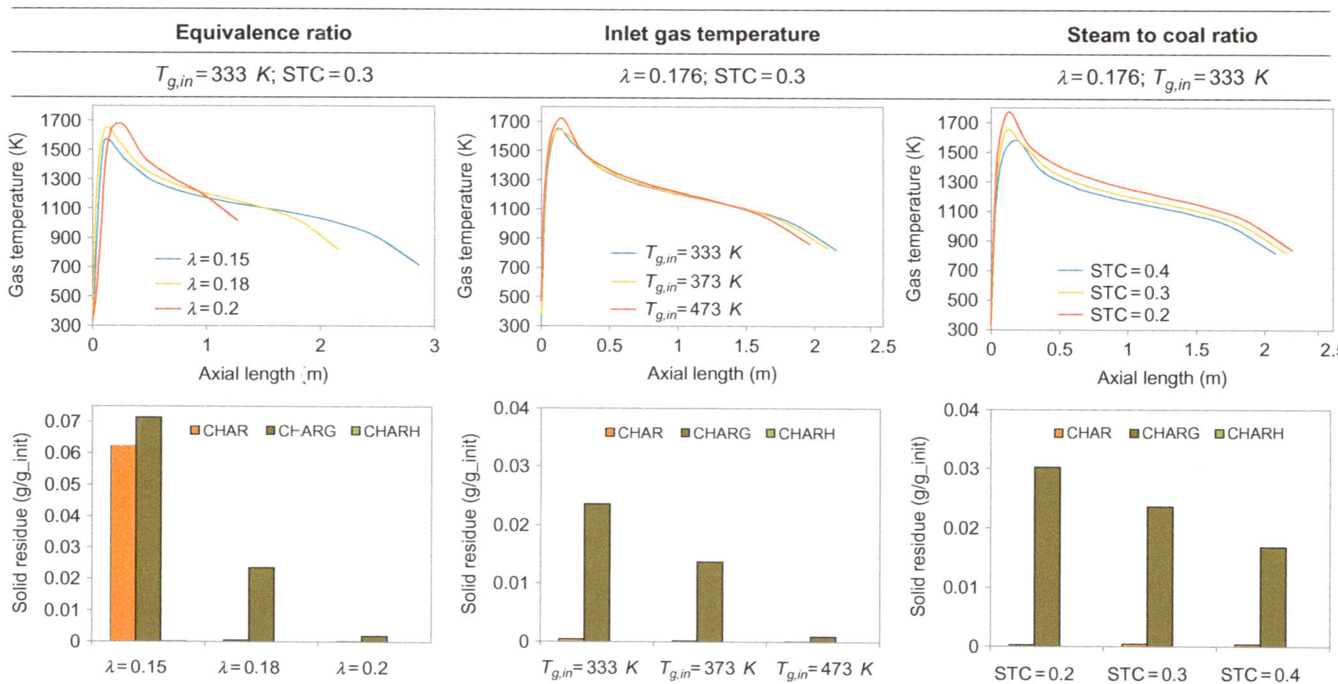

Fig. 50 Sensitivity analysis of equivalence ratio (λ), inlet gas temperature ($T_{g,in}$), and steam to coal ratio (STC) on temperature profiles, gas species, and solid residue.

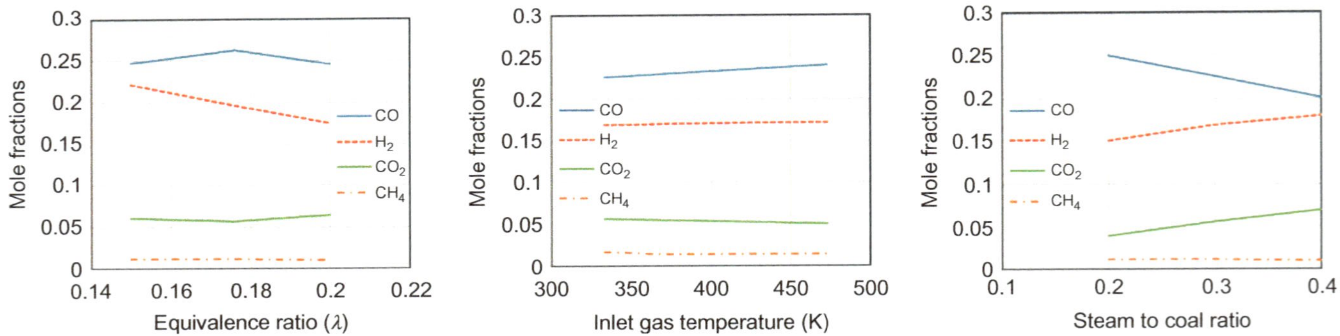

Fig. 51 Sensitivity of dry syngas mole fractions to equivalence ratio, inlet gas temperature, and steam to carbon ratio. Reference conditions: $\lambda = 0.176$, $T_{g,in} = 373K$, and STC = 0.3.

this reason, there is a residual char in the experimental conditions of Grieco and Baldi (see Fig. 9).

A different way to complete the gasification process, without successive oxidation reactions, is to increase the temperature of the inlet gas stream. Axial thermal profiles show a corresponding temperature increase, and globally, an increase in the cold gas efficiency (CGE) is observed. In fact, the change of the inlet temperature from 333K to 473K allows to complete the process, due to the higher reactivity of steam gasification reactions, leading to a negligible carbon content in the solid residue.

The last operating parameter analyzed is the STC. In this case, the increase of steam partial pressure may again result in a higher steam gasification rate. The change of STC from 0.2 up to 0.4 allows to obtain a partial reduction of the carbon content in the residual ash. A significant increase of the H_2/CO ratio is also observed. In fact, CO_2 increases at the expense of CO, due to the water gas shift reaction. For this reason, the CGE remains almost constant in these conditions.

Comparisons between model predictions and experimental data obtained in a pilot-scale coal gasifier were recently discussed by Corbetta et al. (2015), while biomass gasification using low-temperature solar-driven steam supply is reported by Ravaghi-Ardebili et al. (2015).

6.2.1 H_2S Impact on Syngas Production and CO_2 Reduction

There is a relevant environmental concern about coal gasification/combustion for both CO_2 and hydrogen sulfide (H_2S) emissions with related global warming effects. CO_2 is responsible for a great impact on environmental system, without relevant industrial uses due to its thermodynamic stability and low chemical value. Urea and methanol plants worldwide are reusing less than 1% of anthropogenic emissions. At the same time, the tight H_2S legislation threshold of its release into the atmosphere has triggered renewed interest in the modeling of sulfur chemistry (Manenti et al., 2014; Sassi and Gupta, 2008). Moreover, H_2S is a poison for industrial catalysts and its combustion products are responsible for acid rains. Nonetheless, H_2S is an interesting chemical for its hydrogen content and, actually, it is the hydrogen richest molecule after methane (CH_4) and hydrocarbons, methanol (CH_3OH), ammonia (NH_3), and water (H_2O). Consequently, H_2S should deserve more attention than to be considered just a pollutant and then oxidized in Claus plants to obtain elemental sulfur and water (El-Melih et al., 2016). From this different perspective, CO_2 is a potential interesting oxidizing agent and a new reaction pathway called the Acid Gas to Syngas (AG2S™) (Manenti et al., 2015) exploits the oxy-reduction reaction between H_2S and CO_2 (Manenti, 2015; Manenti et al., 2016):

$$2H_2S + CO_2 = CO + H_2 + S_2 + H_2O$$

with relevant benefits with respect to the state of the art of coal gasification. This reaction requires high temperatures ($T > 1000°C$) in a gas-phase regenerative thermal reactor (RTR). High temperatures are necessary to activate the system and reduce by-products. Fig. 52 shows a concept process flow diagram, where CO_2 and H_2S are injected into a common chamber in premixed or unmixed mode, and then react and are transformed into syngas.

The regenerative oxy-reduction between H_2S and CO_2 occurs in the RTR refractory lined chamber, assisted by a minor injection of air or oxygen. RTR effluent with unreacted feed species is quenched and cooled in heat exchangers. A catalytic Unit 1 converts SO_2 into elemental sulfur, analogous to Claus SRU section, cools down the effluent, and acts as a sulfur condenser (S_x). A second catalytic treatment is necessary to convert by-products, such as COS and CS, residual SO_2, and sulfur vapors, into H_2S and CO_2. This improves overall process selectivity on sulfurous species, but this also consumes a portion of CO with CO_2 production. Finally, water is condensed in the separation Unit 2 where unreacted H_2S and CO_2 are recycled to RTR after chemical washing, and the H_2/CO mixture is exported. Process layout described earlier shows the benefit of AG2S™ technology in coal gasification process since it allows to obtain more syngas by reducing pollutant emissions.

A further quantitative example of AG2S™ technology refers to the gasification of Sulcis coal (Bassani et al., 2016). Table 13 shows ultimate analysis of Sulcis coal together with its characterization in terms of the three reference coals (Sommariva et al., 2011). The LHV of this coal is ~20.8 MJ/kg. Table 14 summarizes the operating conditions of inlet and outlet streams of the gasifier, while Fig. 53 shows a comparison between experimental and predicted effluent composition, obtained assuming 10 reactor layers, without particle discretization.

It is assumed that 90% of inlet sulfur leads to the formation of H_2S. The agreement between experimental and predicted data is good. The excess of outlet water could be due either to neglected catalytic effect or to an incomplete attainment of steady conditions of the Sotacarbo pilot plant. Moreover, Fig. 53B shows that the temperature profile along the reactor, which is a relevant indicator of the gasifier operation, is in good agreement with the

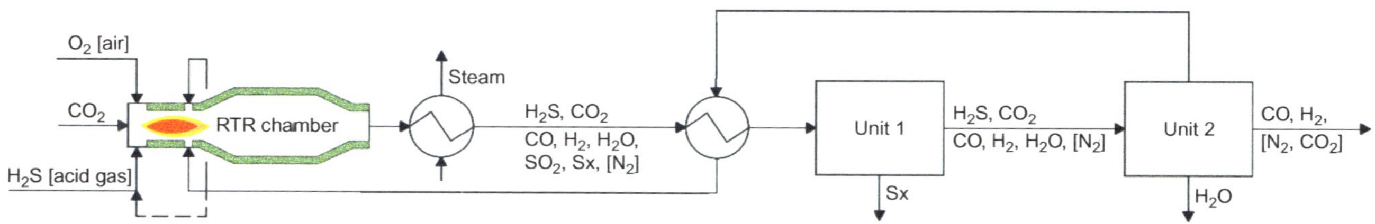

Fig. 52 Concept process flow diagram of AG2S™ technology. *After Manenti G, Molinari L, Manenti F: Syngas from H_2S and CO_2: an alternative, pioneering synthesis route?* Hydrocarb Process *6:1–4, 2016.*

Table 13 Ultimate Analysis and Coal Characterization of Sulcis Coal

Ultimate Analysis

	% C	% H	% N	% S	% O	Moisture	Ash
Composition (wt%)	53.17	3.89	1.29	5.98	6.75	11.51	17.31

Coal Characterization in Terms of Reference Species (Sommariva et al., 2010)

Composition (wt%)	$COAL_1$	$COAL_2$	$COAL_3$	Moisture	Ash
	35.08	18.05	18.05	1.51	17.31

Table 14 Stream Properties and Operating Conditions

				Syngas Effluent	
Operating Parameters	**Coal**	**Air**	**Steam**	**Experimental**	**Predicted**
Mass flow (kg/h)	7.0	8.87	4.20	18.52	17.88
Volume flow (Nm^3/h)	—	6.91	5.23	20.41	20.20
Temperature (°C)	25.0	75.0	120.0	270.0	227.0
Pressure (MPa)	0.14	0.140	0.140	0.107	0.11
Specific heat (kJ/kg K)	0.19	1.01	1.67	1.51	1.55

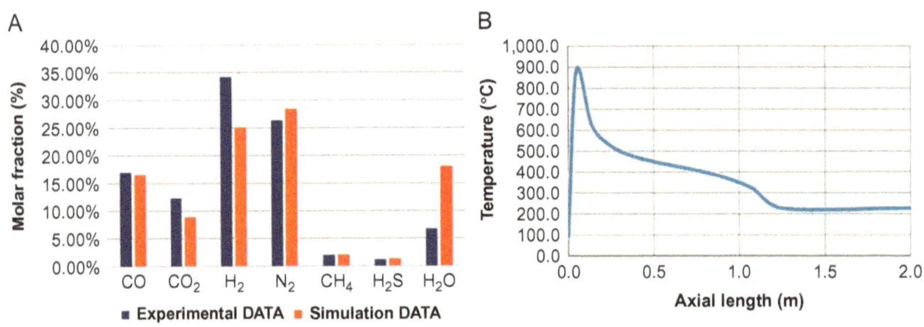

Fig. 53 (A) Outlet syngas composition: comparisons of experimental data with model predictions. (B) Gas-phase temperature along the rector.

experimental peak temperature of ~850°C. The AG2S™ technology was therefore integrated to the coal gasification processes as in Fig. 54 by means of detailed process simulation (Bassani et al., 2016). The raw syngas obtained at the top of the gasifier is usually sent to the chemical washing for purification of raw syngas and acid gas removal. In this case, the raw syngas is

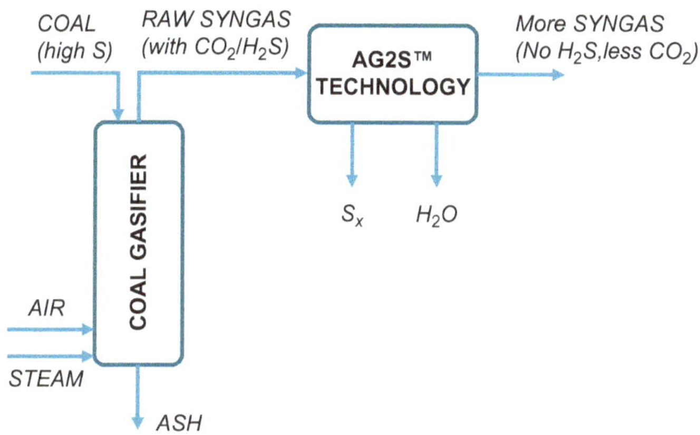

Fig. 54 Integrated AG2S™ and coal gasification process.

directly sent to the chemical washing of Unit 2 of the AG2S™. The acid gases are separated from the syngas and converted into additional syngas, elemental sulfur, and water. This unavoidably leads to a more efficient and less impacting coal gasification. Supposing to use a coal with 10 wt% of sulfur content, it is estimated to obtain about 15 wt% of additional syngas and 10% reduction of CO_2 emissions with a relatively small investment for revamping and a payback in the order of 2 years (Bassani et al., 2016).

6.3 Pyrolysis and Gasification of Polyethylene in a Bubbling Fluidized-Bed Reactor

Mastellone et al. (2007) carried out PE pyrolysis and gasification experiments in a bubbling fluidized-bed (BFB) reactor, with an internal diameter of 102 mm and a height of 1.05 m. Compared to fixed-bed gas–solid reactors, fluidized beds are more effective for the thermal treatments of pyrolysis and gasification of plastic waste. Nitrogen and air/N_2 mixtures were used to fluidize the bed, made of 1440 g of quartz sand. PE was fed by means of a screw-feeder located at the top of the freeboard. The plastic material melts and reaches the bubbling bed in a liquid phase, enveloping sand particles. Experiments were carried out by feeding a mass feed rate of 3 g/min of PE and different equivalent ratios by varying the air/N_2 ratio.

In a BFB, gas bubbles rise, coalesce, and grow up to the top of the bed, promoting a perfect mixing in the emulsion phase where a portion of entering gas fluidizes the inert bed material. The splashing zone just above the bed surface is again well mixed, because of the eruption of gas bubbles. Finally a

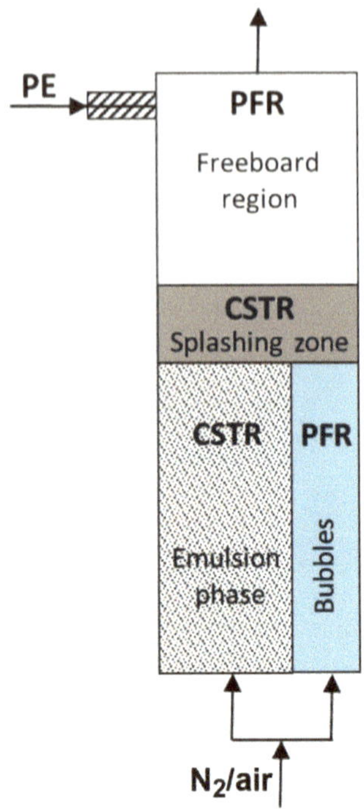

Fig. 55 Simplified modeling framework employed to simulate PE pyrolysis and gasification in a BFB reactor.

plug-flow hydrodynamic regime is present in the freeboard region. Fig. 55 schematically shows the very simplified modeling framework employed to simulate the BFB reactor.

The yields and products obtained from the pyrolysis and gasification experiments in the fluidized-bed reactor are again schematically caused by two contributions. There is the primary devolatilization and pyrolysis of molten plastic, which mainly occurs in the emulsion phase. Then, there are the secondary or successive gas-phase reactions involving the primary volatiles and they occur partially inside the bed but mainly along the freeboard zone (Mastellone et al., 2007). Fig. 56 reports the measured outlet composition obtained from the pyrolysis of PE at three different reactor temperatures. After each test the reactor head was removed in order to verify and clean the bed and the reactor wall. As expected (Dente and Ranzi, 1983), the wall was covered by a significant amount of carbonaceous material (Arena and Mastellone, 2006). Relevant PE dehydrogenation and carbonization during these experiments was also caused by a strong catalytic effect of the iron contained in the reactor wall.

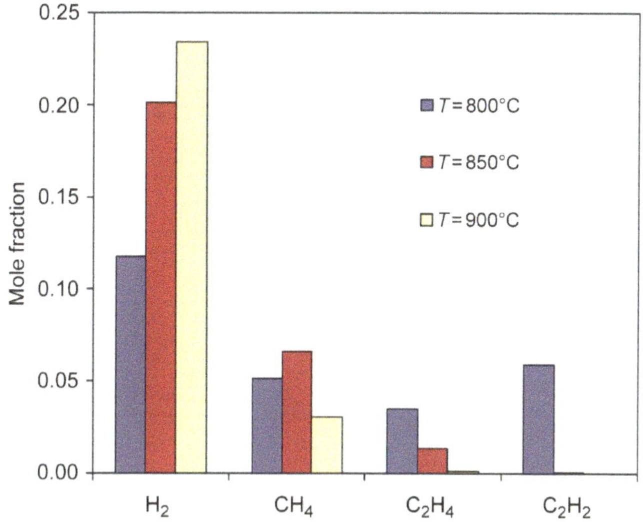

Fig. 56 Polyethylene pyrolysis. Product distribution at three reactor temperatures.

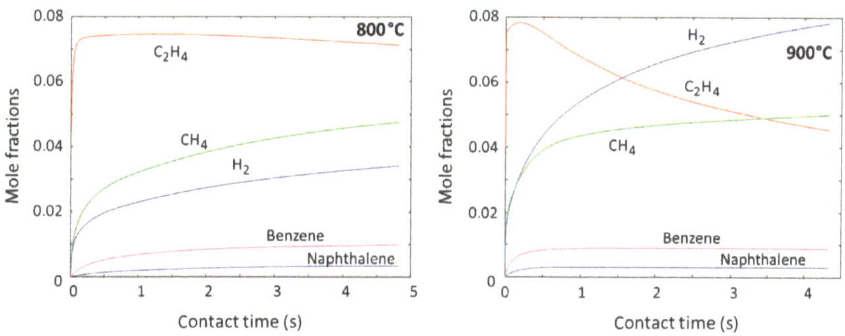

Fig. 57 Polyethylene pyrolysis. Predicted product distribution at 800°C and 900°C.

Fig. 57 shows the predicted product distribution from PE pyrolysis at 800°C and 900°C, at different contact times in the freeboard region. Comparisons between model predictions at these temperatures clearly indicate that higher temperatures favor dehydrogenation reactions and contribute to H_2 formation. C_2H_4 consumption becomes important at 900°C, but still significant amounts of ethylene, benzene, and higher PAH are predicted. A more complete dehydrogenation and the formation of carbon structures, as observed in the experimental tests, can be then justified only on catalytic basis. Temperatures higher than 1150°C would be required to predict similar H_2 production.

Table 15 Polyethylene Gasification: Mole Fractions of Major Gas Products (Experimental Measurements and Model Predictions)

	$\lambda = 0.2$		$\lambda = 0.25$		$\lambda = 0.3$	
	Exp.	**Pred.**	**Exp.**	**Pred.**	**Exp.**	**Pred.**
N_2	0.828	0.839	0.828	0.827	0.84	0.816
H_2	0.084	0.033	0.07	0.034	0.066	0.035
CO_2	0.013	0.008	0.012	0.012	0.016	0.016
CO	0.044	0.062	0.054	0.073	0.054	0.083
CH_4	0.026	0.025	0.027	0.024	0.02	0.024
C_2–C_4	0.005	0.029	0.009	0.026	0.005	0.024
C_6 + (benzene)	—	0.005	—	0.004	—	0.003

Gasification experiments in near isothermal conditions at 850°C were also performed, with equivalence ratios spanning from 0.2 and 0.3. N_2 mole fraction at the reactor outlet was always about 80%. Table 15 shows the experimental measurements and model predictions of mole fractions of major gas products, on dry basis. An increase of λ leads to larger CO and CO_2 formation, with reduction of ethylene, methane, and heavier hydrocarbons. Model predictions of H_2 formation are again significantly lower than the experimental ones as a further confirmation of the presence of a strong catalytic effect, in this reactor and conditions. Further details on these experiments and comparisons can be found in Mastellone et al. (2007).

7. CONCLUSIONS

A comprehensive and unifying mathematical model to describe the pyrolysis, gasification, and combustion of solid fuels is presented and discussed in this Chapter. Emphasis is given to the multicomponent, multiphase, multiscale nature of this system. This challenging problem required several assumptions and kinetic simplifications for both the gas- and solid-phase mechanisms. The characterization of the solid fuels through a limited number of reference components took advantage from the use of van Krevelen diagram. Extensive lumping procedures were applied to describe solid, gas, and tar species in order to reduce the complexity of the overall system. Finally, the coupling of mass and energy transport resistances at particle and reactor scale with the kinetic mechanisms constituted a further difficult aspect. Detailed descriptions of fuel characterization, of kinetic mechanisms, and of particle and reactor balance equations are reported aiming at providing the reader with the useful insights for reproducing the whole set of quoted results. Applications at the particle scale show the possible overshooting of internal temperature during the pyrolysis of thick biomass particles. Further examples illustrate how the start-up procedure can affect the operation behavior of autothermal systems, and demonstrate the possibility of gasification and combustion regimes according to particle geometry and operating conditions. At the reactor scale, the ignition and combustion front in a traveling grate biomass combustor are used as dependent variables to control reactor operation. Coal gasifiers at lab and pilot scale are also analyzed to verify model performances. At the process scale, the enhancing effect of sulfur components in syngas production is also discussed. As already mentioned, these application examples show the flexibility and possibilities as well as the limitations of the proposed approach in the design, simulation, and control of solid fuel pyrolysis, gasification, and combustion units. Selected reference components and relating lumped kinetic models, both for solid fuel pyrolysis and for the secondary gas-phase reactions, are always prone to improvements and extensions. Nevertheless, it seems relevant to observe that this predictive and comprehensive model of solid fuel pyrolysis, gasification, and combustion is already able to provide a wide range of useful applications in a feasible way.

ACKNOWLEDGMENTS

This chapter summarizes the research activities on solid fuel pyrolysis and combustion done at CMIC Department of Politecnico di Milano. The contributions of all the friends, colleagues, and PhD and Master students are gratefully acknowledged. Particularly, the authors acknowledge the very useful discussions, suggestions, and comments of Profs. M. Dente, S. Pierucci, A. Cuoci, and A. Frassoldati.

REFERENCES

Aguado R, Olazar M, San Jose MJ, Aguirre G, Bilbao J: Pyrolysis of sawdust in a conical spouted bed reactor. Yields and product composition, *Ind Eng Chem Res* 39:1925–1933, 2000.

Anca-Couce A: Reaction mechanisms and multi-scale modelling of lignocellulosic biomass pyrolysis, *Prog Energy Combust Sci* 53:41–79, 2016.

Anca-Couce A, Obernberger I: Application of a detailed biomass pyrolysis kinetic scheme to hardwood and softwood torrefaction, *Fuel* 167:158–167, 2016.

Anca-Couce A, Mehrabian R, Scharler R, Obernberger I: Kinetic scheme of biomass pyrolysis considering secondary charring reactions, *Energy Convers Manag* 87: 687–696, 2014.

Antal MJJ, Varhegyi G: Cellulose pyrolysis kinetics: the current state of knowledge, *Ind Eng Chem Res* 34:703–717, 1995.

Antal MJ, Varhegyi G, Jakab E: Cellulose pyrolysis kinetics: revisited, *Ind Eng Chem Res* 37:1267–1279, 1998.

Arena U: Process and technological aspects of municipal solid waste gasification. A review, *Waste Manag* 32(4):625–639, 2012.

Arena U, Mastellone ML: Fluidized bed pyrolysis of plastic wastes. In Schiers J, Kaminsky W, editors: *Feedstock recycling and pyrolysis of waste plastics: converting waste plastics into diesel and other fuels*, Chichester, 2006, John Wiley & Sons Ltd., pp 435–474 (chapter 16).

Ates F, Putun E, Putun AE: Fast pyrolysis of sesame stalk: yields and structural analysis of bio-oil, *J Anal Appl Pyrolysis* 71:779–790, 2004.

Barker-Hemings E, Bozzano G, Dente M, Ranzi E: Detailed kinetics of the pyrolysis and oxidation of anisole, *Chem Eng Trans* 24:61–66, 2011.

Barker-Hemings E, Cavallotti C, Cuoci A, Faravelli T, Ranzi E: A detailed kinetic study of pyrolysis and oxidation of glycerol (propane-1,2,3-triol), *Combust Sci Technol* 184(7–8):1164–1178, 2012.

Bassani A, Pirola C, Maggio E, et al: Acid Gas to Syngas (AG2S™) technology applied to solid fuel gasification: cutting H2S and CO2 emissions by improving syngas production, *Appl Energy*, 2016. http://dx.doi.org/10.1016/j.apenergy.2016.06.040.

Bassilakis R, Zhao Y, Solomon PR, Serio MA: Sulfur and nitrogen evolution in the Argonne coals. Experiment and modeling, *Energy Fuel* 7(6):710–720, 1993.

Bennadji H, Smith K, Shabangu S, Fisher EM: Low-temperature pyrolysis of woody biomass in the thermally thick regime, *Energy Fuel* 27:1453–1459, 2013.

Benson SW: *Thermochemical kinetics*, New York, London, Sydney, Toronto, 1976, John Wiley & Sons.

Brezinsky K, Pecullan M, Glassman I: Pyrolysis and oxidation of phenol, *J Phys Chem A* 102(44):8614–8619, 1998.

Bridgeman TG, Jones JM, Shield I, Williams PT: Torrefaction of reed canary grass, wheat straw and willow to enhance solid fuel qualities and combustion properties, *Fuel* 87(6):844–856, 2008.

Bridgwater AV: Renewable fuels and chemicals by thermal processing of biomass, *Chem Eng J* 91:87–102, 2003.

Bridgwater AV: Review of fast pyrolysis of biomass and product upgrading, *Biomass Bioenergy* 38:68–94, 2012.

Buah WK, Cunliffe AM, Williams PT: Characterization of products from the pyrolysis of municipal solid waste, *Process Saf Environ Prot* 85(5):450–457, 2007.

Burnham AK, Zhou X, Broadbelt LJ: Critical review of the global chemical kinetics of cellulose thermal decomposition, *Energy Fuels* 29:2906–2918, 2015.

Buzzi-Ferraris G: BzzMath 6.0 numerical libraries, Available at: http://homes.chem.polimi.it/gbuzzi/index.htm, 2010.

Buzzi-Ferraris G, Manenti F: A combination of parallel computing and object-oriented programming to improve optimizer robustness and efficiency, *Comput Aided Chem Eng* 28:337–342, 2010.

Buzzi-Ferraris G, Manenti F: BzzMath: library overview and recent advances in numerical methods, *Comput Aided Chem Eng* 30:1312–1316, 2012.

Caballero JA, Conesa JA, Font R, Marcilla A: Pyrolysis kinetics of almond shells and olive stones considering their organic fractions, *J Anal Appl Pyrolysis* 42: 159–175, 1997.

Calonaci M, Grana R, Barker Hemings E, Bozzano G, Dente M, Ranzi E: Comprehensive kinetic modeling study of bio-oil formation from fast pyrolysis of biomass, *Energy Fuels* 24(10):5727–5734, 2010.

Carstensen HH, Dean AM: Development of detailed kinetic models for the thermal conversion of biomass via first principle methods and rate estimation rules. In *Computational modeling in lignocellulosic biofuel production, ACS symposium series*, vol. 1052, 2010, pp 201–243.

Catoire L, Yahyaoui M, Osmont A, et al: Thermochemistry of compounds formed during fast pyrolysis of lignocellulosic biomass, *Energy Fuel* 22:4265–4273, 2008.

Chen JC, Niksa S: Coal devolatilization during rapid transient heating. 1. Primary devolatilization, *Energy Fuel* 6(3):254–264, 1992.

Chouchene A, Jeguirim M, Khiari B, Zagrouba F, Trouvé G: Thermal degradation of olive solid waste: influence of particle size and oxygen concentration, *Resour Conserv Recycl* 54(5):271–277, 2010.

Corbetta M, Frassoldati A, Bennadji H, et al: Pyrolysis of centimeter-scale woody biomass particles: kinetic modeling and experimental validation, *Energy Fuels* 28:3884–3898, 2014.

Corbetta M, Bassani A, Manenti F, et al: Multiscale kinetic modeling and experimental investigation of syngas production from coal gasification in updraft gasifiers, *Energy Fuel* 29:3972–3984, 2015.

Cuoci A, Faravelli T, Frassoldati A, et al: Mathematical modelling of gasification and combustion of solid fuels and wastes, *Chem Eng Trans* 18:989–994, 2009.

Cuoci A, Frassoldati A, Faravelli T, Ranzi E: A computational tool for the detailed kinetic modeling of laminar flames: application to C2H4/CH4 coflow flames, *Combust Flame* 160:870–886, 2013.

Cuoci A, Frassoldati A, Faravelli T, Ranzi E: OpenSMOKE++: an object-oriented framework for the numerical modeling of reactive systems with detailed kinetic mechanisms, *Comput Phys Commun* 192:237–264, 2015.

Curran HJ: Rate constant estimation for C1 to C4 alkyl and alkoxyl radical decomposition, *In J Chem Kinet* 38(4):250–275, 2006.

Debiagi PEA, Pecchi C, Gentile G, et al: Extractives extend the applicability of multistep kinetic scheme of biomass pyrolysis, *Energy Fuel* 29(10):6544–6555, 2015.

Demirbas A: Combustion characteristics of different biomass fuels, *Prog Energy Combust Sci* 30(2):219–230, 2004.

Dente M, Ranzi E: Mathematical modeling of hydrocarbon pyrolysis reactions. In Albright LF, Crynes BL, Corcoran WH, editors: *Pyrolysis: theory and industrial practice*, New York, 1983, Academic Press, pp 133–175.

Dente M, Ranzi E, Goossens AG: Detailed prediction of olefin yields from hydrocarbon pyrolysis through a fundamental simulation model (SPYRO), *Comput Chem Eng* 3:61–75, 1979.

Dente M, Pierucci S, Ranzi E, Bussani G: New improvements in modeling kinetic schemes for hydrocarbons pyrolysis reactors, *Chem Eng Sci* 47(9):2629–2634, 1992.

Dente M, Bozzano G, Faravelli T, Marongiu A, Pierucci S, Ranzi E: Kinetic modelling of pyrolysis processes in gas and condensed phase, *Adv Chem Eng* 32: 51–166, 2007.

Di Blasi C: Modeling and simulation of combustion processes of charring and non-charring solid fuels, *Prog Energy Combust Sci* 19:71–104, 1993.

Di Blasi C: Modeling chemical and physical processes of wood and biomass pyrolysis, *Prog Energy Combust Sci* 34:47–90, 2008.

Di Blasi C. Combustion and gasification rates of lignocellulosic chars, *Prog Energy Combust Sci* 35:121–140, 2009.

Djokic MR, Van Geem KM, Cavallotti C, Frassoldati A, Ranzi E, Marin GB: An experimental and kinetic modeling study of cyclopentadiene pyrolysis: first growth of polycyclic aromatic hydrocarbons, *Combust Flame* 161(11):2739–2751, 2014.

Dou B, Park S, Lim S, Yu TU, Hwang J: Pyrolysis characteristics of refuse derived fuel in a pilot-scale unit, *Energy Fuel* 21:3730–3734, 2007.

Dupont C, Chen L, Cances J, et al: Biomass pyrolysis: kinetic modelling and experimental validation under high temperature and flash heating rate conditions, *J Anal Appl Pyrolysis* 85(1):260–267, 2009.

El-Melih AM, Ibrahim S, Gupta AK, Al Shoaibi A: Experimental examination of syngas recovery from acid gases, *Appl Energy* 164C:64–68, 2016.

Faravelli T, Bozzano G, Scassa C, et al: Gas product distribution from polyethylene pyrolysis, *J Anal Appl Pyrolysis* 52(1):87–103, 1999.

Faravelli T, Pinciroli M, Pisano F, Bozzano G, Dente M, Ranzi E: Thermal degradation of polystyrene, *J Anal Appl Pyrolysis* 60(1):103–121, 2001.
Faravelli T, Bozzano G, Colombo M, Ranzi E, Dente M: Kinetic modeling of the thermal degradation of polyethylene and polystyrene mixtures, *J Anal Appl Pyrolysis* 70(2):761–777, 2003.
Faravelli T, Frassoldati A, Migliavacca G, Ranzi E: Detailed kinetic modeling of the thermal degradation of lignins, *Biomass Bioenergy* 34:290–301, 2010.
Fletcher TH, Hardesty DR: *Compilation of Sandia coal devolatilization data: milestone report (No. SAND-92-8209),* Livermore, CA, 1992, Sandia National Labs.
Fletcher TH, Kerstein AR, Pugmire RJ, Grant DM: Chemical percolation model for devolatilization. 2. Temperature and heating rate effects on product yields, *Energy Fuel* 4(1):54–60, 1990.
Frassoldati A, Cuoci A, Faravelli T, et al: An experimental and kinetic modeling study of n-propanol and iso-propanol combustion, *Combust Flame* 157(1):2–16, 2010.
Frassoldati A, Grana R, Faravelli T, Ranzi E, Oßwald P, Kohse-Höinghaus K: Detailed kinetic modeling of the combustion of the four butanol isomers in premixed low-pressure flames, *Combust Flame* 159(7):2295–2311, 2012.
Frigerio S, Thunman H, Leckner B, Hermansson S: Estimation of gas phase mixing in packed beds, *Combust Flame* 153:137–148, 2008.
Froment GF, Bischoff KB: *Chemical reactor analysis and design,* New York, 1990, John Wiley & Sons.
Gauthier G, Melkior T, Salvador S, et al: Pyrolysis of thick biomass particles: experimental and kinetic modelling, *Chem Eng Trans* 32:601–606, 2013.
Genetti D, Fletcher TH, Pugmire RJ: Development and application of a correlation of 13C NMR chemical structural analyses of coal based on elemental composition and volatile matter content, *Energy Fuel* 13(1):60–68, 1999.
Grana R, Frassoldati A, Faravelli T, et al: An experimental and kinetic modeling study of combustion of isomers of butanol, *Combust Flame* 157(11):2137–2154, 2010.
Gray P, Griffiths JF, Hasko SM, Lignola PG: Oscillatory ignitions and cool flames accompanying the non-isothermal oxidation of acetaldehyde in a well stirred, flow reactor, *Proc Roy Soc Lond A* 374(1758):313–339, 1981. The Royal Society.
Grieco EM, Baldi G: Predictive model for countercurrent coal gasifiers, *Chem Eng Sci* 66(23):5749–5761, 2011.
Groeneveld MJ, Van Swaaij WPM: Gasification of char particles with CO2 and H2O, *Chem Eng Sci* 35:307–313, 1980.
Grønli MG: *A theoretical and experimental study of the thermal degradation of biomass,* Trondheim, Norway, 1996, PhD Thesis NTNU.
Hambly EM: *The chemical structure of coal tar and char during devolatilization,* M.S. Thesis, 1988, Brigham Young University, Provo, UT, USA.
Hercog J, Tognotti L: *Realisation of IFRF solid fuel data base,* 2008. IFRF Doc No. E36/y02.
Himmel ME, Ding SY, Johnson DK, et al: Biomass recalcitrance: engineering plants and enzymes for biofuels production, *Science* 315(5813):804–807, 2007.
Hou Z, Bennett CA, Klein MT, Virk PS: Approaches and software tools for modeling lignin pyrolysis, *Energy Fuel* 24(1):58–67, 2009.
Ince A, Carstensen HH, Reyniers MF, Marin GB: First-principles based group additivity values for thermochemical properties of substituted aromatic compounds, *AICHE J* 61(11):3858–3870, 2015.
Jakab E, Faix O, Till F, Székely T: Thermogravimetry/mass spectrometry study of six lignins within the scope of an international round robin test, *J Anal Appl Pyrolysis* 35:167–179, 1995.
Jupudi RS, Zamansky V, Fletcher TH: Prediction of light gas composition in coal devolatilization, *Energy Fuel* 23(6):3063–3067, 2009.

Kan T, Strezov V, Evans TJ: Lignocellulosic biomass pyrolysis: a review of product properties and effects of pyrolysis parameters, *Renew Sust Energ Rev* 57:1126–1140, 2016.

Kashiwagi T, Nambu H: Global kinetic constants for thermal oxidative degradation of a cellulosic paper, *Combust Flame* 88:345–368, 1992.

Klein MT, Virk PS: Modeling of lignin thermolysis, *Energy Fuel* 22(4):2175–2182, 2008.

Leckner B: Process aspects in combustion and gasification. Waste-to-Energy (WtE) units, *Waste Manag* 37:13–25, 2015.

Lédé J: Cellulose pyrolysis kinetics: an historical review on the existence and role of intermediate active cellulose, *J Anal Appl Pyrolysis* 94:17–32, 2012.

Lewis AD, Fletcher TH: Prediction of sawdust pyrolysis yields from a flat-flame burner using the CPD model, *Energy Fuel* 27:942–953, 2013.

Lindfors C, Kuoppala E, Oasmaa A, Solantausta Y, Arpiainen V: Fractionation of bio-oil, *Energy Fuel* 28(9):5785–5791, 2014.

Maffei T, Senneca O, Ranzi E, Salatino P: Pyrolysis, annealing and char combustion/ oxy-combustion for CFD codes. In *XXXIV* meeting of the Italian Section of the Combustion Institute, 2011, pp 24–26.

Maffei T, Sommariva S, Ranzi E, Faravelli T: A predictive kinetic model of sulfur release from coal, *Fuel* 91(1):213–223, 2012.

Maffei T, Frassoldati A, Cuoci A, Ranzi E, Faravelli T: Predictive one step kinetic model of coal pyrolysis for CFD applications, *Proc Combust Inst* 34(2):2401–2410, 2013a.

Maffei T, Khatami R, Pierucci S, Faravelli T, Ranzi E, Levendis YA: Experimental and modeling study of single coal particle combustion in O2/N2 and oxy-fuel (O2/CO2) atmospheres, *Combust Flame* 160(11):2559–2572, 2013b.

Manenti F: CO2 as feedstock: a new pathway to syngas, *Comput Aided Chem Eng* 37:1049–1054, 2015.

Manenti F, Dones I, Buzzi-Ferraris G, Preisig HA: Efficient numerical solver for partially structured differential and algebraic equation systems, *Ind Eng Chem Res* 48:9979–9984, 2009.

Manenti F, Papasidero D, Bozzano G, Ranzi E: Model-based optimization of sulfur recovery units, *Comput Chem Eng* 66:244–251, 2014.

Manenti F, Pierucci S, Molinari L: *Process for reducing CO2 and producing syngas*, 2015. Patent: WO 2015/015457 A1.

Manenti G, Molinari L, Manenti F: Syngas from H_2S and CO_2: an alternative, pioneering synthesis route? *Hydrocarb Process* 6:1–4, 2016.

Manion JA, Louw R: Rates, products, and mechanisms in the gas-phase hydrogenolysis of phenol between 922 and 1175 K, *J Phys Chem* 93(9):3563–3574, 1989.

Marongiu A, Faravelli T, Bozzano G, Dente M, Ranzi E: Thermal degradation of poly-vinyl chloride, *J Anal Appl Pyrolysis* 70(2):519–553, 2003.

Marongiu A, Faravelli T, Ranzi E: Detailed kinetic modeling of the thermal degradation of vinyl polymers, *J Anal Appl Pyrolysis* 78:343–362, 2007.

Mastellone ML, Arena U, Barbato G, et al: Devolatilization and gasification of plastic wastes in a fluidized bed reactor, *Chem Eng Trans* 11:491–496, 2007.

Matsuoka K, Ma ZX, Akiho H, et al: High-pressure coal pyrolysis in a drop tube furnace, *Energy Fuel* 17(4):984–990, 2003.

Mayes HB, Nolte MW, Beckham GT, Shanks BH, Broadbelt LJ: The alpha–Bet(a) of glucose pyrolysis: computational and experimental investigations of 5-hydroxymethylfurfural and levoglucosan formation reveal implications for cellulose pyrolysis, *ACS Sustain Chem Eng* 2(6):1461–1473, 2014.

Mettler MS, Vlachos DG, Dauenhauer PJ: Top ten fundamental challenges of biomass pyrolysis for biofuels, *Energy Environ Sci* 5:7797–7809, 2012.

Miller RS, Bellan J: A generalized biomass pyrolysis model based on superimposed cellulose, hemicellulose and lignin kinetics, *Combust Sci Technol* 126:97–137, 1997.

Milosavljevic I, Suuberg EM: Cellulose thermal decomposition kinetics: global mass loss kinetics, *Ind Eng Chem Res* 34(4):1081–1091, 1995.
Milosavljevic I, Oja V, Suuberg EM: Thermal effects in cellulose pyrolysis: relationship to char formation processes, *Ind Eng Chem Res* 35:653–666, 1996.
Mitra A, Sarofim AF, Bar-Ziv E: The influence of coal type on the evolution of polycyclic aromatic hydrocarbons during coal devolatilization, *Aerosol Sci Technol* 6(3):261–271, 1987.
Miura K, Mae K, Shimada M, Minami H: Analysis of formation rates of sulfur-containing gases during the pyrolysis of various coals, *Energy Fuel* 15(3):629–636, 2001.
Norinaga K, Shoji T, Kudo S, Hayashi JI: Detailed chemical kinetic modelling of vapour-phase cracking of multi-component molecular mixtures derived from the fast pyrolysis of cellulose, *Fuel* 103:141–150, 2013.
Norinaga K, Yang H, Tanaka R, et al: A mechanistic study on the reaction pathways leading to benzene and naphthalene in cellulose vapor phase cracking, *Biomass Bioenergy* 69:144–154, 2014.
Nowakowska M, Herbinet O, Dufour A, Glaude PA: Detailed kinetic study of anisole pyrolysis and oxidation to understand tar formation during biomass combustion and gasification, *Combust Flame* 161(6):1474–1488, 2014.
Oasmaa A, Sundqvist T, Kuoppala E, et al: Controlling the phase stability of biomass fast pyrolysis bio-oils, *Energy Fuel* 29(7):4373–4381, 2015.
Olah GA: Beyond oil and gas: the methanol economy, *Angew Chem Int Ed* 44(18):2636–2639, 2005.
Olah GA, Goeppert A, Prakash GS: *Beyond oil and gas: the methanol economy,* Weinheim, Germany, 2011, Wiley-VCH Verlag GmbH & Co KGaA.
Oyeyemi VB, Krisiloff DB, Keith JA, Libisch F, Pavone M, Carter EA: Size-extensivity-corrected multireference configuration interaction schemes to accurately predict bond dissociation energies of oxygenated hydrocarbons, *J Chem Phys* 140(4): 044317, 2014a.
Oyeyemi VB, Keith JA, Carter EA: Trends in bond dissociation energies of alcohols and aldehydes computed with multireference averaged coupled-pair functional theory, *J Phys Chem A* 118(17):3039–3050, 2014b.
Park WC, Atreya A, Baum HR: Experimental and theoretical investigation of heat and mass transfer processes during wood pyrolysis, *Combust Flame* 157:481–494, 2010.
Paulsen AD, Mettler MS, Dauenhauer PJ: The role of sample dimension and temperature in cellulose pyrolysis, *Energy Fuel* 27(4):2126–2134, 2013.
Pecullan M, Brezinsky K, Glassman I: Pyrolysis and oxidation of anisole near 1000 K, *J Phys Chem A* 101(18):3305–3316, 1997.
Pelucchi M, Somers KP, Yasunaga K, et al: An experimental and kinetic modeling study of the pyrolysis and oxidation of n-C 3 C 5 aldehydes in shock tubes, *Combust Flame* 162(2):265–286, 2015.
Perry ST: *A global free-radical mechanism for nitrogen release during coal devolatilization based on chemical structure (Doctoral dissertation),* Utah, USA, 1999, Brigham Young University.
Pettinau A, Frau C, Ferrara F: Performance assessment of a fixed-bed gasification pilot plant for combined power generation and hydrogen production, *Fuel Process Technol* 92(10):1946–1953, 2011.
Pettinau A, Ferrara F, Amorino C: Combustion vs. gasification for a demonstration CCS (carbon capture and storage) project in Italy: a techno-economic analysis, *Energy* 50:160–169, 2013.
Pierucci S, Ranzi E: A general mathematical model for a moving bed gasifier. In Braunschweig B, Joulia X, editors: *18th European symposium on computer aided process engineering*, Amsterdam, 2008, Elsevier Science Bv, pp 901–906.
Pohl JH, Sarofim AF: Devolatilization and oxidation of coal nitrogen. In Sixteenth symposium (international) on combustion. Pittsburgh, PA, 1977, The Combustion Institute, pp 491–501.
Ranzi E, Sogaro A, Gaffuri P, Pennati G, Westbrook CK, Pitz WJ: New comprehensive reaction mechanism for combustion of hydrocarbon fuels, *Combust Flame* 99:201–211, 1994a.

Ranzi E, Dente M, Faravelli T, Pennati G: Prediction of kinetic parameters for hydrogen abstraction reactions, *Combust Sci Technol* 95:1–50, 1994b.

Ranzi E, Dente M, Faravelli T, et al: Kinetic modeling of polyethylene and polypropylene thermal degradation, *J Anal Appl Pyrolysis* 40:305–319, 1997.

Ranzi E, Cuoci A, Faravelli T, et al: Chemical kinetics of biomass pyrolysis, *Energy Fuel* 22:4292–4300, 2008.

Ranzi E, Pierucci S, Aliprandi PC, Stringa S: Comprehensive and detailed kinetic model of a traveling grate combustor of biomass, *Energy Fuel* 25:4195–4205, 2011.

Ranzi E, Frassoldati A, Grana R, et al: Hierarchical and comparative kinetic modeling of laminar flame speeds of hydrocarbon and oxygenated fuels, *Prog Energy Combust Sci* 38:468–501, 2012.

Ranzi E, Corbetta M, Manenti F, Pierucci S: Kinetic modeling of the thermal degradation and combustion of biomass, *Chem Eng Sci* 110:2–12, 2014.

Ravaghi-Ardebili Z, Manenti F, Corbetta M, Pirola C, Ranzi E: Biomass gasification using low-temperature solar-driven steam supply, *Renew Energy* 74:671–680, 2015.

Saggese C, Frassoldati A, Cuoci A, Faravelli T, Ranzi E: A wide range kinetic modeling study of pyrolysis and oxidation of benzene, *Combust Flame* 160(7):1168–1190, 2013.

Saggese C, Ferrario S, Camacho J, et al: Kinetic modeling of particle size distribution of soot in a premixed burner-stabilized stagnation ethylene flame, *Combust Flame* 162(9):3356–3369, 2015.

Sarathy SM, Oßwald P, Hansen N, Kohse-Höinghaus K: Alcohol combustion chemistry, *Prog Energy Combust Sci* 44:40–102, 2014.

Sassi M, Gupta AK: Sulfur recovery from acid gas using the Claus process and high temperature air combustion (HiTAC) technology, *Am J Environ Sci* 4(5):502–511, 2008.

Seshadri V, Westmoreland PR: Concerted reactions and mechanism of glucose pyrolysis and implications for cellulose kinetics, *J Phys Chem A* 116:11997–12013, 2012.

Shankar Tumuluru J, Sokhansanj S, Hess JR, Wright CT, Boardman RD: REVIEW: a review on biomass torrefaction process and product properties for energy applications, *Ind Biotechnol* 7(5):384–401, 2011.

Shen D, Jin W, Hu J, Xiao R, Luo K: An overview on fast pyrolysis of the main constituents in lignocellulosic biomass to valued-added chemicals: structures, pathways and interactions, *Renew Sust Energ Rev* 51:761–774, 2015.

Skodras G, Grammelis P, Basinas P, Prokopidou M, Kakaras E, Sakellaropoulos GP: A thermochemical conversion study on the combustion of residue derived fuels, *Water Air Soil Pollut Focus* 9(1–2):151–157, 2008.

Sluiter JB, Ruiz RO, Scarlata CJ, Sluiter AD, Templeton DW: Compositional analysis of lignocellulosic feedstocks. 1. Review and description of methods, *J Agric Food Chem* 58(16):9043–9053, 2010.

Solomon PR, Hamblen DG, Carangelo RM, Serio MA, Deshpande GV: General model of coal devolatilization, *Energy Fuel* 2(4):405–422, 1988.

Solomon PR, Serio MA, Carangelo RM, et al: Analysis of the Argonne premium coal samples by thermogravimetric Fourier transform infrared spectroscopy, *Energy Fuel* 4(3):319–333, 1990.

Sommariva S, Maffei T, Migliavacca G, Faravelli T, Ranzi E: A predictive multi-step kinetic model of coal devolatilization, *Fuel* 89:318–328, 2010.

Sommariva S, Grana R, Maffei T, Pierucci S, Ranzi E: A kinetic approach to the mathematical model of fixed bed gasifiers, *Comput Chem Eng* 35:928–935, 2011.

Stark AK, Bates RB, Zhao Z, Ghoniem AF: Prediction and validation of major gas and tar species from a reactor network model of air-blown fluidized bed biomass gasification, *Energy Fuel* 29(4):2437–2452, 2015.

Tognotti L, Longwell JP, Sarofim AF: The products of the high temperature oxidation of a single char particle in an electrodynamic balance, *Symp Combust* 23:1207–1213, 1991.

Trendewicz A, Evans R, Dutta A, Sykes R, Carpenter D, Braun R: Evaluating the effect of potassium on cellulose pyrolysis reaction kinetics, *Biomass Bioenergy* 74:15–25, 2015.
Van der Stelt MJC, Gerhauser H, Kiel JHA, Ptasinski KJ: Biomass upgrading by torrefaction for the production of biofuels: a review, *Biomass Bioenergy* 35(9):3748–3762, 2011.
van Krevelen DW: Graphical-statistical method for the study of structure and reaction processes of coal, *Fuel* 29:269–284, 1950.
Vinu R, Broadbelt LJ: A mechanistic model of fast pyrolysis of glucose-based carbohydrates to predict bio-oil composition, *Energy Environ Sci* 5:9808–9826, 2012.
Violi A, Truong TN, Sarofim AF: Kinetics of hydrogen abstraction reactions from polycyclic aromatic hydrocarbons by H atoms, *J Phys Chem A* 108(22):4846–4852, 2004.
Wang L, Skreiberg Ø, Gronli M, Specht GP, Antal MJ Jr: Is elevated pressure required to achieve a high fixed-carbon yield of charcoal from biomass? Part 2: the importance of particle size? *Energy Fuel* 27(4):2146–2156, 2013.
Wen CY: Non catalytic heterogeneous solid–fluid reaction models, *Ind Eng Chem* 60(9):34–54, 1968.
Weng J, Jia L, Wang Y, et al: Pyrolysis study of poplar biomass by tunable synchrotron vacuum ultraviolet photoionization mass spectrometry, *Proc Combust Inst* 34: 2347–2354, 2013.
Westerhof RJM, Brilman DWF, van Swaaij WPM, Kersten SRA: Effect of temperature in fluidized bed fast pyrolysis of biomass: oil quality assessment in test units, *Ind Eng Chem Res* 49:1160–1168, 2010.
Williams PT, Besler S: The influence of temperature and heating rate on the slow pyrolysis of biomass, *Renew Energy* 7(3):233–250, 1996.
Williams A, Jones JM, Ma L, Pourkashanian M: Pollutants from the combustion of solid biomass fuels, *Prog Energy Combust Sci* 38:113–137, 2012.
Xu WC, Tomita A: Effect of coal type on the flash pyrolysis of various coals, *Fuel* 66(5):627–631, 1987a.
Xu WC, Tomita A: Effect of temperature on the flash pyrolysis of various coals, *Fuel* 66(5):632–636, 1987b.
Yang HM, Appari S, Kudo S, Hayashi JI, Norinaga K: Detailed chemical kinetic modeling of vapor-phase reactions of volatiles derived from fast pyrolysis of lignin, *Ind Eng Chem Res* 54(27):6855–6864, 2015.
Yildiz G, Pronk M, Djokic M, et al: Validation of a new set-up for continuous catalytic fast pyrolysis of biomass coupled with vapour phase upgrading, *J Anal Appl Pyrolysis* 103:343–351, 2013.
Younan Y, van Goethem MW, Stefanidis GD: A particle scale model for municipal solid waste and refuse-derived fuels pyrolysis, *Comput Chem Eng* 86:148–159, 2016.
Zhang YP: Reviving the carbohydrate economy via multi-product lignocellulose bio-refineries, *J Ind Microbiol Biotechnol* 35:367–375, 2008.
Zhao Y, Serio MA, Bassilakis R, Solomon PR: A method of predicting coal devolatilization behavior based on the elemental composition. In *Symposium (international) on combustion*, vol 25(1), Pittsburgh, PA, 1994, The Combustion Institute, pp 553–560.
Zhou S, Pecha B, van Kuppevelt M, McDonald AG, Garcia-Perez M: Slow and fast pyrolysis of Douglas-fir lignin: importance of liquid-intermediate formation on the distribution of products, *Biomass Bioenergy* 66:398–409, 2014a.
Zhou X, Nolte MW, Mayes HB, Shanks BH, Broadbelt LJ: Experimental and mechanistic modeling of fast pyrolysis of neat glucose-based carbohydrates. 1. Experiments and development of a detailed mechanistic model, *Ind Eng Chem Res* 53(34):13274–13289, 2014b.
Zhou X, Nolte MW, Shanks BH, Broadbelt LJ: Experimental and mechanistic modeling of fast pyrolysis of neat glucose-based carbohydrates. 2. Validation and evaluation of the mechanistic model, *Ind Eng Chem Res* 53(34):13290–13301, 2014c.

CHAPTER TWO

Mechanistic Understanding of Thermochemical Conversion of Polymers and Lignocellulosic Biomass

X. Zhou*, L.J. Broadbelt*, R. Vinu†,1
*Northwestern University, Evanston, IL, United States
†Indian Institute of Technology Madras, Chennai, India
[1]Corresponding author: e-mail address: vinu@iitm.ac.in

Contents

Abstract

Pyrolysis is a promising thermochemical technique to convert polymers such as waste plastics and lignocellulosic biomass to liquid products that are valuable either directly as

Advances in Chemical Engineering, Volume 49
ISSN 0065-2377
http://dx.doi.org/10.1016/bs.ache.2016.09.002

or are potentially upgraded to liquid fuels and fine chemical intermediates. Mechanistically, polymer pyrolysis involves a complex set of free radical, concerted, and/or ionic reactions that occur via numerous competing pathways. Engineering these pathways to produce the required molecules warrant a thorough understanding of kinetics of the reactions under different conditions. In this critical review, after emphasizing the need for resource and energy recovery from polymers, the elementary reactions involved in the pyrolysis of polyolefins to various products are discussed along with an elucidation of detailed kinetic modeling. The reactions involved in oxidative pyrolysis of polymers are also discussed with polystyrene autoxidation as an example case. The influence of catalysts such as zeolites in altering the product distribution from pyrolysis of synthetic polymers is discussed. As lignocellulosic biomass is more complex in structure compared to synthetic polymers, the challenges and methodology involved in modeling the degradation of its basic constituents, viz. cellulose, hemicellulose, and lignin, to various organics are discussed. After elucidating the influence of various concerted reactions involved in cellulose pyrolysis on product yields, the effect of catalysts on biomass fast pyrolysis and bio-oil upgradation are discussed. The review concludes with a note on advantages of copyrolysis of synthetic polymers and biomass to enhance the quality of bio-oil that can be easily converted to biofuel with minimal upgradation.

1. INTRODUCTION

Polymers are indispensable for our everyday living, and a number of natural and synthetic polymers find application in both domestic and industrial sectors such as household appliances, furniture, textile, packaging, agriculture, healthcare, building and construction, medicine, electrical and electronics, automotive and aerospace components, fuel, and energy. Today's human society is knitted with the use of polymers, which is primarily attributed to their superior properties such as light weight, durability, low cost, and ease of production. Moreover, the polymer industry provides jobs, and spurs growth and competitiveness in a global economy. The global production of synthetic polymers (or plastics) rose from 245 million metric tons in 2008 to 311 million metric tons in 2014, with China and Europe contributing nearly 45% of the total share (The Statistics Portal, 2015). Nearly, 8% of the petroleum consumed worldwide is being used for the production of plastics and to power plastic manufacturing processes (Worldwatch Institute, 2016). According to the European data, the major plastics can be ranked in terms of their demand as follows: polypropylene (PP) (19%) > low-density polyethylene (LDPE) (17.5%) > high-density polyethylene (HDPE) (12%) > poly(vinyl chloride) (PVC) (10%) > polystyrene (PS), poly(ethylene terephthalate) (PET), and polyurethanes (PU) (7%)

(Association of Plastics Manufacturers, 2016). The packaging and construction sectors are the key markets for plastics with ca. 40% and 20% demand, respectively. Besides commodity polymers, natural polymers such as lignocellulosic biomass are identified as the most abundant raw materials on earth with annual production of 170 million metric tons (Clarke and Deswarte, 2008). Lignocellulosic biomass, which refers to plant and plant-derived matter produced by photosynthesis, is chemically composed of cellulose, hemicellulose, and lignin, and is the only renewable source of carbon on earth. In fact, only 3% of the 170 million metric tons is utilized for food and nonfood applications (Clarke and Deswarte, 2008). The growth of environmental awareness and dwindling of fossil fuel reserves have ignited interest in utilizing these renewable feedstocks for energy and fuels. The data shown above indicates that the availability of both petroleum-based waste plastics and natural biomass is significant to derive energy, fuels, and chemicals.

1.1 Energy and Resource Recovery From Polymer Wastes

With increase in production and per capita utilization of polymers comes the issue of their disposal. Polymer waste management aims to reduce the accumulation of and reform the waste polymers to valuable products or energy. Landfilling, recycling, and energy recovery are the three options for the treatment of consumed polymer wastes. Landfilling is still a widely practiced technique to get rid of the polymer wastes. Landfilling poses both short-term and long-term risks including (a) excessive land utilization, (b) soil and groundwater contamination, (c) air pollution, (d) disruption of wildlife, and (e) emission of harmful greenhouse gases. Recycling involves either reuse or reformation of plastic wastes. While direct reuse without significant processing is limited to plastic sheets, covers, and containers that are made of PE, PP, PET, and PVC, mechanical reformation of plastics involves a series of steps involving sorting, crushing into flakes, washing, drying, and granulation (Al-Salem et al., 2009). The granules can then be molded to various forms via extrusion and other techniques. Chemical recycling or solvolysis is another technique used to recover monomeric building blocks and/or fine chemicals from polymer wastes. Solvolysis involves treating the classified polymeric wastes with solvents and reagents (or catalysts) to depolymerize the polymer to low molecular weight (LMW) chemicals and oligomers. Based on the type of polymer and the reagent used, a variety of reactions such as glycolysis, methanolysis, hydrolysis, acidolysis, and alcoholysis occur in the liquid phase. One of the disadvantages of solvolysis is the separation of the unreacted polymer, oligomers, and monomers from the reaction medium, which is an intensive downstream operation. The interested reader can refer the reviews of Kumar et al. (2011) and Hamad et al. (2013) for a detailed understanding of chemical recycling techniques.

Incineration or mass burning is a conventional technique to recover energy from waste plastics. It involves combustion of the material in presence of oxygen or air to recover energy that can be used for heating or electric power generation. Owing to the high calorific value of the major consumer plastics (30–40 MJ/kg), good energy recovery is possible via incineration. However, based on the type and composition of the plastic waste, incineration can lead to the emission of toxic volatile organic compounds (VOCs) such as polyaromatic hydrocarbons, furans, dioxins, polychlorinated dibenzo-*p*-dioxins (PCDDs), polychlorinated dibenzo furans (PCDFs), cyano compounds, and halogenated gases (Vinu et al., 2016). In Europe, there is a consistent decline in landfilling, which is compensated by recycling of and energy recovery from plastic wastes. In 2012, out of 25.2 million tons of plastic wastes, nearly 36% and 26% were utilized for energy recovery and recycling, respectively (Association of Plastics Manufacturers, 2016).

Incineration of lignocellulosic biomass obtained from wood cutting, construction debris, and agriculture is also common to recover energy. The use of lignocellulosic agro residues, either alone or in combination, with coal or refuse-derived fuel (RDF) such as dried municipal solid wastes (MSW) in boilers and furnaces is widely employed as a sustainable solution to meet the energy demand. MSW is a heterogeneous and complicated feedstock consisting of plastics, cardboard/paper, cloth, jute, rags, biomass (leaves, wood shavings, residues), metals (iron, aluminum), and inerts (sand, glass, etc). In India, it is estimated that ca. 9% of the MSW from major cities contains plastic wastes (Ojha and Vinu, 2015a). In practice, when MSW is incinerated, a mixture of polymeric wastes with varying degrees of C, H, O, and N composition are subjected to complex thermal reactions that lead to the formation of numerous toxic organic species.

Although incineration has been the conventional processing technique for polymeric wastes, it is considered as a high volume yet low value proposition. Given that polymers are rich in carbon, hydrogen, and various organic functionalities, it is imperative to recover value added chemicals and fuel molecules from them. For example, controlled cracking of hydrogen-rich polymers like PE and PP can yield hydrocarbons of varying carbon

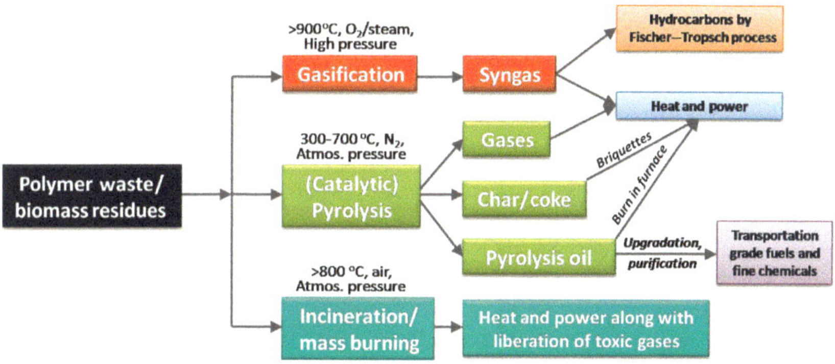

Fig. 1 Various thermochemical techniques to convert waste polymer and lignocellulosic biomass to useful products.

numbers, which are expected to exhibit comparable properties with that of hydrocarbon fuels. This is certainly a high value proposition compared to simply burning them, which also causes environmental pollution. Fig. 1 depicts the thermochemical routes to convert polymers and biomass to valuable products. Gasification and pyrolysis are classified as controlled cracking techniques. Gasification or partial oxidation involves heating the material to very high temperatures (>800°C) in presence of oxygen or steam at high pressures (~0.1–2 MPa) to produce syn gas or producer gas, which is a mixture of carbon monoxide and hydrogen. Owing to high volatile matter content in synthetic polymers (>90%), hydrogen production efficiency of 60–70% is possible with minimal or negligible formation of toxic organics such as dioxins and polyaromatics. The syn gas can be used to either run a gas engine to produce power or it can be subjected to the classic catalytic Fischer–Tropsch process to produce higher alkanes that are valuable as hydrocarbon fuels. The latter process, however, requires two steps to convert the solid polymer to liquid fuels, while pyrolysis yields liquid products in a single step. Unlike synthetic polymer gasification where a high quality of syn gas with minimal contaminants (CO_2, CH_4, C2–C3 hydrocarbons) is produced, biomass gasification results in the production of by-products such as tars originating from the phenolic and oxygenated pyrolysates during biomass pyrolysis. Biomass gasification is well practiced for energy recovery (i.e., heat and electricity generation) from locally available feedstocks (agro residues) in small to medium communities/townships (Bocci et al., 2014; Buragohain et al., 2010). Cogasification of lignocellulosic biomass with polymers (or plastic wastes) is also a promising technique to improve the quality of pyrolysis vapors that can then be combusted to produce a cleaner gas. Al-Salem et al. (2009) reviewed a number of available pilot and commercial scale gasification technologies for treating biomass and plastic wastes. Pyrolysis produces liquid (condensable vapors or oil), solid (char), and gaseous products (noncondensable gases), whose relative yields are determined by the process conditions. Generally, pyrolysis is carried out to produce high yield of oil fraction (or pyrolysis oil) that can be upgraded to transportation grade fuels and other fine chemicals.

1.2 Pyrolysis: A Promising Thermochemical Technique

Pyrolysis involves heating the material to moderate temperatures (300–700°C) in inert, oxygen-free environment. Theoretically, all the volatile matter present in the feedstock can be pyrolyzed to condensable vapor fraction. As synthetic polymers contain high volatile matter with negligible fixed carbon and inorganics like ash, high oil yield (>90 wt.%) can be expected. When commercial polymers containing additives-like fillers, plasticizers, colorants, and flame retardants are pyrolyzed, the yield of char increases due to an intrinsic increase in fixed carbon content. As lignocellulosic biomass contains significant amount of fixed carbon (10–30 wt.% dry basis) and ash (5–10 wt.% dry basis) (Vassilev et al., 2010), complete conversion by pyrolysis is never achieved. The solid char can be value added by chemical treatments to improve its physicochemical properties for use as soil supplements and as adsorbents in environmental applications (Nair and Vinu, 2016a). The solid char can also be briquetted and used as a heating source. The noncondensable gases, based on the type of feedstock, include light gases-like CO, CO_2, H_2, and C1–C5 hydrocarbons (methane, ethane, ethylene, propane, propylene, butanes, butylenes, etc.). Pyrolysis oil from lignocellulosic biomass (also called as bio-oil) is a complex mixture of oxygenated organics including acids, alcohols, carbonyl compounds, dehydrated sugars, furan derivatives, phenolics, and aromatic hydrocarbons (Liu et al., 2014a).

Temperature and residence time are the two key parameters that determine the relative yields of the products. Based on the process temperature and heating rate, pyrolysis can be classified as slow pyrolysis (torrefaction), slow pyrolysis (carbonization), intermediate pyrolysis, and fast pyrolysis. The typical temperature and timescales involved in the above four pyrolysis processes are (~290°C, 10–60 min), (~400°C, hours to days), (~500°C, 10–30 s), (~500°C, 1 s), respectively (Bridgwater, 2012). Importantly, the oil yield increases as the reaction time is decreased at a moderate temperature of 500°C. Long residence time allows the primary volatile species to undergo secondary decomposition reactions both in the gas phase and melt phase to either form completely cracked products such as methane, CO, or CO_2, or form condensed ring aromatics (chars) via repolymerization reactions. High temperature also enhances the pyrolysis severity, and enhances the rate of secondary decomposition reactions, thus reducing the liquid yield. Pyrolysis can, therefore, be considered as a versatile thermo-chemical conversion technique, wherein the yield of desired product can be tailored by altering the reaction conditions. Torrefaction of lignocellulosic biomass results in high char yield (~80 wt.%), carbonization results in ca. 35 wt.% each of gases and char, intermediate pyrolysis results in nearly 50 wt.% of liquid fraction, while fast pyrolysis yields maximum oil (or bio-oil) fraction (ca. 70–80 wt.%) (Bridgwater, 2012). Fast high heating rate of the feedstock (>1000°C/s) and short residence time of the pyrolysates achieved in fast pyrolysis process result in high yield of primary volatile species, which are potentially condensable organic compounds. Residence time plays a crucial role in determining the product yields and the composition of the various constituents in the oil and gas phase. Cozzani et al. (1997) and Westerhout et al. (1998) studied the pyrolysis of PE and PP and showed that at high pyrolysis temperatures (700–800°C) and vapor residence times, tar cracking processes lead to reduction in tar yield with a concomitant increase in gas and char yields. Moreover, the yield of C1–C3 hydrocarbons increased with residence time due to secondary cracking reactions (Cozzani et al., 1997). The vapor residence time can be altered by either varying the length of the reactor (in the case of tubular reactors) or gas flow rate (in the case of semibatch and stirred reactors). Scott and coworkers (Scott et al., 1990) investigated fast pyrolysis of PVC, PS, and LLDPE (linear LDPE) in fluidized bed fast pyrolysis reactor, and reported high yield of monomer recovery from PVC and PS.

Besides the conversion of plastic wastes to useful products, the pyrolysate composition from fast pyrolysis can also yield useful information of the microstructure of polymers, such as sequence isomerism and stereoisomerism (Sugimura et al., 1981; Tsuge and Ohtani, 1997), and the presence of impurities. This is because, the primary pyrolysates carry direct information of the type of bonds and the functional groups that are getting cleaved and reacted, respectively, when subjected to a sudden high temperature environment. Analytical pyrolysis reactors available in the market, viz. micro-pyrolyzer from Frontier Laboratories, Japan, and Pyroprobe® from C.D.S. Analytical, USA, are popular tools to evaluate the purity of polymer samples. The interested reader can refer the handbook of chromatograms and mass spectra of pyrolysates from synthetic polymers by Tsuge et al. (2011). The bio-oil from fast pyrolysis of lignocellulosic biomass contains a range of LMW compounds. While bio-oil has a direct market as a low grade fuel for heating applications in boilers and engines, its value can be improved by upgrading via catalytic hydrodeoxygenation to transportation grade drop-in fuels and fine chemical intermediates (Bridgwater, 2012). Usually catalysts are also added during the fast pyrolysis step to obtain bio-oil of a better quality (i.e., low O content) (Bridgwater, 2012; Carpenter et al., 2014; Chundawat et al., 2011; Kunkes et al., 2008; Mettler et al., 2012c; Pecha and Garcia-Perez, 2015; Ruddy et al., 2014; Vispute et al., 2010). As a promising technology for the production of transportation fuels and multiple commodity chemicals (Alonso et al., 2012; Baker and Elliott, 1988; Bond et al., 2014; Ruddy et al., 2014; Sharma and Bakhshi, 1993c), fast pyrolysis of biomass has recently risen in prominence, and numerous studies have been conducted to understand its complexity. Fig. 2 depicts the major steps involved in fast pyrolysis of biomass. It is evident that biomass fast pyrolysis is a complex process involving the participation of all three phases at different timescales. By using high speed photography, Dauenhauer and coworkers (Teixeira et al., 2011) demonstrated the ejection of primary liquid aerosol particles during cellulose pyrolysis. These fine aerosol particles are shown to contain nonvolatile levoglucosan, anhydro dimers, and cellobiosan. Based on pyrolysis severity,

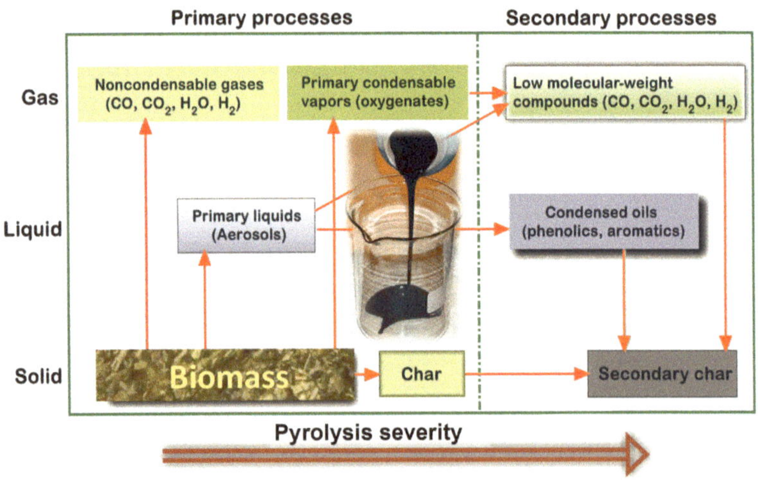

Fig. 2 Products formed during fast pyrolysis of biomass and commercial plastics. Bio-oil is shown as a *dark tarry liquid. Redrawn from Evans RJ, Milne TA: Molecular characterization of the pyrolysis of biomass. 1. Fundamentals,* Energy Fuels *1:123–137, 1987.*

which is determined by the residence time and temperature, various secondary transformations can lead to the formation of condensed aromatics, phenolics, light gases, and secondary char.

This chapter is not intended to be a comprehensive review of all the previous works on fast pyrolysis of synthetic polymers and lignocellulosic biomass. Rather, the present review highlights the mechanism of transformation of these complex feedstocks to various pyrolysis products, and the importance of kinetic models to engineer the pathways for a better understanding of the phenomena. A better understanding of the kinetics is vital to improve the yields and selectivities of the desired products from pyrolysis either by modifying the operating conditions or by using better catalysts. To this end, the various sections of this chapter are classified as follows: Section 2 describes the well-understood free radical mechanism of polymer pyrolysis, specifically olefinic polymers. Emphasis is also given to the mechanism of oxidative pyrolysis of polymers; Section 3 discusses the salient results of catalytic pyrolysis of polymers using different zeolites as catalysts. The ionic mechanism facilitated by the acidic protons is emphasized, and its effect on product distribution is discussed; Section 4 highlights the complexity of biomass structure in terms of its principal components, viz. cellulose, hemicellulose, and lignin, and the challenges involved in modeling the pyrolysis of these components for a successful understanding of whole biomass pyrolysis; Section 5 describes the major steps involved in catalytic pyrolysis of biomass for the production of aromatics, alkenes, and other value added products; Section 6 finally provides mechanistic insights on copyrolysis of biomass with plastics, which is a relatively new technique to produce bio-oil with improved characteristics via enhanced interactions between the intermediates from these two feedstocks.

2. PYROLYSIS OF SYNTHETIC POLYMERS

2.1 Olefinic Polymers

Olefinic polymers like PE, PP, PVC, and PS constitute a major class of commodity polymers due to their wide application in packaging, building and construction, and automotive sectors. Table 1 depicts the major pyrolysates obtained from pyrolysis of a number of synthetic and natural homopolymers. Generally, the common polymers can be structurally classified as follows: vinyl polymers with C and H (PE, PP, PS), halogenated vinyl polymers (PVC, PTFE), O-containing polymers (PET, poly(methyl methacrylate) (PMMA), poly(vinyl alcohol), poly(styrene peroxide) (PSP)), N-containing

Table 1 List of Synthetic and Natural Homopolymers and Their Salient Thermal Degradation Products

S. No.	Polymer	Typical Pyrolysis Products	References
1	Polystyrene	Styrene, α-methyl styrene, styrene dimers, trimers, tetramers, benzene, ethyl benzene, toluene, cumene, naphthalenes, and indenes	Bouster et al. (1989) and Ojha and Vinu (2015a)
2	Polypropylene	2,4-Dimethyl-1-heptene (trimer), methane, ethane, propylene, isobutylene, 2-pentene, 2-methyl-1-pentene, 2,4,6-trimethyl-1-nonene, other C3–C15 alkanes and alkenes	Kiang et al. (1990), Wong and Broadbelt (2001), and Kruse et al. (2003)
3	Polyethylenes (LDPE and HDPE)	Methane, ethylene, ethane, propylene, propane, butane, 1-butenes, 1-pentene, butadiene, C6–C25 alkanes and alkenes	Levine and Broadbelt (2009) and Kannan et al. (2014)
4	Poly(tetrafluoro ethylene)	Tetrafluoro ethylene, hexafluoro propene, cyclic perfluoro butane, and other fluorocarbons	Simon and Kaminsky (1998)
5	Poly(vinyl chloride)	HCl, methane, ethylene, ethane, propylene, propane, benzene, toluene, and PAHs	Marongiu et al. (2003) and Ma et al. (2002)
6	Poly(vinyl acetate)	Benzene, toluene, ethyl benzene, isoxylene, CO, CO_2, and CH_4. Oil quality lies in between gasoline and diesel	Shie et al. (2002)
7	Polyacrylonitrile	HCN, acrylonitrile, acetonitrile, 2-methyl-2-propene nitrile, dimer, trimer, and benzonitrile	Bozi and Blazsó (2009)
8	Poly(ethylene terephthalate)	4-(Vinyl oxycarbonyl) benzoic acid, benzoic acid, vinyl benzoate, benzene, divinyl terephthalate, terephthalic acid, phenyl ethyne, biphenyl, 1,2-ethanediol, dibenzoate, CO, CO_2, and CH_4	Dimitrov et al. (2013)
9	Poly(methyl methacrylate)	Methyl methacrylate, dimer, methyl isobutanol, methane, ethylene, ethane, propylene, CO_2, and CO	Smolders and Baeyens (2004)

Table 1 List of Synthetic and Natural Homopolymers and Their Salient Thermal Degradation Products—cont'd

S. No.	Polymer	Typical Pyrolysis Products	References
10	Poly(styrene peroxide)	Benzaldehyde, formaldehyde, phenyl glycol, and α-hydroxy acetophenone	Vinu et al. (2012)
11	6-Polyamide	ε-Caprolactam, acyclic amides, CO_2, and nitriles	Bozi and Blazsó (2009)
12	Polybutadiene	Butadiene, 4-vinyl cyclohexene, trimer, tetramer and pentamers of butadiene, ethylene, acetylene, ethane, propene, propyne, 1,3-pentadiene, styrene, toluene, and phenyl acetylene	Kiefer et al. (1985)
13	Polyisoprene	Isoprene, dipentene, 2,3-dimethyl cyclopentene, 1,5-dimethyl-5-vinyl cyclohexene, methane, C2–C5 alkenes, $C_{15}H_{24}$, and $C_{16}H_{26}$	Chien and Kiang (1979)
14	Polychloroprene	HCl, chloroprene, dimer, 1,3-butadiene, butenes, chloro butenes, dichloro butane, and dichloro cyclooctadienes	Lehrle et al. (2000)
15	Cellulose	Levoglucosan, formic acid, glycolaldehyde, furfural, 5-hydroxymethyl furfural, dianhydro-glucopyranose, acetol, CO, CO_2, and H_2O	Patwardhan et al. (2009)
16	Hemicellulose (from switch grass)	CO_2, formic acid, water, xylose, dianhydro xylose, CO, acetol, furfural, methyl furan, and acetic acid	Patwardhan et al. (2011a)
17	Lignin (from wheat straw and sarkanda grass by soda pulping process)	Vinyl guaiacol, guaiacol, methyl guaiacol, ethyl guaiacol, isoeugenol, acetovanillone, vanillin, vinyl phenol, ethyl phenol, syringol, methoxy eugenol, acetosyringone, syringaldehyde, phenol, trimethoxy benzene, and CO_2	Nowakowski et al. (2010)

polymers (polyacrylonitrile, polyamide), and elastomers (polyisoprene, polybutadiene, polychlorprene). PS, PMMA, polyisoprene, and nylon-6 are some polymers that yield monomers as the major pyrolysis products via successive depolymerization or end-chain unzipping reactions. However, pyrolysis of polymers like PE and PP do not yield significant amount of monomers, but produce linear hydrocarbons with a variety of chain lengths.

Pyrolysis of olefinic polymers proceeds via free radical mechanism, which is reasonably well established (Bockhorn et al., 1999; Faravelli et al., 1999, 2003; Kruse et al., 2001, 2002, 2003, 2005; Levine and Broadbelt, 2008, 2009; Marongiu et al., 2003, 2007; Poutsma, 2003, 2005, 2009; Vinu and Broadbelt, 2012b; Westerhout et al., 1997a,b). Fig. 3 depicts the elementary reactions involved in the mechanism. These elementary reactions can be generally classified into initiation (radical-forming), propagation (radical-interconverting), and termination (radical-consuming) reactions. Chain fission involves homolytic cleavage of any random C–C bond along the polymer chain, and it forms two radical species. The nature of the radical formed depends on the backbone structure of the polymer. In the case of vinyl polymers like PE and PP, primary and secondary alkyl radicals are formed, whereas in the case of diene polymers (polyisoprene, polybutadiene) with unsaturation in the backbone, allyl, vinyl, and alkyl radicals are formed. Importantly, the bond dissociation energy to form two radicals determines the activation energy of this step. The bond dissociation energy of various C–C bonds follow the trend: $C_{vinyl}–C_{vinyl} > C_{vinyl}–C_{allyl} > C_{alkyl}–C_{alkyl} > C_{alkyl}–C_{allyl} > C_{allyl}–C_{allyl}$ (Vinu and Broadbelt, 2012b). Moreover, tertiary alkyl radicals are more stable than secondary alkyl radicals, which are more stable than primary alkyl radicals. Usually, the activation energy of this initiation step is higher compared to all other elementary reactions (e.g., ca. 90 kcal/mol in the case of HDPE pyrolysis; Levine and Broadbelt, 2009). Therefore, once these radicals are formed, they are rapidly consumed in propagation and termination reactions. Many lumped kinetic models for polymer degradation apply quasi-steady-state approximation for the polymer radical species, such that the rate of formation of polymer free radicals is equal to the rate of their consumption (Madras and McCoy, 1999; Sterling and McCoy, 2001). Chain fission can also occur at the end of the polymer chain to form monomeric free radicals with saturated or unsaturated end group.

Reversible hydrogen abstraction, β-scission, and radical addition form an important class of propagation reactions. Hydrogen abstraction results in interchain hydrogen transfer such that a polymer radical converts to a stable

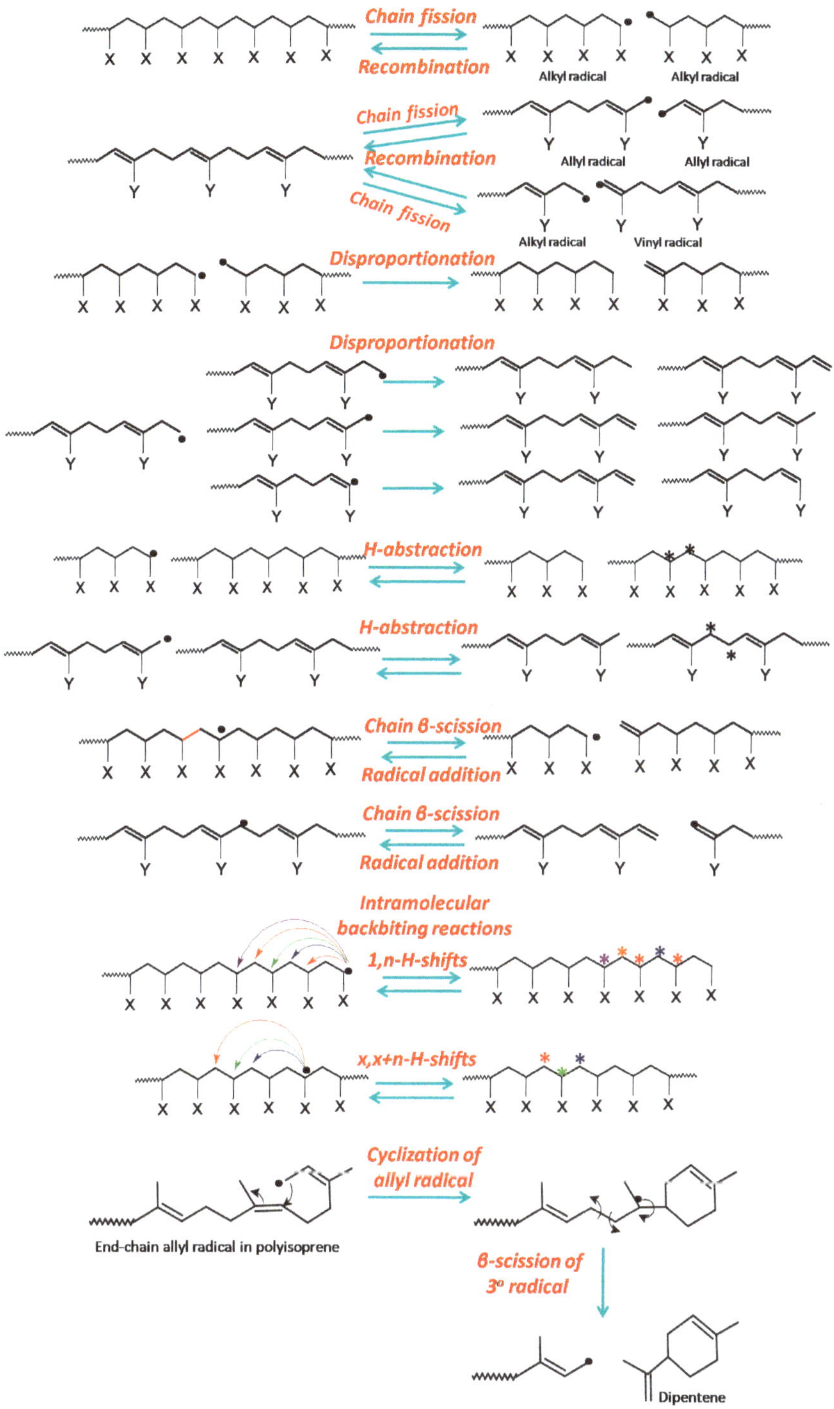

Fig. 3 See legend on next page.

polymer chain, while the stable polymer chain converts to a polymer radical. Based on the position of hydrogen that is transferred, mid-chain or end-chain radicals are formed. Mid-chain β-scission results in molecular weight reduction of polymers and results in the formation of an end-chain polymer radical and a stable polymer with an alkene end. The end-chain polymer radical can further undergo successive β-scission reactions from the chain end to unzip the monomer fragments. This is a major step for the formation of styrene monomer from PS and methyl methacrylate from PMMA (Kruse et al., 2001, 2002; Madras et al., 1996). The formation of isoprene from polyisoprene is also attributed to β-scission of the resonance stabilized allyl radical at the third position from the chain end (Vinu et al., 2016). At pyrolysis temperatures, end-chain β-scission dominates the bimolecular H-abstraction reaction to form high amount of monomers, dimers, and oligomers. However, at moderate temperatures and high reactant concentration corresponding to polymerization, H-abstraction dominates end-chain β-scission. Therefore, Rice–Kossiakoff regime is applicable for polymer pyrolysis, whereas Fabuss–Smith–Satterfield mechanism is applicable for polymerization (Vinu and Broadbelt, 2012b). The formation of C1–C15 alkanes and alkenes during PP pyrolysis, and C8–C23 alkanes and alkenes during HDPE pyrolysis are ascribed to β-scission of the specific end-chain radicals (Fig. 4).

The formation of free radicals at specific positions from the chain end can be due to the following competing reactions, viz. H-abstraction reaction with radical formation at specific end positions, addition of LMW radical to the polymer alkene end, and intramolecular isomerization or backbiting reactions involving 1,*n*- and $x,x+n$-hydrogen shifts (Fig. 3). The occurrence and predominance of intramolecular backbiting reactions are driven by the ring-strain energy of the cyclic transition state, which follows the trend: 1,6-H-shift (0.97 kcal/mol) < 1,5-H-shift (8.35 kcal/mol) < 1,4-H-shift (24.1 kcal/mol) < 1,3-H-shift (25.6 kcal/mol) (Vinu and

Fig. 3 Elementary reactions involved in the pyrolysis of vinyl and diene polymers. The polymer is PE, PP, or PS, when X corresponds to H, alkyl, or phenyl group, respectively. The polymer is polybutadiene or polyisoprene, when Y corresponds to H or alkyl group, respectively. *Asterisks* denote the possible free radical sites. *Redrawn from Vinu R, Broadbelt LJ: Unraveling reaction pathways and specifying reaction kinetics for complex systems,* Annu Rev Chem Biomol Eng *3:29–54, 2012b; Vinu R, Ojha DK, Nair V: Polymer pyrolysis for resource recovery. In Reedijk J, editor:* Elsevier Reference Module in Chemistry, Molecular Sciences and Chemical Engineering. *Waltham, MA, 2016, Elsevier, http://dx.doi.org/10.1016/B978-0-12-409547-2.11641-5.*

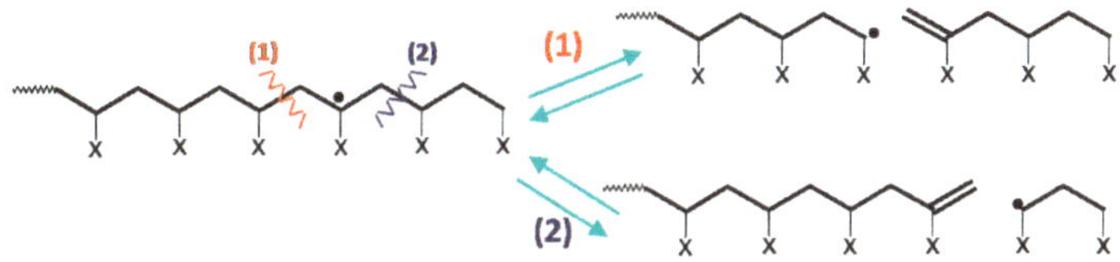

Fig. 4 β-Scission reaction occurring at specific chain-end position leading to the formation of stable alkene end group and a free radical intermediate. The free radical can either undergo further β-scission from the chain end or can abstract a hydrogen from another chain and get stabilized. The group X can be H, CH_3, or C_6H_5. *Redrawn from Kruse TM, Woo OS, Wong H-W, Khan SS, Broadbelt LJ: Mechanistic modeling of polymer degradation: a comprehensive study of polystyrene,* Macromolecules *35:7830–7844, 2002; Kruse TM, Wong HW, Broadbelt LJ: Mechanistic modeling of polymer pyrolysis: polypropylene,* Macromolecules *36:9594–9607, 2003.*

Broadbelt, 2012b). Levine and Broadbelt (2009), by utilizing net rate analysis, showed that $x,x+4$-intramolecular backbiting reactions were the dominant ones for the formation of specific end-chain radicals in HDPE pyrolysis. Similarly, in PS pyrolysis, 1,7- followed by 7,3-H-shift, and a subsequent β-scission was found to be the dominant pathway for the formation of styrene dimer, while direct 1,3-H-shift reaction proceeded at a low rate (Levine and Broadbelt, 2008; Poutsma, 2009). In certain diene polymers like polyisoprene and polybutadiene cyclic compounds are also the major products along with the respective monomers. An example pathway for the formation of dipentene, a cyclic terpene, from polyisoprene is also depicted in Fig. 3. It involves 1,6-cyclization of the end-chain allyl radical to form a tertiary alkyl radical at the end of the polymer chain, which readily undergoes β-scission to form a similar allyl radical and dipentene (Chien and Kiang, 1979; Vinu et al., 2016). The formation of 4-vinyl cyclohexene from polybutadiene also follows a similar pathway.

Termination step can occur via recombination and disproportionation reactions. Recombination of an end-chain radical with a mid-chain polymer radical can result in the formation of branches within the polymer chain. Disproportionation requires at least one participating radical to possess β-hydrogen. Different disproportionation reactions for simple vinyl polymers and diene polymers are depicted in Fig. 3. The mode of termination, i.e., recombination vs disproportionation, depends on the type of polymer and its radicals.

Compared to hydrocarbon polymers like PE, PP, or PS, halogenated polymers like PVC behave differently during pyrolysis. The first step in PVC pyrolysis is dehydrochlorination, which results in the release of large amount of HCl vapors (ca. 50 wt.%) and a significant mass loss (Yu et al., 2016). This occurs by the cleavage of the weakest C–Cl bond to release Cl radicals from the PVC backbone, which then abstracts hydrogen to form HCl (Marongiu et al., 2003). The formation of HCl during polychloroprene pyrolysis, and HCN gas during polyacrylonitrile pyrolysis can also be attributed to a similar mechanism. Ranzi and coworkers (Marongiu et al., 2003) developed a lumped kinetic model involving 250 reactions of 40 species and pseudocomponents to describe the formation of HCl, tarry aromatic compounds, and char from PVC pyrolysis. Crosslinking, bimolecular cyclization, and condensation reactions like Diels–Alder reactions were included in the mechanism to describe the formation of polyaromatic hydrocarbons with 2–10 rings.

2.1.1 Kinetic Modeling

Development of a "predictive" kinetic model of polymer pyrolysis is a challenging task. Owing to the inherent complexity of the structure of the polymer and the reaction types, a number of intermediate free radicals, oligomeric species, and LMW organics are formed. Moreover, polymers of different molecular weights or chain lengths are produced as it degrades, and it is imperative to account for the molecular weight distribution (MWD) of the reacting polymer radicals and stable polymer species. Therefore, simple *n*th-order kinetic models are not suitable to describe the decomposition of the polymer along with the evolution of its pyrolysis products. Mass loss profiles of polymer decomposition under isothermal or dynamic heating in a thermogravimetric analyzer (TGA) are usually subjected to kinetic analysis to evaluate the "apparent" rate parameters, viz. activation energy and frequency factor. These parameters are usually evaluated using single heating rate method of Friedmann, multiple heating rate method of Kissinger, and the classic integral and differential isoconversional methods of Flynn–Wall–Ozawa, Kissinger–Akahira–Sunose, Vyazovkin, and Friedmann (Cooney et al., 1983; Vyazovkin et al., 2011). The apparent rate parameters are used to describe only the mass loss profiles under the limited conditions that were used for fitting the experimental data, and are not sound enough to predict mass loss outside the range. Moreover, they do not account for the change in MWD of the polymers during degradation. Later, continuous distribution kinetic models that account for the MWD of the polymeric and free radical species were adopted to describe polymer degradation (Kodera and McCoy, 1997; Madras and McCoy, 1999; Madras et al., 1996; McCoy and Wang, 1994; Sterling and McCoy, 2001; Ziff and McGrady, 1986). These models, however, lumped all the stable polymeric species and polymeric free radicals, irrespective of their structure, into polymer and radical entities, respectively. The modes of bond cleavage, i.e., random scission or unzipping from the chain end, were incorporated in the stoichiometric kernel in the population balance rate equations for polymeric species. The kernel describes the product distribution when random scission or depolymerization occurred. For example, a Dirac delta function describes the production of LMW products of same molecular weight via end-chain depolymerization reaction.

The general scheme of a polymer (P) undergoing random chain scission to form two free radicals (R), and subsequent end-chain β-scission of R to form LMW product (Q) can be written as follows:

$$P(x)\underset{k_t}{\overset{k_f}{\rightleftarrows}}R(x')+R(x-x')$$

$$R(x)\underset{k_{ra}}{\overset{k_\beta}{\rightleftarrows}}Q(x_s)+R(x-x_s)$$

where x and x' denote the variable molecular weight of the polymer and radical species, and x_s denotes the specific molecular weight of the LMW product. The rate constants of bond fission, termination by recombination, β-scission, and radical addition are denoted by k_f, k_t, k_β, and k_{ra}, respectively. The population balance rate equations for species $P(x)$, $R(x)$, and $Q(x_s)$ can be written as:

$$\frac{\partial p(x)}{\partial t}=-k_f p(x)+k_t\int_0^x r(x')r(x-x')dx'$$

$$\frac{\partial r(x)}{\partial t}=2k_f\int_x^\infty p(x')\Omega(x,x')dx'-2k_t r(x)\int_0^\infty r(x')dx'-k_\beta r(x)$$
$$+k_{ra}\int_0^x r(x'-x_s)q(x_s)dx'+k_\beta\int_x^\infty r(x')\delta(x-(x'-x_s))dx'-k_{ra}r(x)$$
$$\times\int_0^\infty q(x')dx'$$

$$\frac{\partial q}{\partial t}=k_\beta\int_x^\infty r(x')\delta(x'-x_s)dx'-k_{ra}q(x_s)\int_0^\infty r(x')dx'$$

In the above equations, $p(x)$, $r(x)$, and q denote the concentrations, and the random scission stoichiometric kernel, $\Omega(x,x')$, is given by $1/x'$ (Kodera and McCoy, 1997). By applying the moment operation, given by $p^{(n)}(t)=\int p(x,t)x^n dx$, the above integro-differential equations can be converted to the following ordinary differential equations (ODEs), which are simultaneously solved for $n=0$, 1, and 2.

$$\frac{dp^{(n)}}{dt}=-k_f p^{(n)}+k_t\sum_{j=0}^{n} nC_j r^{(j)} r^{(n-j)}$$

$$\begin{aligned}\frac{dr^{(n)}}{dt}=&\,2k_f\frac{p^{(n)}}{n+1}-2k_t r^{(0)} r^{(n)}\\&-k_\beta r^{(n)}+k_{ra}\sum_{j=0}^{n} nC_j r^{(j)} q^{(n-j)}+k_\beta\sum_{j=0}^{n} nC_j x_s^j(-1)^j r^{(n-j)}-k_{ra} r^{(n)} q^{(0)}\end{aligned}$$

$$\frac{dq^{(n)}}{dt}=k_\beta x_s^n r^{(0)}-k_{ra} q^{(n)} r^{(0)}$$

Thus, zeroth, first, and second moments capture the molar concentration, mass concentration, and variance of the distribution, respectively. From these, the number average $(p^{(1)}/p^{(0)})$ and weight average $(p^{(2)}/p^{(1)})$ molecular weights, and polydispersity index $(p^{(2)}p^{(0)}/p^{(1)2})$ can be evaluated.

Broadbelt and coworkers developed robust mechanistic models of polymer pyrolysis by accounting all the possible radical and stable polymer species, and the LMW products (Kruse et al., 2001, 2002, 2003, 2005; Levine and Broadbelt, 2008, 2009; Vinu and Broadbelt, 2012b). They classified the various polymer species based on the identity of the end groups, viz. saturated vs unsaturated. The polymeric radicals were classified based on nature of the radical, i.e., primary, secondary, or tertiary alkyl, benzyl, allyl, vinyl, and whether the radical was present at the end of the polymer chain or at any random position in the mid of the chain. Moreover, in order to account for the formation of hydrocarbons of different carbon chain lengths, specific end-chain radicals at different positions from the chain end were also considered (Kruse et al., 2003; Vinu and Broadbelt, 2012b). By applying the generic family of reactions outlined in Fig. 3 to these different species, a large number of reactions were obtained. Table 2 depicts the typical number of species and reactions involved in various polymer pyrolysis models developed by the Broadbelt group.

It can be observed from Table 2 that when the number of radical and chain-end type increases, as in the case of pyrolysis of mixtures of PS and PP (primary alkyl, secondary alkyl, benzyl) or pyrolysis of polyisoprene (alkyl, allyl, and vinyl), the number of reactions increases enormously.

Table 2 Number of Species and Reactions in Mechanistic Models of Polymer Pyrolysis Available in the Literature

Polymer	Species	Reactions	References
Polystyrene	93	4502	Kruse et al. (2001)
Polypropylene	213	24,480	Kruse et al. (2003)
Blend of polystyrene and polypropylene	277	37,409	Kruse et al. (2005)
High-density polyethylene	151	11,007	Levine and Broadbelt (2009)
Polyisoprene	440	$>10^5$	Vinu and Broadbelt (unpublished data)
Poly(styrene peroxide)	83	949	Vinu et al. (2012)
Cellulose	52	99	Vinu and Broadbelt (2012a)
Cellulose	103	342	Zhou et al. (2014b)
Hemicellulose	123	511	Zhou et al. (2016b)

Importantly, hydrogen abstraction reactions contribute significantly to the overall number of reactions, due to the following possible combinations:

$$
\begin{aligned}
&R_{end,x} + P_y \rightarrow P_x + R_{end,y}\\
&R_{end,x} + P_y \rightarrow P_x + R_{mid,y}\\
&R_{mid,x} + P_y \rightarrow P_x + R_{end,y}\\
&R_{mid,x} + P_y \rightarrow P_x + R_{mid,y}\\
&Q_i^\bullet + P_y \rightarrow Q + R_{end,y}\\
&Q_i^\bullet + P_y \rightarrow Q + R_{mid,y}
\end{aligned}
$$

In the above reaction sequence, R_{end}, R_{mid}, P, $Q^\bullet$, and Q denote end-chain radicals, mid-chain radicals, stable polymer, LMW radical, and stable LMW species, respectively. The stiff set of moment equations for all the species involved in the mechanism were simultaneously solved using numerical solvers to obtain the concentration profiles of stable polymer, intermediates, radicals, and LMW species. The rate parameters for the individual reaction families are usually specified using empirical structure–activity relationships such as (a) Evans–Polanyi for general bond fission, termination, β-scission, and radical addition reactions, and (b) Blowers–Masel for hydrogen abstraction reactions (Vinu and Broadbelt, 2012b). These relationships require the enthalpy of the reaction ($\Delta H_{r\times n}{}^\circ$), which is readily estimated using Benson's group additivity principle (Benson,

1976). The typical range of preexponential factors and energy barriers for a number of elementary free radical reactions involved in polymer pyrolysis are shown in Table 3. The Arrhenius activation energy for a particular reaction family can be determined using the intrinsic barrier, transfer coefficient, and enthalpy of the reaction using the Evans–Polanyi and Blowers–Masel relationships.

Table 3 Summary of Kinetic Rate Parameters (Preexponential Factor and Arrhenius Activation Energy) for the Elementary Steps in Pyrolysis of Polymers

Reaction Family	Recommended Range of *A*, (s^{-1} or L mol^{-1} s^{-1} or L mol^{-1} s^{-1} K^{-n})	Intrinsic Reaction Barrier E_o (kcal mol^{-1})	Transfer Coefficient, α	References
Chain fission/ initiation	10^{16}–10^{17}	2.3, 5.98, 7.88[a]	1.0	Kruse et al. (2002) and Fleischer and Appel (1995)
Recombination	10^{11}	2.3, 5.98, 7.88[a]	0.0	
Disproportionation[b]	5% for alkyl–benzyl	2.3, 5.98, 7.88[a]	0.0	Kruse et al. (2002), Fleischer and Appel (1995), Schreck et al. (1989), Konar (1970), Rossi et al. (1978), and Klein and Kelly (1975)
	5% for alkyl–allyl			
	10% for alkyl–vinyl			
	15% for alkyl–alkyl			
β-scission	10^{13}–10^{14}	11.4	0.76	Kruse et al. (2002) and Poutsma (2000)
Radical addition	10^{7}–10^{8}	11.4	0.24	Kruse et al. (2002) and Poutsma (2000)
Intramolecular Isomerization[c]				
1,3 H-shift	3.8×10^{7}, $n = 0.67$	$E_o = 13.4$, $E_{rs} = 25.6$	0.60	Matheu et al. (2003)

Table 3 Summary of Kinetic Rate Parameters (Preexponential Factor and Arrhenius Activation Energy) for the Elementary Steps in Pyrolysis of Polymers—cont'd

Reaction Family	Recommended Range of A, (s^{-1} or L mol^{-1} s^{-1} or L mol^{-1} s^{-1} K^{-n})	Intrinsic Reaction BarrierE_o (kcal mol^{-1})	Transfer Coefficient, α	References
1,4 H-shift	7.85×10^8, $n=-0.12$	$E_o=13.4$, $E_{rs}=24.1$	0.60	Matheu et al. (2003)
1,5 H-shift	3.67×10^9, $n=-0.60$	$E_o=13.4$, $E_{rs}=8.35$	0.60	Matheu et al. (2003)
1,6 H-shift	2.80×10^7	$E_o=13.4$, $E_{rs}=0.97$	0.60	Matheu et al. (2003)
1,7 H-shift	3.00×10^9	$E_o=13.4$, $E_{rs}=5.0$	0.60	Matheu et al. (2003) and Levine and Broadbelt (2008)
7,3 H-shift	3.67×10^9	$E_a=16.60$	NA	Levine and Broadbelt (2008)
H-abstraction[d]	10^7–10^8	12.0	NA	Poutsma (2000) and Levine and Broadbelt (2008)
H-abstraction from styrene radicals[d]	2.1×10^6	12.0	NA	Kruse et al. (2002)
H-abstraction involving resonance stabilization in the transition state[e]		15.49	0.40	Sabbe et al. (2010)
H-Abstraction by H Atoms				
From allylic C[f]	10^4–10^5	11.56	0.40	Carstensen and Dean (2009)
	$n=2.28$ (allyl CH_3)			
	$n=2.18$ (allyl CHR)			
	$n=1.81$ (allyl CHR_2)			

Continued

Table 3 Summary of Kinetic Rate Parameters (Preexponential Factor and Arrhenius Activation Energy) for the Elementary Steps in Pyrolysis of Polymers—cont'd

Reaction Family	Recommended Range of *A*, (s^{-1} or $L\ mol^{-1}\ s^{-1}$ or $L\ mol^{-1}\ s^{-1}\ K^{-n}$)	Intrinsic Reaction Barrier E_o ($kcal\ mol^{-1}$)	Transfer Coefficient, α	References
From alkylic C[f]	10^4–10^5 $n=1.97$ (1C–H) $n=1.86$ (2C–H) $n=1.75$ (3C–H)	11.25	0.88	Carstensen and Dean (2009)
From cyclic CH_2[f]	10^3–10^4, $n=2.00$	7.73	0.48	Carstensen and Dean (2009)
From cyclic CHR[f]	10^4–10^5, $n=1.88$	9.14	0.50	Carstensen and Dean (2009)
From α-C–H in alcohols[f]		16.28	1.25	Carstensen and Dean (2010)
H-abstraction by CH_3 from α-C–H in alcohols[f]		20.29	1.31	Carstensen and Dean (2009)
Elimination of H_2O from alcohols[f]	1.6×10^4, $n=2.64$	65.2	−0.26	Carstensen and Dean (2010)

[a] E_o values are based on the self-diffusivities of the various polymers in the melt phase. $E_o=2.3$ for PS, PP, and PE; 5.98 for polybutadiene; and 7.88 for polyisoprene (Evans–Polanyi relationship: $E_a=E_o+\alpha\Delta H_{rxn}$).

[b] *A* values for disproportionation are expressed as a percentage of *A* for recombination of different radical types.

[c] $E_a=E_o+E_{rs}+\alpha\Delta H_{rxn}$, where E_{rs} is the ring-strain energy. Rate coefficient is calculated using the modified Arrhenius equation, $k=AT^n\exp(-E_a/RT)$.

[d] E_a is calculated by Blowers–Masel relationship for general H-abstraction reactions. $E_a=0$ (when $\Delta H_{rxn}/4E_o<-1$), $E_o(1+\Delta H_{rxn}/4E_o)^2$ (when $-1\leq\Delta H_{rxn}/4E_o\leq1$), ΔH_{rxn} (when $\Delta H_{rxn}/4E_o>1$).

[e] Includes H-abstraction from allylic radicals to form allylic radicals. Rate constant is calculated as, $k=\kappa A\exp(-E_a/RT)$, where κ is the tunneling coefficient.

[f] *A* values are calculated per H-atom or per C–H bond or –OH group. Rate constant is calculated as, $k=n_H AT^n\exp(-E_a/RT)$, where n_H is the number of equivalent H atoms, or C–H bonds or –OH groups.

Adapted from Vinu R, Broadbelt LJ: Unraveling reaction pathways and specifying reaction kinetics for complex systems. *Annu Rev Chem Biomol Eng* 3:29–54, 2012b.

Mechanistic kinetic models aid in thorough analysis of all the possible pathways that are operative under a particular reaction condition. Using net rate and sensitivity analyses, it is possible to map the major reaction routes for the formation of LMW products. This can be used even outside the experimental operating conditions to predict the product yields and selectivities. Thus, tailoring the product yields can be achieved by using the right catalyst to accelerate the rate of a specific reaction pathway. Broadbelt group has shown that the model yields of various LMW products and molecular weight reduction in PS, PE, and PP pyrolysis match the experimental data well, besides giving valuable insights on dominant reaction pathways (Kruse et al., 2002, 2003; Levine and Broadbelt, 2008, 2009).

Besides the degradation of virgin polymers of a single type, pyrolysis of binary, and ternary polymer mixtures assumes importance from the viewpoint of resource recovery from real polymeric wastes present in MSW. In this case, it is necessary to incorporate the interactions between the intermediates from the two polymer species. This can manifest in the form of cross-hydrogen abstraction and recombination reactions. If the mixing between the melt phases of the two polymers is not good, then the polymers would decompose independently of the other. Faravelli et al. (2003) utilized Flory–Huggins solution theory to predict the extent of mixing of the melt phases of PE and PS, and found that assuming 12.5 wt.% of PE in PS-rich phase and 2.5 wt.% of PS in PE-rich phase resulted in a good match of the kinetic model results with experimental data. Kruse et al. (2005) allowed the diffusion of LMW radicals from PS phase to PP phase and vice versa to incorporate the interaction effects. They found that diffusion of 0.037% of the LMW radicals of PS into PP phase matched the experimental data well, and also captured the enhancement in PP degradation rate by four times.

Kinetic Monte Carlo (KMC) is an alternative to continuous distribution kinetic model to describe polymer pyrolysis. KMC has its origin in the stochastic chemical master equation that determines the reaction probability of a species and its molecular population at a specific future time (Gillespie, 2007). KMC method utilizes an iterative approach to track the reactive species population according to the elementary free radical reaction rules. Compared to the continuum approach that involves solving a number of coupled nonlinear ODEs, this approach is simple to be implemented in a computer program. Nevertheless, bookkeeping the identity of the various reacting species warrants painstaking attention. The Broadbelt group (Vinu et al., 2012) modeled the pyrolysis of PSP using the KMC technique. PSP is a vinyl polyperoxide that decomposes exothermally even at a low temperature of 100°C, and is a good model to understand oxidative degradation of PS. The main products of decomposition of PSP include benzaldehyde and formaldehyde along with minor production of α-hydroxy acetophenone and phenyl glycol. The major end groups in the model included primary alkyl, primary and secondary hydroxide, primary and secondary carbonyl, benzylic carbon, primary and secondary alkoxy radicals, primary alkyl radical, and secondary benzyl radical. Fig. 5 depicts the pathways for the formation of major products from PSP pyrolysis.

Peroxy bond fission is the primary step in the mechanism as the bond dissociation energy of O–O bonds (32–37 kcal/mol) is significantly smaller than that of C–C bonds (87 kcal/mol) (Levine and Broadbelt, 2009; Vinu et al., 2012). The model tracked the different types of O–O linkages attached to the benzyl head group or the alkyl tail group of the polymer. Besides predicting the time evolution of peroxide bonds remaining in the system, the model product yields also matched well with the experimental data. As the complete structure of each and every polymer chain was tracked, the major reactions involved in the formation of minor products were assessed. As shown in Fig. 6, lumped hydrogen abstraction/peroxide β-scission reactions resulted in the formation of the minor products. The radical by-products from this reaction readily forms benzaldehyde and formaldehyde. The model also predicted the formation of unidentified products in experiments like phenyl glyoxal, bibenzyl, and α-benzoyloxy acetophenone.

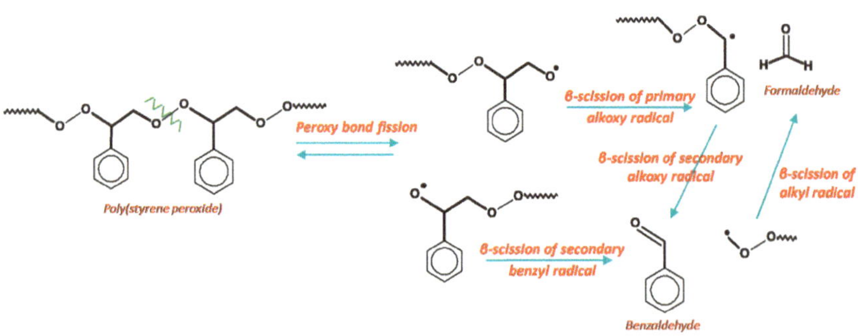

Fig. 5 Elementary pathways involved in the formation of formaldehyde and benzaldehyde from poly(styrene peroxide) pyrolysis. *Redrawn from Vinu R, Levine SE, Wang L, Broadbelt LJ: Detailed mechanistic modeling of poly(styrene peroxide) pyrolysis using kinetic Monte Carlo simulation,* Chem Eng Sci *69:456–471, 2012.*

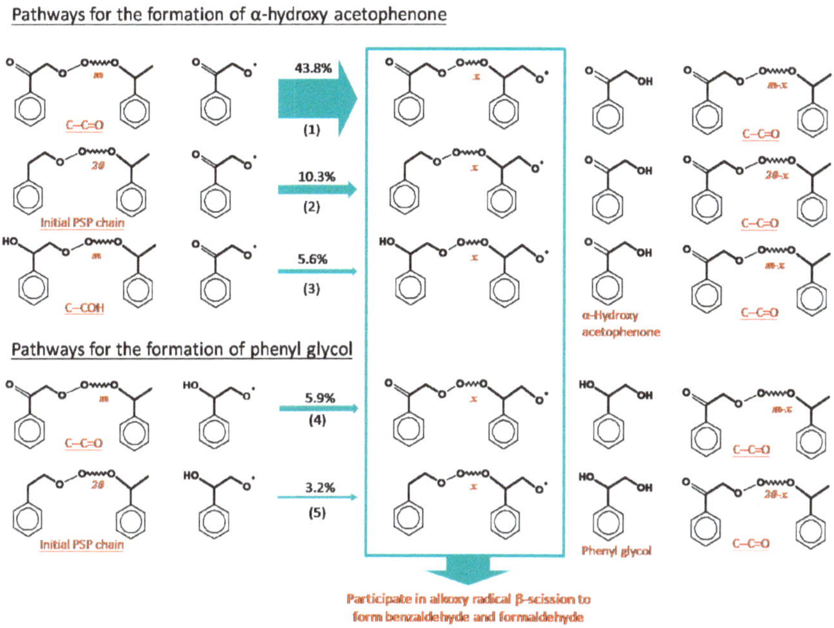

Fig. 6 Major elementary pathways involved in the formation of minor products from PSP pyrolysis, α-hydroxy acetophenone, and phenyl glycol. The *numbers on the arrows* denote the percentage occurrence of a specific reaction. *Redrawn from Vinu R, Levine SE, Wang L, Broadbelt LJ: Detailed mechanistic modeling of poly(styrene peroxide) pyrolysis using kinetic Monte Carlo simulation,* Chem Eng Sci *69:456–471, 2012.*

2.2 Oxidative Pyrolysis

While pyrolysis under inert atmosphere is important to recover valuable resources and monomeric building blocks from polymers, oxidative pyrolysis occurs when polymers are burnt in air or excess oxygen ambience as in combustion/incineration of plastics. More importantly, understanding the fundamental decomposition mechanism of polymers under oxidative environment is essential to design fire/flame retardants. Flame retardants and nanofillers are usually added in polymers to meet the fire safety standards (Dasari et al., 2013). Fig. 7 depicts the elementary reaction steps involved in oxidation of polymers. The mechanism involves steps that are similar to the ones used to model oxidation of hydrocarbon fuels and lubricants (Curran et al., 2002; Pfaendtner and Broadbelt, 2008). Compared to pyrolysis, oxidation reactions involve formation of a number of free radicals such as hydroperoxy, peroxy, alkoxy, and alkyl peroxy radicals. Moreover, the products formed contain different oxygenated functionalities such as alcohols, ketones, aldehydes, carboxylic acids,

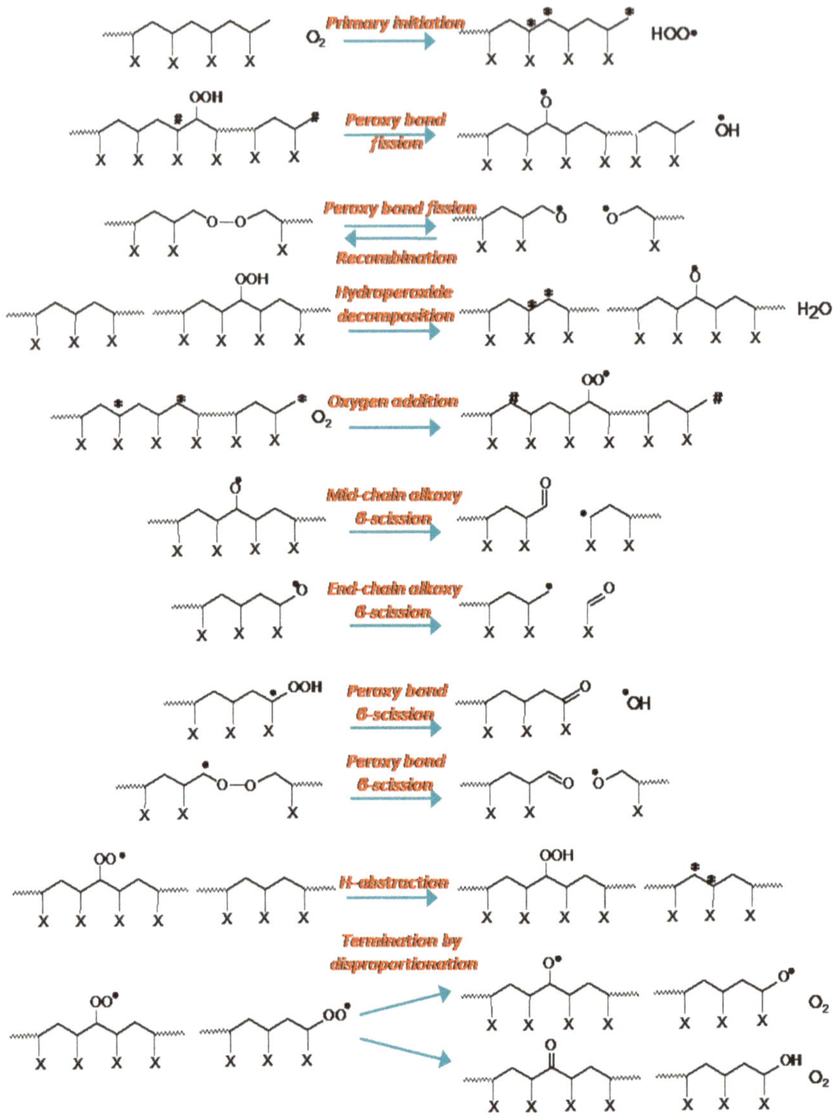

Fig. 7 Elementary reactions involved in oxidative pyrolysis of polymers. The polymer is PE, PP, or PS, when X corresponds to H, alkyl, or phenyl group, respectively. The position in the chain where radicals can form are denoted by *asterisks* (*), while the possible sites where –OOH and –OO˙ can attach are denoted by *hash symbol* (#). *Redrawn from Vinu R, Broadbelt LJ: Unraveling reaction pathways and specifying reaction kinetics for complex systems,* Annu Rev Chem Biomol Eng *3:29–54, 2012b.*

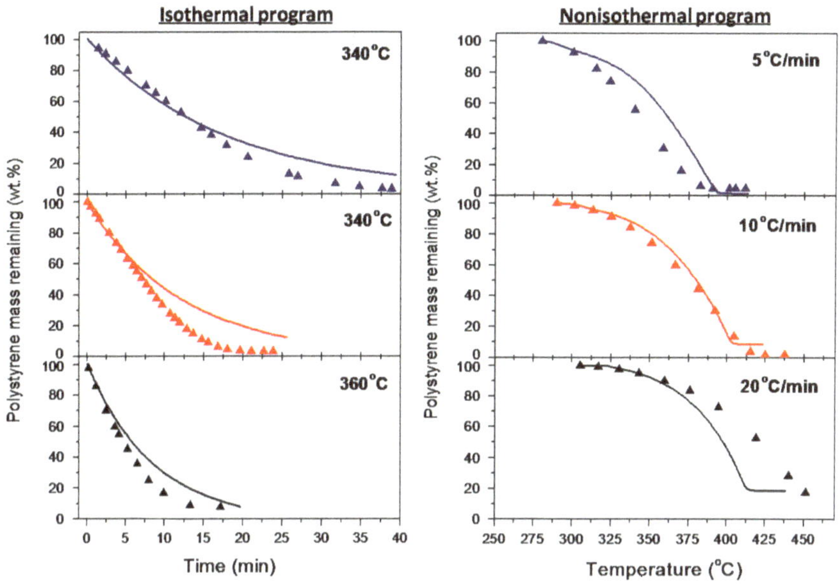

Fig. 8 Experimental and model predictions of isothermal and nonisothermal mass loss profiles of polystyrene ($M_w = 125{,}000$ g/mol) when degraded in air atmosphere. *Points* denote experimental data and *lines* denote mechanistic model prediction. Experimental data are from SDTQ600 (T.A. Instruments) thermogravimetric analyzer.

and esters. Therefore, a detailed kinetic model would involve species and reactions that are more in number compared to pyrolysis, which results in great complexity of the model.

Fig. 8 depicts the mass loss profiles of PS during oxidative pyrolysis in air medium. The experimental mass loss data corresponding to both dynamic and isothermal heating conditions were collected in a TGA. It is evident that the semidetailed model with limited number of 52 species participating in 118 reactions captured the mass loss profiles reasonably well. All possible elementary reactions presented in Fig. 7 were accounted in the kinetic model. The rate coefficients of the elementary steps were similar to that reported by Pfaendtner and Broadbelt (2008).

Importantly, the model also captured reasonably the formation of major and minor oxidation products. The model and experimental yields (in wt.%) of various products obtained under isothermal heating at 350°C after 15 min

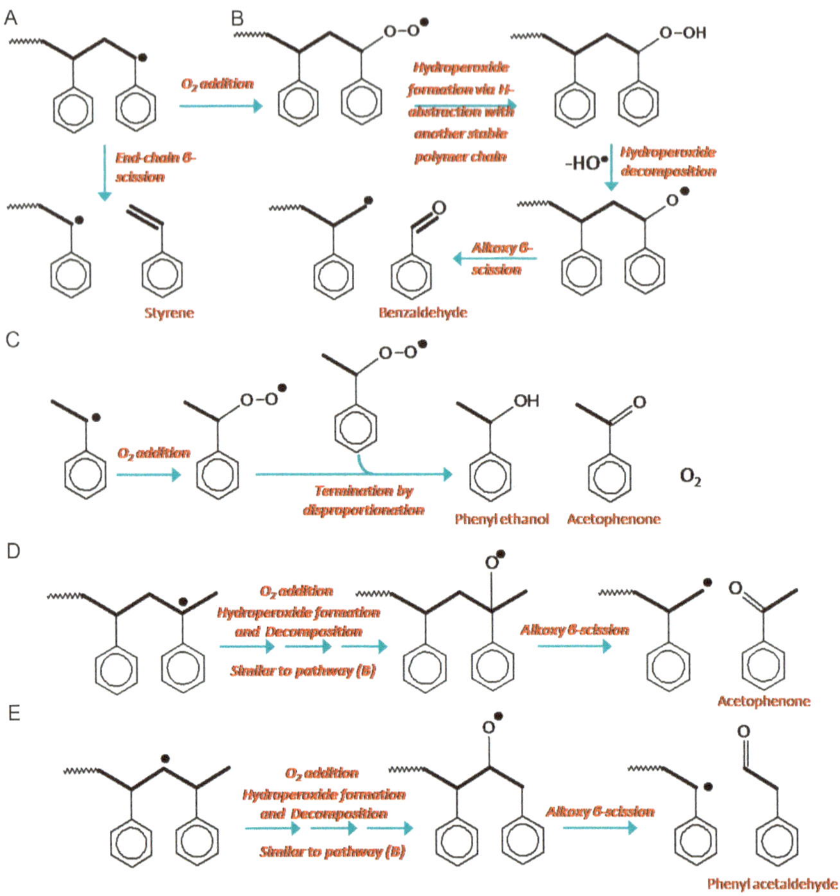

Fig. 9 Major reactions involved in the formation of styrene, benzaldehyde, phenyl ethanol, acetophenone, and phenyl acetaldehyde from oxidative pyrolysis of polystyrene.

are given by: styrene (5.28, 5 ± 0.83), benzaldehyde (1.25, 3 ± 0.97), phenyl ethanol (0.67, 0.98 ± 0.4), phenyl acetaldehyde (0.0008, negligible), and acetophenone (2.4, 0.15 ± 0.03). Based on net rate analysis, the major pathways for the formation of the above products were identified, which are depicted in Fig. 9. It is evident that the formation of a number of LMW compounds proceeds via oxygen addition, hydroperoxide formation, hydroperoxide decomposition, and alkoxy β-scission steps. More studies are required to unravel specific reaction pathways during polymer oxidation/combustion in the presence and absence of flame retardants and other additives.

3. CATALYTIC PYROLYSIS OF SYNTHETIC POLYMERS

Catalytic pyrolysis of polymers is a promising option to tune the overall yields of pyrolysis oil, char, and gases, and the organic composition of the oil fraction. Moreover, the use of catalyst offers benefits in terms of achieving the results at a low operating temperature. Catalysts can either alter the energy barrier of the existing reaction in the noncatalytic pathway or completely change the reaction mechanism to follow a different pathway. Nevertheless, the existence of free radical thermolysis reactions during catalytic pyrolysis cannot be neglected. There are two types of catalytic pyrolysis: (i) in situ catalytic pyrolysis, where the catalyst and the biomass feedstock are mixed in the pyrolysis reactor and the catalytic upgrading happens within the same reactor, and (ii) ex situ catalytic pyrolysis, where the catalysts are placed in a separate reactor for catalytic upgrading of pyrolysis vapors. The composition of pyrolysates differs in both the above configurations. The use of various catalysts like zeolites (HZSM-5, zeolite-Y, zeolite-β, HUSY, mordenite, clinoptilolite), silica-alumina, mesoporous MCM-41 and SBA, metal oxides, $AlCl_3$, super acids (SO_4^{2-}/ZrO_2), FCC catalysts like M/Al_2O_3 (M=Pt, Rh, Ni), X/Al_2O_3 ($X=H_2SO_4$, NaOH), and metal powders are reported for pyrolysis of different polymers. Various key properties of the catalysts influence the pyrolysis oil yield and its composition. These include specific surface area, micropore area, pore size distribution, framework structure, crystalline phase and crystallite size, Lewis vs Brønsted acidity, and presence of basic centers.

The mechanism of pyrolysis in the presence of acidic zeolites is reasonably well established in the literature. The acidic protons, H^+, catalyze the reactions via ionic pathway involving protonation, deprotonation, hydride and methyl shifts, α- and β-pseudo-cyclopropane-branching, methylation, oligomerization, cyclization, and β-scission (Alwahabi and Froment, 2004). The adsorption of the reactant molecule or primary pyrolysate on the active sites of the zeolite is the initial step, followed by protonation and other reactions. Catalytic pyrolysis of PS using acidic and basic catalysts like silica-alumina, HZSM-5, HY and Hβ zeolites, NaOH-coated silica-alumina, and sulfated zirconia is well studied (Lin and White, 1997; Marczewski et al., 2013; Ojha and Vinu, 2015a). In the presence of acid catalysts, the formation of ethylbenzene is predominant over styrene due to hydrogenation of styrene facilitated by acidic protons. However, in presence

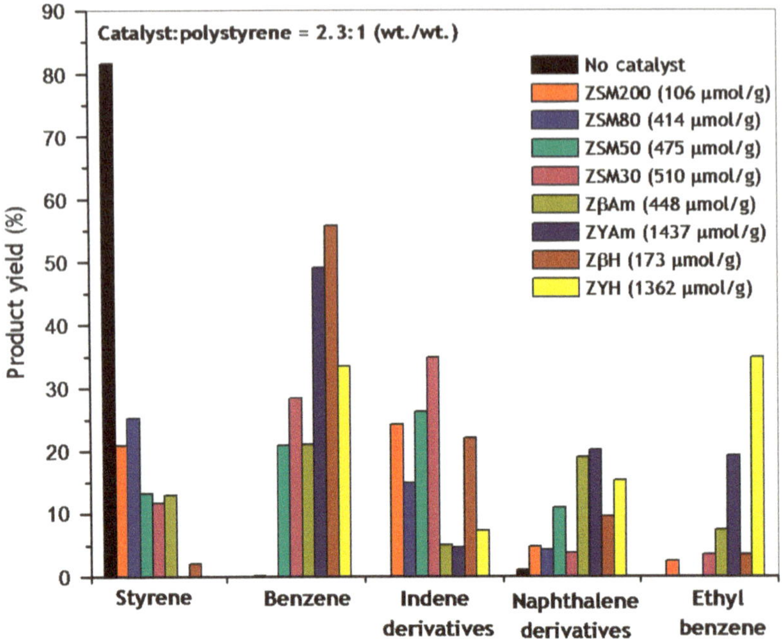

Fig. 10 Effect of various zeolites on the yield of major products from fast pyrolysis of polystyrene at 400°C. The total acidity of the zeolites determined by ammonia-TPD is shown in the *parenthesis* (Ojha and Vinu, 2015a).

of a base catalyst, selective styrene formation takes place (Marczewski et al., 2013). The presence of acidic protons also enhances the yield of indane, indene, and its derivatives. Fig. 10 depicts the yields of various products from in situ catalytic fast pyrolysis (CFP) of PS at 400°C using zeolites of different acidities and framework structures (Ojha and Vinu, 2015a).

Compared to noncatalytic pyrolysis where styrene and dimers are the major products, a number of other products such as benzene, cumene, α-methyl styrene, trimethyl benzene, ethyl benzene, indene, and naphthalene derivatives are also produced in significant quantities in CFP. Importantly, the yield of styrene, dimers, and α-methyl styrene decreased while benzene production increased with increase in Brønsted acidity of the zeolites. Moreover, the yield of naphthalene and methyl naphthalenes increased with pore volume of the catalysts suggesting that cyclization reactions are more pronounced when large surface area is available for the long chain reactive species. The mechanism of transformation of PS to various products via catalytic pyrolysis is depicted in Fig. 11. Based on the position of protonation, i.e., position 1 or position 2 of the aromatic ring, tertiary or

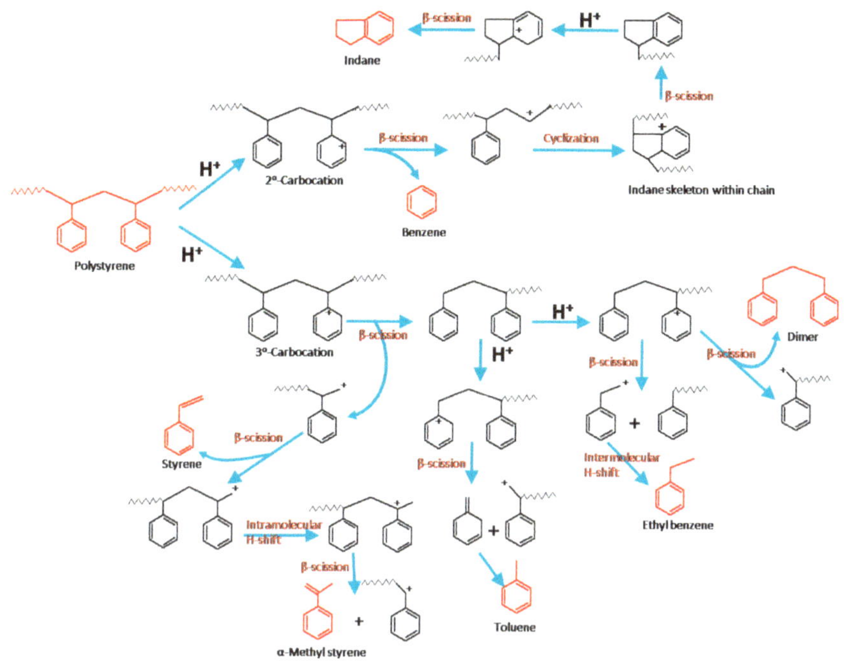

Fig. 11 Mechanism of Brønsted acid catalyzed pyrolysis of polystyrene using zeolites. *Redrawn from Lin R, White RL: Acid-catalyzed cracking of polystyrene,* J Appl Polym Sci *63:1287–1298, 1997; Ojha DK, Vinu R: Resource recovery via catalytic fast pyrolysis of polystyrene using zeolites,* J Anal Appl Pyrol *113:349–359, 2015a.*

secondary carbocation is produced, respectively. These then participate in a number of successive β-scission, cyclization, and H-shift reactions to form various products. The major routes for the formation of benzene and styrene are compared. It is evident that protonation at the ipso position leads to benzene and indane formation. Importantly, it was found from experiments that increasing the pyrolysis temperature in the presence of ZβH led to the formation of styrene with a concomitant decrease in yield of benzene and indene derivatives. This proves that the relative rates of the two pathways at different temperatures play a decisive role in determining the product distribution.

Manos et al. (2000) studied catalytic slow pyrolysis of HDPE using zeolites of different pore sizes. Their results are depicted in Fig. 12. The production of C3–C15 alkanes and heavier hydrocarbons followed the trend: HUSY (heavies, more alkanes) > ZY > Zβ > mordenite > ZSM-5 (light, more alkenes). The large-pore zeolites (ZY, USY, and Zβ) promoted

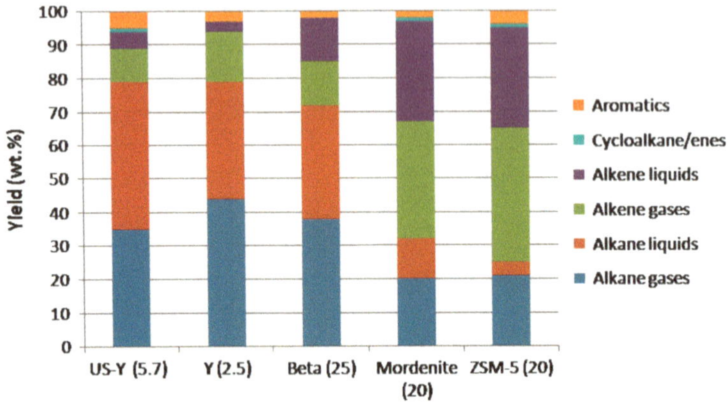

Fig. 12 Yields of various fractions obtained during catalytic pyrolysis of HDPE using zeolites. The heating rate was 5°C/min and the final temperature was 360°C. The Si/Al ratio of the zeolite is shown in the *parenthesis. Redrawn from Manos G, Garforth A, Dwyer J: Catalytic degradation of high-density polyethylene over different zeolitic structures,* Ind Eng Chem Res *39:1198–1202, 2000.*

the formation of heavy hydrocarbons, while the medium-pore zeolites (mordenite and ZSM-5) promoted the formation of lighter products. This shows that the pore size is an important parameter in catalytic pyrolysis. Small pores allow the products to diffuse out of the pore, while larger pores facilitate secondary bimolecular reactions such as oligomerization to form longer chains. Large pores can also sometimes lead to the formation of condensed ring aromatic residues/coke.

High external surface area of the catalyst can be sometimes beneficial to regenerate the catalyst. Serrano et al. (2007) showed that nanocrystalline HZSM-5 with high external surface area could be easily regenerated to regain the activity in LDPE pyrolysis at 340°C. Restoring the activity of catalysts with high micropore area may be difficult, and the catalyst might quickly get deactivated. The form of catalyst also affects the product distribution. For example, ZY catalyst in powder form is reported to promote the formation of olefins, while the same catalyst in pellet form promotes the formation of paraffins and naphthalenes in HDPE pyrolysis (Kumar et al., 2011). Donaj et al. (2012) utilized Ziegler–Natta polymerization catalyst for the degradation of a mixture of polyolefins containing LDPE, HDPE, and PP. They observed nearly 55% improvement in yield of monomers (methane, ethane, ethylene, propane, propylene) at 650°C in presence of catalyst. Nevertheless, the yield was not comparable with stream cracking pyrolysis. Abbas-Abadi et al. (2014) evaluated the effects of temperature, catalyst:polymer ratio, and carrier gas type on product distribution in PP pyrolysis using an FCC catalyst. Hydrogen ambience resulted in high yield of paraffinic products, and 10% catalyst addition resulted in maximum yield of condensable products at 450°C. Catalytic pyrolysis of other polymers like scrap tyre, nylon-6, polyacrylonitrile, and mixed plastics like waste electrical and electronic equipments (WEEE) are also available in the literature (Bozi and Blazsó, 2009; Czernik et al., 1998; Olazar et al., 2008; Santella et al., 2016). A majority of the studies demonstrate that the use of catalyst remarkably enhances the quality of the products, and sometimes, the product yields. For example, (a) 85 wt.% of caprolactam was obtained from nylon-6 depolymerization at 330–360°C using KOH/α-Al_2O_3 (Czernik et al., 1998), (b) ca. 90 wt.% oil with high calorific value (39 MJ/kg) and monoaromatic content was obtained by pyrolysis of WEEE at 400°C using USY and HZSM-5 zeolites (Santella et al., 2016), (c) nearly 20 wt.% of gases (ethylene and propylene) and ca. 12 wt.% of BTX (benzene, toluene, and xylene) were obtained from scrap type fast pyrolysis using HZSM-5 at 450°C (Olazar et al., 2008), and (d) high selectivity of acetonitrile and propene nitrile were achieved when acrylonitrile-based copolymers were fast pyrolyzed using zeolites (Bozi and Blazsó, 2009). Significant opportunities for further research exist in CFP of polymers wherein (a) novel catalyst formulations with tailored microstructure, pore size, and properties, and (b) different reaction conditions can be used to improve the selectivity of fine chemicals and fuel molecules. While the existing studies provide base case scenario of maximum limits of product yields and quality one can expect from pyrolysis of polymers in small-scale laboratory reactors, a deviation in product yields and quality can certainly be expected when the process is scaled up.

4. PYROLYSIS OF BIOMASS

4.1 Composition of Biomass

Lignocellulosic biomass is composed of three major natural polymers: cellulose, hemicellulose, and lignin, together with minor pectins, proteins, and minerals (Vassilev et al., 2010). As shown in Table 4, the abundance of each component varies among different biomass species (softwood, hardwood, agricultural residue, herbaceous biomass, and energy crops), and between the different parts of the same type of biomass (for example, corncob, corn straw, corn stalk, and corn stover). Generally, cellulose is the most abundant natural polymer, accounting for 35–55% of lignocellulosic biomass. Hemicellulose is the second most abundant natural polymer, accounting for 20–35 wt.%, followed by lignin which represents 10–30 wt.% of lignocellulosic biomass. Herbaceous and agricultural biomasses usually have a higher mineral content than wood.

Table 4 Composition Analysis of Biomass

Composition, wt.%	Cellulose	Hemicellulose	Lignin	Ash	Extractives[a]	References
Spruce	45.6	20.0	28.2	0.3	5.9	Taherzadeh et al. (1997)
Pine	46.9	20.3	27.3	0.3	5.1	Taherzadeh et al. (1997)
Birch	47.0	25.9	22.0	0.3	4.7	Taherzadeh et al. (1997)
Beech	45	33	20	0.2	2.0	Di Blasi et al. (2010)
Fir	45	21	18.0	0.5	6.3	Di Blasi et al. (2010)
Corncob	37.6	31.6	20.8	3.2	8.1	Lyu et al. (2015)
Cornstraw	43.1	31.8	11.0	0.8	13.3	Wang et al. (2015c)
Corn stalk	42.7	23.2	17.5	6.8	9.8	Qu et al. (2011)
Corn stover	35.3	28.9	19.9	4.6	6.6	Zhang et al. (2015e)
Rice straw	41.2	30.7	12.2	1.0	14.9	Wang et al. (2015c)
Wheat straw	38.9	21.1	18	9.7	5.5	Carvalheiro et al. (2009)
Sugar cane	51.8	27.6	10.7	0.8	9.1	Dorez et al. (2014)
Miscanthus giganteus	50.34	24.83	12.02	2.67	4.1	Brosse et al. (2012)

[a]Extractives are the material in a biomass sample that is soluble in either water or ethanol during exhaustive extraction (Sluiter et al., 2008).

4.2 Structure of Cellulose, Hemicellulose, and Lignin

4.2.1 Cellulose

Cellulose is a homogeneous, linear polysaccharide with a well-defined structure comprised entirely of β-1,4-linked D-glucopyranose units in a $^4C_1$ conformation. Fig. 13 shows a structural diagram of cellulose, which has a

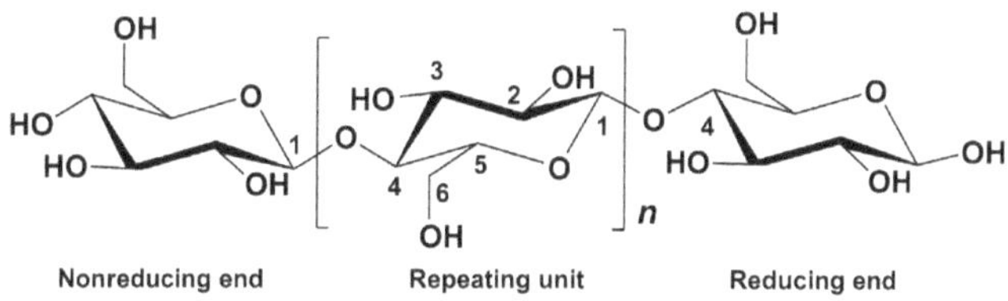

Fig. 13 Structure of cellulose.

nonreducing end, reducing end, and chain length of repeating units ranging from 140 or less to 10,000 or more depending on the source (Hallac and Ragauskas, 2011). The number of glucose units in one cellulose molecular chain is referred to as the degree of polymerization (DP). DP of cellulose is an important structural property that affects solubility, the efficiency of enzymatic hydrolysis of lignocellulosic biomass, and the mechanical properties of lignocellulosic biomass and derived products. The aggregation of hydrogen-bonded cellulose chains within microfibrils creates a highly crystalline structure that gives cellulose its unique properties of mechanical strength and chemical stability. Cellulose microfibrils contain two crystalline forms, cellulose Iα and Iβ. However, both experimental and modeling efforts reveal that there is no evident change in the molecular structure of individual linear chains of cellulose caused by softening or liquefaction during fast pyrolysis. In addition, no significant differences were observed in the product distributions from fast pyrolysis of cellulose with different crystallinities, DP, or feedstock source (Mayes et al., 2014a, 2015; Patwardhan et al., 2009; Zhou et al., 2014b,c). In the native state of biomass, cellulose is intimately associated with lignin, hemicellulose, and other components. However, the interactions of cellulose between these components are much less studied.

4.2.2 Hemicellulose

Hemicellulose is the second most abundant component of lignocellulosic biomass, accounting for 20–35 wt.% of dry biomass. Hemicelluloses are defined as a group of cell wall polysaccharides that are not classified as either cellulose or pectin (Zhou et al., 2016a). The functional groups (building blocks) of hemicellulose include pentoses (xylose and arabinose), hexoses (mannose, glucose, and galactose), hexuronic acids (4-*O*-methyl-D-glucuronic acid, galacturonic acid, and glucuronic acid), small amounts of rhamnose and fucose, and acetyl groups (Fig. 14). These functional groups can assemble into a range of various hemicellulose polysaccharides, such as

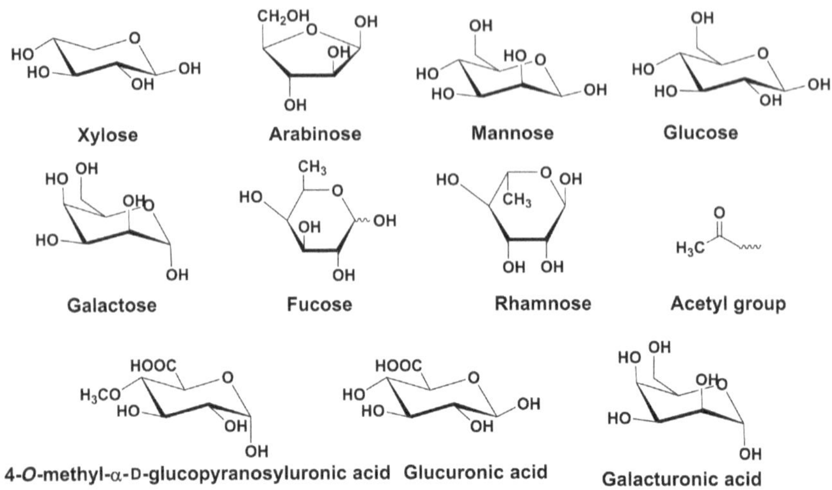

Fig. 14 Building blocks of hemicellulose.

xylans, mannans, xyloglucan, β-1,3;1,4-glucans, and galactans, with diverse structures from linear to highly branched (Ebringerová et al., 2005; Scheller and Ulvskov, 2010; Zhou et al., 2016a). The abundance, detailed chemical composition and structures of these hemicellulose polysaccharides can be very different from each other, and vary widely depending on the biomass source (Obel et al., 2007). In contrast to cellulose, amorphous hemicellulose polysaccharides have a much lower average DP, about 200 for hardwood and 100 for softwood (Ebringerová et al., 2005; Pettersen, 1984). Moreover, in the native state, the xylose, and mannose residues in hemicellulose polysaccharides have acetyl groups attached at position 2 and/or position 3, and the degree of acetylation can range from 0.1 to 0.7 depending on the biomass source and treatment methods utilized (Zhou et al., 2016a).

Major hemicellulose polysaccharides include 4-O-methylglucoronoxylans, arabinoxylans, galactoglucomannans, xyloglucan, and β-1,3;1,4-glucan (Fig. 15). The major hemicellulose polysaccharide of hardwood is 4-O-methylglucoronoxylan (often referred to as glucoronoxylans), accounting for 15–30 wt.%. 4-O-methylglucoronoxylan has a β-1,4 linked xylan backbone with 4-O-methylglucuronic acid and glucuronic acid as primary side groups attached at position 2 of the xylose units of the backbone structure (Dumitriu, 2008). Glucoronoxylans can be ester linked with lignin through the uronic acid groups. Arabinoxylans are the main hemicellulose polysaccharides of herbaceous biomass (Ebringerová et al., 2005; Peng and Wu, 2010; Theander, 1985). Arabinoxylans contain arabinose as the primary side groups,

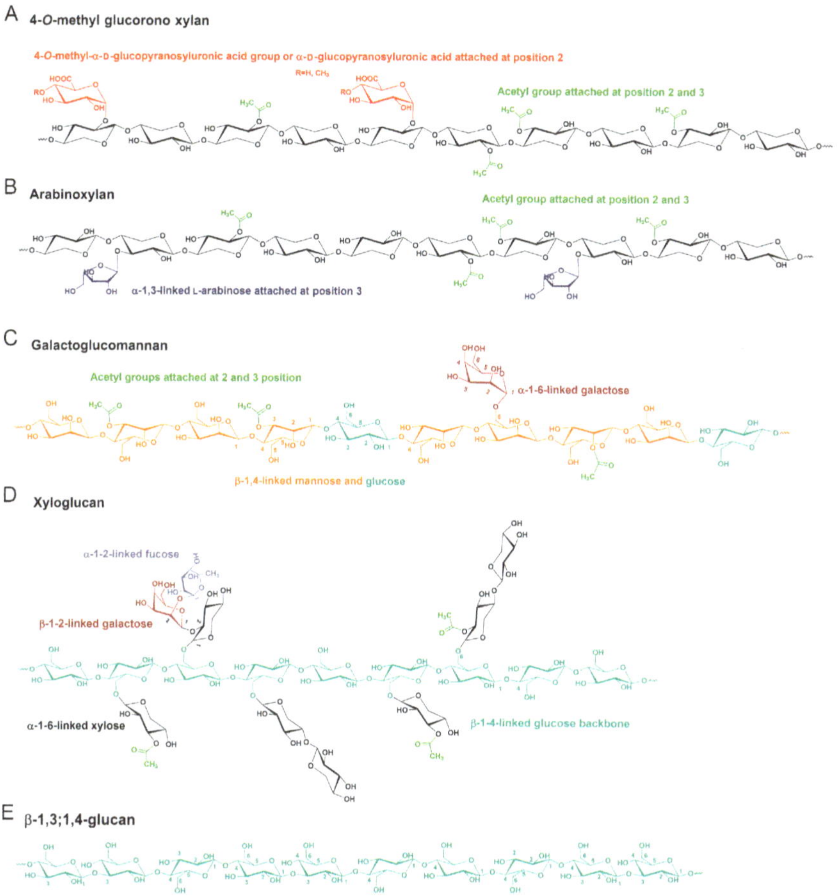

Fig. 15 Structure of major hemicellulose polysaccharides: (A) 4-*O*-methyl-D-glucorono-D-xylan (glucuronoxylan), (B) arabinoxylan xylan, (C) galactoglucomannan, (D) xyloglucan, and (E) β-1,3;1,4-glucan.

which are attached at position 3 of xylose units in the β-1,4-linked xylan backbone. Arabinoxylans can be covalently crosslinked with lignin through these ferulic acid groups. The degree of acetylation and DP of arabinoxylans are generally lower than those of 4-*O*-methylglucoronoxylan. Hemicellulose in all softwoods consists mainly of galactoglucomannans, typically comprising 14–25% of the wood (Desharnais et al., 2011; Donaldson and Knox, 2012; Kusema et al., 2013; Willfor et al., 2005). Galactoglucomannans consist of a β-1,4-linked D-glucopyranose and D-mannopyranose backbone with α-1-6-linked D-galactopyranose groups attached to some of the D-mannopyranose units. The hydroxyl groups bonded to C_2 and C_3 in mannose units are partially substituted by acetyl groups. Xyloglucan has a backbone composed of β-1,4-linked glucose residues, up to 75% of which are substituted with α-1–6 linked xylose side chains. The xylose residues are often capped with a galactose residue occasionally attached to a fucose residue or even arabinose (Ebringerová et al., 2005; Eckardt, 2008; Obel et al., 2007; Ochoa-Villarreal et al., 2012; Popper and Fry, 2004; Xue and Fry, 2012). The primary cell walls of grasses consist of 2–15% of β-1,3;1,4-glucan. β-1,3;1,4-Glucan is an unbranched homopolymer of glucose connected by mixed β-1,4 and β-1,3-linkages (Carpita, 1996). For more detailed structural features, and physicochemical and functional properties of hemicellulose polysaccharides, the interested reader is referred to reviews by Pettersen (1984), Theander (1985), Alén (2000), Ebringerová et al. (2005), Scheller and Ulvskov (2010), and Zhou et al. (2016a).

4.2.3 Lignin

Lignin is the most complex of the three major biopolymers in the cell wall of terrestrial plants. Distinctions between softwood lignin, hardwood lignin, and herbaceous crop lignin (grass lignin) are important (Zeng et al., 2013). Lignin can also be categorized by the isolation methods utilized such as steam explosion, kraft, organosolv, alkaline oxidation, pyrolysis lignins, etc. (Pandey and Kim, 2011). Functionally, lignin reinforces the plant cell walls by bonding with cellulose and enhances the waterproof nature of plant cell walls because of its hydrophobicity and allowing the efficient transport of water in the vascular tissues. Lignin is also deposited in wounds as a barrier against attack by insects and fungi.

Lignin is aromatic in nature, and it is the only large-scale biomass source of aromatic functionality. Lignin is a crosslinked phenolic polymer mainly comprised of three constituent monomers, *p*-hydroxyphenyl (4-hydroxyphenyl, P), guaiacyl (4-hydroxy-3-methoxyphenyl, G), and syringyl (4-hydroxy-3,5-dimethoxyphenyl, S), arranged in a hyperbranched topology with no regular repeating structure (Lebo et al., 2000; Pelzer et al., 2015). As shown in Fig. 16, these monolignol building blocks are hydroxy- and methoxy-substituted phenylpropane units. The relative abundance of monolignols and their linkages, and the detailed molecular structure and MWD of lignin again vary depending on the biomass sources and isolation methods as previously mentioned.

Generally, softwood lignin is comprised almost entirely of G as the dominant monolignol monomer (accounting for 90%) with small quantities of P, while hardwood lignin contains slightly more S than G units with very few P units. Most grass lignins (including lignin of herbaceous and agricultural

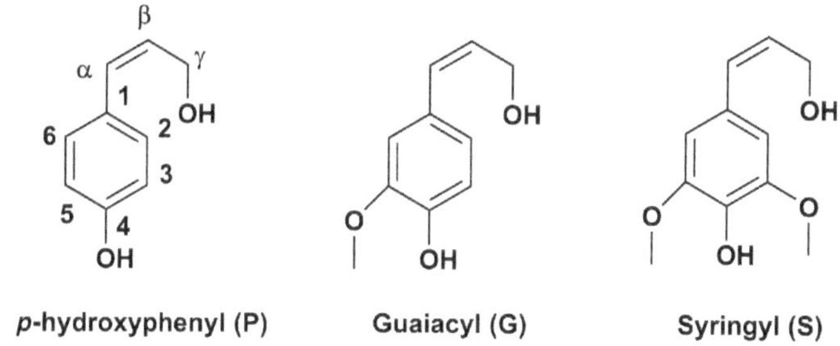

Fig. 16 Monolignol building blocks of lignin: *p*-hydroxyphenyl, guaiacyl, and syringyl.

biomass, and energy crops) are built up of P and G with G as the major component, while few grass lignins also have S units. For example, Zeng et al. (2013) reported that G was the predominant unit in wheat straw cell wall lignin over S and P units.

Another complexity associated with lignin is the variety of interunit linkages identified and quantified in the literature, including β-*O*-4, β-5, β-β, β-1, β-6, α-β, α-*O*-4, α-*O*-γ, γ-*O*-γ, 1-*O*-4, 4-*O*-5, 1-5, 5-5, and 6-5, that irregularly connect these monolignol building blocks (Crestini et al., 2011; Dorrestijn et al., 2000; Zeng et al., 2013). β-*O*-4-Aryl ether (β-*O*-4), α-*O*-4-aryl ether (α-*O*-4), 4-*O*-5-diaryl ether (4-*O*-5), β-5-phenylcoumaran (β-5), 5-5-biphenyl (5-5), β-1-(1,2-diarylpropane) (β-1), and β-β-(resinol) linkages (Fig. 17) were identified to be the seven major linkages in lignin by Dorrestijn et al. (2000). Table 5 summarizes the abundance of these major linkages in lignin from representative sources. The dominant interunit

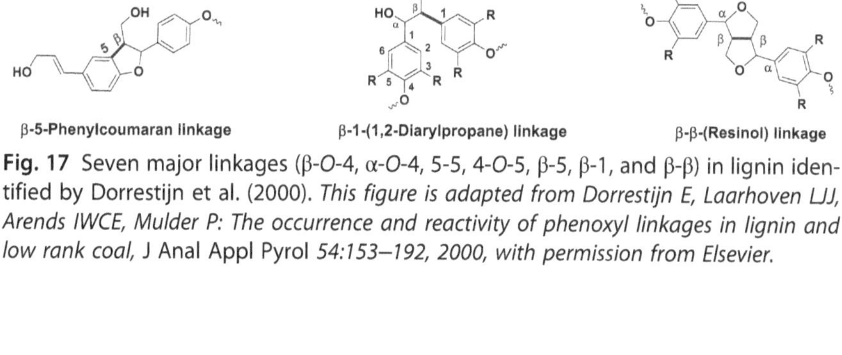

Fig. 17 Seven major linkages (β-*O*-4, α-*O*-4, 5-5, 4-*O*-5, β-5, β-1, and β-β) in lignin identified by Dorrestijn et al. (2000). *This figure is adapted from Dorrestijn E, Laarhoven LJJ, Arends IWCE, Mulder P: The occurrence and reactivity of phenoxyl linkages in lignin and low rank coal,* J Anal Appl Pyrol *54:153–192, 2000, with permission from Elsevier.*

Table 5 The Abundance (in %) of Major Linkages in Lignin (Dorrestijn et al., 2000)

Linkages	Softwood Lignin[a]	Hardwood Lignin[a]	Grass Lignin[b]	Energy Crops Lignin[c]
β-O-4	46	60	77	82–84
β-5	11	6	11	10–11
α-O-4	7	7	2	–
5-5	10	5	3	–
4-O-5	4	7	–	–
β-1	7	7	3	–
β-β	2	3	4	6–7
Others	13	5	–	–

[a]Data from Dorrestijn et al. (2000).
[b]Wheat straw lignin as a representative of grass lignin, data from Sun et al. (2005) and Zeng et al. (2013).
[c]*Miscanthus giganteus* lignin as a representative of energy crops lignin, data from Bauer et al. (2012).

bonding pattern of lignin is ether linkages, with β-O-4 being the most frequent linkage. Hardwood and grass lignins contain about 1.5–2 times more β-O-4-linkages than softwood lignin, respectively (Dorrestijn et al., 2000). The predominant linkages present in *Miscanthus giganteus* (a representative for energy crops) lignin are β-O-4 (82–84%), β-β (6–7%), and β-5 (10–11%) (Bauer et al., 2012).

To date, the exact molecular structure of lignin in its native form in lignocellulosic biomass is not fully understood despite decades of study in part because of the heterogeneity and complex nature of lignin, and in part because of the lack of suitable analytical tools needed to analyze the complex polymer (Crestini et al., 2011; Freudenberg, 1965; Stewart et al., 2009; Vanholme et al., 2010). The lignin structure models in the literature can be broadly classified into two types. One type is the average structure model of lignin. Adler (1957), Freudenberg (1965), Sakakibara (1980), and Glasser et al. (1981) reported their respective lignin structure models and underscored the complexity of lignin structure and the importance of understanding the lignin structure for more efficient utilization of the recalcitrant component of biomass. However, these early efforts have been limited to developing the lignin structure models based on experimentally measurable bulk properties such as monomer composition and bond distribution. Fig. 18 shows the most canonical structural model of (spruce) lignin reported by Freudenberg (1965). Recently, structural representations of lignin considering the role of the biosynthesis mechanism in the generated polymer

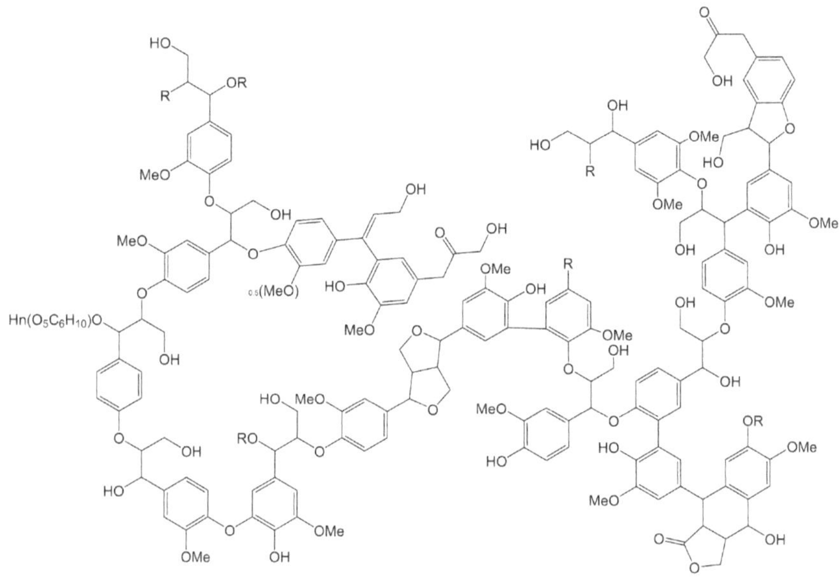

Fig. 18 Lignin structural models for spruce lignin. *Redrawn from Freudenberg K: Lignin: its constitution and formation from p-hydroxycinnamyl alcohols,* Science *148:595–600, 1965.*

have been reported (Stewart et al., 2009). However, these models have been focused on synthetic lignin instead of the native polymer, partially due to an incomplete understanding of the biosynthetic mechanism (van Parijs et al., 2010; Yanez et al., 2016). Stewart et al. (2009) reported representative lignin structural models for fragments of poplar lignin that depict the general features and the main linkage types in their approximate relative frequencies. They discovered that the lignin structural model predicted by NMR analysis is a branched structure, an important feature of native lignin, but very different from the structure of synthesized lignin which was entirely linear with unbranched chains. Furthermore, different structural models have been reported for the same type of lignin (Barta et al., 2010; Stewart et al., 2009). For example, poplar lignin shown in Figs. 19 and 20 reflect that average models have limitations in capturing the variation of the lignin structure based solely on a few bulk properties.

The other type of lignin model is a structure library created by computationally generating a range of diverse molecules rather than using one complex polymer chain that reproduces the observed properties (e.g., monomer and bond distribution) of lignin. Pioneering work was reported by the Klein group in 1988; however, the lignin molecules in this approach exhibit linear topology (Train and Klein, 1988). Recently, the Broadbelt group (Yanez and Broadbelt, 2015a,b; Yanez et al., 2016) has developed

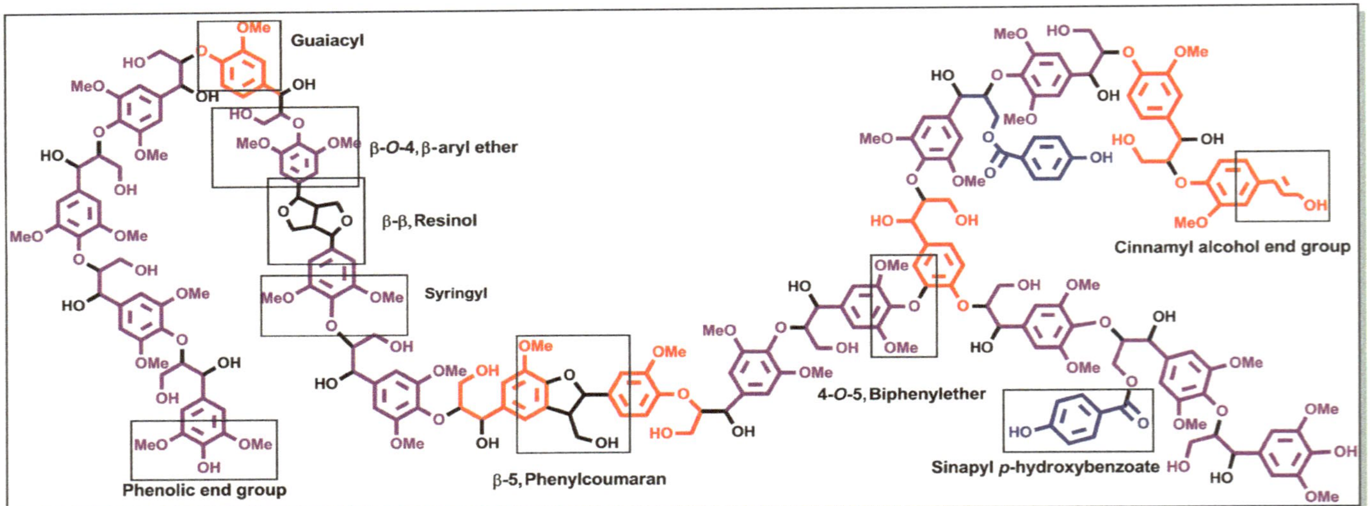

Fig. 19 Structural representation of a lignin polymer from poplar wood. *Adapted from Lochab B, Shukla S, Varma IK: Naturally occurring phenolic sources: monomers and polymers,* RSC Adv *4:21712–21752, 2014, with permission from Royal Society of Chemistry.*

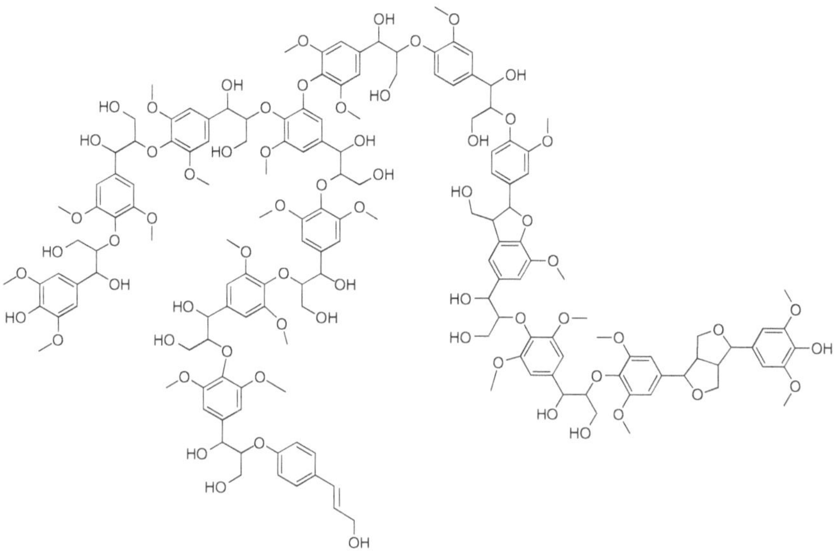

Fig. 20 A hypothetical chemical structural of lignin fragment from poplar sawdust. *Adapted from Baker EG, Elliott DC: Catalytic upgrading of biomass pyrolysis oils. In Bridgwater AV, Kuester JL, editors:* Research in thermochemical biomass conversion, *Netherlands, 1988, Springer, with permission from Royal Society of Chemistry.*

a stochastic method of producing libraries of diverse and complex structural representations with hyperbranched topology (i.e., having branches within branches) that can be generally applied to any type of biomass lignin. The constructed library of lignin molecules agreed well with experimental measurements of monolignol composition (*p*-hydroxyphenyl, guaiacyl, and syringyl), bond distribution (β-O-4, β-5, and 5-5), number-average and weight-average molecular weight, and branching coefficient. The model is also able to make novel predictions of the distribution of free phenolic hydroxyl groups and the dyadic bonding patterns between monolignol building blocks. Broadbelt and coworkers are developing a mechanistic model of lignin pyrolysis using KMC simulations by applying a fast pyrolysis mechanism to a simplified library of structures to predict the product distribution (Yanez and Broadbelt, 2015a,b; Yanez et al., 2016).

4.3 Kinetic Modeling of Biomass Pyrolysis

4.3.1 Global Kinetic Model of Biomass Pyrolysis

Biomass pyrolysis simultaneously yields both condensable and noncondensable gaseous products as well as solid residues called char. The condensable gaseous products are referred to as liquid phase product or bio-oil, which includes

H_2O, LMW alcohols, aldehydes, acids, furans, anhydrosugars, phenols, and aromatics. The noncondensable gaseous products include H_2, CO, CO_2, and light hydrocarbons such as CH_4, C_2H_6, and C_3H_8. The thermal decomposition of biomass during pyrolysis involves an extremely complex reaction network, which consists of hundreds of species and reactions, including the depolymerization of natural polysaccharides and aromatic polymers, reactions of unstable intermediates, and the formation of products through mechanisms such as glycosidic bond cleavage, C–C bond cleavage, hydrolysis, unzipping, dehydration, fragmentation, rearrangement, retro-aldol, and char formation reactions. Over the past 50 years, numerous studies have been conducted to probe the kinetics and mechanism of pyrolysis of biomass and its major components. State-of-the-art reviews have covered different aspects of biomass pyrolysis ranging from the fundamental bench-scale level to the applied reactor and process development level examining pyrolysis parameters, product quality and properties, kinetics and mechanisms, multiscale modeling, technoeconomic analysis, and fundamental challenges. The interested reader is referred to reviews by Antal and Varhegyi (1995), Babu (2008), Bridgwater and Peacocke (2000), Butler et al. (2011), Di Blasi (2008), Graham et al. (1984), Huber et al. (2006), Mohan et al. (2006), Radlein et al. (1991), Venderbosch and Prins (2010), Anca-Couce (2016), Bridgwater (2012), Burnham et al. (2015), Carpenter et al. (2014), Collard and Blin (2014), Kersten and Garcia-Perez (2013), Lede (2012, 2013), Mettler et al. (2012c), Şerbănescu (2014), and White et al. (2011). Here we highlight the kinetic models of biomass pyrolysis with an emphasis on the reaction mechanism.

Great progress in kinetic modeling at both the global kinetic level and the micro-kinetic level has been made to describe the thermal decomposition behavior of biomass and its natural polymers, hemicellulose, lignin, and especially cellulose. The kinetic models reported in the literature for biomass pyrolysis are mostly global models. The first global kinetic scheme of cellulose pyrolysis (Fig. 21) was developed by Broido and his coworkers based on thermogravimetric study (Broido and Nelson, 1975; Kilzer and Broido, 1965). This mechanism included two parallel pathways leading

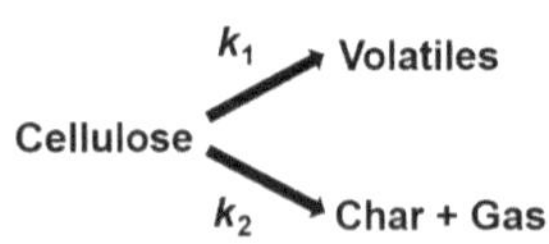

Fig. 21 Broido and Nelson model for cellulose pyrolysis.

Fig. 22 B-S model of cellulose pyrolysis.

to the formation of volatiles, char, and gas. Later, Shafizadeh and coworkers (Bradbury et al., 1979) modified this reaction scheme and reported the most generally accepted kinetic model for cellulose pyrolysis, known as the Broido–Shafizadeh model (B-S model). As shown in Fig. 22, this model describes the formation of an intermediate species, "active cellulose," and its subsequent competing decomposition pathways into volatiles and gas/char.

Subsequently, a variety of kinetic schemes have been developed based on or derived from the pioneering work of Broido and Shafizadeh to describe the pyrolysis of cellulose and biomass, as well as the other two major components: hemicellulose and lignin. The global kinetic models of biomass pyrolysis can be broadly divided into single-component models and three-component models depending on whether biomass is modeled as a single component or the combination of three components: cellulose, hemicellulose, and lignin. Models can also be categorized into one-reaction/stage models and multireaction/stage models. The majority of these global kinetic models were constructed based on the mass loss data obtained from the thermogravimetric analysis of biomass samples.

The simplest model of biomass pyrolysis is the single-component model with a single reaction stage (Fig. 23), in which gas, tar, and char are directly formed from biomass. The number of rate constants that describe the formation of these global products ranges from 1 to 3. Often, the single reaction scheme is governed by a first-order reaction rate, which is expressed in terms of mass conversion, θ, by the following equation (Zaror et al., 1985):

$$\frac{d\theta}{dt} = -(1-\theta)A\exp\left(\frac{-E_a}{RT}\right)$$

where A is the apparent frequency factor, E_a is the apparent activation energy, and θ is the mass conversion of reactant. This is also a common

Fig. 23 Single-component, single-stage kinetic model for biomass pyrolysis.

method to obtain apparent activation energy for pyrolysis of biomass. However, a broad range of E_a from 100 to 300 kJ/mol has been reported for biomass pyrolysis in the literature, and a consensus E_a value of pyrolysis of biomass is yet to be reported. The one-reaction/stage models are able to describe the mass loss curve of biomass during pyrolysis, and the degradation rate of biomass. However, they have limited capability for predicting product yields from different biomass sources or from the same materials under different pyrolysis conditions, where parameters such as sample particle size, heating rate, and temperature are adjustable. Furthermore, one-stage models do not include the secondary reactions of tar.

Another type of global kinetic model for biomass pyrolysis is the multi-reaction/stage model, where the decomposition of biomass or its major components, and the formation of products are described by a series of consecutive/parallel reactions to address both primary and secondary pyrolysis reactions (Chan and Krieger, 1983; Koufopanos et al., 1989, 1991; Radlein et al., 1991; Várhegyi et al., 1997). Fig. 24 depicts a consecutive reaction scheme of biomass pyrolysis which includes competing primary pyrolysis reactions to gas, char, and tar, and secondary pyrolysis reactions of tar at higher temperatures.

Chan and Krieger (1983) reported a multistep reaction scheme for biomass pyrolysis that included considerations for the evaporation of moisture. As shown in Fig. 25, the parallel formation of primary gas competes with the formation of tar, which also experiences a secondary reaction, forming secondary gases and tar. Moreover, the formation of water from the

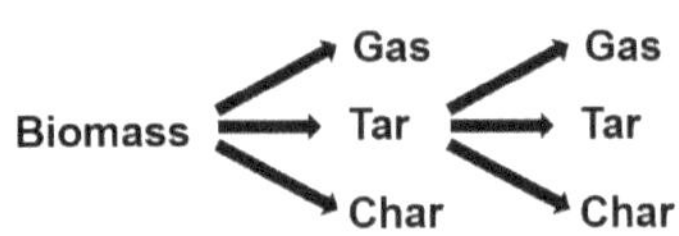

Fig. 24 Single-component, two-stage kinetic model for biomass pyrolysis.

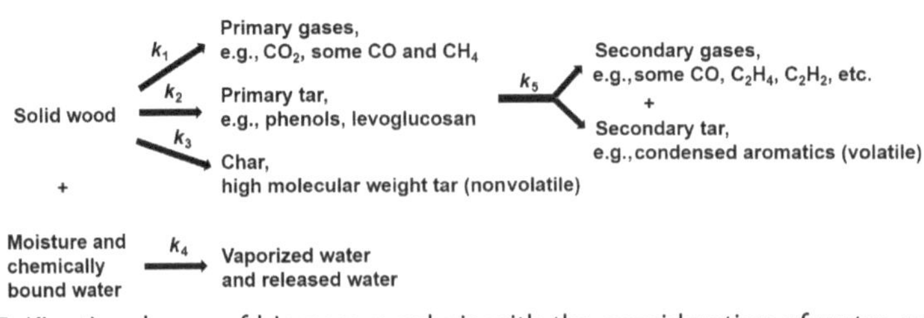

Fig. 25 Kinetic scheme of biomass pyrolysis with the consideration of water evaporation by Chan and Krieger (1983).

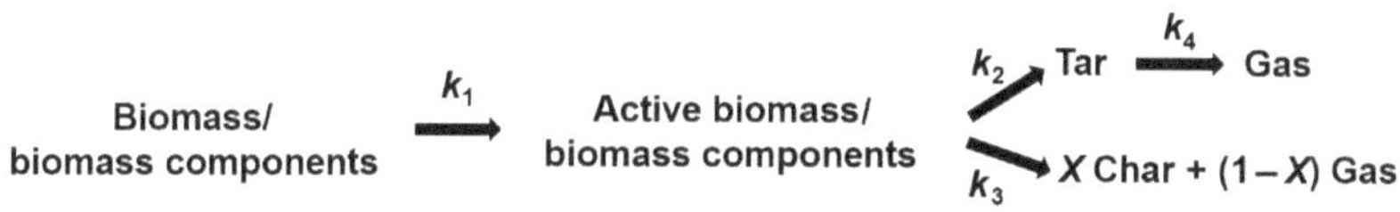

Fig. 26 A general multiple-stage kinetic model of pyrolysis of biomass and its major components. *Redrawn from Miller RS, Bellan J: A generalized biomass pyrolysis model based on superimposed cellulose, hemicellulose and lignin kinetics,* Combust Sci Technol *126:97–137, 1997.*

evaporation of moisture and chemically bound water in wood is also included as the pyrolysis route to describe the physical changes of water during pyrolysis.

Miller and Bellan (1997) proposed a general global reaction scheme for pyrolysis of biomass by extending the B-S model by considering the secondary reactions of tar, as shown in Fig. 26. Biomass feedstock can either be modeled as one component or as a physical mixture of three components: cellulose, hemicellulose, and lignin. The initial step for the formation of active components (e.g., active biomass or its major components) is followed by two parallel routes that lead to the formation of either tar (bio-oil) or char and gases. These parallel routes are governed by distinct kinetic parameters. A subsequent decomposition reaction was included to address the secondary decomposition of tar into gas, which is able to explain the multiple mass loss stages and the increasing yield of gases and the reduction in bio-oil yield observed in experiments at higher temperature. The initiation step that forms active species does not result in any mass variation, which distinguishes this method from one-reaction/stage models. This step enables the model to address the evolution of active species (intermediates) that are formed in the fast pyrolysis of biomass or its major components, leading to the formation of LMW products through further decomposition. There are very subtle changes in the mass loss of the samples observed experimentally during the early stage of biomass pyrolysis, while the mass loss in the early stage shifts the chemical composition of the reacting system dramatically. For example, an initial reaction in cellulose pyrolysis is the formation of a cellulose chain with diminished chain length and another chain species with a levoglucosan-end without any mass change (Vinu and Broadbelt, 2012a; Zhou et al., 2014b,c). A similar scenario is expected to occur in pyrolysis of biomass, which global models fail to capture. However, none of the aforementioned kinetic models of biomass pyrolysis provide a detailed product composition. Moreover, these models are still based on the lumping strategy by which reacting species are grouped into major products based on phase, i.e., solid

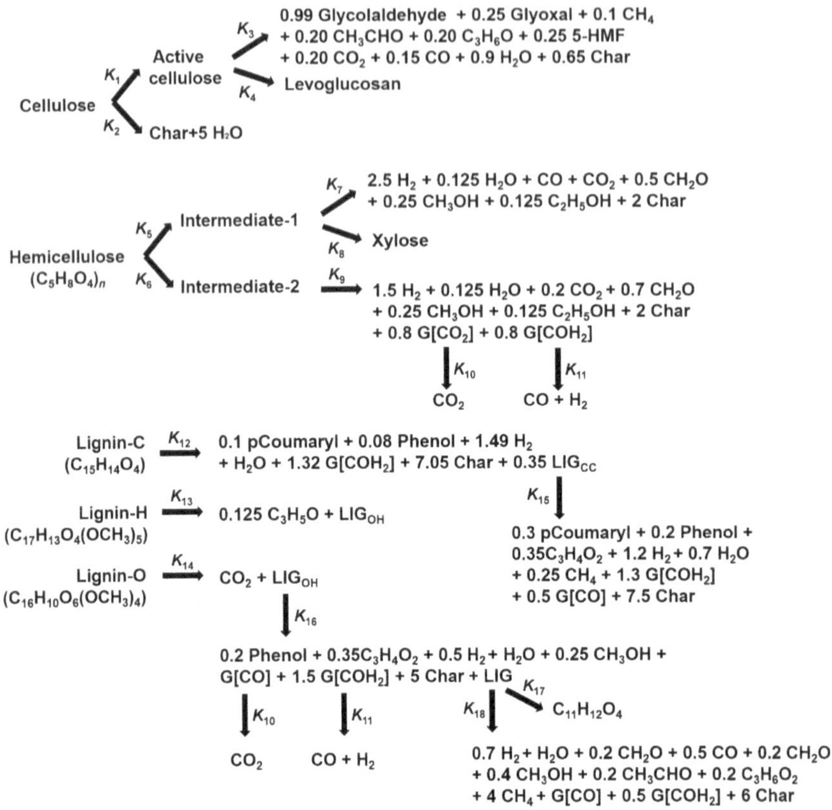

Fig. 27 Three-component, multistage kinetic model of biomass pyrolysis by Ranzi et al. (2008).

biomass, active biomass, volatiles (bio-oil and gas), tar (bio-oil), gas, and char, and are only able to predict the yields of those global lumps without any information on the typical pyrolysis products such as levoglucosan, glycolaldehyde, acetic acid, 5-hydroxymethyl furfural (5-HMF), phenols, aromatics, CO, CO_2, H_2O, and char.

Ranzi et al. (2008) reported the most sophisticated global kinetic model to date for biomass pyrolysis to predict the yields and lumped composition of gas, tar, and solid residue. As shown in Fig. 27, the devolatilization of biomass during fast pyrolysis was modeled as a straightforward combination of three multistep kinetic models for the pyrolysis of each of the three major biomass components: cellulose, hemicellulose, and lignin. To briefly summarize this model's unique contributions, the submodel of cellulose pyrolysis includes one pathway to address the formation of primary char and the loss of water, and another pathway for the formation of active cellulose, which leads to the formation of levoglucosan via a chain-end depolymerization reaction and the formation of other LMW products such as glycolaldehyde, glyoxal, CH_4, acetaldehyde, C_3H_6O, 5-HMF, CO_2, CO, H_2O, and char. For the submodel of hemicellulose pyrolysis, thermal decomposition of hemicellulose (denoted as $(C_5H_8O_4)_n$, xylan) formed two intermediate species, which undergo successive decomposition routes, leading to the formation of xylose, H_2, H_2O, CO, CO_2, HCHO, CH_3OH, C_2H_5OH, and char with independent kinetic parameters. Additionally, the model includes lumped pseudospecies (G[CO_2] and G[COH_2]) which are assumed to be trapped in the solid matrix or melt phase which later release CO_2, and CO and H_2 as gaseous products to address their experimental observation of CO_2, CO, and H_2 released at higher temperatures. The submodel of lignin pyrolysis utilized a combination of three different components: lignin-C (rich in carbon), lignin-O (rich in oxygen), and lignin-H (rich in hydrogen) to model the complex chemical structure of lignin. The decomposition of these initial lignin model components, the reactions of intermediates, and the release of gases are included, leading to the formation of phenol and phenoxy species as the main products of lignin decomposition.

Ranzi and coworkers (Debiagi et al., 2015) further extended their multistep kinetic model of biomass pyrolysis by incorporating several new lumped species to address the presence of hydrophobic and hydrophilic extractives in biomass. Moreover, the model displayed no sensitivity to the different degrees of freedom introduced by new biomass characterization approaches utilized, which greatly widens the applicable range of the model. Anca-Couce et al. (2014) proposed a kinetic scheme for biomass pyrolysis that considered secondary char formation reactions based on the reaction scheme reported by Ranzi et al. (2008). Anca-Couce and Obernberger (2016) further extended their kinetic scheme by Anca-Couce et al. (2014) to model torrefaction of hardwood and softwood.

Ranzi's biomass pyrolysis model and its subsequent modification for more specific biomass materials have an advantage over the existing lumped models by defining stoichiometric coefficients for individual products to handle their different formation rates. Furthermore, these models not only include the main volatilization products but also their subsequent fate in the gas phase. However, there are many factors that are obscured by this type of global model. For example, important products that have been experimentally identified and quantified such as methyl glyoxal, levoglucosan-furanose, acetic acid, furfural, anhydroxylose, dianhydroxylose, and many others are not resolved. Additionally, the chemical composition and structures of hemicellulose and lignin can be very complicated and vary with biomass source, as reviewed in previous sections. The global models cannot capture this complexity when utilizing oversimplified structures for hemicellulose and lignin. For example, hemicellulose was reduced to a representation including only xylan, and lignin was represented by model components assumed from typical β-O-4 linkages as shown in Ranzi's model in Fig. 28.

Generally, global kinetic models are able to explain experimental observations and promote an understanding of the kinetics of biomass pyrolysis to a certain extent. Furthermore, the use of global kinetic models for the complex pyrolysis system simplifies data collection and analysis as well as the numerical implementation, which is attractive for many practical applications. However, global kinetic models of biomass pyrolysis are restricted to a narrow range of operating conditions as reported by the authors, and

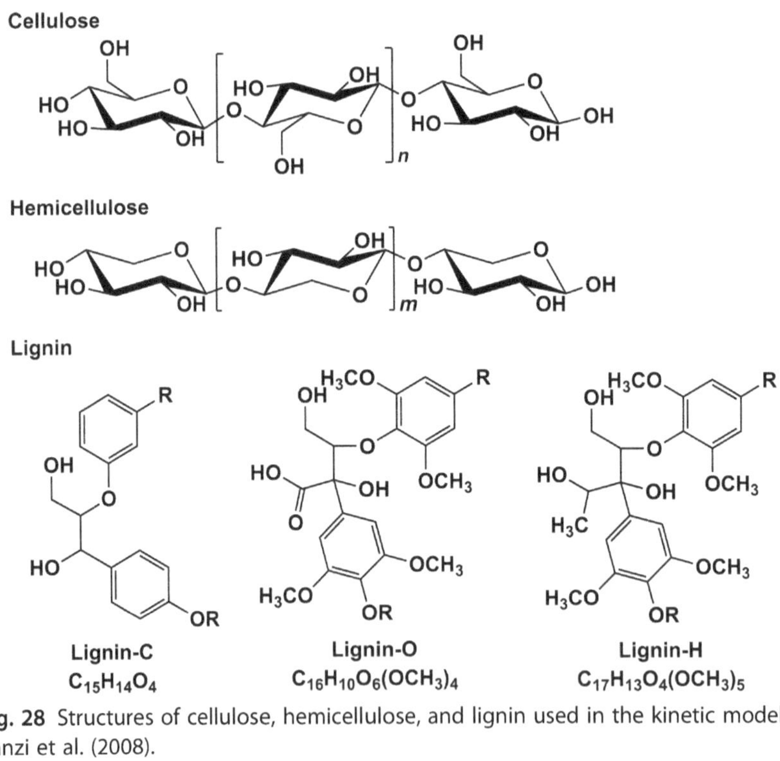

Fig. 28 Structures of cellulose, hemicellulose, and lignin used in the kinetic model by Ranzi et al. (2008).

therefore have very poor potential for extrapolation, which substantially limits the application of the models. Moreover, reactions between lumped pseudospecies were assumed to be irreversible and first order for simplicity. However, fast pyrolysis of cellulose and carbohydrates clearly exhibits sigmoidal reaction character, rather than simply exhibiting first-order reaction character, as recently noted by Burnham et al. (2015). Furthermore, global kinetic models cannot provide detailed information in terms of reaction pathways and the resulting chemical speciation at the mechanistic level.

4.3.2 Mechanistic Modeling of Biomass Pyrolysis

Recently, substantial progress toward the goal of a detailed mechanistic model of complete biomass pyrolysis was made by the Klein group who developed a molecular level kinetic model for biomass gasification, which includes a submodel for biomass pyrolysis (Horton et al., 2016). The model describes the biomass feedstock as a combination of composition models for cellulose, hemicellulose, and lignin, which were developed independently using their in-house software called composition model editor (CME), and based on experimental data from the literature (Horton et al., 2016). Specifically, cellulose was described as a linear polymer of glucose $(C_6H_{10}O_5)_n$, and hemicellulose was represented as a linear xylan structure to reduce the composition model complexity. A crosslinked polymer composition model was developed for lignin based on the attribute identities (including the monomer composition and bonding patterns) as parsed by CME from the Freudenberg model of lignin (Freudenberg, 1965). Pyrolysis chemistry was then applied and the reaction network for biomass pyrolysis was generated automatically using in-house software tools such as the interactive network generator (INGen). The model includes two reaction pathways, hydrolysis and thermolysis for the depolymerization of polysaccharides (cellulose and xylan), which result in conversion of cellulose into glucose and levoglucosan (Fig. 29), and xylan into xylose and anhydroxylopyranose (Fig. 30). These monomeric units can then form light hydrocarbons via thermal cracking, decarbonylation, decarboxylation, and enol–aldehyde tautomerization. The model also includes pathways for the formation of char, higher molecular weight molecules and heavy aromatics via Diels–Alder addition, double bond shift, and dehydrogenation. Similarly, the degradation of lignin in the pyrolysis stage via these same reaction families was also included. This is a breakthrough for modeling biomass pyrolysis at the molecular level, and in understanding the gasification chemistry. However, the submodel for hemicellulose pyrolysis oversimplifies the

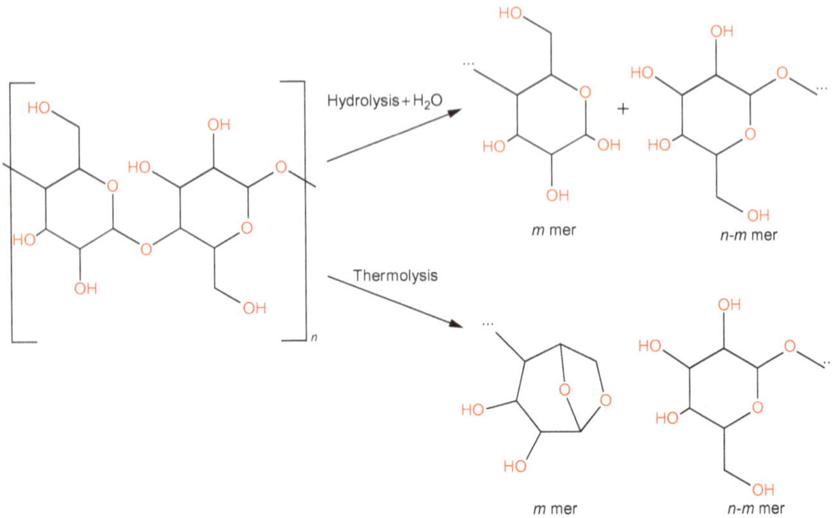

Fig. 29 Depolymerization pathways of cellulose. *Adapted from Horton SR, Mohr RJ, Zhang Y, Petrocelli FP, Klein MT: Molecular-level kinetic modeling of biomass gasification,* Energy Fuels *30:1647–1661, 2016, with permission from American Chemical Society.*

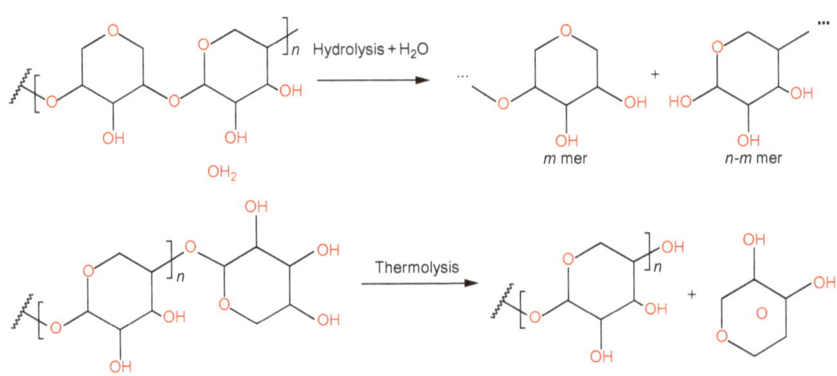

Fig. 30 Depolymerization pathways of hemicellulose (xylan). *Adapted from Horton SR, Mohr RJ, Zhang Y, Petrocelli FP, Klein MT: Molecular-level kinetic modeling of biomass gasification,* Energy Fuels *30:1647–1661, 2016, with permission from American Chemical Society.*

structure of hemicellulose as a linear xylan homopolymer, while in reality it is highly branched. Furthermore, the model focused on the gasification chemistry and many relevant pyrolysis reactions such as dehydration were not included. Detailed reaction pathways for pyrolysis products were also not documented in this work.

As this review has continuously emphasized, fast pyrolysis of biomass is a complex thermal process that breaks a variety of carbohydrates and aromatic polymers into numerous pyrolysis products through many different types of reactions. Given the challenges associated with constructing a complex reaction network to describe the decomposition of whole biomass and the formation of products, it is not surprising that fundamental understanding of biomass pyrolysis in terms of pyrolysis chemistry and reaction mechanisms is lacking. One strategy to tackle this challenge is to critically examine the decomposition pathways of each major component of biomass or model compounds individually, which will lead to a deeper understanding of pyrolysis at a mechanistic level before scaling back to the complex mixtures of starting materials found in biomass.

4.4 Reaction Mechanism of Cellulose Pyrolysis

In 2012, Vinu and Broadbelt (2012a) reported the first mechanistic kinetic model of cellulose pyrolysis. The model was built based on the experimental work of Shanks and coworkers (Patwardhan et al., 2009, 2010, 2011c) using a micropyrolyzer system, isotopic labeling studies of Paine et al. (2007, 2008a,b,c) and Richards and coworkers (Ponder and Richards, 1993; Ponder et al., 1992; Richards, 1987). Many kinetic parameters were derived from the quantum chemical calculations of Mayes and Broadbelt (2012) and other researchers (Hosoya and Sakaki, 2013; Nimlos et al., 2003; Shen et al., 2011). The model was built based on a concerted mechanism, which was demonstrated to be more kinetically favorable than radical or ionic mechanisms, and offered better alignment with experimental findings (Hosoya and Sakaki, 2013; Mayes and Broadbelt, 2012). Later, Zhou et al. (2014b,c) enhanced the model by incorporating updated findings and additional pathways obtained from experiments and quantum chemical calculations. The model focused on two types of reactions: decomposition of cellulose and derived chain species, and reactions of LMW species. The model specified the detailed concerted pathways for the decomposition of cellulose, the reaction of intermediates, the formation of LMW products, and associated kinetic parameters (Vinu and Broadbelt, 2012a; Zhou et al., 2014b,c). As shown in Fig. 31, the model includes the initiation of cellulose chains, yielding levoglucosan-end (LVG-end) chains, and nonreducing-end (NR-end) chains, and the formation of levoglucosan (LVG) from decomposition of the cellulose chain via end-chain initiation and depropagation, which are the dominant pathways for the formation

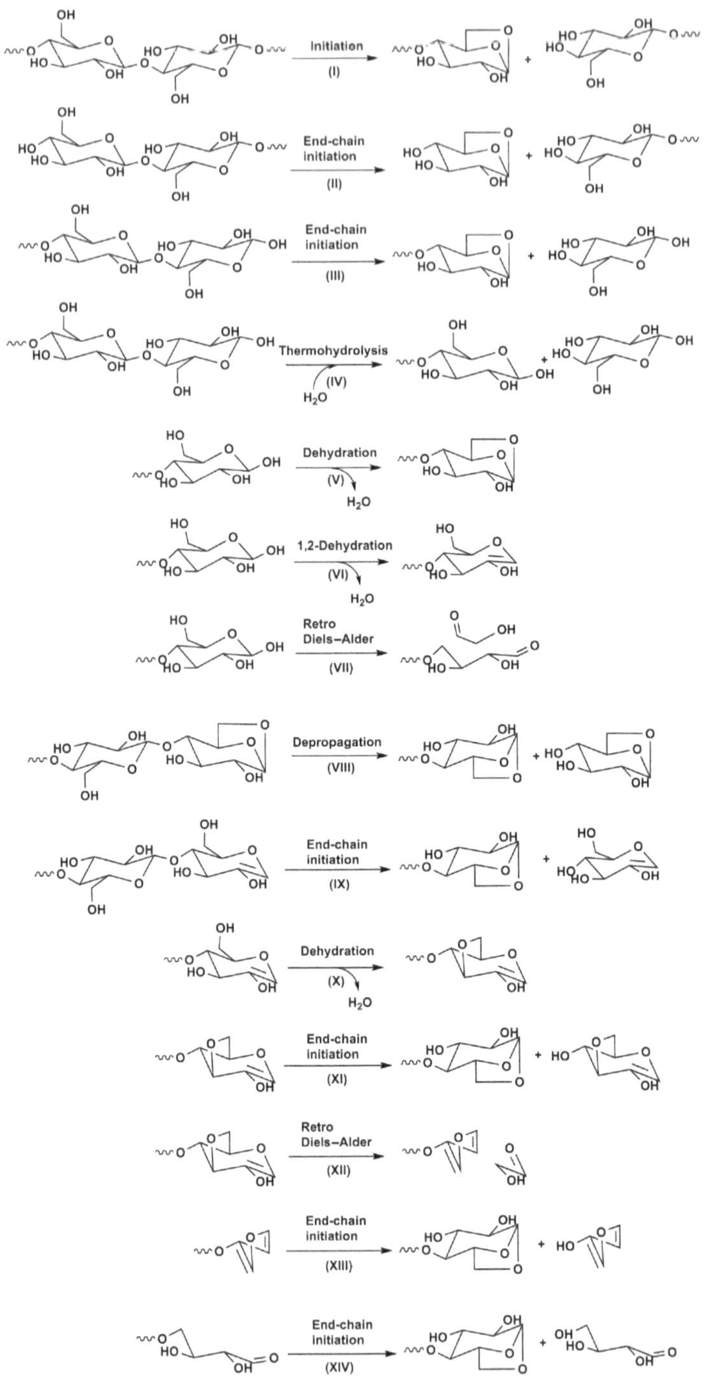

Fig. 31 Decomposition mechanisms of cellulose chains involving end groups. *Adapted from Zhou X, Nolte MW, Mayes HB, Shanks BH, Broadbelt LJ: Experimental and mechanistic modeling of fast pyrolysis of neat glucose-based carbohydrates. 1. Experiments and development of an advanced mechanistic model,* Ind Eng Chem Res *53:13274–13289, 2014b.*

of LVG, the most abundant product. The formation routes of glucose were via end-chain initiation and thermohydrolysis; glucose was a key intermediate leading to the formation of a range of LMW products (LMWP) (Vinu and Broadbelt, 2012a; Zhou et al., 2014b,c). Decomposition of chain species via dehydration, Diels–Alder, and retro Diels–Alder forming anhydro-glucopyranose-end chains, erythrose-end chains, and LMW products were also included. All the chain species with a LMWP-end can undergo end-chain initiation and yield a LVG-end chain and a molecule of the corresponding LMWP, as shown in Fig. 31. Another important type of reaction involving cellulose and its derived chains is mid-chain dehydration which forms a 3,6-anhydro-glucopyranose-mid-chain, eventually leading to the formation of dianhydro-glucopyranose, glycolaldehyde, and char (Fig. 32). The model also explicitly includes decomposition pathways of intermediate glucose and the formation of 67 LMW products via dehydration, ring-opening/closing, retro Diels–Alder, retro aldol, enol–keto tautomerization, isomerization, cyclic/grob fragmentation, and decarbonylation. The formation of char and light gases from dehydrated species was also included in the model. For more details on the reaction mechanism of cellulose pyrolysis, the interested reader is referred to papers by Broadbelt and coworkers (Vinu and Broadbelt, 2012a; Zhou et al., 2014b,c).

The mechanistic model of cellulose pyrolysis was not only able to capture experimental yields of major products such as levoglucosan–pyranose, levoglucosan–furanose, methyl glyoxal, glycolaldehyde, 5-hydroxymethylfurfural, H_2O, CO_2, CO, and char, but also able to well match the yields of minor pyrolysis products such as levoglucosenone, acetone, dihydroxyacetone, and propenal. Net rate analysis revealed that the decomposition of cellulosic chains played a more important role in the formation of levoglucosan and glycolaldehyde than other pyrolysis products. It should be noted that the mechanistic model is able to provide information and insights at the molecular level that a lumped model cannot, and which may be difficult to obtain through experimental methods, such as the dynamic product distribution, the reaction rates of individual reactions, the contributions of different pathways to the formation of individual products during fast pyrolysis, and the pyrolysis timescale. The model provides an improved understanding of the nature of competing reactions in fast pyrolysis in terms of the final yields and underlying chemistry, which could open up avenues to actively control the yield and selectivity for particular pyrolysis products, such as the platform chemical 5-hydroxymethyl furfural (5-HMF), or the pharmaceutical intermediate levoglucosenone, by isolating specific

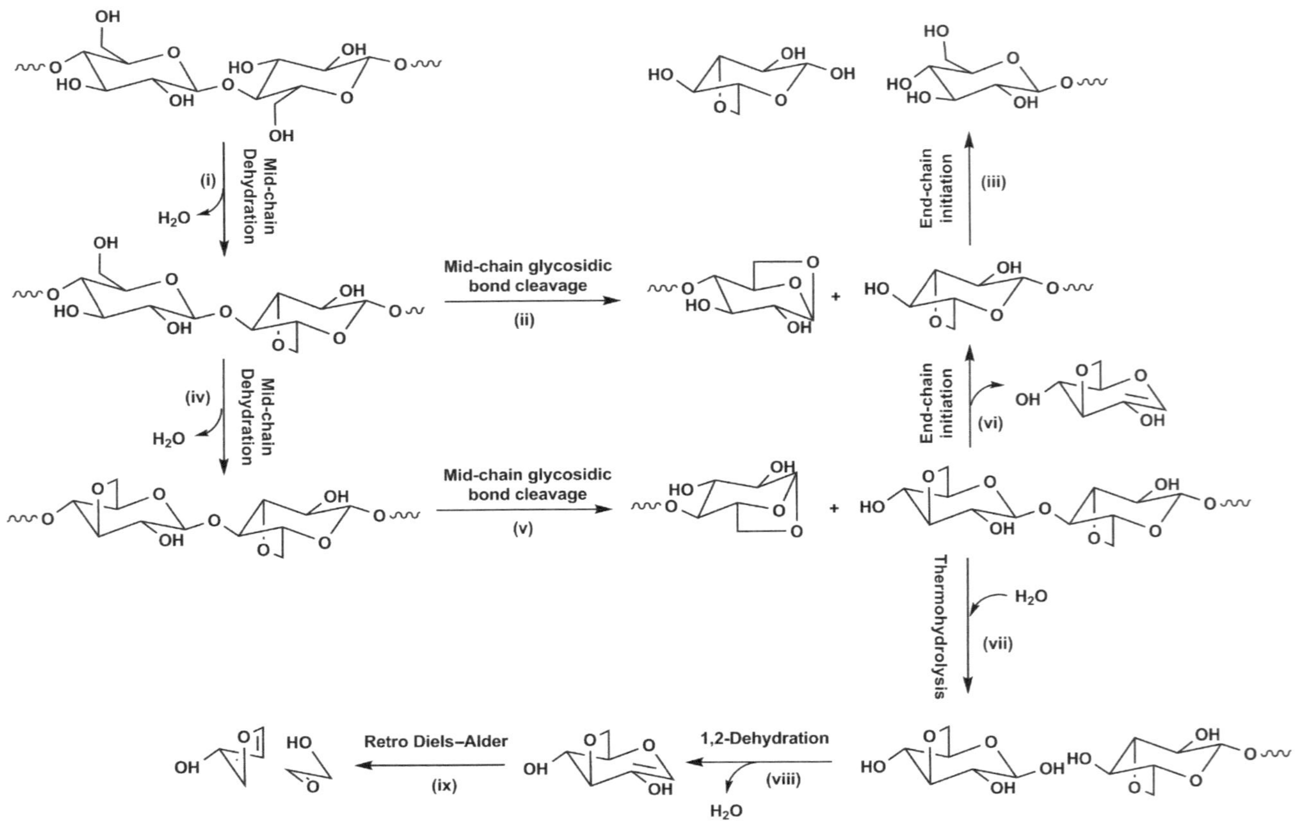

Fig. 32 Decomposition mechanisms of cellulose chains involving mid-groups. *Adapted from Zhou X, Nolte MW, Mayes HB, Shanks BH, Broadbelt LJ: Experimental and mechanistic modeling of fast pyrolysis of neat glucose-based carbohydrates. 1. Experiments and development of an advanced mechanistic model,* Ind Eng Chem Res *53:13274–13289, 2014b.*

reactions through catalysis and optimization of the reaction conditions. More importantly, the mechanistic model of cellulose is extendable to the pyrolysis of other glucose-based carbohydrates and sugars.

Recently, Zhou et al. (2016c,d) extended the mechanistic model of fast pyrolysis of cellulose to address the significant catalytic effects of NaCl on pyrolysis. The model incorporated interactions of Na^+ with cellulosic chains and LMW species, reactions mediated by Na^+ including dehydration, cyclic/Grob fragmentation, ring-opening/closing, isomerization, and char formation, and a degradation network of levoglucosan in the presence of Na^+. Fig. 33 demonstrates reactions in which a sodium ion binds to a LMW species to form a complex, utilizing dehydration of glucose to levoglucosan as a representative example. The complex can then undergo reactions analogous to those in the absence of a sodium ion, albeit governed by different kinetic parameters. In a final elementary step, the sodium ion unbinds from the LMW product. The model that included the interactions of Na^+ with all relevant species and reactions mediated by Na^+ was constructed. Rate coefficients for elementary steps were specified based on Arrhenius parameters. The mechanistic model for sodium-mediated cellulose pyrolysis included 768 reactions of 222 species. 252 reactions between 150 species comprised the additions to the mechanistic model unique to the decomposition of glucose in the presence of NaCl. For the first time, this model revealed that dehydration reactions, especially mid-chain dehydration, that are favored by the presence of NaCl play a vital role in the reduced yield of LVG observed during fast pyrolysis of cellulose (Zhou et al.,

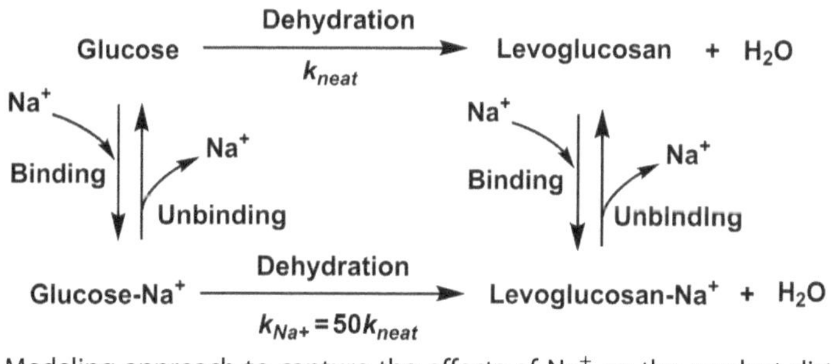

Fig. 33 Modeling approach to capture the effects of Na^+ on the product distribution involved adding interactions of Na^+ with low molecular weight species (e.g., glucose and levoglucosan) and adding parallel reaction pathways (e.g., dehydration) mediated by Na^+ with different kinetic parameters. *Adapted from Zhou X, Nolte MW, Mayes HB, Shanks BH, Broadbelt LJ: Fast pyrolysis of glucose-based carbohydrates with added NaCl, part 1: Experiments and development of a mechanistic model,* AIChE J *62:766–777, 2016c, with permission from Wiley.*

2016c,d). The addition of NaCl to the initial reaction mixture accelerates the thermal decomposition of carbohydrates and substantially reduces the time needed for complete conversion of carbohydrates into LMW pyrolysis products (Zhou et al., 2016c,d).

Mechanistic models produced by the Broadbelt group focused on the primary pyrolysis of cellulose. However, secondary reactions of volatiles in the vapor phase can also occur, especially in the case of long residence times for pyrolysis volatiles. Efforts have also been made to understand these secondary reactions in the vapor phase during cellulose pyrolysis. Norinaga et al. (2013, 2014) experimentally and numerically studied the kinetics of secondary vapor phase cracking of volatiles generated from the fast pyrolysis of cellulose. They developed a detailed chemical kinetic model (DCKM) comprised of more than 500 species and around 8000 elementary step-like reactions, and evaluated it against experimental yields of more than 20 species measured at residence times up to 6 s, and at temperatures ranging from 400 to 900°C. The secondary reactions of volatiles from cellulose pyrolysis yielded light gases such as H_2, CO, CO_2, CH_4, and C_2H_4 as primary products, together with acetaldehyde, acetic acid, acetone, hydroxyl acetone, methanol, C_3 hydrocarbons, furan, benzene, and toluene as minor products. This work, in particular, sought to elucidate the reaction pathways that led to the formation of aromatic species since aromatic compounds such as benzene and toluene were identified as pyrolysis products despite cellulose inherently containing no aromatic structures. Fig. 34 details the proposed reaction pathways for the formation of benzene and naphthalene in the secondary pyrolysis of cellulose at 650°C. They reported that C_3 alkyne and diene were the primary precursors of benzene and the combination of acetylene with propyne or allyl radical formed cyclopentadiene, which was a prominent precursor of naphthalene. Pyrolysis temperature was found to have a major impact on competing pathways during the secondary pyrolysis stage. For example, furan and acrolein are likely important alkyne precursors in cellulose pyrolysis at low temperature, whereas the dehydrogenation of olefins is the major route to alkynes at high temperatures.

4.5 Reaction Mechanism of Hemicellulose Pyrolysis

Due to the escalation in structural complexity, and thus the reaction network, fewer efforts and achievements have been made in the mechanistic modeling of pyrolysis of hemicellulose and lignin when compared to cellulose. However, given the similarities between the polymeric structures of

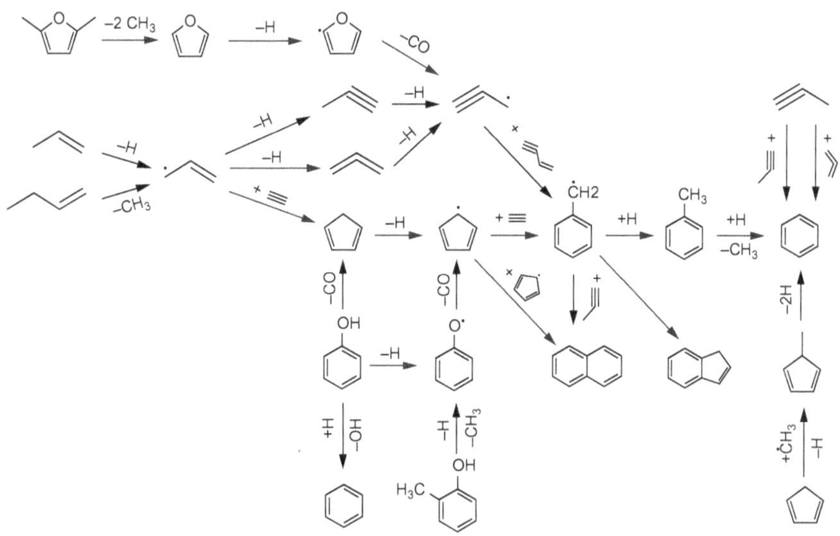

Fig. 34 Reaction pathways for the formation of aromatic hydrocarbons in secondary pyrolysis of cellulose at 650°C and fast pyrolysis of lignin at 700°C. *Blue arrows* indicate dominant pathways and *black arrows* indicate minor pathways. *Redrawn from Norinaga K, Shoji T, Kudo S, Hayashi J: Detailed chemical kinetic modelling of vapour-phase cracking of multi-component molecular mixtures derived from the fast pyrolysis of cellulose,* Fuel *103:141–150, 2013; Norinaga K, Yang H, Tanaka R, et al.: A mechanistic study on the reaction pathways leading to benzene and naphthalene in cellulose vapor phase cracking,* Biomass Bioenergy *69:144–154, 2014; Yang H, Appari S, Kudo S, Hayashi J, Norinaga K: Detailed chemical kinetic modeling of vapor-phase reactions of volatiles derived from fast pyrolysis of lignin,* Ind Eng Chem Res *54:6855–6864, 2015a.*

cellulose and hemicellulose, hemicellulose is speculated to decompose via similar chemical pathways during fast pyrolysis (Patwardhan et al., 2011a).

Fig. 35 shows a postulated reaction scheme for hemicellulose pyrolysis proposed by Patwardhan et al. (2011a), in which there are competing pathways: depolymerization to sugars and anhydrosugars, dehydration to furan and pyran ring derivatives and furanose, as well as pyranose ring breakage to light oxygenated species. Wang et al. (2013a) proposed a reaction scheme for xylan decomposition in which the formation of all pyrolysis products except acetic acid proceeded through an acyclic D-xylose intermediate. Shen et al. (2010) also proposed a reaction scheme (Fig. 36) for the formation of 1,4-anhydro-D-xylopyranose, furfural, acetone, acetic acid, formic acid, CO_2, CO, methanol, etc., from the fast pyrolysis of *O*-acetyl-4-*O*-methylglucurono-xylan. The primary formation pathways for acetic acid and CO_2 were assumed to be the cleavage of acetyl groups off the xylan structure and the decarboxylation reactions of acetyl groups attached on

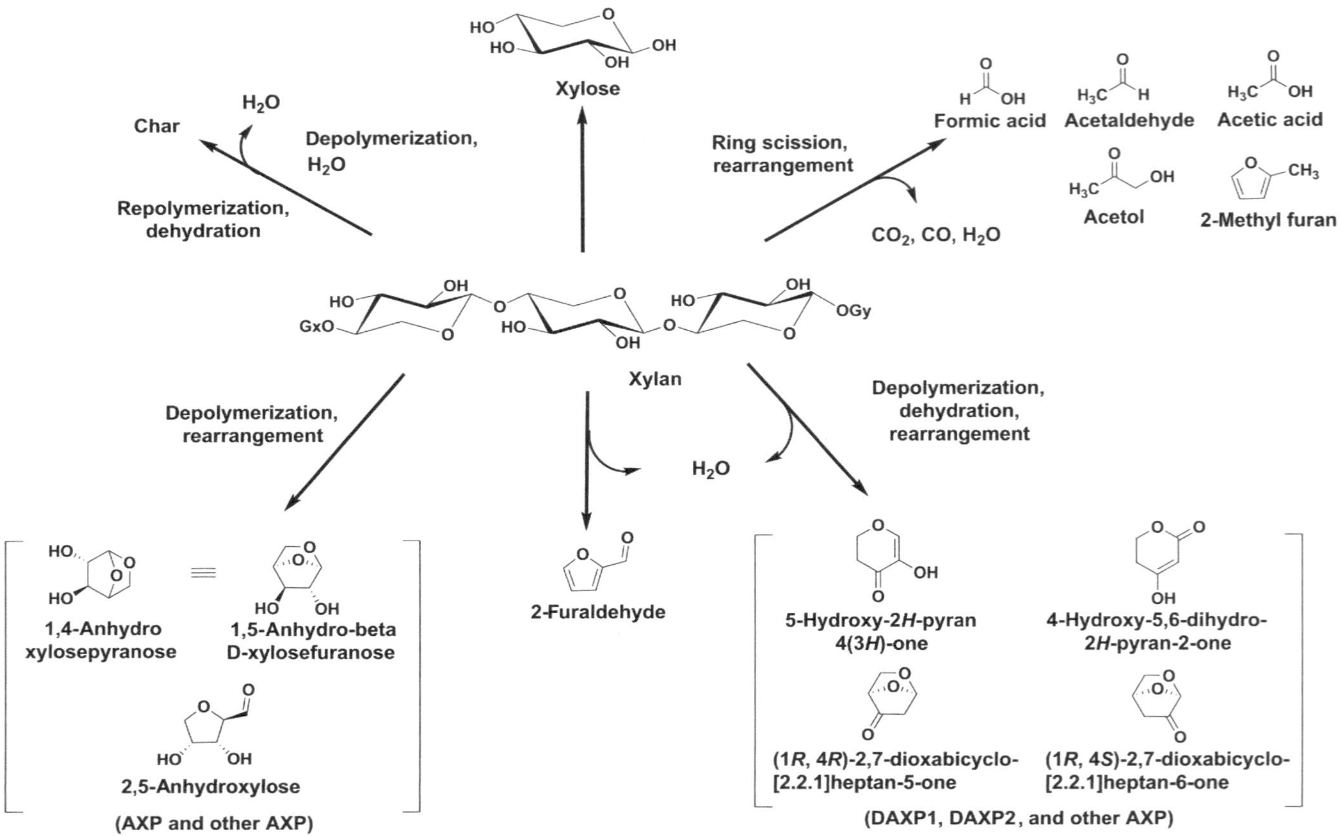

Fig. 35 A postulated reaction scheme for the thermal decomposition of hemicellulose by Shanks and coworkers. *Adapted from Patwardhan PR, Brown RC, Shanks BH: Product distribution from the fast pyrolysis of hemicellulose,* ChemSusChem *4:636–643, 2011a, with permission from Wiley.*

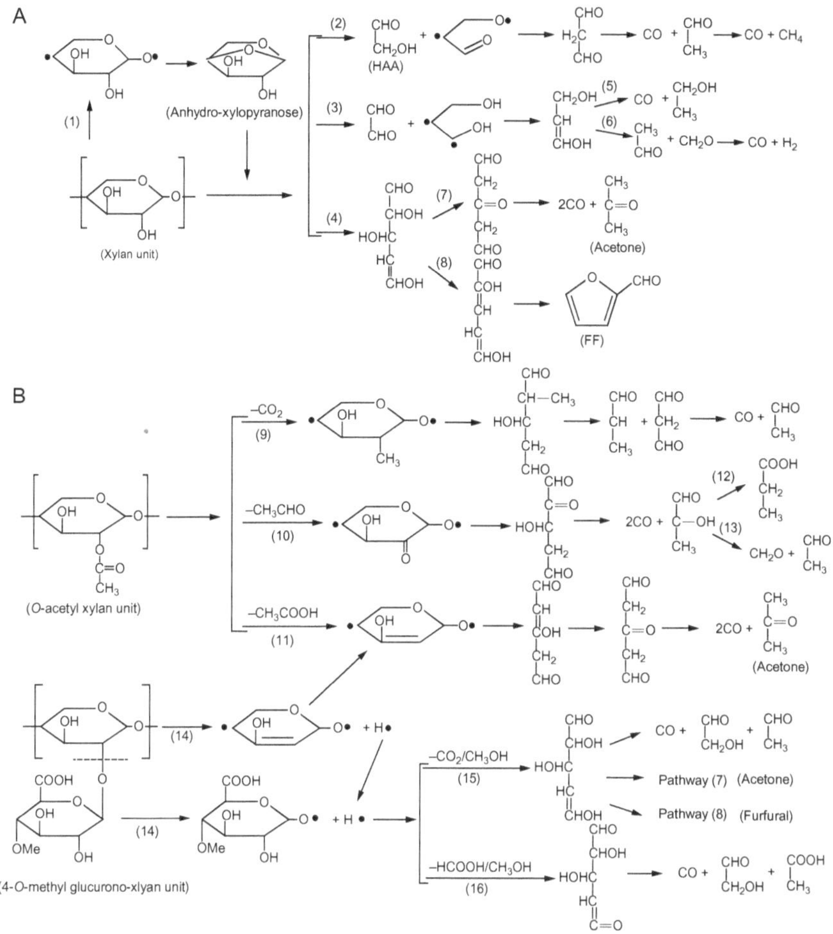

Fig. 36 Postulated reaction scheme for thermal decomposition of (a) *O*-acetyl-4-*O*-methylglucurono-xylan and (b) *O*-acetyl-xylan and 4-*O*-methylglucuronic acid units. *Adapted from Shen DK, Gu S, Bridgwater AV: Study on the pyrolytic behaviour of xylan-based hemicellulose using TG-FTIR and Py-GC-FTIR,* J Anal Appl Pyrol *87:199–206, 2010, with permission from Elsevier.*

the xylan structure, respectively, both of which compete with the formation of CO (Shen et al., 2010). All aldehyde species such as formaldehyde and glycolaldehyde can undergo decarbonylation to form CO and explain the increasing yield of CO with elevated temperature. However, the thermal degradation of xylan via a radical mechanism proposed by Shen et al. (2010) is not likely, as fast pyrolysis of carbohydrates has been increasingly recognized to follow a concerted mechanism (Mayes and Broadbelt,

2012; Mayes et al., 2014b, 2015; Seshadri and Westmoreland, 2012; Vinu and Broadbelt, 2012b; Zhou et al., 2014b, 2016b,c,d). No detailed elementary reaction steps or information at the mechanistic level was reported in these studies. Moreover, none of these proposed decomposition networks/pathways has been validated via kinetic modeling or supported by quantum chemical calculations. Therefore, there is little quantitative information that the speculated reaction pathways can provide to confirm or illuminate the reaction mechanism of hemicellulose pyrolysis.

Recently, the Broadbelt group has developed a mechanistic model for the fast pyrolysis of hemicellulose, specifically hemicellulose extracted from corn stover (with a dominant polysaccharide contribution from arabinoxylans) (Zhou et al., 2016b). The model was built utilizing the reaction family approach which has been applied in the past to construct mechanistic models for fast pyrolysis of glucose-based carbohydrates including cellulose (Broadbelt and Pfaendtner, 2005; Vinu and Broadbelt, 2012a; Zhou et al., 2014b, 2016b,c,d). The model describes the detailed reaction pathways and mechanisms for the decomposition of hemicellulose chain species via initiation, end-chain initiation, dehydration, mid-chain dehydration, hydrolysis, etc., as well as the reactions of intermediates such as xylose, and the formation of 80 pyrolysis products including glycolaldehyde, acetaldehyde, methylglyoxal, furfural, anhydropyranoses, dianhydropyranoses, acetone, acetol, CO_2, CO, H_2O, and char. The model includes more than 500 reactions which are specified in terms of elementary steps and the associated kinetic parameters. This is an important step toward enhancing our fundamental understanding of the fast pyrolysis of hemicellulose. However, this model also made key structural assumptions, focusing on the fast pyrolysis of the hemicellulose polysaccharide arabinoxylans, while in reality composition and structure of native hemicellulose would be much more complicated, as reviewed previously. Therefore, further mechanistic modeling efforts are needed to address the fast pyrolysis of other hemicellulose polysaccharides such as galactoglucomannans, xyloglucans, and β-1,3;1,4-glucans, and their interactions in order to obtain a complete picture of the thermal decomposition of native hemicellulose.

4.6 Reaction Mechanism of Lignin Pyrolysis

Many efforts have also been devoted to unraveling the detailed decomposition pathways of lignin and its model compounds as well as developing kinetic model to describe the pyrolysis chemistry and to predict the pyrolysis product distribution. Both free radical and concerted reaction mechanisms have been proposed to explain the pyrolysis of lignin and its model compounds (Britt et al., 2000; Carstensen and Dean, 2010; Chu et al., 2013; Jarvis et al., 2011; Kawamoto et al., 2008; Kibet et al., 2012; Schlosberg et al., 1983; Vuori and Bredenberg, 1987). For example, Jarvis et al. (2011) reported that concerted reactions dominated over free radical reactions for the pyrolysis of phenethyl phenyl ether (PPE), a lignin model compound, under typical pyrolytic conditions, while Kim et al. (2014) suggested that the pyrolysis of α-O-4 dimeric phenolic compounds followed a free radical reaction mechanism. Indeed, it is possible that the mechanism may change with the structure of the reactant, the reactor and reaction conditions used. Kawamoto et al. (2008) reported that Cγ-structures with different side chain or substituent groups proceeded via different pyrolysis mechanisms. Cγ-deoxy types, especially in phenolic forms, are degraded through radical chain mechanisms; however, introducing an OH group at the Cγ position changes the β-ether cleavage mechanism in phenolic form from a free radical chain to a quinone methide mechanism. Additionally, the radical chain reactions were reported to only be active in a closed-type reactor but not effective in an open-type reactor.

There are two decomposition mechanisms that have been hypothesized in the literature for the primary decomposition of lignin during fast pyrolysis. One is the depolymerization of lignin into monomeric species. Shanks and coworkers (Patwardhan et al., 2011b) reported that fast pyrolysis of lignin primarily resulted in the production of char, gaseous products, and monomeric phenolic compounds with phenol, 4-vinyl phenol, 2-methoxy-4-vinyl phenol, and 2,6-dimethoxy phenol as the major products. Both the pyrolysis of lignin and monolignol compounds resulted in the formation of a remarkable amount of dimeric and other oligomeric compounds formed via reoligomerization of the monomeric pyrolysis products during condensation, which also can be facilitated by the presence of acetic acid. Therefore, it was speculated that depolymerization of lignin forming monomeric compounds is the primary reaction of lignin pyrolysis instead of thermal ejection of oligomeric compounds, as shown in Fig. 37. The oligomeric compounds were formed by the reoligomerization of monomeric compounds during lignin pyrolysis.

Another reaction mechanism is the thermal ejection of lignin forming oligomeric compounds as primary pyrolysis products. For example, Garcia-Perez and coworkers (Zhou et al., 2014a) studied the slow and fast pyrolysis of Douglas-fir lignin using a mesh reactor equipped with microscopy, high

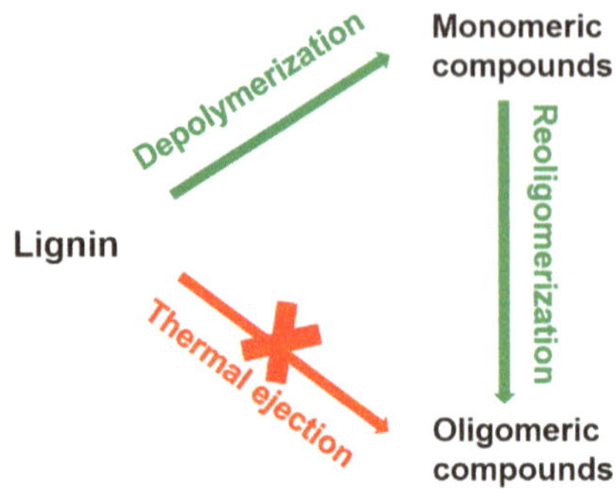

Fig. 37 Reaction mechanism of the formation of oligomers during lignin fast pyrolysis by Patwardhan et al. (2011b).

speed photography, and scanning electron microscopy. They observed the formation of a liquid intermediate phase at the early stage of lignin pyrolysis. Intensified foaming and thermal ejection (atomization) from the surface as vapor or aerosol (referred to as pyrolytic lignin) were also observed at elevated temperatures. The pyrolytic lignin, which was the main product of lignin primary reactions, was identified as mostly oligomeric compounds. The oligomers then underwent secondary degradation into monophenols, pyrolytic lignin derivatives, secondary gases, and char. A scheme representing this reaction mechanism is shown in Fig. 38.

Hough et al. (2015) developed a semidetailed kinetic model for lignin fast pyrolysis, which consisted of elementary and lumped reaction steps to track approximately 100 individual molecules, radicals, lumped heavy species, and functional groups linked to the degrading polymer. The model was able to predict the detailed compositions of gaseous product and tar derived from lignin pyrolysis over a wide range of operating conditions. The model also provided insights into the evolution of functional groups with mechanistic details. However, the model is not able to provide the detailed reaction

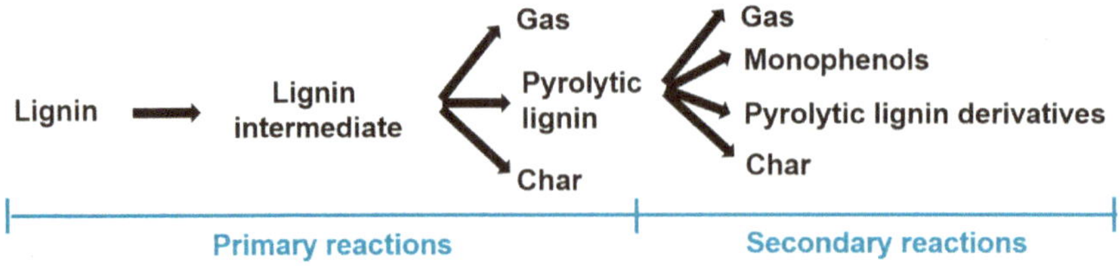

Fig. 38 Reaction mechanism of lignin pyrolysis by Zhou et al. (2014a).

pathways to the final products, and the yields of individual products experimentally observed due to the use of a lumping strategy.

Chu et al. (2013) studied the pyrolysis of β-O-4 type oligomeric lignin model compounds using thermogravimetric analysis and a pyroprobe-GC-MS system. They identified and quantified 25 volatile compounds including vanillin and 2-methoxy-4-methyl phenol in the highest abundance. They proposed a free radical dominant reaction network (Fig. 39) to explain the observed products. As shown in Fig. 39, the initiation step for free radical chain reactions is the homolytic cleavage of the β-O-4 linkage, which generates radicals. The radicals can subsequently abstract hydrogen from other species which have weak C–H or O–H bonding (C_6H_5–OH) to form products. The radicals can also be passed to other species for further reactions leading to chain propagation. Reactions (a) and (b) in Fig. 39 depict example termination reactions. However, secondary degradation via H-abstraction, double bond formation, rearrangement, isomerization, and concerted

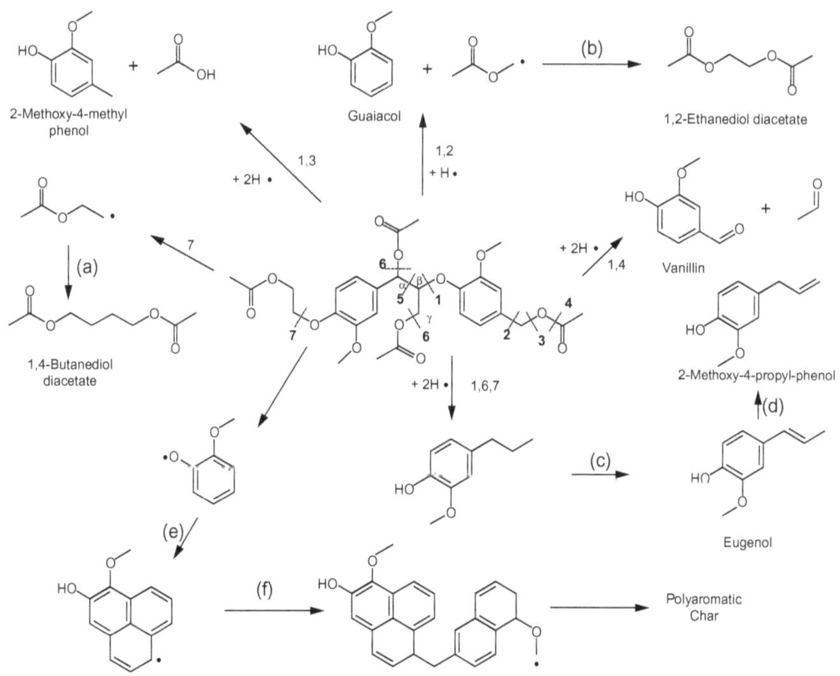

Fig. 39 Reaction mechanism of pyrolysis of oligomeric lignin model compounds with β-O-4 linkages. *Adapted from Chu S, Subrahmanyam AV, Huber GW: The pyrolysis chemistry of a beta-O-4 type oligomeric lignin model compound,* Green Chem *15:125–136, 2013, with permission from Royal Society of Chemistry.*

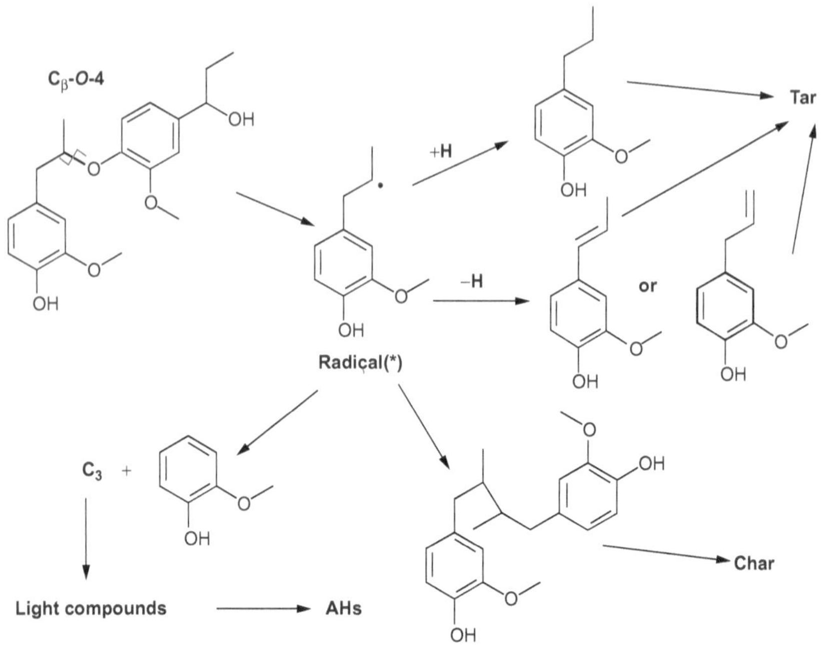

Fig. 40 An example pathway of lignol decomposition during lignin pyrolysis. *Adapted from Yang H, Appari S, Kudo S, Hayashi J, Kumagai S, Norinaga K: Chemical structures and primary pyrolysis characteristics of lignins obtained from different preparation methods,* J Jpn Inst Energy *93:986–994, 2014.*

reactions also occur to diversify the product distribution such as reactions (c) and (d). Char formation is a dominant reaction and is hypothesized to be due to the random repolymerization of radical species such as aromatics, alkanes, and alkenes, followed by the elimination of functional groups such as hydroxyl and methoxyl groups.

Yang et al. (2014) also rationalized a reaction mechanism for lignin decomposition based on their own experimental investigations. As shown in Fig. 40, connections between substituents were suggested to break and generate radicals, leading to the formation of char via polymerization. Moreover, methoxyl groups and aliphatic substituents in lignin enhance char formation by providing active fragments with high tendency toward polymerization. In contrast, the presence of hydrogen in lignin was more likely to suppress char formation by stabilizing active fragments and promote tar formation. Furthermore, Yang et al. (2014) reported that light unsaturates such as olefins, diolefins, and alkynes are important intermediates for the secondary reactions of lignin pyrolysis, and aromatic hydrocarbons from lignin pyrolysis were possibly formed from the recombination of light unsaturates rather than from the natural aromatic structures of lignin. Fig. 34 shows example pathways for the recombination and aromatization of olefins, diolefins, and alkynes that form during fast pyrolysis of lignin even with short residence times of volatiles in the gas phase.

Later, Yang et al. (2015a) reported a DCKM for secondary vapor phase reactions of volatiles derived from lignin fast pyrolysis. The DETCHEM^{BATCH} code, which is designed to simulate the time-dependent homogeneous gas phase reactions in a batch reactor, was used to create the DCKM. Experimental yields of 31 products from the fast pyrolysis of lignin derived from enzymatic hydrolysis of empty fruit bunch (enzymatic hydrolysis lignin—EHL) at 500–950°C were used to evaluate the model. As shown in Fig. 34, analysis of the reaction pathways by Yang et al. (2015a) suggested the importance of the cyclopentadienyl radical (C_5H_5) as an important bridge for the formation of aromatic hydrocarbons from lignin pyrolysis. Specifically, the recombination of cyclopentadienyl radicals, primarily produced by the decomposition of phenols (phenol and cresol) during lignin pyrolysis, was proposed as the dominant route forming naphthalene. The decomposition of methoxy- or hydroxyphenols into 1,3-butadiene led to the formation of vinylacetylene, which reacts with the propargyl radical (C_3H_3) and forms benzyl radical (C_7H_7). Toluene can be formed from the combination of these benzyl radicals (C_7H_7) with H, while indene was mainly formed from C_7H_7 by recombination with C_2H_2. Benzene was produced from two types of reactions: H-substitution reactions of phenol and toluene, and the recombination of small hydrocarbons such as propadiene or propyne with C_3H_3 or C_5H_6 with C_2H_4 with the concomitant formation of a methyl radical.

Although this model was comprised of 8159 elementary reactions of 548 species, it is still not fully mechanistic since the decomposition pathways of phenols (such as guaiacol, syringol, catechol, and their derivatives) were modeled as global reactions. Again, this model is also constructed to a very specific set of conditions since the native lignin structure or lignin obtained from different treatment methods such as organosolv extraction or the Klason procedure have a different composition and structure than EHL (Yang et al., 2014). Moreover, the detailed structure of EHL is not documented. Therefore, continued mechanistic modeling efforts on lignin pyrolysis are needed.

A fundamental understanding of the composition and structure of biomass as well as the mechanistic reaction network and kinetics of biomass pyrolysis at the mechanistic level not only would lead to a deeper fundamental understanding of reaction processes, but also could predict pyrolysis behavior and outcomes to guide the design of efficient pyrolytic reactors, and catalyst innovations for engineering applications. However, many challenges remain in creating comprehensive models capable of making these predictions due to the structural and chemical complexity of biomass, and its decomposition mechanism. Mechanistic models for the pyrolysis of cellulose have enjoyed more success than the other two major biomass components, hemicellulose and lignin, which is mainly due to the greater structural heterogeneity and complexity of hemicellulose, and especially lignin. Moreover, significant experimental and computational efforts devoted to studying cellulose pyrolysis in terms of designing new reaction systems (Krumm et al., 2016; Mettler et al., 2012b; Paulsen et al., 2013, 2014; Teixeira et al., 2011), identification and quantification of intermediates and products (Hilbers et al., 2015; Liu et al., 2013a,b, 2014b; Patwardhan et al., 2009, 2010, 2011c; Wang et al., 2013b, 2014a,b; Yu et al., 2012, 2013), isotopic labeling (Degenstein et al., 2015; Paine et al., 2007, 2008a,b,c), quantum chemical calculations (Agarwal et al., 2012; Hosoya and Sakaki, 2013; Mayes and Broadbelt, 2012; Mayes et al., 2014a,b, 2015; Mettler et al., 2012a; Seshadri and Westmoreland, 2012), and microkinetic modeling (Norinaga et al., 2013, 2014; Vinu and Broadbelt, 2012a; Zhou et al., 2014b, 2016b,c,d) have also contributed remarkably in recent years to unraveling the fundamentals of cellulose pyrolysis. In contrast, comparable efforts have not been made for lignin and, especially, hemicellulose pyrolysis. Consequently, future research efforts should be directed toward mechanistic modeling of pyrolysis of hemicellulose and lignin with special attention given to strategies that effectively capture their heterogeneous structures. Cooperation between experimentalists and modelers will be essential to progress in creating a detailed mechanistic model of fast pyrolysis of model compounds, biomass major components, and ultimately whole biomass.

5. CFP OF BIOMASS

Bio-oil produced by biomass pyrolysis offers several advantages in its potential use as a fuel, beyond its status as a renewable source of energy, as it also burns with low NO_x and SO_x emissions, and reduces net CO_2 input to the atmosphere (Hildebrandt et al., 2009). However, when compared with crude oil, some drawbacks such as poor quality with high oxygen content (up to 60 wt.%) and high water content (10–30 wt.%) make it inferior to

Table 6 Typical Characteristics of Crude Oil and Pyrolysis Bio-Oil

Composition	Crude Oil	Bio-Oil
Water (wt.%)	0.1	10–30
pH	–	2.8–3.8
Density (kg/L)	0.86	1.05–1.25
Viscosity 50°C (cP)	180	40–100
HHV (MJ/kg)	44	16–19
C (wt.%)	83–86	55–65
O (wt.%)	<1	28–60
H (wt.%)	11–14	5–7
S (wt.%)	<4	<0.05
N (wt.%)	<1	<0.4
Solid (wt.%)	0.1	<0.2
H/C	1.5–2.0	0.9–1.5
O/C	≈0	0.3–0.5

Adapted from Dickerson T, Soria J: Catalytic fast pyrolysis: a review. *Energies* 6:514–538, 2013.

conventional hydrocarbon fuels. Additionally, high acidity (pH of 2–4) can lead to potential corrosion problems, its unstable and reactive nature can make it difficult to store, and it expresses a low heating value and high viscosity (Table 6) (Regalbuto, 2009; Yildiz et al., 2015). Therefore, it is important to improve the quality and stability of bio-oil. CFP is an attractive technology that integrates fast pyrolysis of biomass and catalytic upgrading into a single process to produce bio-oil with enhanced quality that could potentially fit into existing infrastructure. In this section, we review the overall comparison of in situ with ex situ CFP of biomass, and recent experimental results from CFP of biomass using zeolite catalysts (Fig. 41).

5.1 Overall Comparison of In Situ and Ex Situ CFP

Although the composition and quality of upgraded bio-oil from CFP can be affected by many factors such as reactor type and configuration (e.g., fixed bed reactor and fluidized bed), biomass source, and properties (e.g., chemical composition, moisture, ash content, and particle size) (Adam et al., 2006), catalyst type and properties (e.g., shape, pore structure, and active sites) (Anand et al., 2016; Jae et al., 2011; Naqvi et al., 2015; Pütün et al.,

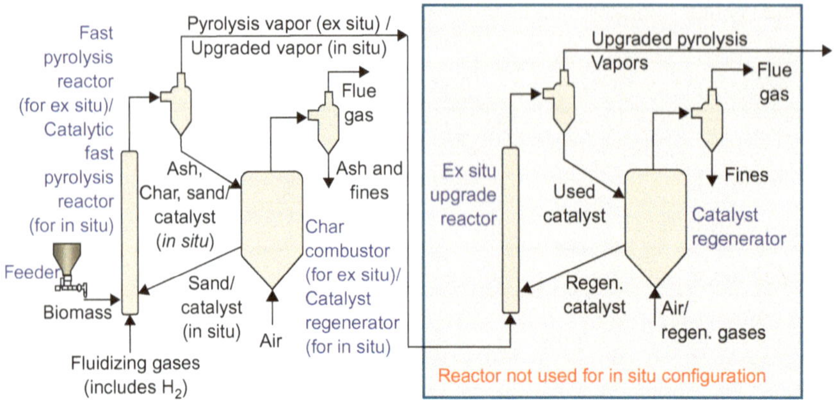

Fig. 41 Simplified process flow diagram for fast pyrolysis and (in situ and ex situ) catalytic fast pyrolysis equipment. *Adapted from Dutta A, Sahir A, Tan E, et al.: Process design and economics for the conversion of lignocellulosic biomass to hydrocarbon fuels thermochemical research pathways with in situ and ex situ upgrading of fast pyrolysis vapors. Technical Report NREL/TP-5100-62455, PNNL-23823, 2015.*

2009), and operating conditions (e.g., heating rate, biomass space velocity, catalyst to feedstock ratio, temperature, and residence time of vapors) (Anand et al., 2016; Liu et al., 2015; Pütün et al., 2009; Zhang et al., 2009, 2012), the bio-oil obtained from both in situ and ex situ CFP consistently exhibits a lower oxygen content (Table 7) than from non-CFP. CFP bio-oil also has a lower content of acids and higher content of LMW hydrocarbons, olefins, aliphatics, aromatics, and phenolics. As shown in Fig. 42, deoxygenation of bio-oil is mainly attributed to the formation of H_2O, CO, and CO_2 through dehydration, decarbonylation, and decarboxylation, respectively (Carlson et al., 2011; Yildiz et al., 2013). Williams and Nugranad (2000) reported deoxygenation that favors the formation of H_2O at low catalyst temperatures, and CO and CO_2 at high catalyst temperatures.

For both in situ and ex situ CFP, the use of catalysts reduced the yield of bio-oil and increased yields of noncondensable gas, water, and solid residue (Mochizuki et al., 2013; Mullen et al., 2011; Naqvi et al., 2015; Williams and Nugranad, 2000; Yildiz et al., 2013; Zhang et al., 2009). The solid residue includes char and coke, which are distinctly defined as a thermal product and catalytic product, respectively (Du et al., 2013; Huber and Corma, 2007). In comparison to non-CFP, similar or slightly decreased amounts of char were formed in CFP. However, a significant amount of coke is produced in CFP. The severe coking not only reduces the overall carbon efficiency to bio-oil, but also rapidly deactivates the catalysts

Table 7 Comparison of Noncatalytic Fast Pyrolysis With In Situ and Ex Situ CFP

	Fast Pyrolysis	In Situ CFP	Ex Situ CFP
Operating Conditions			
Fast pyrolysis temperature (°C)	500	500	500
Catalytic reactor temperature (°C)	–	500	500
Experimental running time (min)	170	100	90
Biomass feed rate (g/h)	200	200	200
Inert gas flow rate (L/h)	120	120	120
Heat carrier flow rate (kg/h)	5.9	–	–
Ratio of sand/catalyst in mixture	–	4:1	1:3
Heat carrier/biomass ratio	26.8	53.8	32.3
Catalyst/biomass ratio		5	5
Yields of Lumped Products (wt.%)			
Liquid	58.9 ± 4.54	50.3 ± 4.39	50.1 ± 1.23
Gas	22.2 ± 2.56	26.2 ± 2.35	23.9 ± 2.43
Char	15.9 ± 1.11	13.6 ± 1.24	15.7 ± 0.97
Coke	2.60 ± 0.14	9.70 ± 0.87	10.1 ± 0.81
Abundant Compounds in Organic Phase (wt.%)			
Levoglucosan	3.54	0.14	0.41
Glycolaldehyde	3.54	0.00	0.00
Acetic acid	7.41	2.93	3.92
Propanoic acid	1.47	0.92	0.53
Furfural	0.59	0.00	0.11
2-Cyclopenten-1-one	0.40	1.36	0.65
1,2-Benzenediol	2.43	6.92	3.15
4-Methyl-1,2-benzenediol	1.76	2.79	1.62
4-Ethylcatechol	0.74	1.15	0.69
2,4-Dimethyl-phenol	0.41	1.19	0.66
4-Methyl-phenol	0.36	1.40	0.82
2-Methyl-phenol	0.30	1.19	0.87

Continued

Table 7 Comparison of Noncatalytic Fast Pyrolysis With In Situ and Ex Situ CFP—cont'd

	Fast Pyrolysis	In Situ CFP	Ex Situ CFP
Phenol	0.27	2.32	1.24
Xylene (m, p)	0.01	4.67	1.54
Toluene	0.02	2.98	0.62
1-Methyl naphthalene	0.02	1.88	0.41
Methyl-1H-indenes	0.01	1.12	0.55
Indene	0.01	1.10	0.53
Indane	0.01	0.52	0.20
1-Hydroxy-2-propanone	2.93	1.22	1.42

Adapted from Yildiz G, Pronk M, Djokic M, et al.: Validation of a new set-up for continuous catalytic fast pyrolysis of biomass coupled with vapour phase upgrading, *J Anal Appl Pyrol* 103:343–351, 2013.

(Carlson et al., 2011; Choi et al., 2015; Horne and Williams, 1995; Jae et al., 2014; Mochizuki et al., 2013; Mullen et al., 2011; Naqvi et al., 2015; Xu et al., 2016; Yildiz et al., 2015). The reduced yield of bio-oil is mainly attributed to the coking and catalytic cracking of pyrolysis oil vapor on the catalyst. Note that this decreased yield in bio-oil also represents a significant loss in carbon yield in bio-oil and energy recovery in CFP products vs non-CFP products (Choi et al., 2015).

In in situ CFP, the catalysts experience intimate contact with both biomass and the derived pyrolysis vapors, which facilitates the catalytic intervention during both the primary reactions of biomass and the secondary reactions of derived pyrolysis vapors. The primary decomposition of biomass could be enhanced by the careful selection of catalysts and the removal of primary products, thus inhibiting the polymerization and secondary charring reactions. However, controlling the reaction severity is challenging at the short timescale of biomass pyrolysis and the short residence time of pyrolysis vapors. For example, the optimal pyrolysis temperature (typically ~500°C) may not be the ideal temperature for the catalytic upgrading of vapors. Often, a large catalyst to biomass ratio is required to achieve a high degree of deoxygenation or a high yield of hydrocarbon products due to the short residence time of the pyrolysis vapor, which simultaneously reduces the effective volume of the pyrolysis reactor. Furthermore, another challenge associated with in situ CFP is that the catalyst is exposed to the multiphase

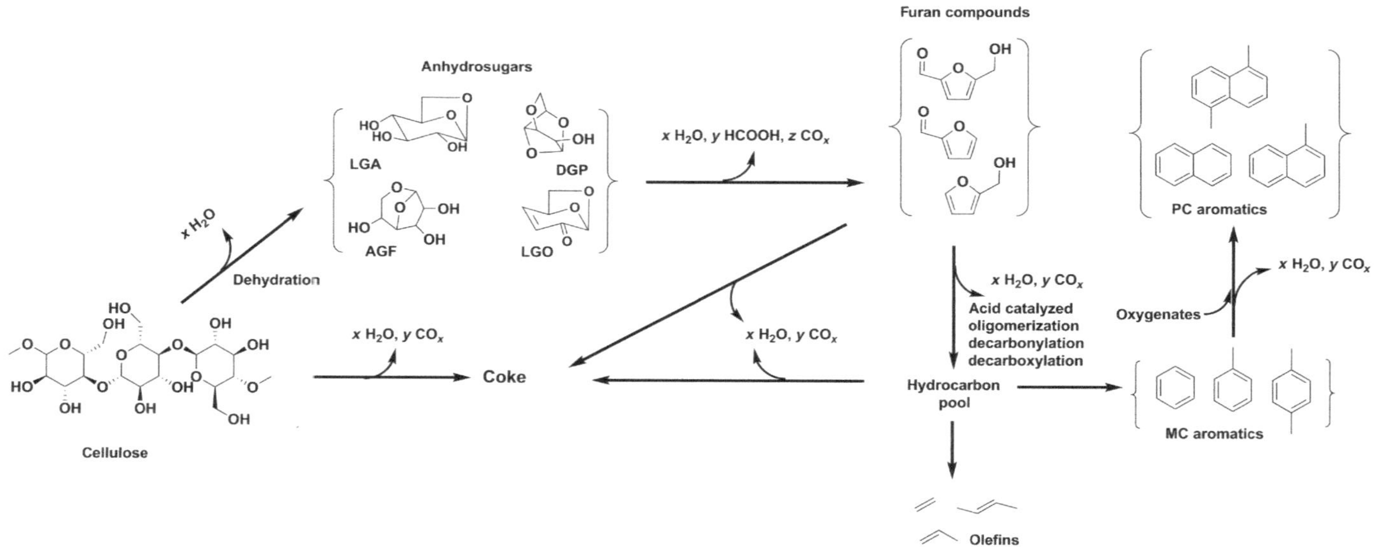

Fig. 42 Overall reaction chemistry for the CFP of cellulose. MC: monocyclic aromatics, PC: polycyclic aromatics. *Adapted from Carlson TR, Cheng YT, Jae J, Huber GW: Production of green aromatics and olefins by catalytic fast pyrolysis of wood sawdust,* Energy Environ Sci *4:145–161, 2011, with permission from Royal Society of Chemistry.*

reacting flows consisting of solid biomass, liquid intermediates, gaseous vapors, and permanent gases, as well as the inherent mineral content of biomass, char, and coke due to the intimate contact/mixing nature of the in situ configuration. The interactions with the multiphase reacting flows may reduce the accessibility of the active sites for oxygenated species to perform the desired chemistry. In particular, the char and coke formed during fast pyrolysis can deposit on the surface of catalysts (e.g., zeolites) and block the pores and active sites, leading to rapid deactivation of the catalysts. Additionally, the mineral matter can not only have a significant effect on the pyrolysis chemistry, for example by forming more char, but also pose a challenge in separation processes in a later stage.

In ex situ CFP, the catalysts are not exposed to the biomass-associated solids (biomass, char, and mineral matter/ash) since a hot gas filter can be applied to remove undesirable solids, making the ex situ configuration a more benign environment for catalyst performance and maintenance. Importantly, the relatively benign environment in a separated reactor allows the catalysts to focus on catalytic chemistries (such as hydrogenation, hydrodeoxygenation, and coupling) and target desired products with less catalyst consumption. Moreover, the independent and flexible control of the catalytic upgrading process can be achieved in an ex situ configuration since it is separated from the initial pyrolysis environment. Reaction parameters such as temperature and residence time can be optimized for biomass pyrolysis and catalytic upgrading separately. Indeed, the flexibility of the ex situ configuration even allows the use of multiple serial catalyst beds. However, higher capital cost due to the use of a hot gas filter for separation and additional reactors for catalytic upgrading, higher operation cost, and longer processing time have precluded this method from achieving more widespread use.

Yildiz et al. (2013) compared in situ CFP and ex situ CFP of pine wood over a heterogeneous ZSM-5-based acidic catalyst with non-CFP. As shown in Table 7, the composition of bio-oil was notably changed by the catalytic treatment. The disappearance of anhydrosugars and aldehydes, decrease in the yields of acids, and the formation of phenols and aromatics were observed for both in situ and ex situ CFP. They concluded that CFP produced bio-oil of higher quality and lower oxygen content than non-CFP, and ex situ CFP consumes less catalyst while in situ CFP produced bio-oil of higher quality (Yildiz et al., 2013).

A technoeconomic assessment of in situ and ex situ CFP by Dutta et al. (2015) concluded that ex situ CFP has less technoeconomic risk than in situ

CFP. Even more importantly, their research demonstrated that production of deoxygenated bio-oil via both in situ and ex situ upgrading of fast pyrolysis vapors followed by hydroprocessing to produce hydrocarbon fuel blendstocks can be cost competitive with petroleum-derived sources (Dutta et al., 2015, 2016). They also comment that a hybrid of in situ configuration using a relatively inexpensive catalyst followed by ex situ upgrading reactors may be economically justifiable (Dutta et al., 2015).

5.2 CFP of Biomass Using Zeolites

This promising technology has led to great efforts in industry and academia in screening and developing catalysts, which can be categorized into several groups including soluble inorganics, metal oxides, microporous materials, mesoporous materials, and supported metal catalysts (Gao et al., 2014). Zeolites are crystalline microporous aluminosilicates have a long history of use and tremendous accomplishment in the catalytic cracking of petroleum into fuels and chemicals. They are also efficient catalysts for aromatic hydrocarbon production with higher selectivity and much less coke production during CFP (Liu et al., 2015; Taarning et al., 2011; Xu et al., 2016). Based on the review of CFP of biomass using different catalysts by Gao et al. (2014), and catalytic conversion of biomass using zeolites by Taarning et al. (2011), here we highlight the salient experimental results of CFP of biomass using zeolite catalysts, with an emphasis on recent experimental efforts. The interested reader is referred to more comprehensive reviews of CFP of biomass by Taarning et al. (2011), Dickerson and Soria (2013), Gao et al. (2014), and Wan and Wang (2014).

5.2.1 In Situ Catalytic Pyrolysis of Biomass Using Zeolites

Naqvi et al. (2015) studied in situ CFP of paddy husk in a drop-type fixed bed pyrolyzer in the presence of a medium-pore framework zeolite (MCM-22) and its delaminated counterpart with a higher external surface area (ITQ-2). The ITQ-2 catalyst produced slightly higher yield of bio-oil with lower oxygenate content and higher aromatic content than the MCM-22 catalyst due to its thinner delaminated structure, more active acidic sites and larger external surface area, facilitating the access of oxygenates to the catalytic sites.

Cole and Lee (2015) developed an analytical platform that couples gas chromatography (GC) to dopant-assisted atmospheric pressure chemical ionization (APCI), and high-resolution time-of-flight mass spectrometry (TOF-MS) to evaluate in situ CFP of cellulose. The use of ZSM-5 catalyst resulted in the formation of fully deoxygenated hydrocarbon compounds (Cole and Lee, 2015), while the use of ZY catalyst converted the highly oxygenated compounds to various oxygenated compounds with lower oxygen content. Zeolites of higher acidity yielded more water and aromatic hydrocarbons. The analytical platform allowed direct chemical composition analysis of GC-separated molecules, regardless of their presence in the database, and identified 142 compounds from the CFP of cellulose in contrast with 38 species by traditional Py-GC-EI-MS analysis.

Liu et al. (2015) studied the effect of operating conditions, and optimized them for the production of aromatic hydrocarbons, xylene, benzene, and toluene, from in situ CFP of duckweed with an HZSM-5 catalyst in a micropyrolyzer. The maximized carbon yields of 27.2 mol% of aromatic hydrocarbon, 6.2 mol% of xylene, 5.5 mol% of benzene, and 8.0 mol% of toluene were obtained. Uemura et al. (2015) compared the production of arenes, which were mainly composed of benzene, toluene, and naphthalene, from in situ CFP of Japanese cedar using different zeolites (HZSM-5, Mordenite, Y-type-1, Y-type-2, and β-Type). The highest carbon yield of 26% was obtained with HZSM-5.

Park et al. (2015) tested HZSM-5 with varying compositional ratios of SiO_2/Al_2O_3 from 23 to 280 for in situ CFP. The HZSM-5 with a lower SiO_2/Al_2O_3 ratio showed a higher catalytic activity because of the increased strength of Brønsted acid sites, which catalyze the conversion of carbohydrates to aromatics. The presence of HZSM-5 considerably enhanced the yield of aliphatic and aromatic hydrocarbons in bio-oil, and significantly reduced the fractions of phenolics and acids from CFP (Park et al., 2015).

In situ CFP experiments of torrefied hemicellulose, lignin, and cellulose over HZSM-5 in a semibatch pyroprobe reactor by Zheng et al. (2015) demonstrated that torrefaction can cause a reduction in the aromatic yield and increase in the BTX selectivities from CFP of torrefied hemicellulose and lignin, while little impact was observed for cellulose.

Jin et al. (2015) reported that HZSM-5 was efficient at increasing the proportion of monoaromatics in the bio-oil with minimal coke production, while Hβ and HY significantly increased the proportion of light phenolics and PAHs from in situ CFP of Geodae–Uksae. The use of three microporous zeolites, HZSM-5, Hβ, and HY, yielded improved bio-oil with more phenolics, monoaromatics, and polycyclic aromatic hydrocarbons (PAHs), and fewer acids, oxygenates, and furans. However, these benefits balance an overall decreased yield of bio-oil and increased yields of CO_2, CO, and light hydrocarbons (C_1–C_4) also caused by catalytic effects promoting deoxygenation and cracking reactions (Jin et al., 2015).

Recently, Anand et al. (2016) tested different zeolites, including ZSM5, Zβ, and ZY, for in situ CFP of spirulina algae using Py-GC/MS (pyrolysis coupled with gas chromatograph/mass spectrometry) at 350–800°C. They revealed that the formation of nitriles, especially C_2–C_4 nitriles, via the dehydration of amides can be significantly promoted by high acidity ZY and Zβ catalysts. Moreover, all the three tested zeolite catalysts promoted the formation of mono aromatics, PAHs, and indoles, yielding BTX, cyclobutane, dimethyl cyclopropane, and acetonitrile as the major organic products. This study also demonstrated that tailoring the reaction conditions (such as temperature, catalyst to biomass ratio, etc.) of CFP can be a potential strategy for selective production of hydrocarbons from microalgae.

A recent study by Hoff et al. (2016) discovered that the crystallinity and framework aluminum site accessibility are critical to achieve high aromatic yields from CFP of biomass. In this work, they demonstrated that amorphous silica-alumina surface species, even present in small concentrations, can greatly impact the diffusion of bulky reactants, lower the amount of Al in framework sites, and consequently, alter the Brønsted acid site density and strength. Based on those findings, a highly crystalline zeolite catalyst with microporous micron-sized crystals was designed, and achieved 32% yield of aromatic hydrocarbons from CFP of cellulose. This was 12% higher than the aromatic hydrocarbon yield obtained for commercial ZSM-5 tested under the same conditions.

5.2.2 Ex Situ Catalytic Pyrolysis of Biomass Using Zeolites

During ex situ CFP, primary vapors from fast pyrolysis are passed over catalysts for upgrading. Horne and Williams (1995) studied the ex situ CFP of wood in a fixed bed of ZSM-5 catalyst at 550°C, producing upgraded bio-oil mainly consisting of light and heavy aromatics, oxygenated phenolics, and neutral compounds. The deactivation of the catalyst became apparent as the pyrolysis time increased, resulting in the decreased yield of monocyclic aromatic hydrocarbons, and more oxygenated products. Nguyen et al. (2013) utilized Na-faujasite and H-faujasite zeolites for CFP of pine wood in both in situ and ex situ modes, and found that the latter method yielded bio-oil of a better quality (i.e., oxygen removal degree) than the former.

Pütün et al. (2009) tested different catalysts including clinoptilolite (a natural zeolite, NZ), ZSM-5, and HY for ex situ upgrading of the pyrolysis vapors with varying catalyst/feed ratios and different catalyst bed temperatures ranging from 350 to 500°C. Natural zeolite was determined to be very effective in aliphatic production, whereas ZSM-5 effectively produced gas and aromatics. HY showed a higher tendency toward the formation of coke and tar. Imran et al. (2014) observed a high extent of deoxygenation in bio-oil with a neutral pH using $Na_2CO_3/\gamma\text{-}Al_2O_3$ catalyst. As a result of complete removal of carboxylic acids, the heating value of bio-oil was high (~36 MJ/kg).

Recently, Wang et al. (2015b) reported that the functionality and molecular size of oxygenates can strongly affect the pathway of oxygen rejection and the formation of hydrocarbons during ex situ CFP. More hydrocarbons and less coke were formed from oxygenates of smaller molecular size, while oxygenates of larger molecular size yielded more external coke due to their relatively lower diffusion rates into pore channels. No significant interactions between glucose-derived oxygenates during CFP over HZSM-5 catalyst were observed in this study.

Mukarakate et al. (2015) studied ex situ CFP over HZSM-5 with added steam. They found that phenols can also be formed by reactions of water molecules with aromatic intermediates generated during biomass pyrolysis rather than simply from lignin decomposition. Additionally, steam also reduced the formation of nonreactive carbon on zeolite, decreasing the deactivation rate of the catalyst. Fig. 43 shows a reaction schematic for the integrated hydroprocessing and zeolitic upgrading of pyrolysis oil by Vispute et al. (2010).

Research efforts are also focused on ex situ catalytic reforming of pyrolysis oil rather than pyrolysis vapors for the production of syn gas (Leijenhorst et al., 2014), and the successive catalytic upgrading of pyrolysis oil (Adjaye and Bakhshi, 1995a,b; Elliott and Wang, 2015; Jackson, 2013; Kastner et al., 2015; Li et al., 2014b; Sharma and Bakhshi, 1993a,b; Vitolo et al., 1999, 2001; Wang et al., 1997, 1998) for the production of hydrocarbons. Li et al. (2014b) demonstrated that the stability and quality of bio-oil derived from the fast pyrolysis of woody biomass can be significantly enhanced by the combination of esterification and hydrotreatment of bio-oil over a Pd/C catalyst, which does not favor the formation of large aromatic ring species and coke. For more information on the catalytic hydrodeoxygenation of pyrolysis bio-oil, interested readers are referred to the review by Elliott (2015).

Recently, Wang et al. (2015a) reported a novel transformation of bio-oil derived from fast pyrolysis of biomass into jet range hydrocarbons. As shown in Fig. 44, the catalytic cracking of bio-oil produced low carbon aromatics

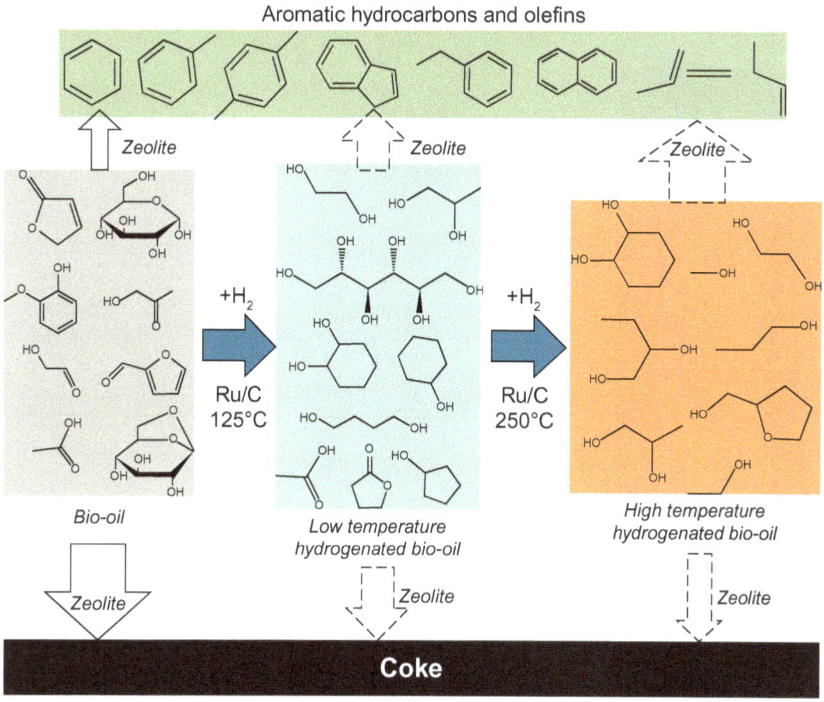

Fig. 43 Reaction schematic for the integrated hydroprocessing and zeolitic upgrading of pyrolysis oil. *Solid arrows*: pyrolysis oil is directly passed over the zeolite catalyst; *long-dashed arrows*: pyrolysis oil is hydrogenated over Ru/C at 398 K and then passed over the zeolite catalyst; *short-dashed arrows*: pyrolysis oil is first hydrogenated over Ru/C at 398 K, then over Pt/C at 523 K, and then passed over the zeolite catalyst. The width of the *vertical arrows* represents the product carbon yield from a particular field. In addition to the product streams shown in the figure, oxygen is removed at the zeolite stage as a mixture of CO, CO_2, and H_2O. Boosting the hydrogen content of the zeolite feed (left to right) increases the thermal stability of the feed, resulting in a reduction in the amount of coke and an increase in the yields of aromatic hydrocarbons and olefins. The addition of hydrogen also raises the proportion of oxygen lost as water relative to CO and CO_2 and thereby further raises the proportion of carbon incorporated into marketable compounds. *Redrawn from Mukarakate C, McBrayer JD, Evans TJ, et al.: Catalytic fast pyrolysis of biomass: the reactions of water and aromatic intermediates produces phenols.* Green Chem *17:4217–4227, 2015; Vispute TP, Zhang H, Sanna A, Xiao R, Huber GW: Renewable chemical commodity feedstocks from integrated catalytic processing of pyrolysis oils,* Science *330:1222–1227, 2010.*

and light olefins, alkylation of which generated C_8–C_{15} aromatic hydrocarbons. Lastly, the hydrogenation of C_8–C_{15} aromatics produced C_8–C_{15} cyclic alkanes.

More recently, Zhang et al. (2016) studied the catalytic conversion of bio-oil model compounds (acetol, methylglyoxal, 5-HMF, acetaldehyde,

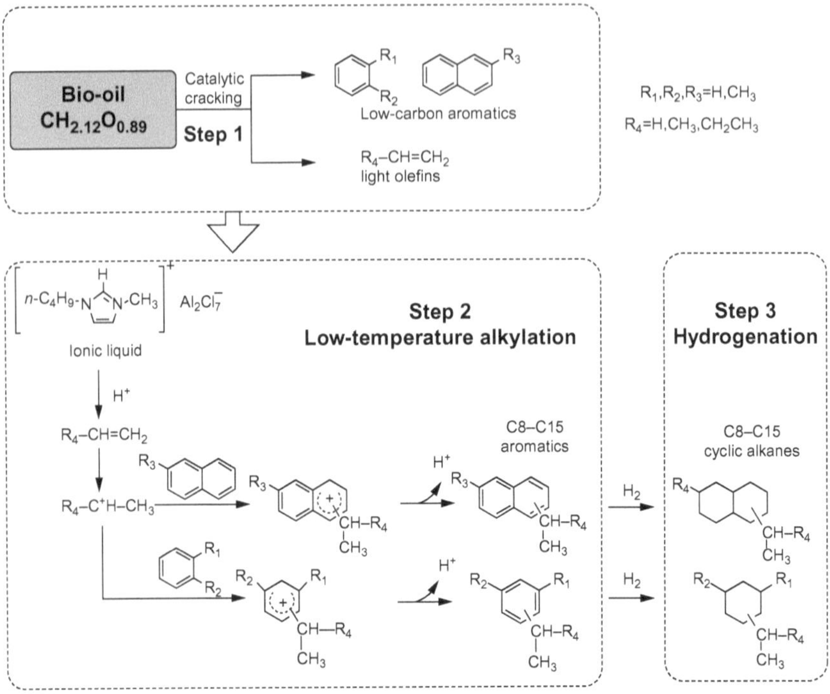

Fig. 44 Reaction routes designed for transformation of bio-oil to jet and diesel fuel range hydrocarbons by Wang et al. (2015a). *This figure is adapted from Wang JC, Bi PY, Zhang YJ, et al.: Preparation of jet fuel range hydrocarbons by catalytic transformation of bio-oil derived from fast pyrolysis of straw stalk,* Energy *86:488–499, 2015a, with permission from Elsevier.*

acetone, methyl vinyl ketone, furfural, 5-methylfurfural) over different acid/base catalysts including ground serpentine, acid-treated serpentine, silica-alumina, sulfated zirconia, and MgO. The acid–base bifunctional MgO/sulfated zirconia catalyst, which contains sulfated zirconia (a strong acid) as the acidic catalyst and MgO as the basic catalyst, displayed the highest activity in aldol condensation reactions, and the best deoxygenation performance. Deoxygenation reactions were promoted by the cooperative catalysis between the closely located acid and base sites.

5.2.3 Modified Zeolites for CFP

Zeolites have been widely studied as catalysts for the CFP of biomass largely due to their ability to deoxygenate bio-oil and form olefins and aromatic hydrocarbons. However, the drawbacks of CFP with zeolites are the low yield of hydrocarbons, carbon loss due to formation of light gases and coke, and rapid catalyst deactivation. Recent efforts have been made to improve hydrocarbon yields and minimize coke formation by modifying zeolite catalysts for CFP of biomass. Li et al. (2014a) reported that carefully controlled desilication of a zeolite can improve the conversion of lignocellulose to valuable aromatic hydrocarbons and decrease the formation of undesired coke. Zhang et al. (2015a,b,c,d) reported a reduction in coke formation by modifying the internal and external acid sites of zeolites. Vichaphund et al. (2013, 2014a,b, 2015) reported that modifying the HZSM-5 catalyst with the addition of metals (Co, Ni, Mo, Ga, and Pd) enhanced the selectivity (up to 97%) toward aromatics, including BTX, while simultaneously reducing the formation of unfavorable naphthalenes, and undesirable oxygenated and N-containing compounds from in situ CFP of Jatropha residues. Rezaei et al. (2015) showed that interesting effects between key zeolite characterization parameters, such as pore structure and acidity, could be exploited to design more efficient catalysts. Liu et al. (2016) reported the addition of CaO into the HZSM-5 catalyst increased both the bio-oil and aromatic yields. An optimal conversion of 35.77 wt.% of feedstock into aromatics was obtained when the CaO to HZSM-5 ratio was 1:4. Xu et al. (2016) designed zeolites to limit the mass diffusion of bulky molecules and prevent the accumulation of carbonaceous coke.

5.3 Other Catalysts for CFP

Other catalysts such as MoO_3 (Nolte et al., 2016), metal phosphide (Griffin et al., 2016), nano-NiO (Chen et al., 2016), molybdenum supported on KIT-5 mesoporous silica (Budhi et al., 2015), MgO, γ-Al_2O_3, and silica-alumina (Choi et al., 2015), CeO_2, Nb_2O_5, SiO_2, high surface area SiO_2, Si-MCM-48, and Al-Fe-MCM48 (Nieva et al., 2015), Pd/C (Ye et al., 2016), TiO_2, CeO_2, and ZrO_2 (Nair and Vinu, 2016b) have also been reported in the literature for hydrodeoxygenation, hydrocarbon, and phenolics production from CFP of biomass.

5.4 CFP With Cofeeding

Many studies have been conducted to improve the outcome of CFP of biomass by cofeeding. Carlson et al. (2011) reported that recycling/cofeeding olefins (primarily ethylene and propylene) into CFP of pine wood sawdust and furan using ZSM-5 as a catalyst increased the yield of aromatics. Zhang et al. (2012) demonstrated the petrochemical yield from the CFP of biomass can be enhanced by cofeeding alcohols that have a high hydrogen to carbon effective ratio. Li et al. (2014c) reported that Diels–Alder reactions of cellulose-derived furans with linear α-olefins greatly enhances aromatic production and reduces coke formation in the catalytic copyrolysis of cellulose/pine wood with plastics. Yang et al. (2015b) reported that steam cofeeding reversibly lowered yields of aromatics, char/coke, and identifiable oxygenated species. The cofeeding of 10–33.3 wt.% LDPE with corn fermentation residues for CFP increased the yield of hydrocarbons from 30.4 to 40.3–52.1 C%, in which BTX, ethylene and propylene account for 66.8–71.1% of the hydrocarbon products (Li et al., 2016). Naik et al. (2015) demonstrated that the coprocessing of hydrodeoxygenated fast pyrolysis oil (HDO) with vacuum gas oil (VGO) over a Pd/Al_2O_3 catalyst led to higher yield of liquefied petroleum gases (LPG), while the coprocessing of fast pyrolysis oil (FPO) with VGO yielded more gasoline and light cycle oil (LCO). Cofeeding of torrefied wood with plastics was reported by Hassan et al. (2016) to increase the production of aromatics from CFP. The introduction of PS resulted in the formation of an additional 20 aromatic hydrocarbon compounds.

5.5 Catalyst Deactivation

A major obstacle of CFP is the rapid deactivation of the catalyst (Carlson et al., 2011; Horne and Williams, 1995; Jae et al., 2014; Mochizuki et al., 2013; Xu et al., 2016; Yildiz et al., 2015). This deactivation is primarily caused by the deposition of coke, which covers the active sites and blocks the pores of the catalyst during CFP. There are two origins of coke: thermal and catalytic (Du et al., 2013). The catalytic coke is usually deposited inside the zeolite channels of the crystal, while the thermal coke is often deposited on the outer surface of the catalyst (meso-, macropores). Characterization of char and coke from the (catalytic) fast pyrolysis of biomass and cellulose showed that neither are graphitic-type carbons (Du et al., 2013). Char possesses a much greater content of carbonyl groups and significantly more Raman defects than coke. Consequently, the formation of char is typically associated with the formation of CO and CO_2 via decarbonylation and decarboxylation. Both coke and char coexist in the solid residue of CFP; however, the coke is the dominant carbon form of the solid residue (ratio of coke to char is 7/3) (Du et al., 2013). Besides coke formation, the deposition of metals, accumulation of ash, and interaction with S-containing and N-containing species can also poison the catalyst or significantly reduce its activity and selectivity.

Progress has been made toward developing strategies that regenerate catalysts in CFP. Jae et al. (2014) reported that spray-dried ZSM-5 catalyst was stable throughout 30 reaction/regeneration cycles for CFP of pine sawdust. However, catalyst regeneration by high temperature oxidative treatment may result in the loss of active (acid) sites, and changes the structure and composition of the catalysts, as noted by Yildiz et al. (2014, 2015).

6. COPYROLYSIS OF SYNTHETIC POLYMERS WITH BIOMASS

The earlier sections have demonstrated the challenges, salient results, and opportunities in understanding both non-CFP and CFP of synthetic polymers and lignocellulosic biomass. Recently, copyrolysis of polymers and biomass has emerged as a popular technique to recover chemicals from both these feedstocks. This technique involves treating oxygen-rich biomass with hydrogen-rich plastics to obtain pyrolysis oil of better quality in terms of lower oxygen content. Importantly, as a result of lower oxygen content, total acid number of the oil decreases, which improves its storage stability. Besides this, the calorific value of the copyrolysis bio-oil is high, viscosity is low, and water content is low compared to pyrolysis bio-oil produced from only biomass (Bhattacharya et al., 2009). The density of the copyrolysis bio-oil lies in between that of bio-oil from individual polymer and biomass pyrolysis. In a modern biorefinery based on CFP, copyrolysis can be a promising approach to produce high quality bio-oil (high H/O ratio) in the initial step so that hydrogen requirement in the subsequent upgradation by catalytic hydrodeoxygenation process is significantly lowered. This technique also paves way to utilize partially segregated MSW, which predominantly contains biomass residues and waste plastics, in fast pyrolysis process.

A number of studies on (fast) copyrolysis of biomasses (such as cellulose, xylan, lignin, switch grass, pine, beech, and spruce woods) with polymers (PS, PE, PP, PET, PVC, and tyres) are available in the literature (Dorado et al., 2014, 2015; Jakab et al., 2001; Li et al., 2014a,b,c; Marin et al., 2002; Martínez et al., 2014; Ojha and Vinu, 2015a,b; Sharypov et al., 2002, 2003; Suriapparao et al., 2014; Zhang et al., 2015f; Zhou et al., 2015). A majority of the studies show that the bio-oil yield and its stability are enhanced by mixing a proportion of polymers with biomass. While independent thermal decomposition behavior is exhibited by the mixture, Sharypov and coworkers (Marin et al., 2002; Sharypov et al., 2002, 2003) showed that biomass:polymer composition and the type of feedstock significantly affects the product yields. Importantly, when the polymer fraction was more in the mixture the product yields were nonadditive, which signifies the interactions between the two components during pyrolysis. As the yield of heavy alkanes increased during copyrolysis, it was thought that the interactions between the polymer and biomass were in the form of free radical exchange. Polyolefins can readily donate hydrogen to biomass radicals to stabilize the degradation products. Similarly, the biomass-derived radicals were expected to react with polymeric intermediates, as high yield of 2-alkenes and low yield of aromatics were observed when the fraction of biomass was more in the mixture (Sharypov et al., 2003). Jakab et al. (2001) studied the effects of wood, cellulose, lignin, and activated charcoal on decomposition of PS and PE, and proposed that the interactions are facilitated by the formation of char from biomass pyrolysis, which initiates hydrogenation reactions of alkadienes. Williams and coworkers (Zhou et al., 2015) observed a positive effect of mixing cellulose, hemicellulose, and lignin with PVC during pyrolysis. They observed that the interactions led to decrease in production of HCl and PAHs, which was attributed to the catalyzing effect of biomass char.

Suriapparao et al. (2014) performed extensive TGA experiments of copyrolysis of cellulose and PP to understand the interaction effects. They evaluated apparent activation energies of thermal decomposition using various methods like First Kissinger method and isoconversional Kissinger–Akahira–Sunose (KAS) method. Even though cellulose and PP decomposed in distinct temperature regimes, it was observed that the activation energy of decomposition of cellulose (158 ± 3 kJ/mol) was not affected by the presence of PP (20–80 wt.%), while the presence of cellulose significantly decreased the activation energy of PP decomposition (206–120 kJ/mol on addition of 20–80 wt.% of cellulose). Furthermore, from KAS method, it was ascertained that the apparent activation energy sharply decreased when PP decomposition was about to begin after completion of cellulose pyrolysis. This prompted them to propose that the primary volatiles generated from cellulose (comprising of levoglucosan, anhydrosugars, and cellulose dimers) might react with the melt phase of PP to form products of interaction. The presence of PP along with cellulose also shifted the exotherm corresponding to char formation to higher temperatures, which shows that char formation can be reduced at nominal pyrolysis temperatures (400–450°C). In order to investigate the specific products of interactions under fast pyrolysis conditions, Ojha and Vinu (2015b) utilized Pyroprobe® reactor coupled with GC/MS and FT-IR (Fourier transform infrared spectroscopy) to probe (a) the decomposition products at different cellulose: PP composition and temperature, and (b) the evolution of major functional groups with pyrolysis time. Linear chain C8–C20 alcohols were observed to be the major decomposition products along with hydrocarbons from PP pyrolysis. A maximum of ~36% alcohols were observed from PP-rich mixture at 600°C. The experimental yields of individual pyrolysate groups did not match the predicted value based on additive rule. Importantly, the estimated heating values of the bio-oil from copyrolysis were in the range of 36–40 MJ/kg. The heating value also exhibited a linear increase with sum total concentration of alcohols, hydrocarbons, and aromatics in the pyrolysate. The presence of PP lowered the time of maximum evolution of major functional groups like C–OH, $-CH_2-$, and C=O, which shows that the reaction timescale is lowered in fast copyrolysis.

While it was believed that linear chain alcohols are formed via simple recombination of hydroxyl radicals from cellulose with primary, secondary, and tertiary alkyl hydrocarbon radicals from PP, it was recently shown via DFT calculations that this is not a possible pathway (Ojha et al., 2016). The energy barrier involved in the cleavage of C–OH bond in cellulose to form hydroxyl radicals is very high (>90 kcal/mol) compared to other concerted reactions involved in cellulose fast pyrolysis (40–60 kcal/mol) (Zhou et al., 2014b). DFT calculations performed using propylene trimer, ˙OH and water molecules showed that the water of dehydration from cellulose pyrolysis (with an activation energy of ca. 50 kcal/mol) readily donates a ˙OH group to PP radicals and increases the availability of hydrogen radicals in the system for H-abstraction reactions (Fig. 45). This reaction had relatively low activation energy of 10–14 kcal/mol based on the nature of PP radical (primary, secondary, or tertiary).

Recent works are also focused on catalytic fast copyrolysis of biomass and polymers, and all the studies show that the presence of HZSM-5 zeolite enhances the yield of aromatic hydrocarbons like BTX, ethyl benzene, and naphthalene derivatives (Dorado et al., 2014, 2015; Li et al., 2014c; Zhang et al., 2015f). Importantly, the mixture of cellulose with LDPE is reported to yield maximum aromatic hydrocarbons (Li et al., 2014c). This was verified using ^{13}C isotopic labeling studies of Dorado et al. (2015) who showed that alkyl benzenes derived from cellulose and PE fast copyrolysis had carbon incorporation from both sources. Moreover, naphthalenes were found to be the products of interaction when PS or PET was copyrolyzed with biomass. The plausible mechanism of formation of BTX from catalytic fast copyrolysis is depicted in Fig. 46.

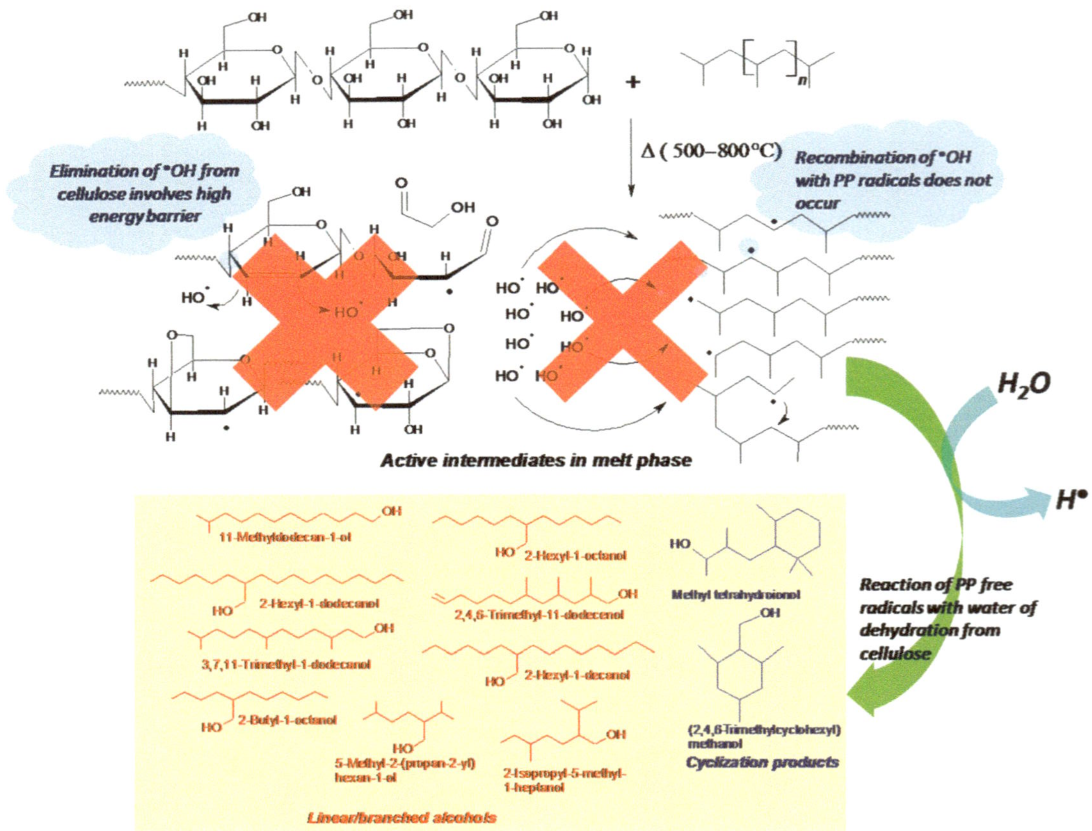

Fig. 45 Major steps involved in the formation of long chain alcohols from fast copyrolysis of cellulose and polypropylene. *Redrawn from Ojha DK, Vinu R: Fast co-pyrolysis of cellulose and polypropylene using Py-GC/MS and Py-FT-IR,* RSC Adv *5:66861–66870, 2015b; Ojha DK, Shukla S, Sachin RS, Vinu R: Understanding the interactions between cellulose and polypropylene during fast co-pyrolysis via experiments and DFT calculations,* Chem Eng Transact *50:67–72, 2016.*

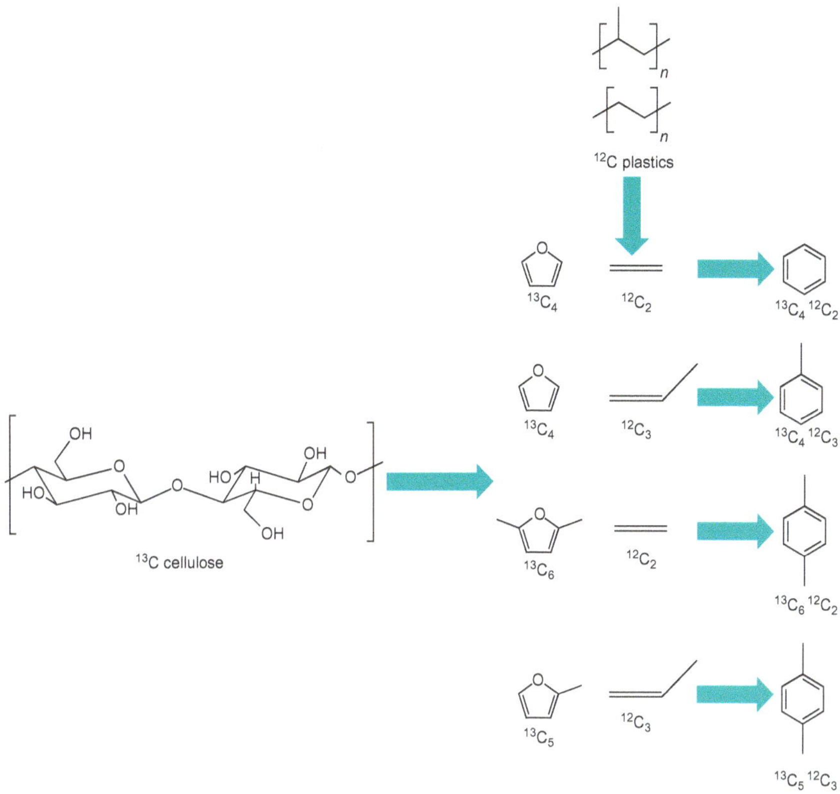

Fig. 46 Major pathways involved in the formation of BTX from catalytic fast copyrolysis of cellulose with polyolefins. *Adapted from Dorado C, Mullen CA, Boateng AA: Origin of carbon in aromatic and olefin products derived from HZSM-5 catalyzed co-pyrolysis of cellulose and plastics via isotopic labeling,* Appl Catal B: Environ *162:338–345, 2015, with permission from Elsevier.*

7. CONCLUSIONS

This review has showcased the importance of understanding the mechanism of pyrolysis of polymers and lignocellulosic biomass. While the mechanism and kinetics of pyrolysis of virgin polyolefins is well established and validated, a better understanding of kinetics is required for pyrolysis of mixed polymers that mimic real plastic wastes (Yu et al., 2016), and catalytic pyrolysis of polymers using shape selective and commercially viable catalysts. In the front of lignocellulosic biomass pyrolysis, significant success is achieved in the development of mechanistic models for cellulose pyrolysis, while similar models for hemicellulose and lignin are emerging. Nevertheless, integrating the models of cellulose, hemicellulose, and lignin pyrolysis by incorporating the interactions between these components, and the intermediates formed from them is a complex task. Moreover, including the catalytic effect of ash and other extractives in the model is a challenge. CFP of biomass that combines fast pyrolysis and catalytic upgrading into one process is an attractive technology for converting lignocellulosic biomass into bio-oil for the production of renewable hydrocarbons (e.g., aromatics, olefins) and chemicals (e.g., phenolics). Developing a fundamental understanding of the underlying chemistry of the CFP process via experiments, quantum chemistry modeling of reaction rates, and mechanistic modeling would guide the refinement of CFP processes. As outlined by Dutta et al. (2015, 2016), advances in catalysis research and development will be necessary to continue to improve the quality and yield of the bio-oil. Specifically, development of catalysts that (a) reduce coke and non-condensable gas formation, (b) induce vapor phase coupling to produce diesel/jet fuel range via deeper hydrodeoxygenation and hydrogenation, and (c) are cheaper with longer lifetimes are essential to achieve better carbon efficiency, and also tailor the product distribution from the CFP of biomass. While the initial results on fast copyrolysis of biomass with polymers are promising, continued fundamental research is important to better understand the interactions between the pyrolysis intermediates from these two different feedstocks.

ACKNOWLEDGMENTS

X.Z. and L.J.B. are grateful for financial support by the National Science Foundation (CBET-1435228), the Institute for Sustainability and Energy at Northwestern (ISEN), and the Department of Energy (DOE) Office of Energy Efficiency and Renewable Energy (EERE) through the Office of Biomass Program, grant number DEEE0003044. The authors thank collaborators at ExxonMobil Research and Engineering Company for useful discussions. R.V. thanks Department of Science and Technology, India (SR/S3/CE/074/2012), and National Center for Combustion Research and Development (NCCRD), IIT Madras, for financial support.

REFERENCES

Abbas-Abadi MS, Haghighi MN, Yeganeh H, McDonald AG: Evaluation of pyrolysis process parameters on polypropylene degradation products, *J Anal Appl Pyrol* 109: 272–277, 2014.

Adam J, Antonakou E, Lappas A, et al: *In situ* catalytic upgrading of biomass derived fast pyrolysis vapours in a fixed bed reactor using mesoporous materials, *Micropor Mesopor Mat* 96:93–101, 2006.

Adjaye JD, Bakhshi NN: Production of hydrocarbons by catalytic upgrading of a fast pyrolysis bio-oil. Part I: conversion over various catalysts, *Fuel Process Technol* 45:161–183, 1995a.
Adjaye JD, Bakhshi NN: Production of hydrocarbons by catalytic upgrading of a fast pyrolysis bio-oil. Part II: comparative catalyst performance and reaction pathways, *Fuel Process Technol* 45:185–202, 1995b.
Adler E: Structural elements of lignin, *Ind Eng Chem* 49:1377–1383, 1957.
Agarwal V, Dauenhauer PJ, Huber GW, Auerbach SM: Ab initio dynamics of cellulose pyrolysis: nascent decomposition pathways at 327 and 600°C, *J Am Chem Soc* 134: 14958–14972, 2012.
Alén R: Structure and chemical composition of wood. In Stenius P, editor: *Forest products chemistry*, Helsinki, 2000, Fapet Oy.
Alonso DM, Wettstein SG, Dumesic JA: Bimetallic catalysts for upgrading of biomass to fuels and chemicals, *Chem Soc Rev* 41:8075–8098, 2012.
Al-Salem SM, Lettieri P, Baeyens J: Recycling and recovery routes of plastic solid waste (PSW): a review, *Waste Manage* 29:2625–2643, 2009.
Alwahabi SM, Froment GF: Single event kinetic modeling of the methanol-to-olefins process on SAPO-34, *Ind Eng Chem Res* 43:5098–5111, 2004.
Anand V, Sunjeev V, Vinu R: Catalytic fast pyrolysis of *Arthrospira platensis* (spirulina) algae using zeolites, *J Anal Appl Pyrol* 118:298–307, 2016.
Anca-Couce A: Reaction mechanisms and multi-scale modelling of lignocellulosic biomass pyrolysis, *Progr Energy Combust Sci* 53:41–79, 2016.
Anca-Couce A, Obernberger I: Application of a detailed biomass pyrolysis kinetic scheme to hardwood and softwood torrefaction, *Fuel* 167:158–167, 2016.
Anca-Couce A, Mehrabian R, Scharler R, Obernberger I: Kinetic scheme of biomass pyrolysis considering secondary charring reactions, *Energ Convers Manage* 87:687–696, 2014.
Antal MJ, Varhegyi G: Cellulose pyrolysis kinetics: the current state of knowledge, *Ind Eng Chem Res* 34:703–717, 1995.
Association of Plastics Manufacturers: Plastics—the Facts 2014/2015. An analysis of European plastics production, demand and waste data (website). http://www.plasticseurope.org/documents/document/20150227150049-final_plastics_the_facts_2014_2015_260215.pdf. Accessed June, 2016.
Babu BV: Biomass pyrolysis: a state-of-the-art review, *Biofuel Bioprod Bior* 2:393–414, 2008.
Baker EG, Elliott DC: Catalytic upgrading of biomass pyrolysis oils. In Bridgwater AV, Kuester JL, editors: *Research in thermochemical biomass conversion*, Netherlands, 1988, Springer.
Barta K, Matson TD, Fettig ML, Scott SL, Iretskii AV, Ford PC: Catalytic disassembly of an organosolv lignin via hydrogen transfer from supercritical methanol, *Green Chem* 12:1640–1647, 2010.
Bauer S, Sorek H, Mitchell VD, Ibáñez AB, Wemmer DE: Characterization of Miscanthus giganteus lignin isolated by ethanol organosolv process under reflux condition, *J Agr Food Chem* 60:8203–8212, 2012.
Benson SW: *Thermochemical kinetics: methods for the estimation of thermochemical data and rate parameters,* ed 2, New York, 1976, John Wiley and Sons.
Bhattacharya P, Steele PH, Hassan EBM, Mitchell B, Ingram L, Pittman CU Jr., : Wood/plastic copyrolysis in an auger reactor: chemical and physical analysis of the products, *Fuel* 88:1251–1260, 2009.
Bocci E, Sisinni M, Moneti M, Vecchione L, Di Carlo A, Villarini M: State of art of small scale biomass gasification power systems: a review of the different typologies, *Energy Procedia* 45:247–256, 2014.
Bockhorn H, Hornung A, Hornung U, Jakobströer P: Modeling of isothermal and dynamic pyrolysis of plastics considering non-homogeneous temperature distribution and detailed degradation mechanism, *J Anal Appl Pyrol* 49:53–74, 1999.

Bond JQ, Upadhye AA, Olcay H, et al: Production of renewable jet fuel range alkanes and commodity chemicals from integrated catalytic processing of biomass, *Energy Environ Sci* 7:1500–1523, 2014.

Bouster C, Vermande P, Veron J: Evolution of the product yield with temperature and molecular weight in the pyrolysis of polystyrene, *J Anal Appl Pyrol* 15:249–259, 1989.

Bozi J, Blazsó M: Catalytic modification of pyrolysis products of nitrogen-containing polymers over Y zeolites, *Greenchem* 11:1638–1645, 2009.

Bradbury AGW, Sakai Y, Shafizadeh F: A kinetic model for pyrolysis of cellulose, *J Appl Polym Sci* 23:3271–3280, 1979.

Bridgwater AV: Review of fast pyrolysis of biomass and product upgrading, *Biomass Bioenergy* 38:68–94, 2012.

Bridgwater AV, Peacocke GVC: Fast pyrolysis processes for biomass, *Renew Sust Energy Rev* 4:1–73, 2000.

Britt PF, Buchanan AC, Cooney MJ, Martineau DR: Flash vacuum pyrolysis of methoxy-substituted lignin model compounds, *J Org Chem* 65:1376–1389, 2000.

Broadbelt LJ, Pfaendtner J: Lexicography of kinetic modeling of complex reaction networks, *AIChE J* 51:2112–2121, 2005.

Broido A, Nelson MA: Char yield on pyrolysis of cellulose, *Combust Flame* 24:263–268, 1975.

Brosse N, Dufour A, Meng XZ, Sun QN, Ragauskas A: Miscanthus: a fast-growing crop for biofuels and chemicals production, *Biofuel Bioprod Bior* 6:580–598, 2012.

Budhi S, Mukarakate C, Iisa K, et al: Molybdenum incorporated mesoporous silica catalyst for production of biofuels and value-added chemicals via catalytic fast pyrolysis, *Green Chem* 17:3035–3046, 2015.

Buragohain B, Mahanta P, Moholkar VS: Biomass gasification for decentralized power generation: the Indian perspective, *Renew Sustain Energ Rev* 14:73–92, 2010.

Burnham AK, Zhou X, Broadbelt LJ: Critical review of the global chemical kinetics of cellulose thermal decomposition, *Energy Fuels* 29:2906–2918, 2015.

Butler E, Devlin G, Meier D, McDonnell K: A review of recent laboratory research and commercial developments in fast pyrolysis and upgrading, *Renew Sust Energy Rev* 15:4171–4186, 2011.

Carlson TR, Cheng YT, Jae J, Huber GW: Production of green aromatics and olefins by catalytic fast pyrolysis of wood sawdust, *Energy Environ Sci* 4:145–161, 2011.

Carpenter D, Westover TL, Czernik S, Jablonski W: Biomass feedstocks for renewable fuel production: a review of the impacts of feedstock and pretreatment on the yield and product distribution of fast pyrolysis bio-oils and vapors, *Green Chem* 16:384–406, 2014.

Carpita NC: Stucture and biogenesis of the cell walls of grasses, *Annu Rev Plant Physiol Plant Mol Biol* 47:445–476, 1996.

Carstensen H-H, Dean AM: Rate constant rules for the automated generation of gas-phase reaction mechanisms, *J Phys Chem A* 113:367–380, 2009.

Carstensen H-H, Dean AM: Development of detailed kinetic models for the thermal conversion of biomass via first principle methods and rate estimation rules. In Nimlos MR, Crowley MF, editors: *Computational modeling in lignocellulosic biofuel production*, 2010, pp 201–243. ACS Symp. Series, Vol. 1052.

Carvalheiro F, Silva-Fernandes T, Duarte L, Gírio F: Wheat straw autohydrolysis: process optimization and products characterization, *Appl Biochem Biotechnol* 153:84–93, 2009.

Chan R, Krieger BB: Modeling of physical and chemical processes during pyrolysis of a large biomass pellet with experimental-verification, *Abstr Pap Am Chem Soc* 186, 1983. 84-Fuel.

Chen J, Liu C, Wu SB: Catalytic fast pyrolysis of alcell lignin with nano-NiO, *Bioresources* 11:663–673, 2016.

Chien JCW, Kiang JKY: Polymer reactions—X. Thermal pyrolysis of poly(isoprene), *Eur Polym J* 15:1059–1065, 1979.
Choi YS, Lee KH, Zhang J, Brown RC, Shanks BH: Manipulation of chemical species in bio-oil using *in situ* catalytic fast pyrolysis in both a bench-scale fluidized bed pyrolyzer and micropyrolyzer, *Biomass Bioenerg* 81:256–264, 2015.
Chu S, Subrahmanyam AV, Huber GW: The pyrolysis chemistry of a beta-O-4 type oligomeric lignin model compound, *Green Chem* 15:125–136, 2013.
Chundawat SPS, Beckham GT, Himmel ME, Dale BE: Deconstruction of lignocellulosic biomass to fuels and chemicals, *Annu Rev Chem Biomol Eng* 2:121–145, 2011.
Clarke J, Deswarte F: The biorefinery concept—an integrated approach. In Clarke J, Deswarte F, editors: *Introduction to chemicals from biomass*, UK, 2008, John Wiley and Sons Ltd.
Cole DP, Lee YJ: Effective evaluation of catalytic deoxygenation for *in situ* catalytic fast pyrolysis using gas chromatography-high resolution mass spectrometry, *J Anal Appl Pyrol* 112:129–134, 2015.
Collard F-X, Blin J: A review on pyrolysis of biomass constituents: mechanisms and composition of the products obtained from the conversion of cellulose, hemicelluloses and lignin, *Renew Sustainable Energy Rev* 38:594–608, 2014.
Cooney JD, Day M, Wiles DM: Thermal degradation of poly(ethylene terephthalate): a kinetic analysis of thermogravimetric data, *J Appl Polym Sci* 28:2887–2902, 1983.
Cozzani V, Nicolella C, Rovatti M, Tognotti L: Influence of gas-phase reactions on the product yields obtained in the pyrolysis of polyethylene, *Ind Eng Chem Res* 36:342–348, 1997.
Crestini C, Melone F, Sette M, Saladino R: Milled wood lignin: a linear oligomer, *Biomacromolecules* 12:3928–3935, 2011.
Curran HJ, Gaffuri P, Pitz WJ, Westbrook CK: A comprehensive modeling study of iso-octane oxidation, *Combust Flame* 129:253–280, 2002.
Czernik S, Elam CC, Evans RJ, Meglen RR, Moens L, Tatsumoto K: Catalytic pyrolysis of nylon-6 to recover caprolactam, *J Anal Appl Pyrol* 46:51–64, 1998.
Dasari A, Yu Z-Z, Cai G-P, Mai Y-M: Recent developments in the fire retardance of polymeric materials, *Prog Polym Sci* 38:1357–1387, 2013.
Debiagi PEA, Pecchi C, Gentile G, et al: Extractives extend the applicability of multistep kinetic scheme of biomass pyrolysis, *Energy Fuels* 29:6544–6555, 2015.
Degenstein JC, Murria P, Easton M, et al: Fast pyrolysis of ^{13}C-labeled cellobioses: gaining insights into the mechanisms of fast pyrolysis of carbohydrates, *J Org Chem* 80:1909–1914, 2015.
Desharnais L, Du F, Brosse N: Optimization of galactoglucomannans and acidic arabinans recovery in softwood, *Ind Eng Chem Res* 50:14217–14220, 2011.
Di Blasi C: Modeling chemical and physical processes of wood and biomass pyrolysis, *Progr Energy Combust Sci* 34:47–90, 2008.
Di Blasi C, Branca C, Galgano A: Biomass screening for the production of furfural via thermal decomposition, *Ind Eng Chem Res* 49:2658–2671, 2010.
Dickerson T, Soria J: Catalytic fast pyrolysis: a review, *Energies* 6:514–538, 2013.
Dimitrov N, Krehula LK, Siročić AP, Murgić ZH: Analysis of recycled PET bottles products by pyrolysis-gas chromatography, *Polym Degrad Stabil* 98:972–979, 2013.
Donaj PJ, Kaminsky W, Buzeto F, Yang W: Pyrolysis of polyolefins for increasing the yield of monomers' recovery, *Waste Manage* 32:840–846, 2012.
Donaldson LA, Knox JP: Localization of cell wall polysaccharides in normal and compression wood of radiata pine: relationships with lignification and microfibril orientation, *Plant Physiol* 158:642–653, 2012.
Dorado C, Mullen CA, Boateng AA: H-ZSM5 catalyzed co-pyrolysis of biomass and plastics, *ACS Sustainable Chem Eng* 2:301–311, 2014.

Dorado C, Mullen CA, Boateng AA: Origin of carbon in aromatic and olefin products derived from HZSM-5 catalyzed co-pyrolysis of cellulose and plastics via isotopic labeling, *Appl Catal B: Environ* 162:338–345, 2015.

Dorez G, Ferry L, Sonnier R, Taguet A, Lopez-Cuesta JM: Effect of cellulose, hemicellulose and lignin contents on pyrolysis and combustion of natural fibers, *J Anal Appl Pyrol* 107:323–331, 2014.

Dorrestijn E, Laarhoven LJJ, Arends IWCE, Mulder P: The occurrence and reactivity of phenoxyl linkages in lignin and low rank coal, *J Anal Appl Pyrol* 54:153–192, 2000.

Du SC, Valla JA, Bollas GM: Characteristics and origin of char and coke from fast and slow, catalytic and thermal pyrolysis of biomass and relevant model compounds, *Green Chem* 15:3214–3229, 2013.

Dumitriu S: *Polysaccharides: structural diversity and functional versatility*, ed 4, New York, 2008, Marcel Dekker.

Dutta A, Sahir A, Tan E, et al: *Process design and economics for the conversion of lignocellulosic biomass to hydrocarbon fuels thermochemical research pathways with* in situ *and* ex situ *upgrading of fast pyrolysis vapors*, Technical Report NREL/TP-5100-62455, PNNL-23823, 2015.

Dutta A, Schaidle JA, Humbird D, Baddour FG, Sahir A: Conceptual process design and techno-economic assessment of *ex situ* catalytic fast pyrolysis of biomass: a fixed bed reactor implementation scenario for future feasibility, *Top Catal* 59:2–18, 2016.

Ebringerová A, Hromádková Z, Heinze T: Hemicellulose. In Heinze T, editor: *Polysaccharides i*, Berlin, Heidelberg, 2005, Springer.

Eckardt NA: Role of xyloglucan in primary cell walls, *Plant Cell* 20:1421–1422, 2008.

Elliott DC: Biofuel from fast pyrolysis and catalytic hydrodeoxygenation, *Curr Opin Chem Eng* 9:59–65, 2015.

Elliott DC, Wang HM: Hydrocarbon liquid production via catalytic hydroprocessing of phenolic oils fractionated from fast pyrolysis of red oak and corn stover, *ACS Sustainable Chem Eng* 3:892–902, 2015.

Faravelli T, Bozzano G, Scassa C, et al: Gas product distribution from polyethylene pyrolysis, *J Anal Appl Pyrol* 52:87–103, 1999.

Faravelli T, Bozzano G, Colombo M, Ranzi E, Dente M: Kinetic modeling of the thermal degradation of polyethylene and polystyrene mixtures, *J Anal Appl Pyrol* 70:761–777, 2003.

Fleischer G, Appel M: Chain length dependence of the self-diffusion of polyisoprene and polybutadiene in the melt, *Macromolecules* 28:7281–7283, 1995.

Freudenberg K: Lignin: its constitution and formation from p-hydroxycinnamyl alcohols, *Science* 148:595–600, 1965.

Gao HL, Liu HT, Pang B, et al: Production of furfural from waste aqueous hemicellulose solution of hardwood over ZSM-5 zeolite, *Bioresour Technol* 172:453–456, 2014.

Gillespie DT: Stochastic simulation of chemical kinetics, *Annu Rev Phys Chem* 58:35–55, 2007.

Glasser WG, Glasser HR, Morohoshi N: Simulation of reactions with lignin by computer (SIMREL). 6. Interpretation of primary experimental analysis data ("analysis program"), *Macromolecules* 14:253–262, 1981.

Graham RG, Bergougnou MA, Overend RP: Fast pyrolysis of biomass, *J Anal Appl Pyrol* 6:95–135, 1984.

Griffin MB, Baddour FG, Habas SE, Ruddy DA, Schaidle JA: Evaluation of silica-supported metal and metal phosphide nanoparticle catalysts for the hydrodeoxygenation of guaiacol under *ex situ* catalytic fast pyrolysis conditions, *Top Catal* 59:124–137, 2016.

Hallac BB, Ragauskas AJ: Analyzing cellulose degree of polymerization and its relevancy to cellulosic ethanol, *Biofuel Bioprod Bior* 5:215–225, 2011.

Hamad K, Kaseem M, Deri F: Recycling of waste from polymeric materials: an overview of the recent works, *Polym Degrad Stabil* 98:2801–2812, 2013.

Hassan E, Elsayed I, Eseyin A: Production high yields of aromatic hydrocarbons through catalytic fast pyrolysis of torrefied wood and polystyrene, *Fuel* 174: 317–324, 2016.

Hilbers TJ, Wang ZH, Pecha B, et al: Cellulose-lignin interactions during slow and fast pyrolysis, *J Anal Appl Pyrol* 114:197–207, 2015.

Hildebrandt D, Glasser D, Hausberger B, Patel B, Glasser BJ: Producing transportation fuels with less work, *Science* 323:1680–1681, 2009.

Hoff TC, Gardner DW, Thilakaratne R, et al: Tailoring ZSM-5 zeolites for the fast pyrolysis of biomass to aromatic hydrocarbons, *ChemSusChem* 9:1473–1482, 2016.

Horne PA, Williams PT: The effect of zeolite ZSM-5 catalyst deactivation during the upgrading of biomass-derived pyrolysis vapours, *J Anal Appl Pyrol* 34:65–85, 1995.

Horton SR, Mohr RJ, Zhang Y, Petrocelli FP, Klein MT: Molecular-level kinetic modeling of biomass gasification, *Energy Fuels* 30:1647–1661, 2016.

Hosoya T, Sakaki S: Levoglucosan formation from crystalline cellulose: importance of a hydrogen bonding network in the reaction, *ChemSusChem* 6:2356–2368, 2013.

Hough B, Pfaendtner J, Schwartz DT: *Application of a semi-detailed kinetic model for lignin fast pyrolysis, 2015, American Institute of Chemical Engineers Annual Meeting*, Utah, 2015, Salt Lake City.

Huber GW, Corma A: Synergies between bio- and oil refineries for the production of fuels from biomass, *Angew Chem Int Ed* 46:7184–7201, 2007.

Huber GW, Iborra S, Corma A: Synthesis of transportation fuels from biomass: chemistry, catalysts, and engineering, *Chem Rev* 106:4044–4098, 2006.

Imran A, Bramer EA, Seshan K, Brem G: High quality bio-oil from catalytic flash pyrolysis of lignocellulosic biomass over alumina-supported sodium carbonate, *Fuel Proc Technol* 127:72–79, 2014.

Jackson MA: Ketonization of model pyrolysis bio-oil solutions in a plug-flow reactor over a mixed oxide of Fe, Ce, and Al, *Energy Fuels* 27:3936–3943, 2013.

Jae J, Tompsett GA, Foster AJ, et al: Investigation into the shape selectivity of zeolite catalysts for biomass conversion, *J Catal* 279:257–268, 2011.

Jae J, Coolman R, Mountziaris TJ, Huber GW: Catalytic fast pyrolysis of lignocellulosic biomass in a process development unit with continual catalyst addition and removal, *Chem Eng Sci* 108:33–46, 2014.

Jakab E, Blazsó M, Faix O: Thermal decomposition of mixtures of vinyl polymers and lignocellulosic materials, *J Anal Appl Pyrol* 58–59:49–62, 2001.

Jarvis MW, Daily JW, Carstensen HH, et al: Direct detection of products from the pyrolysis of 2-phenethyl phenyl ether, *J Phys Chem A* 115:428–438, 2011.

Jin SH, Lee HW, Ryu C, Jeon JK, Park YK: Catalytic fast pyrolysis of Geodae-Uksae 1 over zeolites, *Energy* 81:41–46, 2015.

Kannan P, Shoaibi AA, Srinivasakannan C: Temperature effects of the yield of gaseous olefins from waste polyethylene via flash pyrolysis, *Energy Fuels* 28:3363–3366, 2014.

Kastner JR, Hilten R, Weber J, McFarlane AR, Hargreaves JSJ, Batra VS: Continuous catalytic upgrading of fast pyrolysis oil using iron oxides in red mud, *RSC Adv* 5:29375–29385, 2015.

Kawamoto H, Ryoritani M, Saka S: Different pyrolytic cleavage mechanisms of β-ether bond depending on the side-chain structure of lignin dimers, *J Anal Appl Pyrol* 81: 88–94, 2008.

Kersten S, Garcia-Perez M: Recent developments in fast pyrolysis of ligno-cellulosic materials, *Curr Opin Biotechnol* 24:414–420, 2013.

Kiang JKY, Uden PC, Chien JCW: Polymer reactions—Part VII: thermal pyrolysis of polypropylene, *Polym Degrad Stabil* 2:113–127, 1990.

Kibet J, Khachatryan L, Dellinger B: Molecular products and radicals from pyrolysis of lignin, *Environ Sci Technol* 46:12994–13001, 2012.

Kiefer JH, Wei HC, Kern RD, Wu CH: The high temperature pyrolysis of 1,3-butadiene: heat of formation and rate of dissociation of vinyl radical, *Int J Chem Kinetics* 17:225–253, 1985.
Kilzer FJ, Broido A: Speculation on the nature of cellulose pyrolysis, *Pyrodynamics* 2:151–163, 1965.
Kim KH, Bai X, Brown RC: Pyrolysis mechanisms of methoxy substituted α-O-4 lignin dimeric model compounds and detection of free radicals using electron paramagnetic resonance analysis, *J Anal Appl Pyrol* 110:254–263, 2014.
Klein R, Kelly RD: Combination and disproportionation of allylic radicals at low temperatures, *J Phys Chem* 79:1780–1784, 1975.
Kodera Y, McCoy BJ: Distribution kinetics of radical mechanisms: reversible polymer decomposition, *AICHE J* 43:3205–3214, 1997.
Konar RS: The disproportionation and combination of alkyl radicals, *Int J Chem Kinet* 2:419–422, 1970.
Koufopanos CA, Maschio G, Lucchesi A: Kinetic modeling of the pyrolysis of biomass and biomass components, *Can J Chem Eng* 67:75–84, 1989.
Koufopanos CA, Papayannakos N, Maschio G, Lucchesi A: Modeling of the pyrolysis of biomass particles—studies on kinetics, thermal and heat-transfer effects, *Can J Chem Eng* 69:907–915, 1991.
Krumm C, Pfaendtner J, Dauenhauer PJ: Millisecond pulsed films unify the mechanisms of cellulose fragmentation, *Chem Mater* 28:3108–3114, 2016.
Kruse TM, Woo OS, Broadbelt LJ: Detailed mechanistic modeling of polymer degradation: application to polystyrene, *Chem Eng Sci* 56:971–979, 2001.
Kruse TM, Woo OS, Wong H-W, Khan SS, Broadbelt LJ: Mechanistic modeling of polymer degradation: a comprehensive study of polystyrene, *Macromolecules* 35:7830–7844, 2002.
Kruse TM, Wong HW, Broadbelt LJ: Mechanistic modeling of polymer pyrolysis: polypropylene, *Macromolecules* 36:9594–9607, 2003.
Kruse TM, Levine SE, Wong H-W, et al: Binary mixture pyrolysis of polypropylene and polystyrene: a modeling and experimental study, *J Anal Appl Pyrol* 73:342–354, 2005.
Kumar S, Panda AK, Singh RK: A review on tertiary recycling of high-density polyethylene to fuel, *Resour Conserv Recy* 55:893–910, 2011.
Kunkes EL, Simonetti DA, West RM, Serrano-Ruiz JC, Gärtner CA, Dumesic JA: Catalytic conversion of biomass to monofunctional hydrocarbons and targeted liquid-fuel classes, *Science* 322:417–421, 2008.
Kusema BT, Tonnov T, Maki-Arvela P, et al: Acid hydrolysis of O-acetyl-galactoglucomannan, *Catal Sci Technol* 3:116–122, 2013.
Lebo SE, Gargulak JD, McNally TJ: Lignin. In Kirk-Othmer, editor: *Kirk–Othmer encyclopedia of chemical technology*, New York, 2000, John Wiley & Sons, Inc., pp 2474–2481
Lede J: Cellulose pyrolysis kinetics: an historical review on the existence and role of intermediate active cellulose, *J Anal Appl Pyrol* 94:17–32, 2012.
Lede J: Biomass fast pyrolysis reactors: a review of a few scientific challenges and of related recommended research topics, *Oil Gas Sci Technol: Rev IFP* 68:801–814, 2013.
Lehrle RS, Dadvand N, Parsons IW, Rollinson M, Horn IM, Skinner AR: Pyrolysis-GC-MS used to study the thermal degradation of polymers containing chlorine. III. Kinetics and mechanisms of polychloroprene pyrolysis. Selected ion current plots used to evaluate rate constants for the evolution of HCl and other degradation products, *Polym Degrad Stabil* 70:395–407, 2000.
Leijenhorst EJ, Wolters W, van de Beld B, Prins W: Autothermal catalytic reforming of pine-wood-derived fast pyrolysis oil in a 1.5 kg/h pilot installation: performance of monolithic catalysts, *Energy Fuels* 28:5212–5221, 2014.
Levine SE, Broadbelt LJ: Reaction pathways to dimer in polystyrene pyrolysis: a mechanistic modeling study, *Polym Degrad Stabil* 93:941–951, 2008.

Levine SE, Broadbelt LJ: Detailed mechanistic modeling of high-density polyethylene pyrolysis: low molecular weight product distribution, *Polym Degrad Stabil* 94:810–822, 2009.
Li J, Li XY, Zhou GQ, et al: Catalytic fast pyrolysis of biomass with mesoporous ZSM-5 zeolites prepared by desilication with NaOH solutions, *Appl Catal A: Gen* 470: 115–122, 2014a.
Li X, Gunawan R, Wang Y, et al: Upgrading of bio-oil into advanced biofuels and chemicals part iii Changes in aromatic structure and coke forming propensity during the catalytic hydrotreatment of a fast pyrolysis bio-oil with Pd/C catalyst, *Fuel* 116:642–649, 2014b.
Li XY, Li J, Zhou GQ, et al: Enhancing the production of renewable petrochemicals by co-feeding of biomass with plastics in catalytic fast pyrolysis with ZSM-5 zeolites, *Appl Catal A: Gen* 481:173–182, 2014c.
Li XY, Li GY, Li J, et al: Producing petrochemicals from catalytic fast pyrolysis of corn fermentation residual by-products generated from citric acid production, *Renew Energ* 89:331–338, 2016.
Lin R, White RL: Acid-catalyzed cracking of polystyrene, *J Appl Polym Sci* 63:1287–1298, 1997.
Liu D, Yu Y, Wu H: Evolution of water-soluble and water-insoluble portions in the solid products from fast pyrolysis of amorphous cellulose, *Ind Eng Chem Res* 52:12785–12793, 2013a.
Liu DW, Yu Y, Wu HW: Differences in water-soluble intermediates from slow pyrolysis of amorphous and crystalline cellulose, *Energy Fuels* 27:1371–1380, 2013b.
Liu C, Wang H, Karim AM, Sun J, Wang Y: Catalytic fast pyrolysis of lignocellulosic biomass, *Chem Soc Rev* 43:7594–7623, 2014a.
Liu D, Yu Y, Hayashi J, Moghtaderi B, Wu H: Contribution of dehydration and depolymerization reactions during the fast pyrolysis of various salt-loaded celluloses at low temperatures, *Fuel* 136:62–68, 2014b.
Liu GY, Wright MM, Zhao QL, Brown RC: Catalytic fast pyrolysis of duckweed: effects of pyrolysis parameters and optimization of aromatic production, *J Anal Appl Pyrol* 112:29–36, 2015.
Liu SY, Xie QL, Zhang B, et al: Fast microwave-assisted catalytic co-pyrolysis of corn stover and scum for bio-oil production with CaO and HZSM-5 as the catalyst, *Bioresource Technol* 204:164–170, 2016.
Lyu G, Wu S, Zhang H: Estimation and comparison of bio-oil components from different pyrolysis conditions, *Front Energy Res* 3(28), 2015.
Ma S, Lu J, Gao J: Study of the low temperature pyrolysis of PVC, *Energy Fuels* 16:338–342, 2002.
Madras G, McCoy BJ: Distribution kinetics for polymer mixture degradation, *Ind Eng Chem Res* 38:352–357, 1999.
Madras G, Smith JM, McCoy BJ: Degradation of poly(methyl methacrylate) in solution, *Ind Eng Chem Res* 35:1795–1800, 1996.
Manos G, Garforth A, Dwyer J: Catalytic degradation of high-density polyethylene over different zeolitic structures, *Ind Eng Chem Res* 39:1198–1202, 2000.
Marczewski M, Kaminska E, Marczewska H, Godek M, Rokicki G, Sokołowski J: Catalytic decomposition of polystyrene. The role of acid and basic active centers, *Appl Catal B: Environ* 129:236–246, 2013.
Marin N, Collura S, Sharypov V, et al: Copyrolysis of wood biomass and synthetic polymer mixtures. Part II: characterisation of the liquid phases, *J Anal Appl Pyrol* 65:41–55, 2002.
Marongiu A, Faravelli T, Bozzano G, Dente M, Ranzi E: Thermal degradation of poly(vinyl chloride), *J Anal Appl Pyrol* 70:519–553, 2003.
Marongiu A, Faravelli T, Ranzi E: Detailed kinetic modeling of the thermal degradation of vinyl polymers, *J Anal Appl Pyrol* 78:343–362, 2007.

Martínez JD, Veses A, Mastral AM, et al: Co-pyrolysis of biomass with waste tyres: upgrading of liquid bio-fuel, *Fuel Proc Technol* 119:263–271, 2014.

Matheu DM, Dean AM, Grenda JM, Green WH Jr: Mechanism generation with integrated pressure dependence: a new model for methane pyrolysis, *J Phys Chem A* 107: 8552–8565, 2003.

Mayes HB, Broadbelt LJ: Unraveling the reactions that unravel cellulose, *J Phys Chem A* 116:7098–7106, 2012.

Mayes HB, Nolte MW, Beckham GT, Shanks BH, Broadbelt LJ: The alpha-bet(a) of glucose pyrolysis: computational and experimental investigations of 5-hydroxymethylfurfural and levoglucosan formation reveal implications for cellulose pyrolysis, *ACS Sustainable Chem Eng* 2:1461–1473, 2014a.

Mayes HB, Tian J, Nolte MW, et al: Sodium ion interactions with aqueous glucose: insights from quantum mechanics, molecular dynamics, and experiment, *J Phys Chem B* 118:1990–2000, 2014b.

Mayes HB, Nolte MW, Beckham GT, Shanks BH, Broadbelt LJ: The alpha-bet(a) of salty glucose pyrolysis: computational investigations of 5-hydroxymethylfurfural and levoglucosan formation reveal cellulose pyrolysis catalytic action by sodium ions, *ACS Catal* 5:192–202, 2015.

McCoy BJ, Wang M: Continuous-mixture fragmentation kinetics: particle size reduction and molecular cracking, *Chem Eng Sci* 49:3773–3785, 1994.

Mettler MS, Mushrif SH, Paulsen AD, Javadekar AD, Vlachos DG, Dauenhauer PJ: Revealing pyrolysis chemistry for biofuels production: conversion of cellulose to furans and small oxygenates, *Energy Environ Sci* 5:5414–5424, 2012a.

Mettler MS, Paulsen AD, Vlachos DG, Dauenhauer PJ: The chain length effect in pyrolysis: bridging the gap between glucose and cellulose, *Green Chem* 14:1284–1288, 2012b.

Mettler MS, Vlachos DG, Dauenhauer PJ: Top ten fundamental challenges of biomass pyrolysis for biofuels, *Energy Environ Sci* 5:7797–7809, 2012c.

Miller RS, Bellan J: A generalized biomass pyrolysis model based on superimposed cellulose, hemicellulose and lignin kinetics, *Combust Sci Technol* 126:97–137, 1997.

Mochizuki T, Chen S-Y, Toba M, Yoshimura Y: Pyrolyzer–GC/MS system-based analysis of the effects of zeolite catalysts on the fast pyrolysis of Jatropha husk, *Appl Catal A: Gen* 456:174–181, 2013.

Mohan D, Pittman CU, Steele PH: Pyrolysis of wood/biomass for bio-oil: a critical review, *Energy Fuels* 20:848–889, 2006.

Mukarakate C, McBrayer JD, Evans TJ, et al: Catalytic fast pyrolysis of biomass: the reactions of water and aromatic intermediates produces phenols, *Green Chem* 17:4217–4227, 2015.

Mullen CA, Boateng AA, Mihalcik DJ, Goldberg NM: Catalytic fast pyrolysis of white oak wood in a bubbling fluidized bed, *Energy Fuels* 25:5444–5451, 2011.

Naik DV, Kumar V, Prasad B, et al: Catalytic cracking of jatropha-derived fast pyrolysis oils with VGO and their NMR characterization, *RSC Adv* 5:398–409, 2015.

Nair V, Vinu R: Peroxide-assisted microwave activation of pyrolysis char for adsorption of dyes from waste water, *Bioresour Technol* 216:511–519, 2016a.

Nair V, Vinu R: Production of guaiacols via catalytic fast pyrolysis of alkali lignin using titania, zirconia and ceria, *J Anal Appl Pyrol* 119:31–39, 2016b.

Naqvi SR, Uemura Y, Yusup S, Sugiura Y, Nishiyama N: *In situ* catalytic fast pyrolysis of paddy husk pyrolysis vapors over MCM-22 and ITQ-2 zeolites, *J Anal Appl Pyrol* 114:32–39, 2015.

Nguyen TS, Zabeti M, Lefferts L, Brem G, Seshan K: Catalytic upgrading of biomass pyrolysis vapours using faujasite zeolite catalysts, *Biomass Bioenergy* 48:100–110, 2013.

Nieva ML, Volpe MA, Moyano EL: Catalytic and catalytic free process for cellulose conversion: fast pyrolysis and microwave induced pyrolysis studies, *Cellul* 22:215–228, 2015.

Nimlos MR, Blanksby SJ, Ellison GB, Evans RJ: Enhancement of 1,2-dehydration of alcohols by alkali cations and protons: a model for dehydration of carbohydrates, *J Anal Appl Pyrol* 66:3–27, 2003.

Nolte MW, Zhang J, Shanks BH: *Ex situ* hydrodeoxygenation in biomass pyrolysis using molybdenum oxide and low pressure hydrogen, *Green Chem* 18:134–138, 2016.

Norinaga K, Shoji T, Kudo S, Hayashi J: Detailed chemical kinetic modelling of vapour-phase cracking of multi-component molecular mixtures derived from the fast pyrolysis of cellulose, *Fuel* 103:141–150, 2013.

Norinaga K, Yang H, Tanaka R, et al: A mechanistic study on the reaction pathways leading to benzene and naphthalene in cellulose vapor phase cracking, *Biomass Bioenergy* 69:144–154, 2014.

Nowakowski DJ, Bridgwater AV, Elliott DC, Meier D, de Wild P: Lignin fast pyrolysis: results from an international collaboration, *J Anal Appl Pyrol* 88:53–72, 2010.

Obel N, Neumetzler L, Pauly M: Hemicelluloses and cell expansion. In Verbelen JP, Vissenberg K, editors: *The expanding cell*, Berlin, Heidelberg, 2007, Springer.

Ochoa-Villarreal M, Aispuro-Hernández E, Martínez-Téllez MA, Vargas-Arispuro I: Plant cell wall polymers: Function, structure and biological activity of their derivatives. In Gomes ADS, editor: *Polymerization*, Rijeka, Croatia, 2012, InTech.

Ojha DK, Vinu R: Fast co-pyrolysis of cellulose and polypropylene using Py-GC/MS and Py-FT-IR, *RSC Adv* 5:66861–66870, 2015b.

Ojha DK, Vinu R: Resource recovery via catalytic fast pyrolysis of polystyrene using zeolites, *J Anal Appl Pyrol* 113:349–359, 2015a.

Ojha DK, Shukla S, Sachin RS, Vinu R: Understanding the interactions between cellulose and polypropylene during fast co-pyrolysis via experiments and DFT calculations, *Chem Eng Transact* 50:67–72, 2016.

Olazar M, Arabiourrutia M, López G, Aguado R, Bilbao J: Effect of acid catalysts on scrap tyre pyrolysis under fast heating conditions, *J Anal Appl Pyrol* 82:199–204, 2008.

Paine JB III, Pithawalla YB, Naworal JD, Thomas CE: Carbohydrate pyrolysis mechanisms from isotopic labeling. Part 1: the pyrolysis of glycerin: discovery of competing fragmentation mechanisms affording acetaldehyde and formaldehyde and the implications for carbohydrate pyrolysis, *J Anal Appl Pyrol* 80:297–311, 2007.

Paine JB III, Pithawalla YB, Naworal JD: Carbohydrate pyrolysis mechanisms from isotopic labeling. Part 2. The pyrolysis of D-glucose: general disconnective analysis and the formation of C_1 and C_2 carbonyl compounds by electrocyclic fragmentation mechanisms, *J Anal Appl Pyrol* 82:10–41, 2008a.

Paine JB III, Pithawalla YB, Naworal JD: Carbohydrate pyrolysis mechanisms from isotopic labeling: part 3. The pyrolysis of D-glucose: formation of C_3 and C_4 carbonyl compounds and a cyclopentenedione isomer by electrocyclic fragmentation mechanisms, *J Anal Appl Pyrol* 82:42–69, 2008b.

Paine JB III, Pithawalla YB, Naworal JD: Carbohydrate pyrolysis mechanisms from isotopic labeling: part 4. The pyrolysis of D-glucose: the formation of furans, *J Anal Appl Pyrol* 83:37–63, 2008c.

Pandey MP, Kim CS: Lignin depolymerization and conversion: a review of thermochemical methods, *Chem Eng Technol* 34:29–41, 2011.

Park YK, Yoo ML, Jin SH, Park SH: Catalytic fast pyrolysis of waste pepper stems over HZSM-5, *Renew Energ* 79:20–27, 2015.

Patwardhan PR, Satrio JA, Brown RC, Shanks BH: Product distribution from fast pyrolysis of glucose-based carbohydrates, *J Anal Appl Pyrol* 86:323–330, 2009.

Patwardhan PR, Satrio JA, Brown RC, Shanks BH: Influence of inorganic salts on the primary pyrolysis products of cellulose, *Bioresour Technol* 101:4646–4655, 2010.

Patwardhan PR, Brown RC, Shanks BH: Product distribution from the fast pyrolysis of hemicellulose, *ChemSusChem* 4:636–643, 2011a.

Patwardhan PR, Brown RC, Shanks BH: Understanding the fast pyrolysis of lignin, *ChemSusChem* 4:1629–1636, 2011b.

Patwardhan PR, Dalluge DL, Shanks BH, Brown RC: Distinguishing primary and secondary reactions of cellulose pyrolysis, *Bioresour Technol* 102:5265–5269, 2011c.

Paulsen AD, Mettler MS, Dauenhauer PJ: The role of sample dimension and temperature in cellulose pyrolysis, *Energy Fuels* 27:2126–2134, 2013.

Paulsen AD, Hough BR, Williams CL, et al: Fast pyrolysis of wood for biofuels: spatiotemporally resolved diffuse reflectance in situ spectroscopy of particles, *ChemSusChem* 7:652, 2014.

Pecha B, Garcia-Perez M: Chapter 26—Pyrolysis of lignocellulosic biomass: oil, char, and gas. In Dahiya A, editor: *Bioenergy*, Boston, 2015, Academic Press.

Pelzer AW, Sturgeon MR, Yanez AJ, et al: Acidolysis of α-O-4 aryl-ether bonds in lignin model compounds: a modeling and experimental study, *ACS Sustainable Chem Eng* 3:1339–1347, 2015.

Peng Y, Wu S: The structural and thermal characteristics of wheat straw hemicellulose, *J Anal Appl Pyrol* 88:134–139, 2010.

Pettersen RC: The chemical composition of wood. In *The chemistry of solid wood*, Washington, DC, 1984, American Chemical Society, pp 1–56.

Pfaendtner J, Broadbelt LJ: Mechanistic modeling of lubricant degradation. 1. Structure–reactivity relationships for free-radical oxidation, *Ind Eng Chem Res* 47:2886–2896, 2008.

Ponder GR, Richards GN: Pyrolysis of some ^{13}C-labeled glucans: a mechanistic study, *Carbohydr Res* 244:27–47, 1993.

Ponder GR, Richards GN, Stevenson TT: Influence of linkage position and orientation in pyrolysis of polysaccharides: a study of several glucans, *J Anal Appl Pyrol* 22:217–229, 1992.

Popper ZA, Fry SC: Primary cell wall composition of pteridophytes and spermatophytes, *New Phytol* 164:165–174, 2004.

Poutsma ML: Fundamental reactions of free radicals relevant to pyrolysis reactions, *J Anal Appl Pyrol* 54:5–35, 2000.

Poutsma ML: Reexamination of the pyrolysis of polyethylene: data needs, free-radical mechanistic considerations, and thermochemical kinetic simulation of initial product-forming pathways, *Macromolecules* 36:8931–8957, 2003.

Poutsma ML: Comparison of literature models for volatile product formation from the pyrolysis of poluisobutylene at mild conditions: data analysis, free-radical mechanistic considerations, and simulation of initial product-forming pathways, *J Anal Appl Pyrol* 73:159–203, 2005.

Poutsma ML: Further considerations of the sources of the volatiles from pyrolysis of polystyrene, *Polym Degrad Stabil* 94:2055–2064, 2009.

Pütün E, Uzun BB, Pütün AE: Rapid pyrolysis of olive residue. 2. Effect of catalytic upgrading of pyrolysis vapors in a two-stage fixed-bed reactor, *Energy Fuels* 23:2248–2258, 2009.

Qu TT, Guo WJ, Shen LH, Xiao J, Zhao K: Experimental study of biomass pyrolysis based on three major components: hemicellulose, cellulose, and lignin, *Ind Eng Chem Res* 50:10424–10433, 2011.

Radlein D, Piskorz J, Scott DS: Proceedings of the 9th international conference on fundamentals aspects, analytical techniques, processes and applications of pyrolysisfast pyrolysis of natural polysaccharides as a potential industrial process, *J Anal Appl Pyrol* 19:41–63, 1991.

Ranzi E, Cuoci A, Faravelli T, et al: Chemical kinetics of biomass pyrolysis, *Energy Fuels* 22:4292–4300, 2008.

Regalbuto JR: Cellulosic biofuels—got gasoline? *Science* 325:822–824, 2009.

Rezaei PS, Shafaghat H, Daud WMAW: Suppression of coke formation and enhancement of aromatic hydrocarbon production in catalytic fast pyrolysis of cellulose over different zeolites: effects of pore structure and acidity, *RSC Adv* 5:65408–65414, 2015.

Richards GN: Glycolaldehyde from pyrolysis of cellulose, *J Anal Appl Pyrol* 10:251–255, 1987.

Rossi M, King KD, Golden DM: The equilibrium constant and rate constant for allyl radical recombination in the gas phase, *J Am Chem Soc* 101:1223–1230, 1978.

Ruddy DA, Schaidle JA, Ferrell JR III, Wang J, Moens L, Hensley JE: Recent advances in heterogeneous catalysts for bio-oil upgrading via "ex situ catalytic fast pyrolysis": catalyst development through the study of model compounds, *Green Chem* 16:454–490, 2014.

Sabbe MK, Vandeputte AG, Reyniers M-F, Waroquier M, Marin GB: Modeling the influence of resonance stabilization on the kinetics of hydrogen abstractions, *Phys Chem Chem Phys* 12:1278–1298, 2010.

Sakakibara A: A structural model of softwood lignin, *Wood Sci Technol* 14:89–100, 1980.

Santella C, Cafiero L, Angelis DD, Marca FL, Tuffi R, Ciprioti SV: Thermal and catalytic pyrolysis of a mixture of plastics from small waste electrical and electronic equipment (WEEE), *Waste Manage* 54:143–152, 2016.

Scheller HV, Ulvskov P: Hemicelluloses, *Annu Rev Plant Biol* 61:263–289, 2010.

Schlosberg RH, Szajowski PF, Dupre GD, et al: Pyrolysis studies of organic oxygenates, *Fuel* 62:690–694, 1983.

Schreck VA, Serelis AK, Solomon DH: Self-reactions of 1,3-diphenylpropyl and 1,3,5-triphenylpentyl radicals: models for termination in styrene polymerization, *Aust J Chem* 42:375–393, 1989.

Scott DS, Czernik SR, Piskorz J, Radlein DSAG: Fast pyrolysis of plastic wastes, *Energy Fuels* 4:407–411, 1990.

Şerbănescu C: Kinetic analysis of cellulose pyrolysis: a short review, *Chem Pap* 68:1–14, 2014.

Serrano DP, Aguado J, Rodríguez JM, Peral A: Catalytic cracking of polyethylene over nanocrystalline HZSM-5: catalyst deactivation and regeneration study, *J Anal Appl Pyrol* 79:456–464, 2007.

Seshadri V, Westmoreland PR: Concerted reactions and mechanism of glucose pyrolysis and implications for cellulose kinetics, *J Phys Chem A* 116:11997–12013, 2012.

Sharma RK, Bakhshi NN: Catalytic conversion of fast pyrolysis oil to hydrocarbon fuels over HZSM-5 in a dual reactor system, *Biomass Bioenerg* 5:445–455, 1993a.

Sharma RK, Bakhshi NN: Catalytic upgrading of fast pyrolysis oil over HZSM-5, *Can J Chem Eng* 71:383–391, 1993b.

Sharma RK, Bakhshi NN: Catalytic upgrading of pyrolysis oil, *Energy Fuels* 7:306–314, 1993c.

Sharypov VI, Marin N, Beregovtsova NG, et al: Co-pyrolysis of wood biomass and synthetic polymer mixtures. Part I: influence of experimental conditions on the evolution of solids, liquids and gases, *J Anal Appl Pyrol* 64:15–28, 2002.

Sharypov VI, Beregovtsova NG, Kuznetsov BN: Co-pyrolysis of wood biomass and synthetic polymers mixtures. Part III: characterisation of heavy products, *J Anal Appl Pyrol* 67:325–340, 2003.

Shen DK, Gu S, Bridgwater AV: Study on the pyrolytic behaviour of xylan-based hemicellulose using TG-FTIR and Py-GC-FTIR, *J Anal Appl Pyrol* 87:199–206, 2010.

Shen C, Zhang IY, Fu G, Xu X: Pyrolysis of D-glucose to acrolein, *Chin J Chem Phys* 24:249–252, 2011.

Shie JL, Chen YH, Chang CY, Lin JP, Lee DJ, Wu CH: Thermal pyrolysis of poly(vinyl alcohol) and its major products, *Energy Fuels* 16:109–118, 2002.

Simon CM, Kaminsky W: Chemical recycling of polytetrafluoroethylene by pyrolysis, *Polym Degrad Stabil* 62:1–7, 1998.

Sluiter A, Ruiz R, Scarlata C, Sluiter J, Templeton D: *Determination of extractives in biomass, laboratory analytical procedure (LAP),* Technical Report NREL/TP-510-42619, 2008.

Smolders K, Baeyens J: Thermal degradation of PMMA in fluidized beds, *Waste Manage* 24:49–857, 2004.

Sterling WJ, McCoy BJ: Distribution kinetics of thermolytic macromolecular reactions, *AIChE J* 47:2289–2303, 2001.

Stewart JJ, Akiyama T, Chapple C, Ralph J, Mansfield SD: The effects on lignin structure of overexpression of ferulate 5-hydroxylase in hybrid poplar, *Plant Physiol* 150:621–635, 2009.

Sugimura Y, Nagaya T, Tsuge T: Pyrolysis–gas chromatographic studies on head-to-head polystyrene, *Macromolecules* 14:520–523, 1981.

Sun X-F, Sun R, Fowler P, Baird MS: Extraction and characterization of original lignin and hemicelluloses from wheat straw, *J Agr Food Chem* 53:860–870, 2005.

Suriapparao DV, Ojha DK, Ray T, Vinu R: Kinetic analysis of co-pyrolysis of cellulose and polypropylene, *J Therm Anal Calorim* 117:1441–1451, 2014.

Taarning E, Osmundsen CM, Yang XB, Voss B, Andersen SI, Christensen CH: Zeolite-catalyzed biomass conversion to fuels and chemicals, *Energy Environ Sci* 4:793–804, 2011.

Taherzadeh MJ, Eklund R, Gustafsson L, Niklasson C, Liden G: Characterization and fermentation of dilute-acid hydrolyzates from wood, *Ind Eng Chem Res* 36:4659–4665, 1997.

Teixeira AR, Mooney KG, Kruger JS, et al: Aerosol generation by reactive boiling ejection of molten cellulose, *Energy Environ Sci* 4:4306–4321, 2011.

The Statistics Portal (website). http://www.statista.com/statistics/282732/global-production-of-plastics-since-1950/. Accessed November, 2015.

Theander O: Cellulose, hemicellulose and extractives. In Overend RP, Milne TA, Mudge LK, editors: *Fundamentals of thermochemical biomass conversion*, Netherlands, 1985, Springer, pp 35–60.

Train PM, Klein MT: Chemical modeling of lignin. In *Pyrolysis oils from biomass*, , Washington, DC, 1988, American Chemical Society, pp 241–263.

Tsuge S, Ohtani H: Structural charcaterization of polymeric materials by pyrolysis-GC/MS, *Polym Degrad Stabil* 58:109–130, 1997.

Tsuge S, Ohtani H, Watanabe C: *Pyrolysis-GC/MS handbook of synthetic polymers,* Oxford, UK, 2011, Elsevier.

Uemura K, Appari S, Kudo S, Hayashi J, Einaga H, Norinaga K: *In-situ* reforming of the volatiles from fast pyrolysis of ligno-cellulosic biomass over zeolite catalysts for aromatic compound production, *Fuel Process Technol* 136:73–78, 2015.

van Parijs FRD, Morreel K, Ralph J, Boerjan W, Merks RMH: Modeling lignin polymerization. I. Simulation model of dehydrogenation polymers, *Plant Physiol* 153:1332–1344, 2010.

Vanholme R, Demedts B, Morreel K, Ralph J, Boerjan W: Lignin biosynthesis and structure, *Plant Physiol* 153:895–905, 2010.

Várhegyi G, Antal MJ Jr, Jakab E, Szabó P: Kinetic modeling of biomass pyrolysis, *J Anal Appl Pyrol* 42:73–87, 1997.

Vassilev SV, Baxter D, Andersen LK, Vassileva CG: An overview of the chemical composition of biomass, *Fuel* 89:913–933, 2010.

Venderbosch RH, Prins W: Fast pyrolysis technology development, *Biofuel Bioprod Bior* 4:178–208, 2010.

Vichaphund S, Aht-Ong D, Sricharoenchaikul V, Atong D: Catalytic upgrading of Jatropha waste fast pyrolysis vapors over synthesized HZSM-5 using analytical Py-GC/MS, *J Biobased Mater Bio* 7:252–258, 2013.

Vichaphund S, Aht-Ong D, Sricharoenchaikul V, Atong D: Characteristic of fly ash derived-zeolite and its catalytic performance for fast pyrolysis of Jatropha waste, *Environ Technol* 35:2254–2261, 2014a.

Vichaphund S, Aht-Ong D, Sricharoenchaikul V, Atong D: Effect of synthesis time on physical properties and catalytic activities of synthesized HZSM-5 on the fast pyrolysis of Jatropha waste, *Res Chem Intermediat* 40:2395–2406, 2014b.

Vichaphund S, Aht-ong D, Sricharoenchaikul V, Atong D: Production of aromatic compounds from catalytic fast pyrolysis of Jatropha residues using metal/HZSM-5 prepared by ion-exchange and impregnation methods, *Renew Energ* 79:28–37, 2015.

Vinu R, Broadbelt LJ: A mechanistic model of fast pyrolysis of glucose-based carbohydrates to predict bio-oil composition, *Energy Environ Sci* 5:9808–9826, 2012a.

Vinu R, Broadbelt LJ: Unraveling reaction pathways and specifying reaction kinetics for complex systems, *Annu Rev Chem Biomol Eng* 3:29–54, 2012b.

Vinu R, Levine SE, Wang L, Broadbelt LJ: Detailed mechanistic modeling of poly(styrene peroxide) pyrolysis using kinetic Monte Carlo simulation, *Chem Eng Sci* 69:456–471, 2012.

Vinu R, Ojha DK, Nair V: Polymer pyrolysis for resource recovery. In Reedijk J, editor: *Elsevier Reference Module in Chemistry, Molecular Sciences and Chemical Engineering*, Waltham, MA, 2016, Elsevier. http://dx.doi.org/10.1016/B978-0-12-409547-2.11641-5.

Vispute TP, Zhang H, Sanna A, Xiao R, Huber GW: Renewable chemical commodity feedstocks from integrated catalytic processing of pyrolysis oils, *Science* 330:1222–1227, 2010.

Vitolo S, Seggiani M, Frediani P, Ambrosini G, Politi L: Catalytic upgrading of pyrolytic oils to fuel over different zeolites, *Fuel* 78:1147–1159, 1999.

Vitolo S, Bresci B, Seggiani M, Gallo MG: Catalytic upgrading of pyrolytic oils over HZSM-5 zeolite: behaviour of the catalyst when used in repeated upgrading–regenerating cycles, *Fuel* 80:17–26, 2001.

Vuori AI, Bredenberg JB: Thermal chemistry pathways of substituted anisoles, *Ind Eng Chem Res* 26:359–365, 1987.

Vyazovkin S, Burnham AK, Criado JM, Pérez-Maqueda LA, Popescu C, Sbirrazzuoli N: ICTAC Kinetics Committee recommendations for performing kinetic computations on thermal analysis data, *Thermochim Acta* 520:1–19, 2011.

Wan SL, Wang Y: A review on *ex situ* catalytic fast pyrolysis of biomass, *Front Chem Sci Eng* 8:280–294, 2014.

Wang D, Czernik S, Montane D, Mann M, Chornet E: Biomass to hydrogen via fast pyrolysis and catalytic steam reforming of the pyrolysis oil or its fractions, *Ind Eng Chem Res* 36:1507–1518, 1997.

Wang DN, Czernik S, Chornet E: Production of hydrogen from biomass by catalytic steam reforming of fast pyrolysis oils, *Energy Fuels* 12:19–24, 1998.

Wang S, Liang T, Ru B, Guo X: Mechanism of xylan pyrolysis by Py-GC/MS, *Chem Res Chin Univ* 29:782–787, 2013a.

Wang ZH, McDonald AG, Westerhof RJM, et al: Effect of cellulose crystallinity on the formation of a liquid intermediate and on product distribution during pyrolysis, *J Anal Appl Pyrol* 100:56–66, 2013b.

Wang Z, Pecha B, Westerhof RJM, et al: Effect of cellulose crystallinity on solid/liquid phase reactions responsible for the formation of carbonaceous residues during pyrolysis, *Ind Eng Chem Res* 53:2940–2955, 2014a.

Wang ZH, Zhou S, Pecha B, Westerhof RJM, Garcia-Perez M: Effect of pyrolysis temperature and sulfuric acid during the fast pyrolysis of cellulose and Douglas fir in an atmospheric pressure wire mesh reactor, *Energy Fuels* 28:5167–5177, 2014b.

Wang JC, Bi PY, Zhang YJ, et al: Preparation of jet fuel range hydrocarbons by catalytic transformation of bio-oil derived from fast pyrolysis of straw stalk, *Energy* 86:488–499, 2015a.

Wang K, Zhang J, Brent SH, Brown RC: Catalytic conversion of carbohydrate-derived oxygenates over HZSM-5 in a tandem micro-reactor system, *Green Chem* 17:557–564, 2015b.

Wang SR, Ru B, Dai GX, Sun WX, Qiu KZ, Zhou JS: Pyrolysis mechanism study of minimally damaged hemicellulose polymers isolated from agricultural waste straw samples, *Bioresour Technol* 190:211–218, 2015c.

Westerhout RWJ, Waanders J, Kuipers JAM, van Swaaij WMP: Kinetics of the low-temperature pyrolysis of polyethene, polypropene, and polystyrene modeling, experimental determination, and comparison with literature models and data, *Ind Eng Chem Res* 36:1955–1964, 1997a.

Westerhout RWJ, Balk RHP, Meijer R, Kuipers JAM, van Swaaij WMP: Examination and evaluation of the use of screen heaters for the measurement of the high temperature pyrolysis kinetics of polyethene and polypropene, *Ind Eng Chem Res* 36:3360–3368, 1997b.

Westerhout RWJ, Kuipers JAM, van Swaaij WMP: Experimental determination of the yield of pyrolysis products of polyethene and polypropene. Influence of reaction conditions, *Ind Eng Chem Res* 37:841–847, 1998.

White JE, Catallo WJ, Legendre BL: Biomass pyrolysis kinetics: a comparative critical review with relevant agricultural residue case studies, *J Anal Appl Pyrol* 91:1–33, 2011.

Willfor S, Sundberg A, Hemming J, Holmbom B: Polysaccharides in some industrially important softwood species, *Wood Sci Technol* 39:245–258, 2005.

Williams PT, Nugranad N: Comparison of products from the pyrolysis and catalytic pyrolysis of rice husks, *Energy* 25:493–513, 2000.

Wong H-W, Broadbelt LJ: Tertiary resource recovery from waste polymers via pyrolysis: neat and binary mixture reactions of polypropylene and polystyrene, *Ind Eng Chem Res* 40:4716–4723, 2001.

Worldwatch Institute, Washington, DC: *Global plastic production rises, recycling lags* (website). http://www.worldwatch.org/global-plastic-production-rises-recycling-lags-0. Accessed June, 2016.

Xu MZ, Mukarakate C, Robichaud DJ, Nimlos MR, Richards RM, Trewyn BG: Elucidating zeolite deactivation mechanisms during biomass catalytic fast pyrolysis from model reactions and zeolite syntheses, *Top Catal* 59:73–85, 2016.

Xue X, Fry SC: Evolution of mixed-linkage (1→3, 1→4)-β-D-glucan (MLG) and xyloglucan in equisetum (horsetails) and other monilophytes, *Ann Bot* 109:873–886, 2012.

Yanez AJ, Broadbelt LJ: Exploring diverse lignin populations: complex structural representations and fast pyrolysis modeling, 2015, *American Institute of Chemical Engineers Annual Meeting*, Utah, 2015a, Salt Lake City.

Yanez AJ, Broadbelt LJ: Toward mechanistic modeling of lignin deconstruction: a stochastic method to produce libraries of structural representations of lignin (gce180), *19th Annual Green Chemistry & Engineering Conference*, Boston, 2015b, MA.

Yanez AJ, Li W, Mabon R, Broadbelt LJ: A stochastic method to generate libraries of structural representations of lignin, *Energy Fuels* 30:5835–5845, 2016.

Yang H, Appari S, Kudo S, Hayashi J, Kumagai S, Norinaga K: Chemical structures and primary pyrolysis characteristics of lignins obtained from different preparation methods, *J Jpn Inst Energy* 93:986–994, 2014.

Yang H, Appari S, Kudo S, Hayashi J, Norinaga K: Detailed chemical kinetic modeling of vapor-phase reactions of volatiles derived from fast pyrolysis of lignin, *Ind Eng Chem Res* 54:6855–6864, 2015a.

Yang H, Coolman RJ, Karanjkar P, et al: The effect of steam on the catalytic fast pyrolysis of cellulose, *Green Chem* 17:2912–2923, 2015b.

Ye XN, Lu Q, Li WT, et al: Selective production of nicotyrine from catalytic fast pyrolysis of tobacco biomass with Pd/C catalyst, *J Anal Appl Pyrol* 117:88–93, 2016.

Yildiz G, Pronk M, Djokic M, et al: Validation of a new set-up for continuous catalytic fast pyrolysis of biomass coupled with vapour phase upgrading, *J Anal Appl Pyrol* 103: 343–351, 2013.

Yildiz G, Lathouwers T, Toraman HE, et al: Catalytic fast pyrolysis of pine wood: effect of successive catalyst regeneration, *Energy Fuels* 28:4560–4572, 2014.

Yildiz G, Ronsse F, Venderbosch R, van Duren R, Kersten SRA, Prins W: Effect of biomass ash in catalytic fast pyrolysis of pine wood, *Appl Catal B: Environ* 168:203–211, 2015.

Yu Y, Liu DW, Wu HW: Characterization of water-soluble intermediates from slow pyrolysis of cellulose at low temperatures, *Energy Fuels* 26:7331–7339, 2012.

Yu Y, Liu D, Wu H: Formation and characteristics of reaction intermediates from the fast pyrolysis of NaCl- and $MgCl_2$-loaded celluloses, *Energy Fuels* 28:245–253, 2013.

Yu J, Sun L, Ma C, Qiao Y, Yao H: Thermal degradation of PVC: a review, *Waste Manage* 48:300–314, 2016.

Zaror CA, Hutchings IS, Pyle DL, Stiles HN, Kandiyoti R: Secondary char formation in the catalytic pyrolysis of biomass, *Fuel* 64:990–994, 1985.

Zeng J, Helms GL, Gao X, Chen S: Quantification of wheat straw lignin structure by comprehensive NMR analysis, *J Agr Food Chem* 61:10848–10857, 2013.

Zhang H, Xiao R, Huang H, Xiao G: Comparison of non-catalytic and catalytic fast pyrolysis of corncob in a fluidized bed reactor, *Bioresour Technol* 100:1428–1434, 2009.

Zhang HY, Carlson TR, Xiao R, Huber GW: Catalytic fast pyrolysis of wood and alcohol mixtures in a fluidized bed reactor, *Green Chem* 14:98–110, 2012.

Zhang B, Zhong ZP, Chen P, Ruan R: Microwave-assisted catalytic fast pyrolysis of biomass for bio-oil production using chemical vapor deposition modified HZSM-5 catalyst, *Bioresour Technol* 197:79–84, 2015a.

Zhang B, Zhong ZP, Song ZW, Ding K, Chen P, Ruan R: Optimizing anti-coking abilities of zeolites by ethylene diamine tetraacetie acid modification on catalytic fast pyrolysis of corn stalk, *J Power Sources* 300:87–94, 2015b.

Zhang B, Zhong ZP, Wang XB, Ding K, Song ZW: Catalytic upgrading of fast pyrolysis biomass vapors over fresh, spent and regenerated ZSM-5 zeolites, *Fuel Process Technol* 138:430–434, 2015c.

Zhang B, Zhong ZP, Xie QL, Chen P, Ruan R: Reducing coke formation in the catalytic fast pyrolysis of bio-derived furan with surface modified HZSM-5 catalysts, *RSC Adv* 5:56286–56292, 2015d.

Zhang J, Choi YS, Yoo CG, Kim TH, Brown RC, Shanks BH: Cellulose-hemicellulose and cellulose-lignin interactions during fast pyrolysis, *ACS Sustainable Chem Eng* 3:293–301, 2015e.

Zhang B, Zhong Z, Ding K, Song Z: Production of aromatic hydrocarbons from catalytic co-pyrolysis of biomass and high density polyethylene: analytical Py-GC/MS study, *Fuel* 139:622–628, 2015f.

Zhang J, Wang KG, Nolte MW, Choi YS, Brown RC, Shanks BH: Catalytic deoxygenation of bio-oil model compounds over acid–base bifunctional catalysts, *ACS Catal* 6:2608–2621, 2016.

Zheng AQ, Jiang LQ, Zhao ZL, et al: Impact of torrefaction on the chemical structure and catalytic fast pyrolysis behavior of hemicellulose, lignin, and cellulose, *Energy Fuels* 29:8027–8034, 2015.

Zhou S, Pecha B, van Kuppevelt M, McDonald AG, Garcia-Perez M: Slow and fast pyrolysis of Douglas-fir lignin: importance of liquid-intermediate formation on the distribution of products, *Biomass Bioenergy* 66:398–409, 2014a.

Zhou X, Nolte MW, Mayes HB, Shanks BH, Broadbelt LJ: Experimental and mechanistic modeling of fast pyrolysis of neat glucose-based carbohydrates. 1. Experiments and development of an advanced mechanistic model, *Ind Eng Chem Res* 53:13274–13289, 2014b.

Zhou X, Nolte MW, Shanks BH, Broadbelt LJ: Experimental and mechanistic modeling of fast pyrolysis of neat glucose-based carbohydrates. 2. Validation and evaluation of the mechanistic model, *Ind Eng Chem Res* 53:13290–13301, 2014c.

Zhou H, Wu C, Onwudili JA, Meng A, Zhang Y, Williams PT: Effect of interactions of PVC and biomass components on the formation of polycyclic aromatic hydrocarbons (PAH) during fast co-pyrolysis, *RSC Adv* 5:11371–11377, 2015.

Zhou X, Li W, Mabon R, Broadbelt LJ: A critical review on hemicellulose pyrolysis, *Energy Technol*, 2016. http://dx.doi.org/10.1002/ente.201600327.

Zhou X, Nolte MW, Mayes HB, Shanks BH, Broadbelt LJ: Fast pyrolysis of glucose-based carbohydrates with added NaCl, part 1: experiments and development of a mechanistic model, *AIChE J* 62:766–777, 2016c.

Zhou X, Nolte MW, Shanks BH, Broadbelt LJ: Fast pyrolysis of glucose-based carbohydrates with added NaCl, part 2: validation and evaluation of the mechanistic model, *AIChE J* 62:778–791, 2016d.

Zhou X, Li W, Mabon R, Broadbelt LJ: A mechanistic model of fast pyrolysis of hemicellulose extracted from cornstover, 2016b.

Ziff RM, McGrady ED: Kinetics of polymer degradation, *Macromolecules* 19:2513–2519, 1986.

CHAPTER THREE

Steam Cracking and EDC Furnace Simulation

Y. Zhang, G. Hu, W. Du[1], F. Qian[1]
Key Laboratory of Advanced Control and Optimization for Chemical Processes of Ministry of Education, East China University of Science and Technology, Shanghai, China
[1]Corresponding author: e-mail address: wldu@ecust.edu.cn; fqian@ecust.edu.cn

Contents

Abstract

Industrial cracking furnaces for ethene and ethylene dichloride (EDC) production are highly energy intensive. Much effort has been exerted toward achieving higher thermal efficiency and product yield. Numerical modeling has shown to be useful in furnace design and optimization. The advantage of numerical modeling over experiments is that a wider operating range can be simulated, allowing theoretical optimum to be found as guidance for practical. Industrial steam cracking furnaces have been extensively studied via numerical simulations over the past few decades, leading to many interesting results that offer deep insights into steam cracking processes. In this chapter, progress in the modeling of industrial cracking furnaces for both ethene and EDC production that has taken place in the last 5 years is reviewed. The discussion on ethene cracking furnaces includes fully coupled furnace-reactor simulations for the furnaces, the investigation of radiative heat transfer, the heat transfer and flow boiling of complex hydrocarbon feedstock in convection sections, and the integrated operation and cyclic scheduling optimization of cracking furnace systems. For the EDC cracking furnace, the effect of the fuel gas allocation factor, CCl_4 concentration, on the run length of the furnace is presented. In addition to the recent advances in cracking furnace modeling, the problems encountered by the current furnace design in fulfilling the new requirements of higher efficiency and

Advances in Chemical Engineering, Volume 49
ISSN 0065-2377
http://dx.doi.org/10.1016/bs.ache.2016.09.003

lower emission of NO_x and CO_2 are detailed. The alternative approaches, which may provide a better solution for this issue, are given, with a potential research focus on cracking furnace design and optimization that may occur in the future.

1. INTRODUCTION

Ethene and propene are the most important building blocks for the chemical industry, and their derivatives are widely used in daily life. They are mainly produced by thermal cracking of hydrocarbons mixed with steam in large-scale furnaces. A typical arrangement of an industrial steam cracking furnace is depicted in Fig. 1. The furnace consists of mainly two sections: i.e., the convection section and the radiant section. Hydrocarbon feedstock is first introduced into the convection section and heated to the incipient cracking temperature, which ranges from 500°C to 680°C depending on the type of feedstock. During the preheating, dilution steam is mixed with the feedstock to reduce the partial pressure of the hydrocarbon, which favors the ethene and propene yields. The diluted feedstock then enters the tubular reactors suspended vertically in the middle of the radiant section (firebox), where the cracking of hydrocarbon occurs. The heat required by the cracking reactions is supplied by fuel gas combustion in the radiant section outside the reactor coils. Two types of burners are often used in

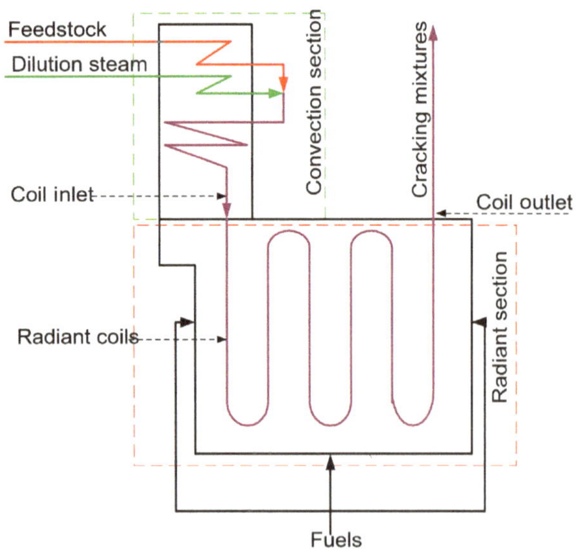

Fig. 1 Typical arrangement of an industrial steam cracking furnace.

the firebox: Floor burners located on the bottom of the furnace create turbulent, diffusive, long flames, while radiant burners, which are also called *sidewall burners*, use premixed fuel and air and thus produce much smaller flames. After cracking in a very short time (about 0.1–0.5 s, or even in milliseconds in some furnaces), the process gas leaves the reactor coil at 750–890°C and is cooled rapidly to prevent degradation of the highly reactive products by secondary reactions.

As the core unit of an ethene plant, steam cracking furnaces are very energy-intensive and consume about 8% of the sector's total primary energy (Ren et al., 2006). To meet the rapid increase in the world's ethene and propene demand, it is desired to produce these light olefins in steam cracking furnaces at an even larger scale, which requires modern thermal cracking units to be designed to operate not only efficiently, but also in an environmentally friendly manner. Numerical modeling, as a powerful tool for process design and optimization, has been widely used to investigate the steam-cracking process on both the process gas side and the fire side in the past several decades.

For the reactor side, a lot of efforts have been made toward the modeling of cracking reactions. The development of the cracking mechanisms has evolved from molecular reaction models (Ghassabzadeh et al., 2009; Kumar and Kunzru, 1985) to the more advanced free radical reaction models (Ranzi et al., 2004; Sabbe et al., 2011; Sun and Saeys, 2011; Van Geem et al., 2006; van Goethem et al., 2001). As detailed feedstock composition is needed by the free radical-based models, molecular reconstruction of complex feedstocks using commercial indices such as specific density and average molecular mass *n*-paraffins, iso-paraffins, olefins, naphthenes, and aromatics (PIONA) analysis was also widely studied (Charon-Revellin et al., 2010; Hudebine and Verstraete, 2004; Pyl et al., 2010; Van Geem et al., 2007a; Verstraete et al., 2004, 2010). The aforementioned research has led to commercial software used in industrial settings, such as COILSIM1D (Van Geem et al., 2007b) and SPYRO (van Goethem et al., 2001). In addition, many studies of reactor design optimization have been conducted by multidimensional reactor simulations using the detailed reaction kinetics (e.g., Heynderickx and Froment, 1998; Plehiers et al., 1990; Reyniers et al., 2015; Schietekat et al., 2014; Van Cauwenberge et al., 2015).

Research on the fire side of the furnace initially focused on radiative heat transfer modeling (Detemmerman and Froment, 1998; Heynderickx and Nozawa, 2004, 2005; Heynderickx et al., 2001; Oprins et al., 2001) using the zone method proposed by Hottel and Sarofim (1970). With the development of computational fluid dynamics (CFD) and the drastic increase in computational power over the last several decades, radiative transfer equations (RTEs) can nowadays be solved together with the fluid dynamics in the entire flow domain. This has resulted in much research on fire side simulation to investigate flow pattern, fuel combustion, radiative heat transfer, and the NO_x and CO emissions of the furnaces (Habibi et al., 2007a,b; Lan et al., 2007; Oprins and Heynderickx, 2003; Stefanidis et al., 2006, 2007, 2008).

Another important material used in the chemical industry is the vinyl chloride monomer (VCM), the precursor of the world's second-most-used plastic, polyvinyl chloride (PVC). It is produced by thermal cracking of ethylene dichloride (EDC) in industrial furnaces in a manner similar to the steam-cracking process. The EDC cracking furnace also consists of a convection section and a radiant section. The ethene dichloride is preheated in the former up to around 670K and cracked in the latter to produce VCM. To prevent secondary reactions, the process gas leaving the radiant section is quickly quenched and then sent to the separation units. The configuration of the EDC cracking furnace differs from that of the steam cracking furnace in two main aspects. First, the typical arrangement of the reactor tubes in EDC crackers is horizontal instead of vertical, as seen in steam cracking furnaces. Second, the furnace is equipped with only radiant burners on its side wall, as opposed to the steam cracking furnace, where floor burners are more often used.

Despite of the different configurations on the fire side, the underlying combustion and radiative heat transfer in the two types of cracking furnaces are basically the same. Hence, the fire side models dealing with the steam cracking furnaces are also applicable for EDC cracking furnace modeling. This has led to much less research on the fire side of the EDC cracking furnaces than that of the ethene cracking furnaces. Therefore, more attention has been focused on the free-radical mechanism of EDC thermal cracking over the last several decades, giving rise to a number of detailed and simplified reaction networks (e.g., Ashmore et al., 1982a,b; Barton, 1949; Barton and Howlett, 1949; Choi et al., 2001; Huybrechts and Wouters, 2002; Ranzi et al., 1992; Schirmeister et al., 2009).

In the last 5 years, more studies on ethene and EDC cracking have been conducted by fully coupled furnace-reactor simulations of practically installed industrial furnaces. Many interesting results have been obtained, which give some insight into the interaction between the process side and the fire side and allow better design and operation of the cracking furnaces.

In this chapter, we present a review of the most recent discoveries in numerical modeling of industrial cracking furnaces for ethene and EDC production. The chapter is organized as follows. Section 2 deals with the ethene cracking furnace simulations. Fully coupled furnace-reactor simulation is discussed in Section 2.1, with detailed information on both the process side and the fire side. Section 2.2 describes the influence of radiative heat transfer on reactor simulations. The heat transfer and two-phase flow boiling of hydrocarbon feedstock in the convection section are presented in Section 2.3, followed by the optimization and scheduling of cracking furnace systems in Section 2.4. Section 3 is dedicated to the modeling of EDC cracking furnaces. A comprehensive coupled furnace-reactor simulation is introduced in Section 3.1, and the study of the furnace run length is presented in Section 3.2. Section 4 summarizes this chapter and offers some closing remarks.

2. ETHENE STEAM CRACKING FURNACE

2.1 Comprehensive Coupled Furnace-Reactor Simulation Using CFD

Due to the lack of research on ethene cracking furnaces with the bottom and sidewall heat-suppliers, coupled simulations of a 100-kt/a industrial SL-II naphtha furnace were conducted to study the flow, combustion, heat transfer in furnace, and thermal cracking reactions in reactor coils. Fig. 2, taken from Hu et al. (2011), illustrates the configuration of one-sixth of an ethene furnace. This furnace adopts a two-pass branch and changed-diameter reactor tube design. The reactor is the "4-1" type—i.e., four inlet tubes are linked with an outlet tube. For the firebox modeling, the standard k-ε model was applied to turbulence simulation. The finite-rate/eddy-dissipation model, which can simulate both nonpremixed combustion and premixed combustion, was used to fuel gas combustion. The radiative heat transfer process was described by the discrete ordinates (DO) model, and the flue gas radiative properties in terms of the absorption coefficients were calculated using the Weighted-Sum-of-Gray-Gases model (WSGGM) proposed by Smith et al. (1982). For the steam cracking kinetics, a molecular model containing 22 reactions proposed by Kumar and Kunzru (1985) and a radical reactions COILSIM1D model (Van Geem et al., 2007b) were employed. The finite rate chemistry model was employed to examine the transfer reaction process. The procedure of the coupled simulation is as follows: Initially, an estimated external tube

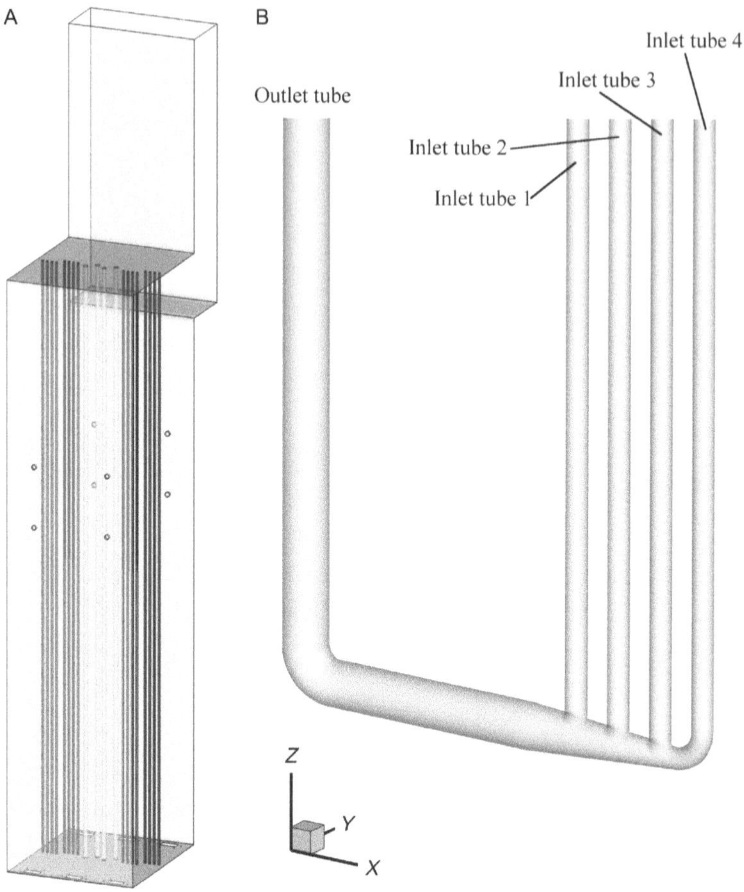

Fig. 2 Schematic diagram of cracking furnace: (A) front view of the considered segment of cracking furnace and (B) SL-II coil configuration. *Adapted with permission from Hu G, Wang H, Qian F, et al.: Comprehensive CFD simulation of product yields and coking rates for a floor- and wall-fired naphtha cracking furnace,* Ind Eng Chem Res *50(24):13672–13685, 2011. Copyright 2011 ACS Publications.*

metal-temperature profile was assigned as the boundary condition of the furnace simulation by a user-defined function (UDF) in FLUENT. Then the combustion and heat transfer inside the furnace were calculated to obtain the heat flux profile along the reactor tube, which was used in the next step to perform reactor simulation, which yielded an updated tube skin temperature profile. This iterative loop was continued until the maximal difference in the tube metal temperature (TMT) of two successive iterations reached a predefined threshold value (e.g., 1K). The simulation provided for the first time detailed information about the concentration, velocity, and temperature fields for both the reactor side and the fire side in this type of furnace. The simulation results agreed well with industrial data and provided a theoretical basis for the optimization of the geometrical structure and operational parameters of the cracking furnace.

2.1.1 Flow, Temperature, and Concentration Fields of the Fire Side

Fig. 3 shows the flue gas velocity field at various heights in the plane $x=0.525$ m along the width of the furnace. A large recirculation zone exists near the reactor tubes due to the entrainment effect of the high-velocity jet of the floor burners. The height of this recirculation zone in the middle in the y-direction is about 9 m. The flue gas velocity decreases gradually, and the velocity gradient is small due to the diffusion and recirculation of flue gas between two rows of floor burners. The inner Fig. 3C shows the local enlarged diagram of the other two recirculation zones in the crossover

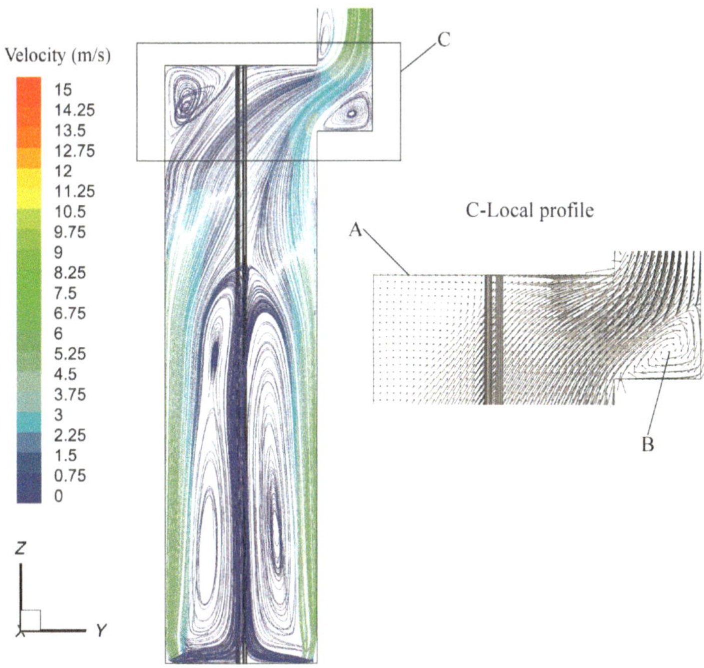

Fig. 3 Flue gas velocity vectors in the bridge section. *Reprinted with permission from Hu G, Wang H, Qian F, et al.: Comprehensive CFD simulation of product yields and coking rates for a floor- and wall-fired naphtha cracking furnace,* Ind Eng Chem Res *50(24):13672–13685, 2011. Copyright 2011 ACS Publications.*

section. With the rise of flue gas along the furnace height, the flue gas encounters obstacles at the top, leading to a change in the flue gas flow direction and a separation of the boundary layer. Hence a recirculation zone A appears at the corner of the top of the furnace. When the flue gas enters the convection section, it is forced to make a quick turn where the cross section becomes small. This causes the flue gas velocity to increase, creating another recirculation zone B at the corner of the convection section inlet.

The complexity of the flue gas velocity distribution determines the complexity of the flue gas concentration distribution in the furnace, affecting the flue gas temperature distribution in the furnace. Fig. 4A–D shows the flue gas temperature contours at different cross sections along the furnace width, which are 0.225, 1.655, 2.015, and 3.445 m, respectively. The cross sections depicted in Fig. 4A and D are located above the bottom burner of both sides of the furnace, and the other two sections shown in Fig. 4B and C are located on both sides of the reactor tubes. There are clear temperature gradients in the height and width directions of the furnace. The flue gas temperature is high near the furnace wall region, the central furnace, and the sidewall burners region, while the flue gas temperature is low on the furnace bottom and top, and near the reactor tubes region. In the width direction,

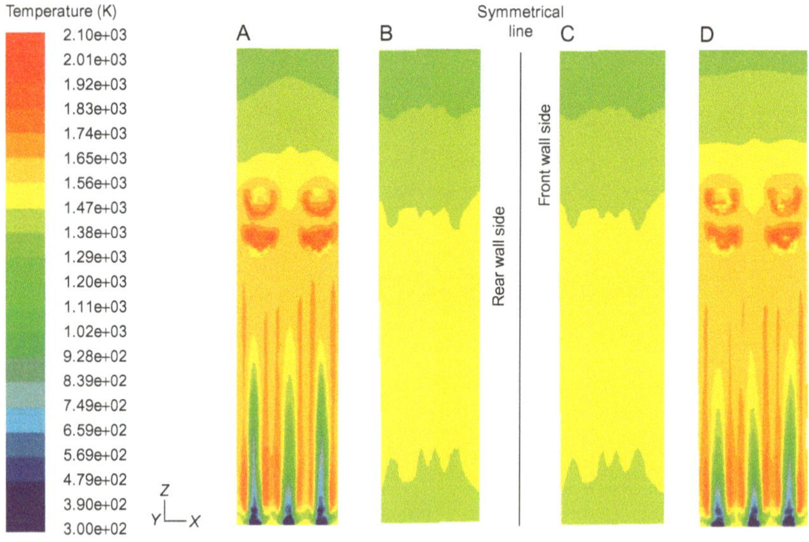

Fig. 4 Flue gas temperature contours at the different sections along the furnace width: (A) 0.225 m; (B) 1.655 m; (C) 2.015 m; (D) 3.445 m. *Reprinted with permission from Guihua H, Honggang W, Feng Q. Numerical simulation on flow, combustion and heat transfer of ethylene cracking furnaces,* Chem Eng Sci *66(8):1600–1611, 2011. Copyright 2011 Elsevier.*

temperature distribution of the flue gas on two sides of the furnace is not completely symmetrical when the reactor tubes are taken as a symmetrical line. Near the first row of sidewall burners on both sides of the furnace, the high-temperature zone on the front wall side does not occur on the rear wall side. The reason for this phenomenon is that the furnace outlet is located on the front wall side, and the resulting asymmetric flow pattern extends the high-temperature plane on the rear wall side to the front wall side. In the height direction, the flue gas temperature at the bottom of the furnace is low. This is mainly caused by the influence of the jet of the bottom fuel gas and air. That is, the fuel gas ejected from the burners has a very high velocity, so it is not completely mixed with the ambient air at lower elevations of the furnace. With the development of jets, the fuel gas and air further mix and an intense combustion reaction occurs, which leads to a large amount of heat being released. Therefore, the flue gas temperature gradually increases with the increase in the furnace height, reaching its highest value at about 5–7 m. With increasing elevation, the fuel gas and air are gradually consumed and the reactor tubes continuously absorb the heat, leading to temperature drop of the flue gas. While at higher elevations where the sidewall burners are located, the flue gas temperature begins to rise again. Therefore, the sidewall burners enlarge the zone of high-temperature flue gas at the top of the furnace and effectively complement the lack of heat at the top of furnace. Moreover, the bottom burners are arranged near the furnace wall so as to avoid the flame produced by the burners directly impinging upon the reactor tubes.

Fig. 5 shows the flue gas temperature profiles in different zones along the furnace height. The top two curves (solid and dashed-dotted lines) represent the average flue gas temperature profiles of the floor burner zone, on the front wall side and rear wall side, respectively. The middle curve (dotted line) is the average flue gas temperature profile of the whole furnace zone. The bottom two curves (dashed and double dotted-dashed lines) denote the average flue gas temperature profiles of the zone near the reactor tubes on the front wall side and rear wall side, respectively. Comparing the top two curves, the flue gas temperature above the floor burners of the front wall side is higher than that of the rear wall side when the furnace height is less than 6 m. Near the radiant burners, the flue gas temperature above the floor burners of the front wall side is lower than that of the rear wall side. The main reason for this is the asymmetric flue gas velocity distribution. The flue gas temperature on the left tube (the rear wall side) is significantly higher than that on the right tube (the front wall side), especially above the radiation

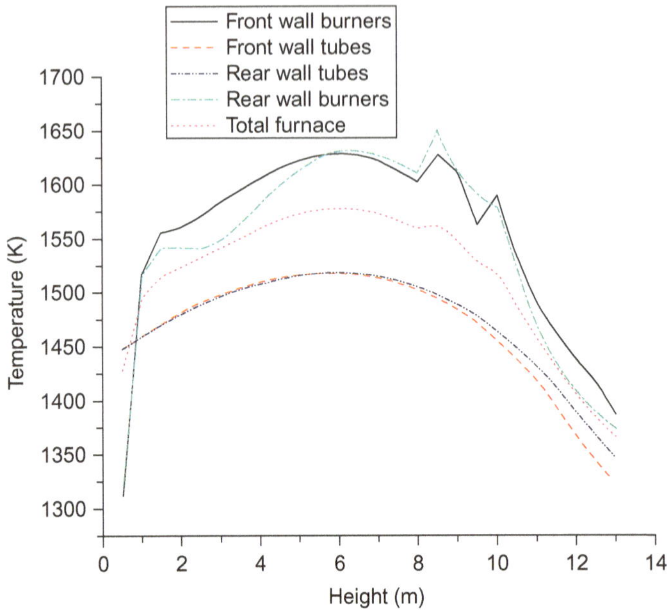

Fig. 5 Flue gas temperature profiles in different zones along the furnace height. *Reprinted with permission from Hu G, Wang H, Qian F, et al.: Comprehensive CFD simulation of product yields and coking rates for a floor- and wall-fired naphtha cracking furnace,* Ind Eng Chem Res *50(24):13672–13685, 2011. Copyright 2011 ACS Publications.*

burners. This is because the high-temperature flue gas from the radiation burners of the rear wall side should have been parallel to the wall for flowing upward, but the high-temperature flue gas flows to the front wall side of the furnace, away from the original route, due to the asymmetry of flue gas velocity distribution and the obstruction effect of recirculation of the top of the furnace. Thus, there are a large, high-temperature surface at the rear wall side of the row of reactor tubes and a high-temperature flue gas flow that contacts with the tube walls on this side.

In Fig. 6, the mass fraction profile of the flue gas above the floor burners along the furnace height is given. One can see that CO_2 and H_2O concentrations rapidly increase below 2.5 m in the furnace height direction, while those of O_2 and CO gradually decrease because of fuel gas consumption. The profiles of CO_2, H_2O, and O_2 concentration in both the front wall side burners and rear wall side burners follow similar trends. In the radiant burner zone (i.e., 8–10-m height of the furnace), some fluctuations of the concentration of O_2, H_2O, O_2, and CO are observed. Near the radiant burners, the CO_2 and H_2O concentrations are low and the O_2 concentration is high

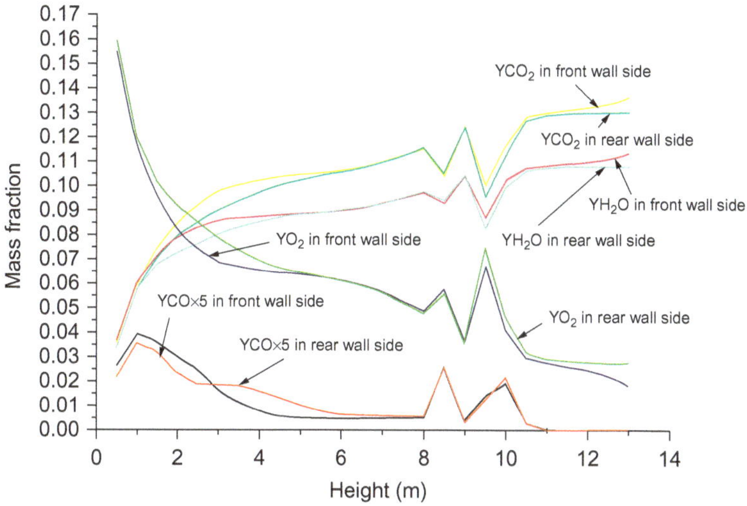

Fig. 6 Mass fraction profiles of flue gas above the floor burners along the furnace height. *Reprinted with permission from Hu G, Wang H, Qian F, et al.: Comprehensive CFD simulation of product yields and coking rates for a floor- and wall-fired naphtha cracking furnace,* Ind Eng Chem Res *50(24):13672–13685, 2011. Copyright 2011 ACS Publications.*

because of combustion of fuel gas in the radiation burners. In the floor burner zone on the front wall side, the CO_2 and H_2O concentrations continuously increase along the furnace height, while the O_2 concentration decreases. In the floor burner zone of the rear wall side, the concentrations of CO_2, H_2O, and O_2 basically keep constant along the furnace height. The reason for this is that the outlet of the furnace is located on the front wall side, and all the flue gases must pass through the central reactor tubes from the outlet into the convection section, which causes the asymmetrical flue gas flow in the top of the furnace. In the crossover section, the concentration of oxygen is low because a large amount of O_2 with low concentration in the center of the furnace mixes with a small amount of O_2 with high concentration of the front wall side, while on the rear wall side, the O_2 with high concentration forms a "dead zone" at the corner of the top of the furnace (see Fig. 3).

Fig. 7 shows the O_2, CO_2, CO mass concentration contours at different cross sections along the furnace length. The O_2 mass concentration is high in the jet core zone because both the fuel gas and air have a high flow velocity, where the fuel combustion does not have time to be fully carried out and little oxygen is consumed. Outside the jet core zone, the mixing and combustion of the fuel gas and air is completed; therefore, the oxygen is gradually

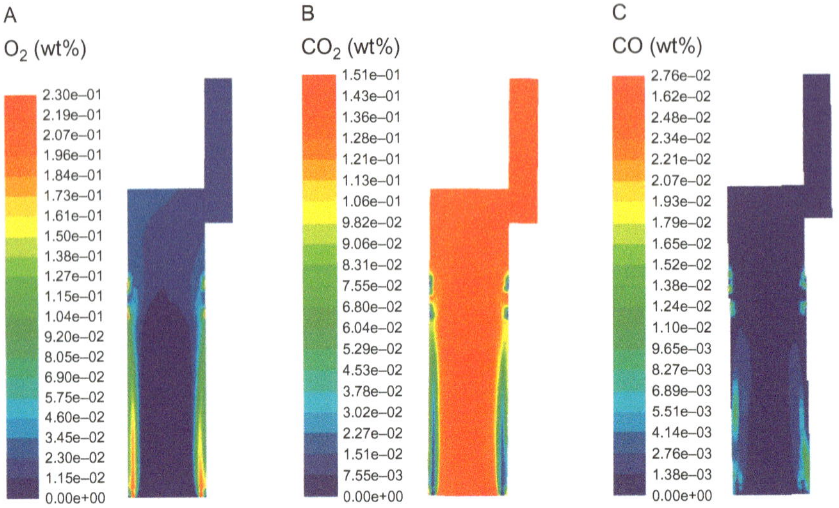

Fig. 7 Mass fraction distribution of O_2, CO_2, and CO at $x=0.525$ m along the length of the firebox. (A) O_2; (B) CO_2; (C) CO. *Adapted with permission from Hu G, Wang H, Qian F, et al.: Coupled simulation of an industrial naphtha cracking furnace equipped with long-flame and radiation burners,* Comput Chem Eng *38:24–34, 2012. Copyright 2012 Elsevier.*

consumed, resulting in less O_2 mass concentration. O_2 mass concentration decreases due to the combustion of the fuel gas along the furnace height direction. The CO_2 mass concentration above the bottom burners is less than that outside the bottom burners, and CO_2 mass concentration increases with the increase in furnace height at the jet core zone. The explanation is as follows: In the jet core zone of the bottom burners, incomplete mixing of the fuel gas and air leads to an inadequate combustion, and thus a small quantity of CO_2 is produced. Outside the jet core zone, full combustion condition leads to a large quantity of CO_2 generation and a high CO_2 mass concentration. The CO mass concentration contours at different sections along the furnace length and width. CO mass concentration focuses above the auxiliary nozzles outside the air door of the bottom burners zone, while it is relatively small above the primary nozzles inside the air door. Because of the poor mixing with air, the fuel gas ejected from the auxiliary nozzles generates more CO due to partial oxidation. While the fuel gas ejected from the primary nozzles is completely surrounded by air, achieving better mixing with small amount of CO produced. In addition, CO mass concentration is high at the furnace bottom due to incomplete combustion of the fuel gas, and it gradually decreases to zero with the furnace height due to sufficient combustion of the fuel gas.

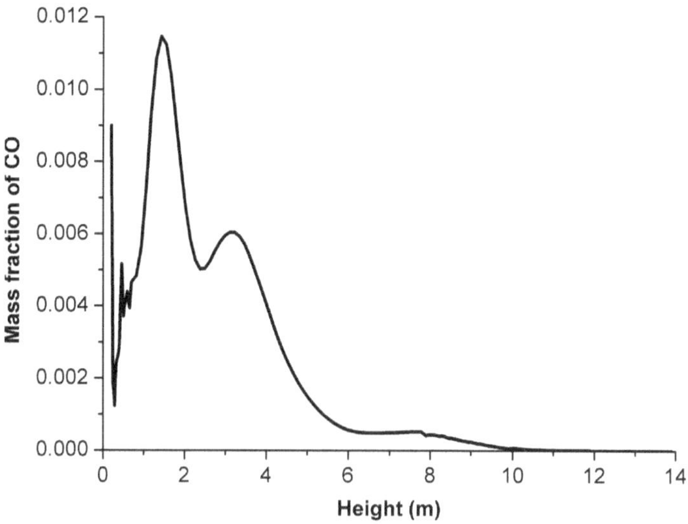

Fig. 8 CO mass concentration distribution in plane $x = 1.565$ m above a bottom burner. *Reprinted with permission from Guihua H, Honggang W, Feng Q. Numerical simulation on flow, combustion and heat transfer of ethylene cracking furnaces,* Chem Eng Sci *66(8):1600–1611, 2011. Copyright 2011 Elsevier.*

Fig. 8 shows the CO mass concentration distribution in plane $x = 1.565$ m above the bottom burners. CO mass concentration is relatively high at a height of 0–2 m above the burners. When the height is more than 2 m, CO mass concentration gradually decreases and reduces to about zero at the furnace height of 6 m. It is well known that the termination position of combustion flame is the end of oxidation reaction; therefore, the flame height can be estimated by CO content distribution. In other words, the place where CO is depleted, which is at the height of about 6 m in this study, is the flame height.

2.1.2 Process Gas Temperature, Velocity, Pressure, and Concentration Profiles of the Reactor Tube Side

The molecular reaction kinetic model treats the naphtha feedstock as a single pseudocomponent, and the pyrolysis process is simplified as a primary reaction and many secondary reactions. Therefore, to describe different naphtha feedstocks, the stoichiometric coefficients of the primary reaction have to be adjusted accordingly, but this does not affect the secondary reaction. The pyrolysis reaction kinetic model established by this method has the advantages of less computational cost, so the method is widely used in industry. Hu et al. (2011) adopted a coupled furnace-reactor simulation for an industrial SL-II naphtha cracking furnace fired by both the floor and wall burners. Both the process gas side and the fire side were simulated using the CFD approach. The molecular kinetic model of Kumar and Kunzru (1985) was used to simulate the naphtha cracking reactions in the reactor.

The process gas temperature profiles along the axial and radial directions of the reactor are depicted in Fig. 9. In the first part of the inlet tube, the temperature of the process gas rapidly increases because of the low-process gas temperature and limited heat consumption. The temperature rapidly increases from 883K at the entrance of the inlet tube to 1039K at the tube length of 6.5 m. With increasing process gas temperature, the reaction rate becomes larger and the absorbed heat is mainly used for the reaction. Hence, the slope of the process gas temperature is reduced. At the tube bend, where the diameter changes, the temperature of the process gas experiences a small decline because of a reduction of the absorbed heat due to the decrease in specific surface area, while the heat requirement for reaction inside the tubes remains high. In the following section, the reaction rate decreases and the absorbed heat is again mainly used for the temperature increase of the process gas inside the outlet tube, making the process gas temperature increase again.

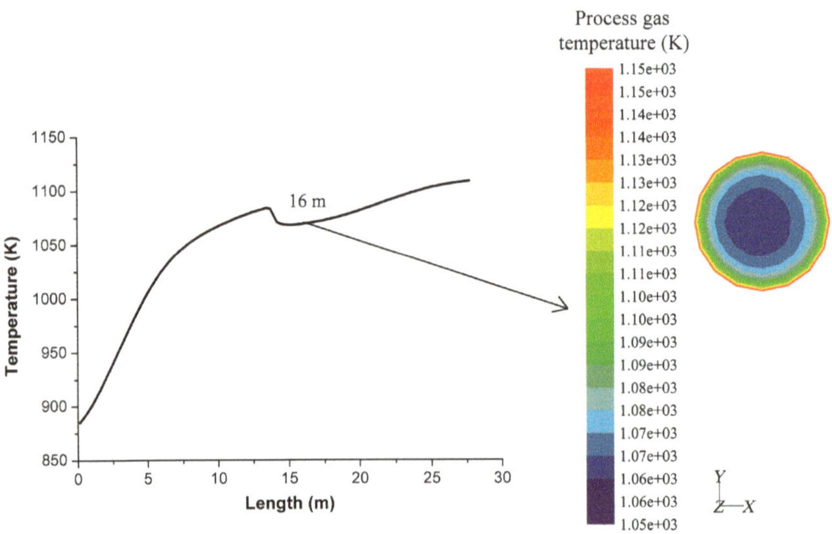

Fig. 9 Process gas temperature profiles along the axial and radial directions of the reactor. *Adapted with permission from Hu G, Wang H, Qian F, et al.: Comprehensive CFD simulation of product yields and coking rates for a floor- and wall-fired naphtha cracking furnace,* Ind Eng Chem Res *50(24):13672–13685, 2011. Copyright 2011 ACS Publications.*

It can be seen that there is significant radial temperature gradients in the order of 100K, which agrees well with values reported by Van Geem et al. (2004) and Lan et al. (2007). The flow pattern directly affects the heat transfer inside the reactor tube. The velocity distribution in the radial direction of the reactor tube affects the temperature distribution in the radial direction. The temperature variation is small in 90% of the radical region from the center of the reactor tube and rapid near the wall, as shown in Fig. 9. This is because laminar flow of the boundary layer leads to a steep temperature gradient near the wall of the reactor tube.

Fig. 10 shows that the process gas velocity gradually increases along the length of the inlet tube as a result of the process gas expansion caused by cracking of the molecules into smaller pieces. In the inlet tube, the process gas velocity increases from 76 m/s at the entrance of the inlet tube to 146 m/s at the changed-diameter site, and the velocity steeply decreases from 146 to 130 m/s because of the sudden increase in the diameter of the second pass. Obviously, the increase of the process gas velocity is smaller than that in the inlet tube. On the one hand, the diameter of the outlet tube is bigger than that of the inlet tube, while on the other hand, the volume expansion rate of the process gas is smaller as a significant portion of the

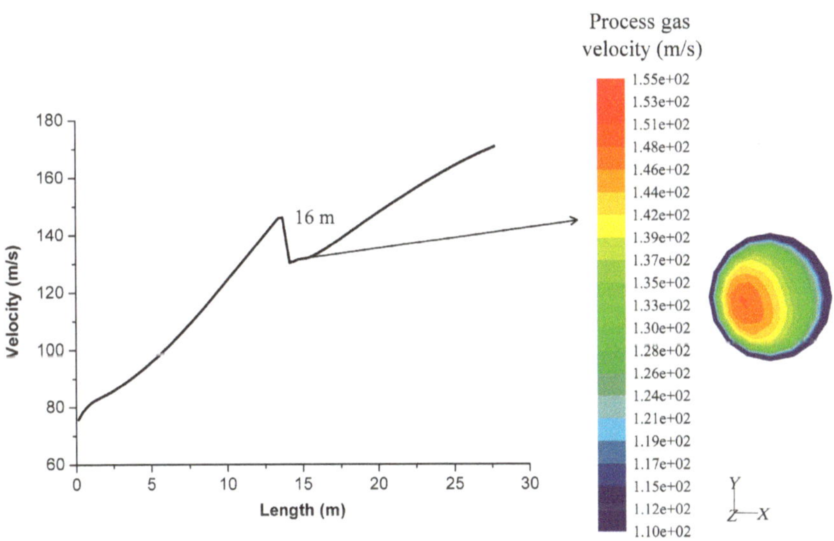

Fig. 10 Process gas velocity profiles along the axial and radial directions of the reactor. *Adapted with permission from Hu G, Wang H, Qian F, et al.: Comprehensive CFD simulation of product yields and coking rates for a floor- and wall-fired naphtha cracking furnace,* Ind Eng Chem Res *50(24):13672–13685, 2011. Copyright 2011 ACS Publications.*

feedstock was already converted into smaller molecules. Based on the obtained average velocity, the calculated residence time of the process gas inside the reactor tube is 0.24 s, which is in good agreement with the industrial data (0.25 s). In this figure, a gentle change of the velocity in 90% radial from center of the reactor tube is shown, followed by a sharp velocity decrease close to the tube wall. The variation of the velocity in the radial direction implies a distribution of residence time along the radius. This is also because of the laminar flow of the boundary condition, which leads to a steep velocity gradient near the wall of the reactor tube.

Fig. 11 shows the pressure profile of process gas. The pressure in the inlet tubes decreases with the reactor length. It is expected that the pressure drop in the first pass is larger than that in the second pass due to the increased diameter from the inlet tube to the outlet tube. Therefore, the slope of the pressure profile changes at the bend of the reactor tube, which is the connecting point of the inlet and the outlet tubes. Based on the simulated pressure profile, the pressure drop over the coil is 0.0444 MPa, which agrees well with the industrial data (0.0433 MPa), indicating that the calculated pressure results are reasonable.

The profiles of the main product yield along the reactor length are depicted in Fig. 12. The C_2H_4 yield initially slowly increases because of

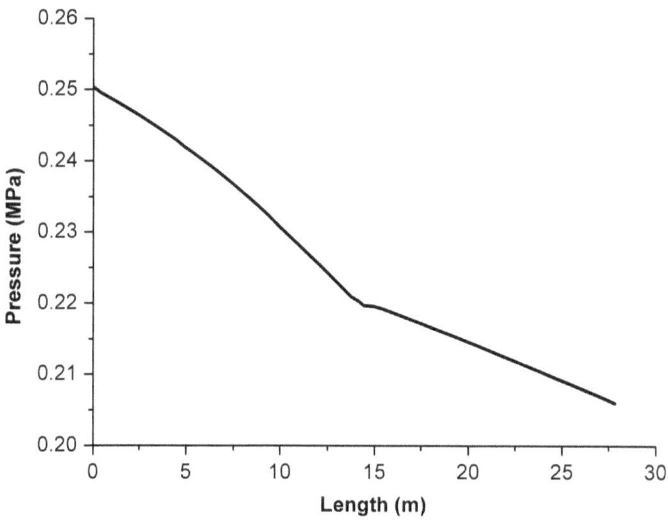

Fig. 11 Process gas pressure profile along the reactor length. *Adapted with permission from Hu G, Wang H, Qian F, et al.: Comprehensive CFD simulation of product yields and coking rates for a floor- and wall-fired naphtha cracking furnace,* Ind Eng Chem Res *50(24):13672–13685, 2011. Copyright 2011 ACS Publications.*

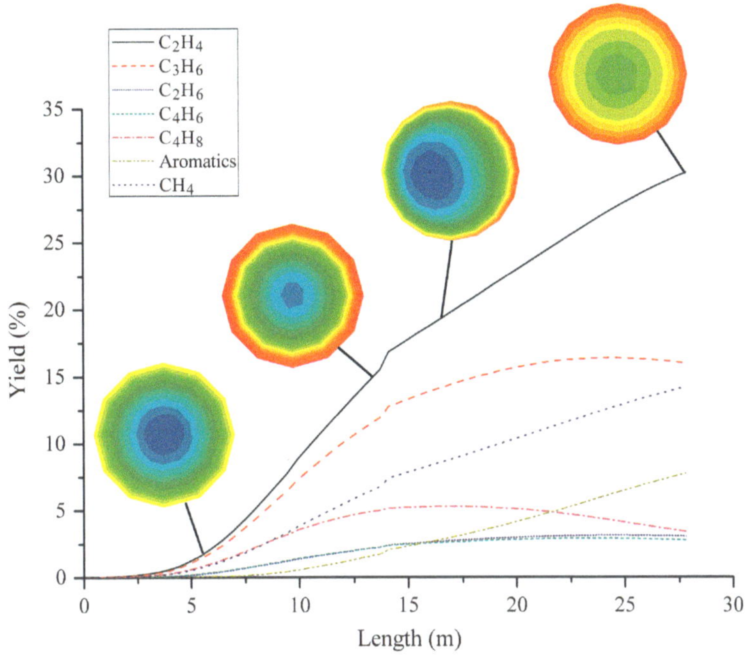

Fig. 12 Product yield profile along the tubular reactor.

the low process gas temperature. In the second half of the first pass, the process gas temperature reaches the value for the rate of cracking reaction to increase substantially, giving rise to more and more ethene. The C_2H_4 yield increases more slowly when the process gas enters into the outlet tube as a result of the decreased reactant. The C_3H_6 yield has a different shape and undergoes a maximum because at higher temperatures, the formed C_3H_6 will be converted to C_2H_4 and CH_4. Also, additional reactions will transfer part of formed C_3H_6 into heavier components such as aromatics. For the yields of other components like C_2H_6, C_4H_6, and C_4H_8, the trends are similar to that of C_3H_6, while the yield of CH_4 increases monotonically along the axial reactor coordinate.

The CFD simulation of the cracking coil also allows one to address the importance of the radial nonuniformities of species concentrations in the coil. Fig. 13A shows the mass fraction of ethene at the height of 3 m in the second pass reactor tube and reveals that there are strong radial nonuniformities at this position. The mass fraction of ethene gradually increases from center of the reactor tube to the tube wall, but the difference of mass fraction between the center and the wall region is approximately 1%. On the

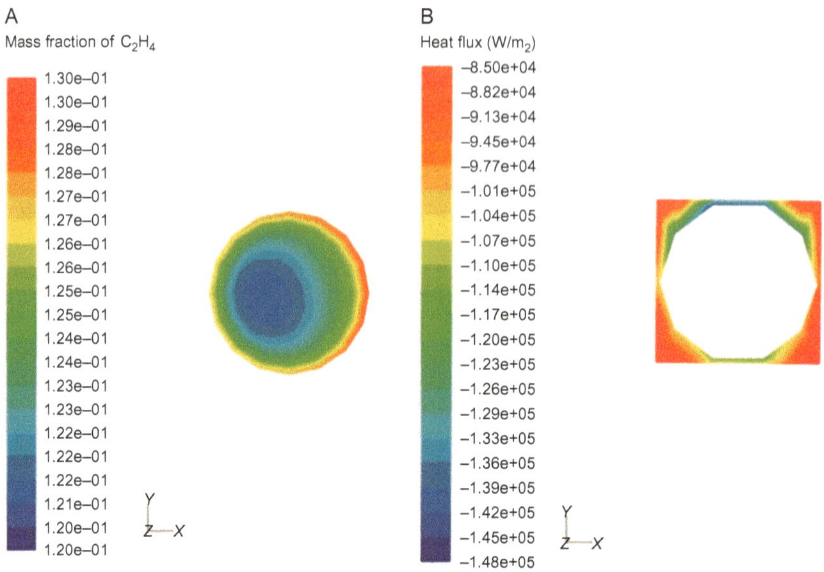

Fig. 13 Radial profiles of (A) ethene mass fraction; (B) Circumferential heat flux at a height of 3 m on the second pass of the reactor coil. *Adapted with permission from Hu G, Wang H, Qian F, et al.: Comprehensive CFD simulation of product yields and coking rates for a floor- and wall-fired naphtha cracking furnace,* Ind Eng Chem Res *50(24):13672–13685, 2011. Copyright 2011 ACS Publications.*

one hand, this is caused by nonuniformities of the temperature and velocity, as can be seen in Figs. 9 and 10. On the other hand, the circumferential non-uniformities of the heat flux profile (as reported by Heynderickx et al., 1992) are also responsible, as shown in Fig. 13B.

Although the molecular reaction model is relatively simple and less computationally expensive, the stoichiometric coefficients of the primary reaction always must be adjusted for different feedstocks and operation conditions to ensure good agreement with the simulated results and industrial data. Based on the theory of (Rice, 1931), the free radical reaction kinetic model describes the reaction process according to the detailed composition of the feedstock. The product distribution is obtained by solving all kinds of reaction differential equations. Because the free radical mechanism has strictly characterized the process of pyrolysis reaction, this model has strong adaptability and high precision and reliability. Hu et al. (2012) performed coupled reactor-furnace simulations for a 100-kt/a SL-II naphtha cracking furnace containing both long-flame and radiation burners. The CFD approach was used to simulate the flow, combustion, and radiative heat

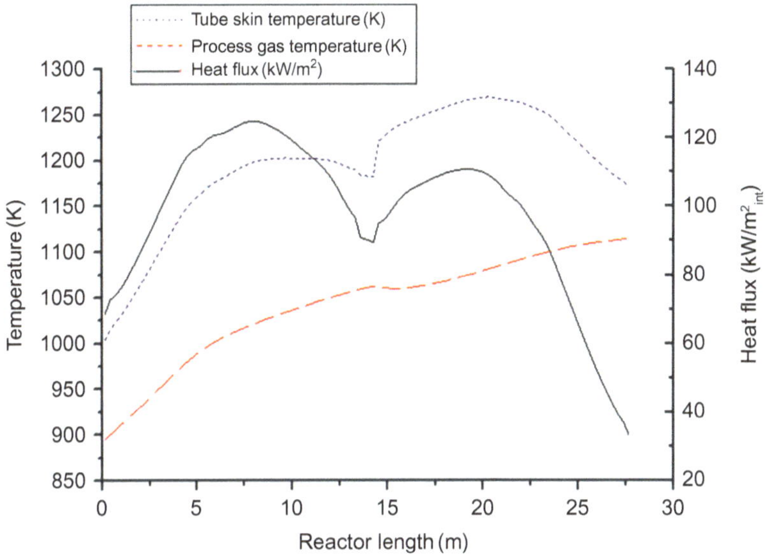

Fig. 14 Profile of the tube skin and process gas temperature and heat flux along the reactor length. *Adapted with permission from Hu G, Wang H, Qian F, et al.: Coupled simulation of an industrial naphtha cracking furnace equipped with long-flame and radiation burners,* Comput Chem Eng *38:24–34, 2012. Copyright 2012 Elsevier.*

transfer in the furnace. The software packages COILSIM1D and SimCO were used to account for the cracking process in the reactor coils.

Fig. 14 shows the heat flux profile and the TMT and process gas temperature profiles along the reactor length. As mentioned previously, the maximum flue gas-temperature is located in a zone at 4–6 m from the furnace floor. Hence, the TMT and the heat flux reach their maximum in this zone. It can be seen in Fig. 14 that the TMT in the first pass increases from 1004K at the entrance to a first maximum of 1202K at an axial tube length position of about 10 m, and then falls to 1183K at the exit of the first pass. The TMT of the second pass increases from 1220K at the entrance of the second pass to the second maximum, 1269K, at a tube length of about 20 m, and then falls to 1171K at the exit of the second pass. As mentioned previously, the maximum external wall temperature is relatively low for this medium-severity operation because of the proper positioning of the burners and convection section. Hence, a longer run length can be expected with these types of furnaces.

The heat flux profile is determined by the cracking reactions in the reactor coil, the process gas temperature profile inside the tube, and the TMT profile. The heat flux is relatively low in the beginning of the inlet tube due to low flue gas-temperature. Hence, in the first section of the reactor, almost no cracking reaction occurs. The absorbed heat leads to a rapid increase of the process gas temperature. With increasing process gas temperature inside the tube, both the reaction rate of the cracking reactions and the heat absorbed by the process gas increase. Hence, the heat flux reaches the first peak at the tube length of about 8 m, where the flue gas-temperature reaches its maximum. With increasing conversion, the rise of the process gas temperature becomes less steep because the absorbed heat is consumed to maintain the endothermic cracking reactions. When reaching the second pass of the coil, the process gas reenters the high-temperature flue gas region. In the first section of the second pass where the reactor diameter increases, the process gas temperature remains almost constant. Afterward, the increasing heat input from the furnace to the coil makes results in another increase in the process gas temperature. The heat flux increases until the maximum is reached at a tube length of about 19 m. Fig. 14 also shows that the average heat flux in the first pass is higher than that of the second pass. This can be explained by the smaller driving force for heat transfer in the second pass because of the higher temperature of the process gas in this section of the reactor.

Table 1 compares simulation results and industrial data for the coupled reactor-furnace simulation. High light olefin yields are obtained because of the combination of short residence times with the rapid heating of process gas in the split coil design. The simulated temperature of outlet flue gas, excess oxygen ratio, coil outlet temperature (COT), residence time, coil inlet pressure, and coil pressure drop are in good agreement with the industrial data. Also, the yields of the main products ethene and propene agree well with the yields measured in the industrial cracker. For the heavier aromatics, the agreement is not perfect, but this can be caused by the way the yields have been determined in the industrial cracker. Generally, part of the heavier products is lost in the separation section before they can be analyzed (Van Geem et al., 2008). This shows the potential of the coupled furnace-reactor simulation in providing quantitative results for a furnace equipped with both long flame burners and radiation burners.

In conclusion, for the first time, an industrial naphtha cracking furnace consisting of both long flame burners and radiation burners was simulated using CFD for the furnace side and detailed kinetics on the process gas side. Good agreement was obtained between industrial data and simulated light olefin yields via a coupled simulation of furnace and reactor, therefore

Table 1 Comparison of Calculated Results With Industrial Data

Items	Industrial Data	Simulation Results
Temperature of outlet flue gas (K)	1397	1384
Excess oxygen ratio (vol%)	2.00	1.95
Temperature of outlet process gas (K)	1108.5	1111.1
Residence time (s)	0.24	0.27
Coil inlet pressure (MPa)	0.27	0.25
Coil pressure drop (MPa)	0.07	0.05
Yield (wt%)		
Hydrogen	0.86	0.65
Methane	13.81	13.82
Acetylene	0.37	0.39
Ethene	29.69	29.56
Ethane	3.72	3.70
Propylene	15.53	15.17
Propane	0.53	0.44
1,3-Butadiene	4.20	4.49
Sum of butenes	4.21	3.59
Benzene	4.62	5.63
Toluene	3.83	3.24
Xylene	1.28	3.04

Reprinted with permission from Hu G, Wang H, Qian F, et al.: Coupled simulation of an industrial naphtha cracking furnace equipped with long-flame and radiation burners, *Comput Chem Eng* 38:24–34, 2012. Copyright 2012 Elsevier.

providing a proper theoretical basis for gaining better insight into the behavior of industrial steam cracking furnaces. Very high ethene and propene yields can be obtained due to the long run length as a result of the specific design of the furnace. The optimal design of a steam cracking furnace requires the combination of coil design, the positioning of the convection section, and the positioning of long-flame and radiation burners, and can be addressed by coupling CFD simulations on the fire gas side with fundamental kinetic models on the process gas side.

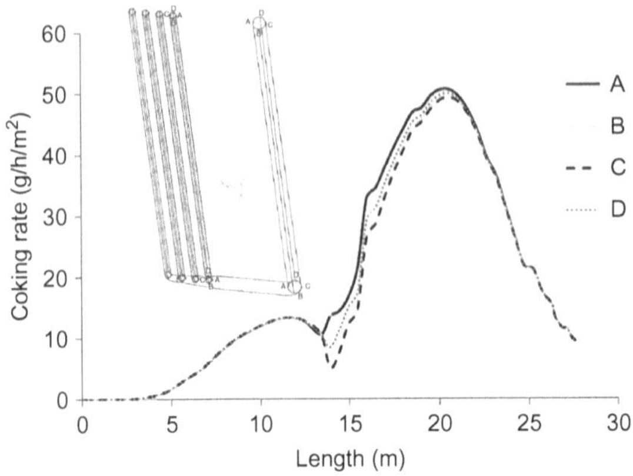

Fig. 15 Simulated coking rate along the length of the reactor at for four different circumferential positions (A, B, C, and D) in the coils. *Reprinted with permission from Hu G, Wang H, Qian F, et al.: Comprehensive CFD simulation of product yields and coking rates for a floor- and wall-fired naphtha cracking furnace,* Ind Eng Chem Res *50(24):13672–13685, 2011. Copyright 2011 ACS Publications.*

2.1.3 Coking Rate

Fig. 15 shows the coking rate profile calculated based on the internal TMT at four circumferential positions. For all four of these coking curves, two maxima can be observed to correspond to the maxima in the heat flux and the external and internal wall temperature profiles. The maximum coking rate is observed at approximately the middle of the second pass of the reactor. In the first pass, the difference in coking rate for the four circumferential positions remains small, and only in the second pass do some differences become visible. The circumferential differences in temperature and concentrations of the coke precursors are responsible for the nonuniform coking rates, and that will lead to nonuniform coke layers on the reactor wall. Local excessive coking can cause significant damage to the reactor material of the tubes. In the present reactor design, circumferential temperature differences of over 20°C have been simulated. However, the difference in circumferential temperature remains limited to 4°C where the coking rate reaches its maximum value. The latter implies that the simulated maximum coking rate does only slightly depend on the circumferential position for the present reactor. Fig. 15 further shows the strong relation between the coking rate and the wall temperature. It is obvious that reducing the maximum wall temperature will significantly reduce the coking tendency of the coil. One approach is increasing turbulence using a mixing element radiant tube (MERT) type of reactor (Gyorffy et al., 2009).

In summary, the coupled reactor-furnace simulation of a 100-kt/a SL-II ethene naphtha cracking furnace containing both floor and wall burners has been discussed in this section. Current studies have excessively simplified the structure of cracking furnace and the combustion model of wall burners, which was considerably different in the actual situation of the cracking furnace, and this affected the accuracy of the computation. This discussion looked at using the CFD method to carry out computer simulation studies on the process gas side and the fire side in the cracking furnace. The simulation provides for the first time detailed information about the profiles of temperature, velocity, and concentration of the process gas, as well as those of temperature, velocity, and pressure distributions of the flue gas, TMTs, and heat flux profiles along the reactor tube for these types of furnaces. A good agreement between simulation and industrial product yields has been obtained without any tuning of the kinetics, indicating that the proposed approach can be used as a guide for further optimization of geometries and operating parameters of naphtha cracking furnaces with burners located both in the floor and in the wall.

2.2 Investigation of Radiative Heat Transfer Process in Ethene Cracking Furnace

2.2.1 Comparison of Radiation Models

In past decades, many researchers have used the Lobo–Evans method (Lobo, 1974), Belokon's method (Jegla, 2006), and zone methods (Plehiers et al., 1990; Ramana Rao et al., 1988) for simulating industrial steam cracking furnaces. The fuel combustion was not simulated rigorously; instead, a predefined heat release rate was imposed to estimate the composition and temperature of the flue gas. Furthermore, convective heat transfer to the reactor tubes was often ignored. More particularly, for the simulation of steam cracking furnaces, different radiation models have been evaluated by many researchers over the past two decades. For instance, Habibi et al. (2007a,b) assessed the influence of the different radiation models on the temperature, concentration, and radiative properties of the flue gas mixture. However, the effect of radiation models on the reactor side, which has to be studied by coupled furnace-reactor simulations, was omitted. Thus, we performed coupled furnace-reactor simulations of an industrial naphtha cracking furnace with a 130-kt/a capacity. CFD was used to study on the furnace side, and the one-dimensional (1D) reactor model COILSIM1D was used to simulate on the process gas side in the cracking furnace. For the first time, different radiation models were evaluated in coupled furnace-reactor simulations to predict run lengths. The influence of the adopted radiation model on the flue gas flow, radiative properties, product yields, heat flux profiles to the reactor tubes, external TMT and process gas temperature profiles, and run length is analyzed. Accordingly, the best radiation model for the simulation of steam cracking furnaces is determined.

Fig. 16 shows one-eighth of the simulated naphtha cracking furnace. Fuel gas is uniformly distributed over 64 floor burners and divided per burner over two types of fuel inlets. The primary fuel inlets lie inside the air inlet, while the secondary burners are outside the air inlet, as shown in Fig. 16C. All the reactors comprise two passes: an inlet and an outlet pass. The inlet passes enter the furnace from the top and bend in an alternating fashion toward both walls through an S-type bend close to the furnace floor. U-type bends connect each inlet pass to its outlet pass, realigning all the reactors in a single row.

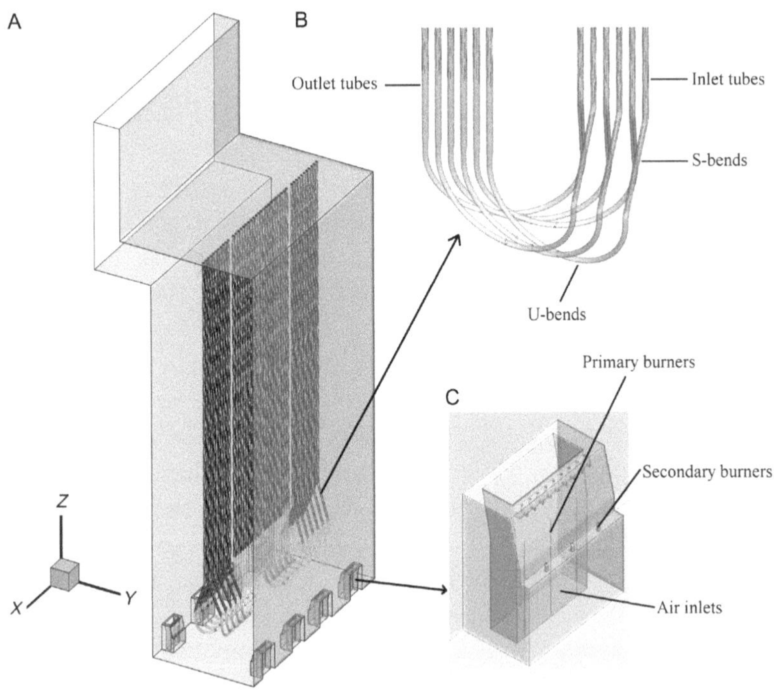

Fig. 16 Schematic of the naphtha cracking furnace: (A) overall geometry of the furnace; (B) detail of the reactor tube bends; (C) details of a burner. *Reprinted with permission from Hu G, Schietekat CM, Zhang Y, et al.: Impact of radiation models in coupled simulations of steam cracking furnaces and reactors,* Ind Eng Chem Res *54(9):2453–2465, 2015. Copyright 2015 ACS Publications.*

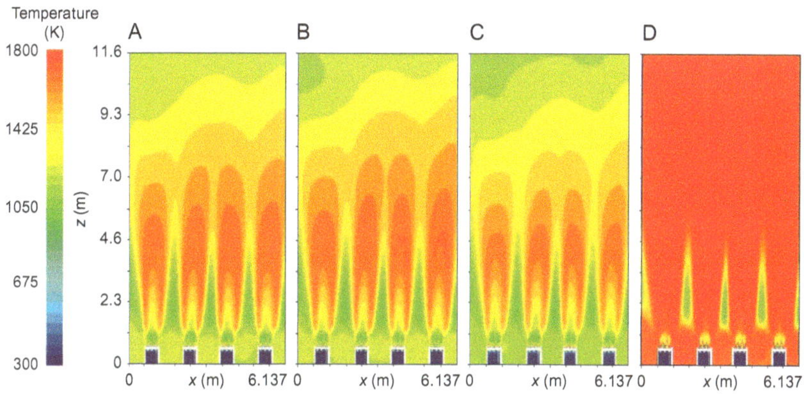

Fig. 17 Flue gas temperature in a vertical cross section at $y = 2.75$ m for different radiation models: (A) DOM; (B) DTRM; (C) P-1; (D) adiabatic. *Reprinted with permission from Hu G, Schietekat CM, Zhang Y, et al.: Impact of radiation models in coupled simulations of steam cracking furnaces and reactors,* Ind Eng Chem Res *54(9):2453–2465, 2015. Copyright 2015 ACS Publications.*

Fig. 17 shows the flue gas temperature–predicted different radiation models such as (A) discrete ordinates model (DOM), (B) discrete transfer radiation model (DTRM), (C) P-1, and (D) adiabatic in a vertical cross section along the furnace length at $y = 2.75$ m. The flue gas temperature of the adiabatic simulation is unrealistically high due to the absence of radiative heat transfer. For DOM, DTRM, and P-1 models, clear flame shapes are simulated. The results of the DOM and DTRM simulations are very similar, while slightly lower flue gas temperatures can be observed in the results obtained by the P-1 model.

Fig. 18 shows the computed contours of the optical thickness in the plane $x = 2.3$ m for the different radiation models. In this work, the optical thickness is defined as the product of the local mean absorption coefficient and the local integral turbulent length scale ($k^{3/2}/\varepsilon$). For all the models, the optical thickness in the jet core zones of the burners is rather low due to the locally low absorption coefficient. Outside the jet core, larger values of the optical thickness are simulated due to the abundance of H_2O and CO_2. On the upper left side, close to the reactors, low velocities and accompanying low kinetic energy levels explain the low optical thickness. A similar reasoning holds for the two recirculation zones in the outlet. The P-1 model is applicable at optical thicknesses above 1 (Habibi et al., 2007a). Hence, it is clear that the P-1 model is used outside its applicability range in some regions in the furnace.

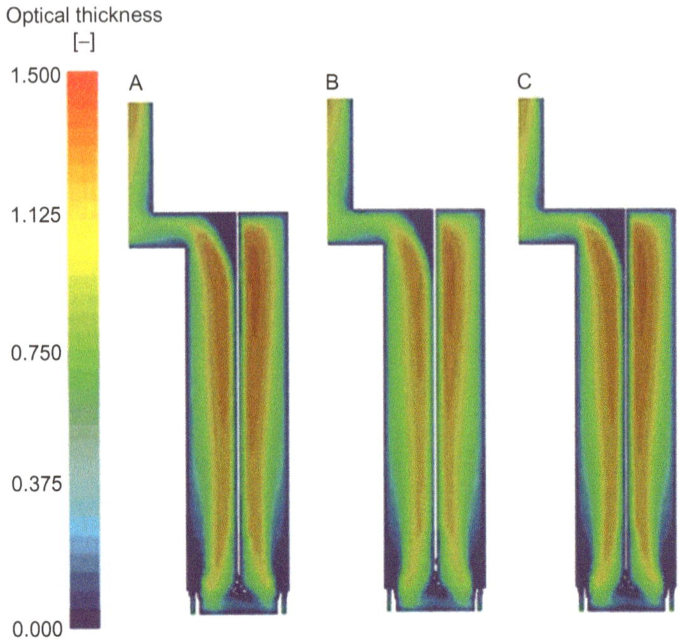

Fig. 18 Optical thickness in a vertical cross section at $x=2.3$ m for different radiation models: (A) DOM; (B) DTRM; (C) P-1. *Reprinted with permission from Hu G, Schietekat CM, Zhang Y, et al.: Impact of radiation models in coupled simulations of steam cracking furnaces and reactors,* Ind Eng Chem Res *54(9):2453–2465, 2015. Copyright 2015 ACS Publications.*

Fig. 19 shows (A) the heat flux to the reactors, (B) the external TMT, and (C) the process gas temperature along the reactor axial position at the start of run (SOR) conditions for DOM, DTRM, and P-1. The results of the adiabatic simulation are not shown. As can be observed from Fig. 19A, the heat flux profiles for DOM and DTRM are very similar; both show two maxima (i.e., one in each pass). With the P-1 model, a higher heat input to the first pass of the reactors is simulated compared to DOM and DTRM, while a lower heat input to the second pass of the reactors is simulated. The higher heat flux to the first pass with the P-1 model leads to a higher process gas temperature at the inlet of the second pass of the reactor compared to DOM and DTRM. This leads to a lower driving force for heat transfer from the furnace to the second pass, resulting in the lower heat flux to the second pass with the P-1 model than for DOM and DTRM. The reactor external wall temperature profiles depicted in Fig. 19B basically show the same trends as the heat flux profile. DOM and DTRM show

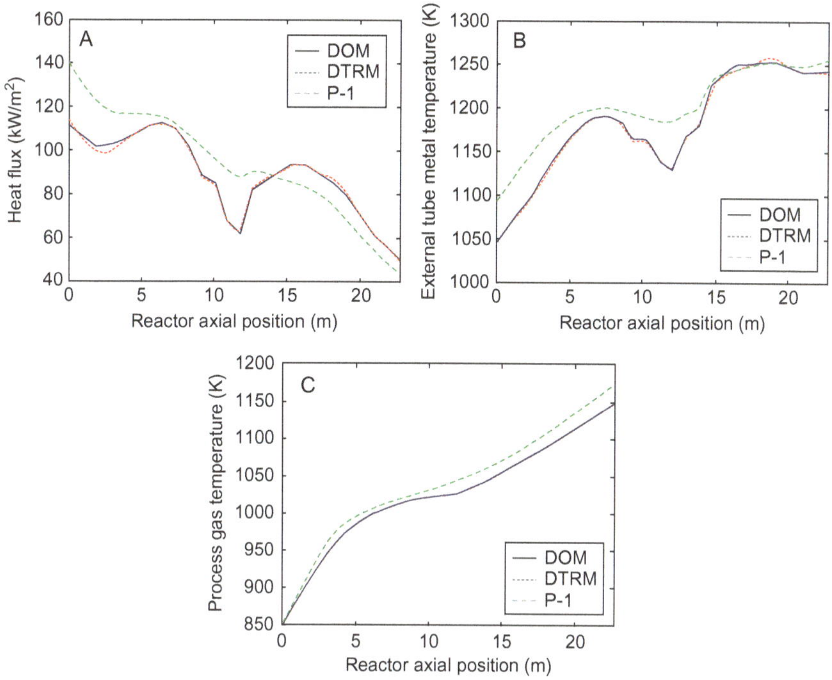

Fig. 19 (A) Heat flux per tube internal surface area; (B) SOR external wall temperature; (C) process gas temperature as a function of axial position with different radiation models. *Reprinted with permission from Hu G, Schietekat CM, Zhang Y, et al.: Impact of radiation models in coupled simulations of steam cracking furnaces and reactors,* Ind Eng Chem Res *54(9):2453–2465, 2015. Copyright 2015 ACS Publications.*

very similar results because of the similar heat flux profiles. With the P-1 model, higher TMTs are simulated in the first pass, whereas the profile flattens out in the second pass because of the lower heat input. In Fig. 19C, the process gas temperature profile for the P-1 model increases faster than for DOM and DTRM because of the higher heat flux to the first pass. However, after 15 m, the increase in temperature is highly similar because of the similar heat input. The similar increase in temperature after 15 m for P-1 compared to DOM and DTRM (notwithstanding the lower heat input) is explained by the higher cracking severity with P-1. Because the cracking severity is higher, the heat of the reaction is lower because of the higher reaction rates of exothermic reactions that form aromatics.

Fig. 20 shows the ethene and propene yield along the reactor axial position. Because of the higher process gas temperature for P-1, cracking starts faster and results in a higher cracking severity. This leads to a higher ethene

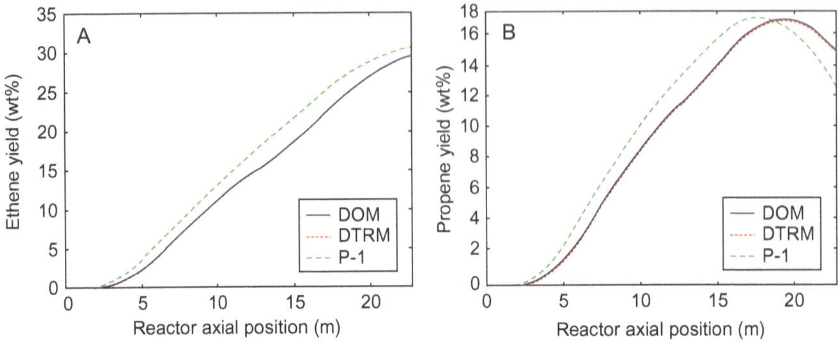

Fig. 20 (A) Ethene yield; (B) propene yield profile as a function of the axial position with different radiation models. *Reprinted with permission from Hu G, Schietekat CM, Zhang Y, et al.: Impact of radiation models in coupled simulations of steam cracking furnaces and reactors,* Ind Eng Chem Res *54(9):2453–2465, 2015. Copyright 2015 ACS Publications.*

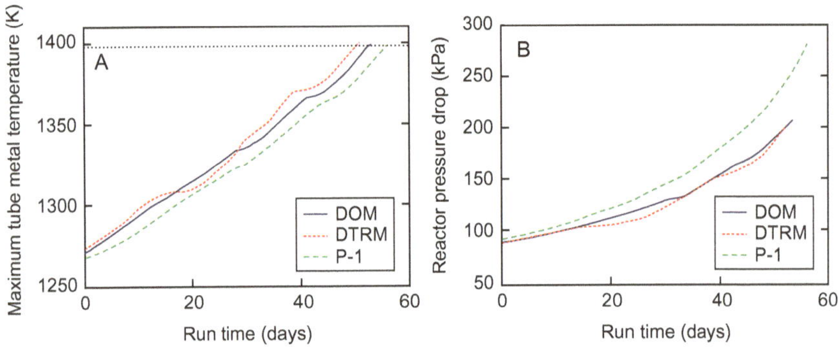

Fig. 21 (A) Maximum TMT; (B) reactor pressure drop as a function of run time. *Reprinted with permission from Hu G, Schietekat CM, Zhang Y, et al.: Impact of radiation models in coupled simulations of steam cracking furnaces and reactors,* Ind Eng Chem Res *54(9):2453–2465, 2015. Copyright 2015 ACS Publications.*

yield and a shift of the propene maximum upstream of the reactor because of the secondary reactions converting propene to ethene more upstream of the reactor.

Fig. 21 shows the maximum external TMT and the pressure drop as a function of run time for the three simulations. The operational limits are reached after 53, 51, and 56 days of operation when using DOM, DTRM, and P-1, respectively. These are all slightly lower than the design run length of 60 days. Despite the higher cracking severity in the P-1 simulation compared to DOM and DTRM simulations (i.e., 0.42 as opposed to 0.52), a longer run length is simulated in the P-1 case. This shows the importance

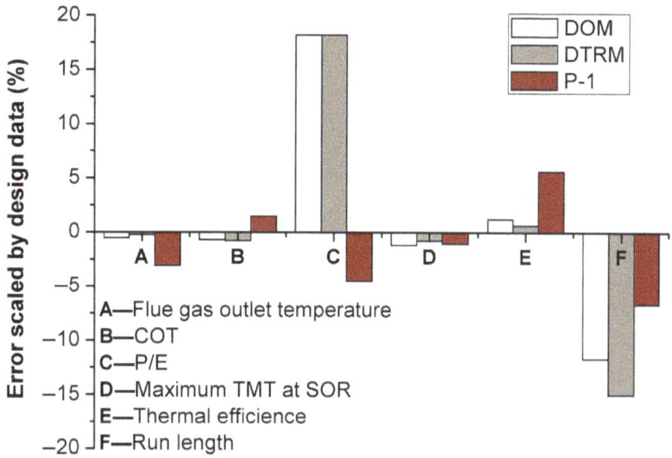

Fig. 22 Comparison of simulation results with different radiation models (scaled by design data).

of a correct shape of the heat flux profile for an accurate run length prediction, justifying the detailed and computationally intensive simulations performed, analyzed, and presented in this chapter.

Fig. 22 compares the most important simulation results obtained using various radiation models with the plant design data. The results of the adiabatic method deviate too much from the reasonable values and hence are not presented here. It can be seen that DOM and DTRM yield much smaller errors than the P-1 model, except for P/E and run length. This is due to the higher total heat input predicted by the P-1 model, which produces higher cracking severity and shorter run length. The thermal efficiency compared in Fig. 22 is defined as the ratio of the radiative heat transfer to the reactors to the total heat transfer to the coils. The higher total heat transfer to the reactors with the P-1 model results in a higher thermal efficiency of the furnace. The SOR thermal efficiency of the furnace with DOM and DTRM is within 0.6% of the thermal efficiency of the plant design data. However, the COT and the maximum SOR external wall temperature are somewhat less than the design data. This results in a lower simulated cracking severity, i.e., a higher propene-to-ethene ratio. Comparing the simulated results of all radiation models to the design data, DOM and DTRM perform best. However, the current implementation of DTRM in ANSYS FLUENT 14.0 is not compatible with parallel processing. Therefore, DOM is the recommended radiation model to use for the simulation of steam cracking furnaces.

In summary, coupled furnace-reactor simulations of an industrial naphtha cracking furnace with 130-kt/a capacity were performed. For the first time, different radiation models were evaluated in coupled furnace-reactor simulations for the purpose of predicting run lengths. Adopting the adiabatic model resulted in unrealistically high flue gas temperatures due to the neglect of radiative heat transfer, which is the main contribution of heat transfer in the furnace. The results with the DOM and DTRM radiation models are very similar. The flue gas temperature with DOM is higher than with the P-1 radiation model, resulting in higher incident radiation. Calculation of the optical thickness shows that it is less than 1 in some regions of the furnace. The P-1 model is valid only for an optical thickness above 1 (Habibi et al., 2007a), which explains the discrepancies with DTRM and DOM. With the P-1 model, a higher total heat input to the reactors is simulated compared to DOM and DTRM. Also, the heat flux profile along the reactor axial length is different for P-1 than for DOM and DTRM. Comparing the simulated results of all radiation models to the plant design data, DOM and DTRM perform best. However, as DOM has a broader application range than DTRM, DOM is the recommended radiation model to use for the simulation of steam cracking furnaces, particularly when run-length simulations are carried out.

2.2.2 Flue Gas Nongray Radiative Properties

As the heat required for cracking reactions is mainly transferred by radiation from the flue gas and the furnace refractory, it is essential to study their radiative properties using accurate models. In principle, both the flue gas and the surface show strong nongray behavior in radiative heat transfer; i.e., the absorption coefficient of the flue gas and the surface emissivity is dependent on the wavelength and varies a lot over the spectrum. However, simulations taking nongray properties into account are too expensive computations to be performed, especially for large-scale industrial furnaces. So far, the most widely used models in industrial applications still treat the flue gas and furnace refractory as gray; i.e., assuming that the gas and the surface emit and absorb radiation over the entire spectrum.

Normally, the surface shows much less nongray behavior than the flue gas does, and it can be treated as gray without introducing significant errors. The only exception is when the surface radiative properties have to be calculated accurately, such as in the research conducted by Heynderickx and Nozawa (2004, 2005) and Stefanidis et al. (2008). Nongray radiative properties models were used in these studies, as the gray models are incapable of evaluating the effect of high-emissivity coating on furnace thermal efficiency improvement (Heynderickx and Nozawa, 2004). On the other hand, it has been reported that treating the flue gas as gray in combustion systems may lead to underprediction of the flame temperature by up to 100K or even more (Modest, 2013). This was also observed by Stefanidis et al. (2007), who compared gray and nongray radiative models in a steam cracking furnace. As a consequence of temperature underprediction, the furnace thermal efficiency was 5% overpredicted by the gray model, which is quite large considering the scale of the industrial furnace.

Although the aforementioned research clearly shows that the flue gas radiative properties have a great impact on the furnace simulation, the effect still needs to be studied quantitatively. Therefore, we preformed coupled furnace-reactor simulations of an industrial naphtha cracking furnace with 130 kt/a capacity, as shown in Fig. 16. The nongray gas radiative model was developed based on the exponential wide band model (EWBM) proposed by Edwards (1976), which consists of 11 absorption bands for the H_2O and CO_2 gas mixture. To reduce the computational cost, only five bands suggested by Stefanidis et al. (2007) were taken into account in nongray model development and are listed as follows:

- H_2O: 6.3 μm, 2.7 μm
- CO_2: 15 μm, 4.3 μm, 2.7 μm

As there are spectral windows between the selected absorption bands, the total number of spectral intervals in which the RTE needs to be solved is nine. Therefore, the developed nongray model is referred to as the *nine-band model.* The nine-band model was validated against the full EWBM and a correlation developed by Leckner (1972) based on spectrally integrated emissivities. Fig. 23A shows the total emissivity as a function of temperature for H_2O and CO_2, respectively. It can be seen that the agreement between the emissivity predicted by the EWBM and Leckner's correlation is generally good for both H_2O and CO_2. The nine-band model also gives quite reasonable values, even with some omitted absorption bands. Large deviation only occurs at lower temperatures for H_2O due to the lack of the H_2O rotational band at 71 μm in the nine-band model. However, it is expected to have less impact on the radiative heat transfer calculation because the flue gas temperature always exceeds 1200K in steam cracking furnaces. To confirm this, simulations of a test furnace were performed and the predicted heat flux profiles using Leckner's correlation and the nine-band model were compared. As shown in Fig. 23B, the difference in the heat flux profiles is negligible, indicating that the nine-band model is suitable to account for the flue gas radiative properties.

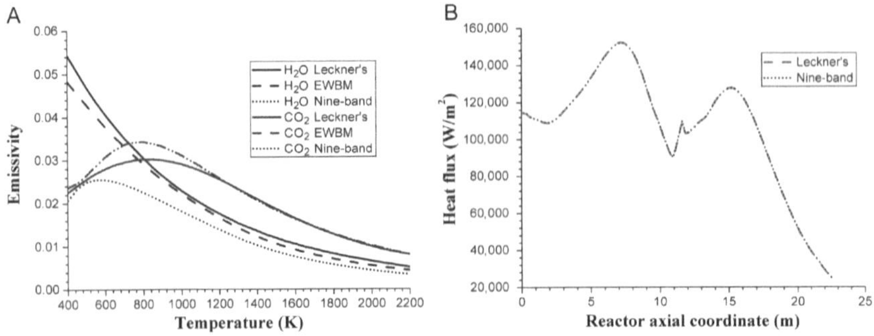

Fig. 23 (A) Emissivities of H_2O and CO_2 as a function of temperature. (B) Averaged heat flux (W/m^2) over all reactors in the test furnace as a function of the reactor axial coordinate (m). *Adapted with permission from Zhang Y, Qian F, Schietekat CM, et al.: Impact of flue gas radiative properties and burner geometry in furnace simulations,* AIChE J *61(3):936–954, 2015. Copyright 2015 John Wiley and Sons.*

For the purpose of evaluating the influence of nongray gas treatment on the simulation of the steam cracking furnace, the gray implementation of the WSGGM proposed by Smith et al. (1982) was used as the gray gas model to be compared with the nine-band nongray gas model. A fully coupled furnace-reactor simulation was performed using the gray and the nongray model, respectively. The commercial software ANSYS FLUENT 14.0 was used for the fire side modeling with the nine-band model implemented as a UDF. The reactor simulations were performed using COILSIM1D (Van Geem et al., 2007b) developed in the Laboratory for Chemical Technology.

The flue gas temperature contours at $x=0.778$ m of the two simulation cases are compared in Fig. 24. It can be clearly seen that the nongray gas model leads to higher flame temperature in the furnace, which is consistent with the previous discussion. On the other hand, the difference in the flame shape between the two cases is quite small, indicating that the effect of the flue gas radiative properties on the combustion process is negligible. This is obvious because flue gas appears only when large amounts of fuel and air are already consumed; hence, it mainly affects the temperature field of the region where combustion is completed. The conclusion is further confirmed by Fig. 25A. The average flue gas temperature profiles calculated from the gray and the nongray cases show little difference at furnace elevations below 2.5 m, where the turbulent mixing of fuel and air occurs. Fig. 25A also shows that the predicted flue gas outlet temperature of the nongray gas model is about 70K higher than that of the gray gas model, which means that more heat is taken away by the flue gas; therefore, the furnace efficiency

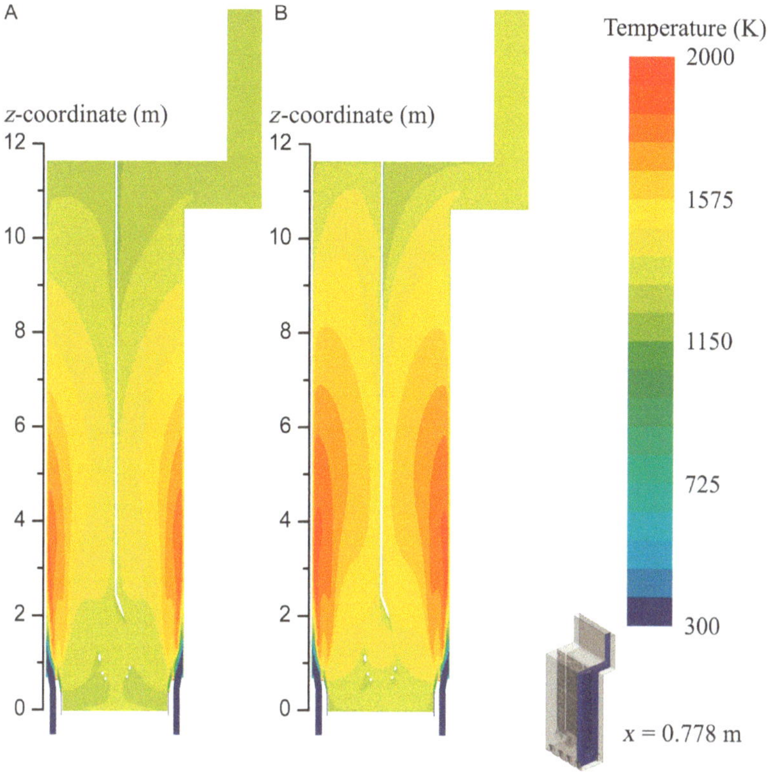

Fig. 24 Temperature field in a vertical cross section at the center of the burner set 1 (i.e., at $x = 0.778$ m): (A) gray gas model; (B) nongray gas model. *Adapted with permission from Zhang Y, Qian F, Schietekat CM, et al.: Impact of flue gas radiative properties and burner geometry in furnace simulations,* AIChE J *61(3):936–954, 2015. Copyright 2015 John Wiley and Sons.*

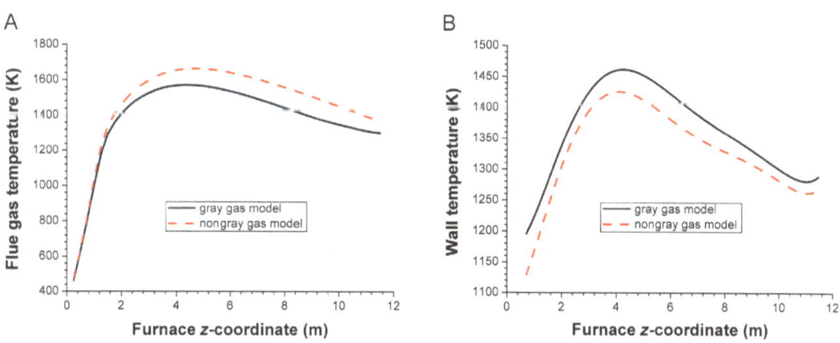

Fig. 25 (A) Mixing-cup averaged flue gas temperature and (B) area-averaged furnace refractory wall temperature as a function of furnace z-coordinate (m) above the burner at $x = 0.778$ m. *Adapted with permission from Zhang Y, Qian F, Schietekat CM, et al.: Impact of flue gas radiative properties and burner geometry in furnace simulations,* AIChE J *61(3):936–954, 2015. Copyright 2015 John Wiley and Sons.*

decreases about 3.6% in the nongray case. As the furnace refractory contributes more than the flue gas in radiative heat transfer to the reactor coils, the temperature of the furnace refractory wall in the nongray case is also expected to be lower than that in the gray case, as shown in Fig. 25B.

As mentioned previously, coupled furnace-reactor simulations were performed to study the impact of the flue gas radiative properties on the cracking reactor coils. The COTs and maximum TMTs of all 22 reactor coils are depicted in Fig. 26. It should be noted here that all reactor coils were coupled with the furnace to yield more accurate results. The distribution of the COT and the maximum TMT over the reactor coils is due to the different spatial locations of the reactor coils relative to the burners and the shadow effects (Zhang et al., 2015). Despite this uneven distribution, the nongray gas model generally results in a much lower COT and a maximum TMT, which are about 44K and 38K, respectively. This clearly shows the importance of using a nongray gas model in steam cracking furnace simulations for two reasons: First, cracking severity is quite sensitive to COT, and 44-K overprediction of COT by the gray gas model will introduce unacceptable large errors; and second, maximum TMT is used as the operational limit for steam cracking furnaces, so the inaccurate gray gas model will also have a negative impact on the run-length simulation.

To provide a better understanding of the nongray radiative heat transfer behavior of the flue gas, a top view of the incident radiation fields in the spectral window and the absorption band is shown in Fig. 27. Since the flue gas does not participate in the radiative heat transfer in the spectral window, the incident radiation is determined only by the temperature of the furnace

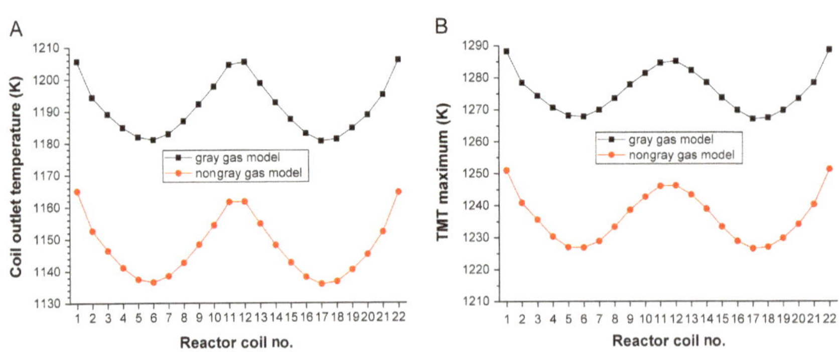

Fig. 26 The distribution of (A) coil outlet temperature; (B) maximum external TMT over all 22 reactors. *Adapted with permission from Zhang Y, Qian F, Schietekat CM, et al.: Impact of flue gas radiative properties and burner geometry in furnace simulations,* AIChE J *61(3):936–954, 2015. Copyright 2015 John Wiley and Sons.*

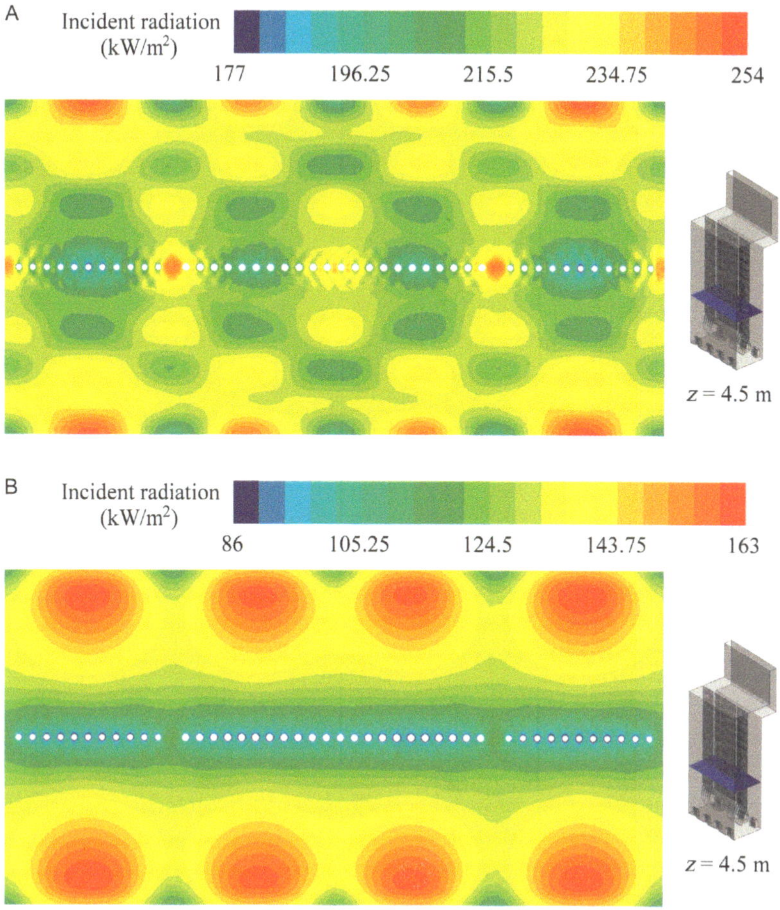

Fig. 27 Incident radiation field calculated by the nongray gas model in a horizontal plane at $z=4.5$ m in (A) spectral window; (B) absorption band. *Adapted with permission from Zhang Y, Qian F, Schietekat CM, et al.: Impact of flue gas radiative properties and burner geometry in furnace simulations,* AIChE J *61(3):936–954, 2015. Copyright 2015 John Wiley and Sons.*

refractory and the reactor tube wall, as well as the view factors between the surfaces. Hence, the maximum incident radiation appears at the furnace refractory right above the burners. While for the absorption band, flue gas acts as the main contributor of the radiation source. Therefore, the maximum shifts about 0.2 m away from the furnace refractory and located at the place of the flame center. As the radiation emitted from the furnace refractory and the flame is reabsorbed by the flue gas closer to the reactor coils at a relatively lower temperature, the distribution of the incident radiation in the absorption band is more uniform than that in the spectral window.

Finally, the computational cost of using different radiation models and radiative properties models is presented. Because the simulations in Sections 2.2.1 and 2.2.2 were performed on different computer infrastructures, it is not fair to directly compare the wall times for those cases. However, all the radiation models in Section 2.2.1 were combined with gray gas implementation of WSGGM, so they can be considered comparable to the gray gas model discussed in Section 2.2.2. The computational cost was scaled by the case that uses the combination of DOM and WSGGM as it is presented in both studies. Taking the central processing unit (CPU) time for DOM as 1 unit, it is shown that DTRM (11 units) is the most time consuming model for gray gas implantation comparing to P-1 (0.82 units) and Adiabatic (0.65 units). Interestingly, the CPU time does not increase a lot when using nongray gas models (1.67 units). This is because the RTE was solved once in every 10 iterations of the conservation equations and nongray implantation requires only four more equations to solve with every cell. However, the data file for the nongray gas model is more than five times larger than that of the gray gas model.

2.3 Convection Section

In most of the existing studies of steam cracking furnaces, the convection section was omitted or simplified as a single unit without considering the details in fluid flow of a complex hydrocarbon mixture and the corresponding heat transfer process, especially for two-phase flow boiling. For example, Aspen Plus and other software programs were used to establish the mathematical model of the convection section and simulate the flow, mass transfer, and heat transfer in the tube of the convection section. With the development of CFD and computer technology, research in the microscopic phenomena of the convection section has attracted more and more attention. The most recent work on the convection section using CFD was done in studies by De Schepper et al. (2009a,b, 2010) and Verhees et al. (2016), in which the two-phase flow boiling, the coke formation on the convection tubes, and the coupled effect of the process side and the flue gas side were comprehensively studied. However, this research still has some deficiencies: (1) a uniform heat flux profile was imposed on the tube wall when simulating the flow boiling process in the evaporation section, which is apparently not the case for industrial convection sections and will lead to some errors in prediction of heat transfer and feedstock vaporization; and (2) the evaporation section was not accounted for in coupled CFD simulations of the convection section due to its highly computational cost—in this case, the effects of coupling between the process side and the flue gas side were not fully considered for the entire convection section. Therefore, we carried out complete coupled simulations of the convection chamber and tubes with dual-stage steam feed mixing of an industrial ethene cracking furnace using CFD for the first time.

A schematic diagram of a convection section with dual-stage steam feed mixing is shown in Fig. 28. This section consists of eight tube sections, which are the feed preheater (FPH), economizer (ECO), high-temperature coil I (HTC-I), high-temperature coil II (HTC-II), high-pressure steam

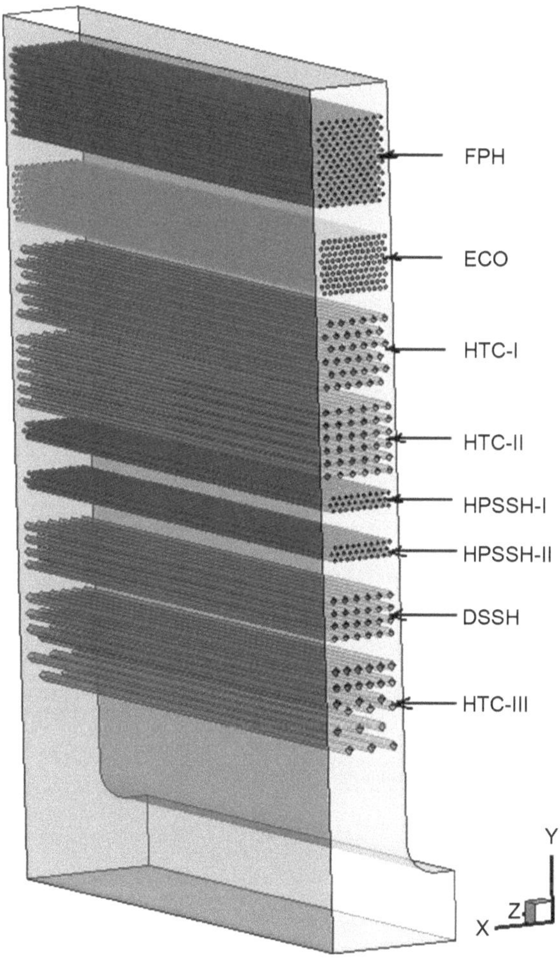

Fig. 28 Schematic diagram of the convection section with dual-stage steam feed mixing. *Reprinted with permission from Hu G, Yuan B, Zhang L, et al.: Coupled simulation of convection section with dual stage steam feed mixing of an industrial ethylene cracking furnace,* Chem Eng J *286:436–446, 2016. Copyright 2016 Elsevier.*

superheater I(HPSSH-I), high-pressure steam superheater II (HPSSH-II), dilution steam superheater (DSSH) and high-temperature coil III (HTC-III). Feedstock is heated in the FPH and changes gradually from liquid to gas. Boiler feed water is heated in the ECO. The first steam flux is carried out before the feedstock enters the HTC-I. The second steam flux enters the DSSH and is mixed with the HTC-II outlet, and then enters the HTC-III to be heated.

2.3.1 Convection Chamber

Fig. 29 shows flue gas velocity vector field at cross section $z=5$ m. At the corner, both close to and away from the entrance to the convection chamber,

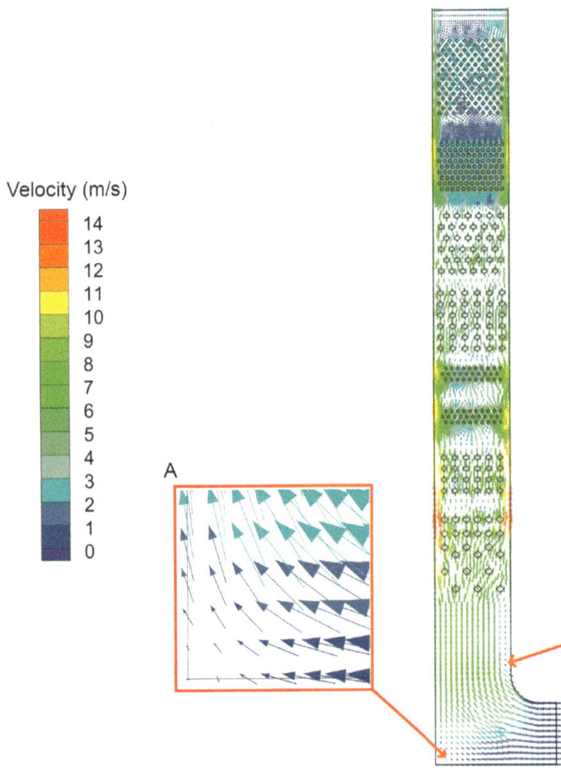

Fig. 29 Flue gas velocity vector filed at cross section $z=5$ m. (A) velocity vectors enlarged diagram at the corner away from the entrance to the convection chamber; (B) velocity vectors enlarged diagram at corner near the entrance to the convection chamber. *Reprinted with permission from Hu G, Yuan B, Zhang L, et al.: Coupled simulation of convection section with dual stage steam feed mixing of an industrial ethylene cracking furnace,* Chem Eng J *286:436–446, 2016. Copyright 2016 Elsevier.*

there are clearly two recirculation zones. The recirculation zone makes high-temperature flue gas remain longer in the convection chamber. At the same time, it causes local overheating in the furnace wall of the convection chamber, thus not being conducive to the stable operation of the cracking furnace. When high-temperature flue gas enters the convection chamber, its velocity is larger and will scour the furnace wall of the opposite side. In the convection chamber bottom, flue gas velocity distribution along the x axial direction is not uniform. This is because the structure of the convection chamber is asymmetrical. With the increase of height, the flue gas passes through the staggered tube bundles and its velocity field along the x axial direction gradually becomes uniform. While arriving at the outlet of convection chamber, the flue gas velocity is basically consistent along the x axial direction.

Fig. 30 shows flue gas temperature profiles along the x and z axial directions. At the height of 3.10 m, the flue gas temperature at $x=0.1$ m is smaller than that of $x=1.0$ m. This is because the flue gas velocity of $x=0.1$ m, which is near the entrance to the convection chamber, is less than that of $x=1.0$ m, leading to a smaller heat transfer coefficient and thus a lower flue gas temperature. The flue gas of upward flow carries out heat exchange with the tubes, and the flue gas temperature field also gradually becomes uniform. In Fig. 30, the nonuniformity of the flue gas temperature along the x axial direction gradually reduces

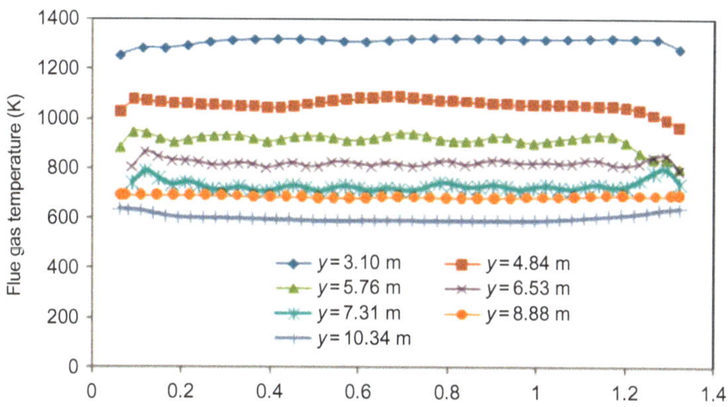

Fig. 30 Flue gas temperature distribution along the x axial direction in the convection chamber. *Reprinted with permission from Hu G, Yuan B, Zhang L, et al.: Coupled simulation of convection section with dual stage steam feed mixing of an industrial ethylene cracking furnace,* Chem Eng J *286:436–446, 2016. Copyright 2016 Elsevier.*

with increasing elevation. Uniform flue gas temperature distribution is achieved at $y=8.88$ m, indicating that the staggered tubes facilitate the heat transfer. It is also shown that the flue gas temperature gradually decreases with increasing height.

2.3.2 The Simulation Results of the Tube Section

The mass fraction profile of an FPH straight tube is shown in Fig. 31. Because of the long tube length, in order to show the calculation results clearly, the x-, y-, and z-axes are adopted as the proportion of 5:5:1. The flow pattern change of hydrocarbon feedstock has a close relationship with the TMT. In the tube, the liquid near the tube wall first reaches the saturation temperature, and then it forms a bubble that distributes in a continuous liquid phase called the *bubble flow,* as shown in Fig. 31A. With the influence of gravity, the bubble will move to the top of the tube, and thus the liquid is gradually divided by a long, thin bubble. At this time, the fluid will become chaotic, forming an elongated bubble flow, as shown in Fig. 31B. When a lot of bubbles gather on the top of the tube and form a continuous gas phase, a clear interface between the gas phase and the liquid phase can be observed. Then, the liquid phase lies below the gas phase, forming the stratified flow, as shown in Fig. 31C. When most of the liquid

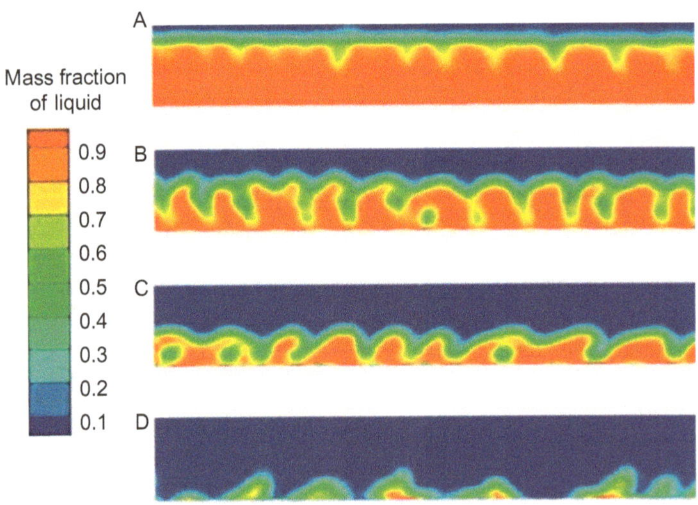

Fig. 31 Mass fraction profile of FPH straight tube: (A) bubble flow; (B) elongated bubble flow; (C) stratified flow; (D) spray flow. *Reprinted with permission from Hu G, Yuan B, Zhang L, et al.: Coupled simulation of convection section with dual stage steam feed mixing of an industrial ethylene cracking furnace,* Chem Eng J *286:436–446, 2016. Copyright 2016 Elsevier.*

phase finishes vaporization, a small amount of liquid phase disperses in the continuous gas phase, forming the spray flow, as shown in Fig. 31D. In the vaporized part of the feedstock, the TMT will sharply increase, and thus the TMT in this place will be higher than that near the tube wall. The evolution of the flow patterns shown in Fig. 31 is in general agreement with that of a two-phase flow in a horizontal evaporator tube (Kattan et al., 1998).

In view of the similarity of the simulation results of different tube sections, this work will take the simulation results of the HTC-I section as an example to describe and analyze. Fig. 32 shows external tube heat flux profiles of different passes along the tube axial position in HTC-I. The heat flux at the two ends of the tube is obviously lower than that of the other location. This is because the two ends of the tube lie near the furnace wall, where the flue gas temperature is low, leading to a low difference between the flue gas temperature and TMT, in turn resulting in a low heat flux. One can also see from this figure that in Pass 1, Pass 3, and Pass 5, the heat flux increases along the tube axial position, while in Pass 2, Pass 4, and Pass 6, the heat flux decreases along the tube axial position.

Fig. 33A shows the process gas temperature profiles of different passes along the tube axial position in HTC-I. The process gas temperature increases from Pass 1 to Pass 6 because the process gas absorbs the heat transferred by the flue gas. Fig. 33B shows the external TMT profiles of different passes along tube axial position in HTC-I. The external TMT increases from

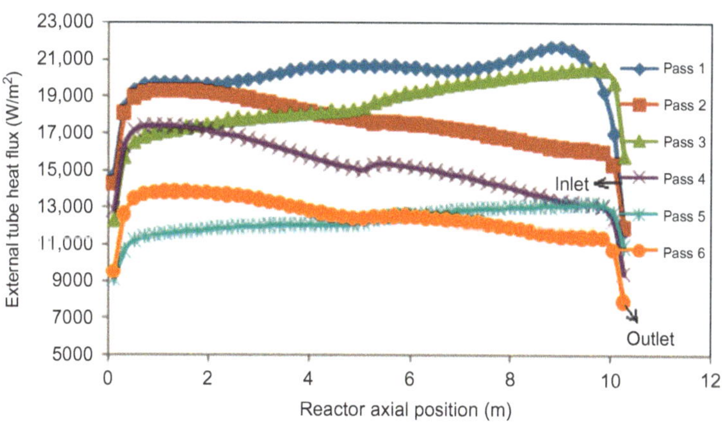

Fig. 32 External tube heat flux profiles of different passes in HTC-I. *Reprinted with permission from Hu G, Yuan B, Zhang L, et al.: Coupled simulation of convection section with dual stage steam feed mixing of an industrial ethylene cracking furnace,* Chem Eng J *286:436–446, 2016. Copyright 2016 Elsevier.*

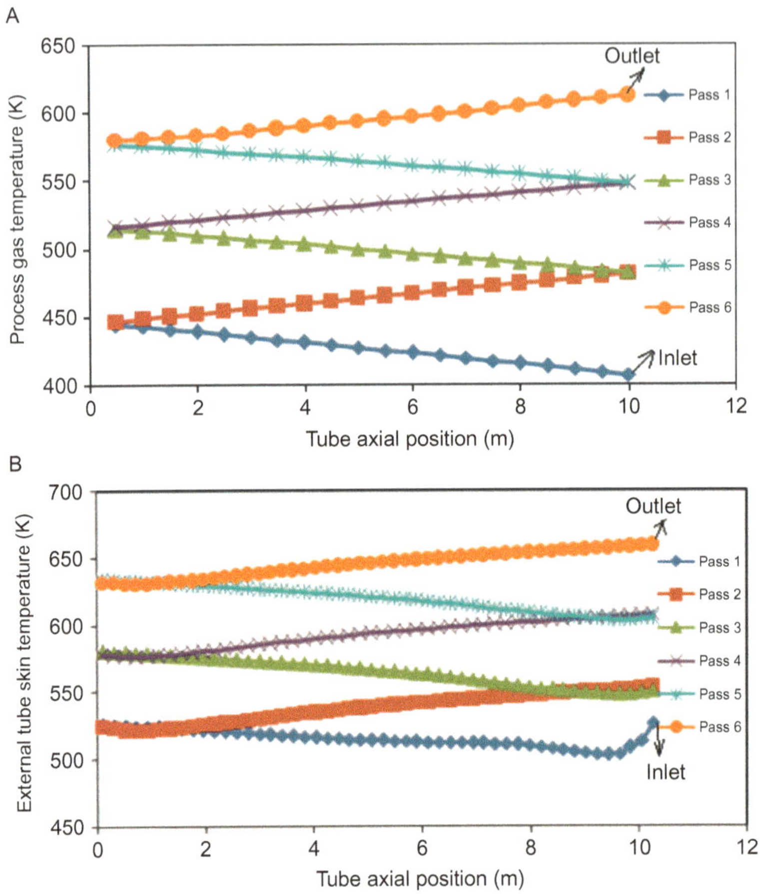

Fig. 33 (A) Process gas temperature profiles and (B) external tube skin temperature of different passes in HTC-I. *Reprinted with permission from Hu G, Yuan B, Zhang L, et al.: Coupled simulation of convection section with dual stage steam feed mixing of an industrial ethylene cracking furnace,* Chem Eng J *286:436–446, 2016. Copyright 2016 Elsevier.*

Pass 1 to Pass 6, except near the two ends of the tubes (i.e., in the crossover position of two tube passes) where the external TMT stays stable. This is because the two ends of the tubes connect with the bend tubes that are outside the convection section, which are not heated by flue gas, leading to a lower TMT there than inside the convection section.

In conclusion, a complete coupled convection chamber/tube simulation of an industrial cracking furnace with dual-stage steam feed mixing was carried out for the first time in this study. Detailed simulation results, including the flue gas velocity and temperature fields in the convection

Table 2 Comparison of Simulation Results and Industrial Data

Items	Industrial Data	Simulation Results
Exhaust gas temperature (K)	385.15	388.65
Outlet temperature of FPH (K)	430.15	437.41
Outlet temperature of ECO (K)	523.15	527.78
Outlet temperature of HTC-I (K)	609.15	615.75
Outlet temperature of HTC-II (K)	677.15	681.61
Outlet temperature of HPSSH-I (K)	695.15	698.66
Outlet temperature of HPSSH-II (K)	793.15	792.44
Outlet temperature of DSSH (K)	844.15	843.18
Outlet temperature of HTC-III (K)	884.15	887.38

Hu G, Yuan B, Zhang L, et al.: Coupled simulation of convection section with dual stage steam feed mixing of an industrial ethylene cracking furnace, *Chem Eng J* 286:436–446, 2016. Copyright 2016 Elsevier.

chamber and the process gas temperature, TMT, and heat flux profiles in the tubes, were obtained. As shown in Table 2, the simulation results are in good agreement with the industrial data, which explains the accuracy of the model and indicates that it could be used for further research. A complete coupled convection chamber/tube simulation of an industrial cracking furnace with dual-stage steam feed mixing was proposed for the first time. Because the structure of the convection chamber is asymmetrical, flue gas velocity distribution along the width direction is not uniform. With the increase of height, the flue gas flow field along the width direction gradually becomes uniform. This shows that staggered tubes facilitate the exchange of heat. Two recirculation zones occur at the corner both close to and away from the entrance to the convection chamber, which will cause a longer residence time of the flue gas and local overheating in the furnace wall of the convection chamber. The process gas temperature, TMT, and heat flux profiles are different along the axial and radial directions. The changes of flow pattern from bubble flow to spray flow are effected by gravity and centrifugal force during evaporation.

2.4 Optimization and Scheduling of Steam Cracking Furnace Systems

The aim of establishing accurate models for steam cracking furnaces is to carry out studies on design and operation optimization based on these models. Moreover, multiple ethene furnaces are usually run in parallel in a cracking plant, with different feedstocks available from the upstream tanks. Thus, finding an optimal scheduling of the entire cracking furnace system has great significance for the purpose of improving the economic potential of the cracking plant. However, the computational cost of the rigorous models involving detailed kinetics is enormous due to the complexity of the steam cracking process, making it infeasible to carry out optimization using these models. Therefore, surrogate models, which can be considered to be a regression of a set of input-output data obtained by evaluating the complex rigorous models, are often used for this purpose. This so-called model of the model (Singh et al., 2013) has the advantage of easy and fast access, and it has been widely used in chemical process control, optimization, and scheduling. Surrogate models for industrial steam cracking furnaces usually take operating variables as inputs, including feedstock flow rate, COT, and dilution steam. While the outputs of the models are the results which are of interest for the purpose of furnace operation and optimization, such as product yield, TMT, coke thickness, etc. In the following section, the recent progress in surrogate model development is discussed, followed by multiobjective optimization and integrated operation and cyclic scheduling for an industrial ethene cracking furnace system based on the established surrogate model.

2.4.1 ANN-Based Surrogate Model

Artificial neural networks (ANNs) are the most widely used surrogate model compared to other surrogate models such as the response surface methodology and Kriging interpolation. Due to its excellent performance in machine learning and simple structure, ANN has been applied in many areas of chemical engineering (Eason and Cremaschi, 2014; Nuchitprasittichai and Cremaschi, 2013; Zendehboudi et al., 2014). However, in most of the previous studies, the structure of the ANN was predetermined. As the accuracy of the ANN-based surrogate model is strongly dependent on its structure, an inappropriately selected structure may lead to great error in the surrogate model predictions. In addition, the data set selected for the model training also has a great impact on the final surrogate model. As Eason and Cremaschi (2014) have reported, ANN may exhibit undesirable oscillations between the sample points and highly nonlinear behavior may be missed by the variance-based data set sampling method often used in previous studies. To overcome these deficiencies, we proposed a new approach combining automatic ANN structure determination with a prediction error–based, mixed adaptive sampling method for the surrogate model development (Jin et al., 2016).

In this work, a three-layer, feed-forward ANN was used due to its clear mathematical expression. As there is only one hidden layer between the input and the output layers, the ANN structure determination is equivalent to the selection of the number of hidden nodes. A five-step procedure was proposed to calculate the proper number of hidden nodes in the ANN based on its effective number of parameters estimated by Bayesian regularization (Hagan et al., 2014). For the adaptive sampling method, the prediction error–based mixed adaptive sampling method was used to prevent overfitting and underfitting that may occur in the variance-based method of Eason and Cremaschi (2014).

To test the performance of the two methods (i.e., the automatic ANN structure determination and the prediction error-based mixed adaptive sampling), three challenge functions were used as the models of complex processes: one 2D Shekel function on domain $[0,10]^2$, one 2D Ackley function on domain $[2,2]^2$, and one 2D Peaks on domain $[3,3]^2$. Four surrogate models were established using the following combinations of methods: fixed ANN structure and variance-based adaptive sampling by Eason and Cremaschi (2014), referred to as "EC_fixed"; improved version of the "EC_fixed" model with automatic ANN structure determination, referred to as "EC_free"; improved version of the "EC_fixed" model using prediction error–based mixed adaptive sampling, referred to as "JIN_fixed"; and a combination of both the automatic ANN structure determination and the prediction error–based mixed adaptive sampling, referred to as "JIN_free."

Due to space limitations, only the 2D Shekel function is discussed here, with the graphical representation shown in Fig. 34. This function has a highly nonlinear dip, which makes it a good candidate for evaluating the ANN-based surrogate model.

The responses of the four surrogate models are depicted in Fig. 35. It can be clearly seen that the automatic ANN structure determination and the prediction error–based mixed adaptive sampling method significantly improve the performance of the established surrogate model. Comparing Fig. 35A and B, we can conclude that the ANN structure used by Eason and Cremaschi (2014) failed to capture the dip of the 2D Shekel function, indicating the necessity of the automatically determined ANN structure, especially for a process model whose behavior is not known a priori for a proper structure to be selected. Using the prediction error-based mixed adaptive sampling method also helps to overcome the problem; however, the surrogate model shows strong oscillation due to the inappropriate structure of the ANN, as shown in Fig. 35C. The predicted minimum value of the 2D

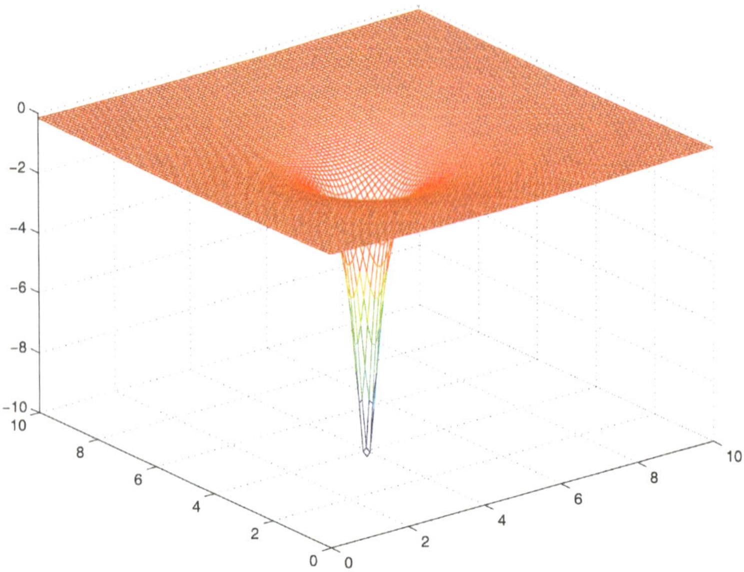

Fig. 34 2D Shekel function on the domain of $[0,10]^2$. *Reprinted with permission from Jin Y, Li J, Du W, et al.: Adaptive sampling for surrogate modelling with artificial neural network and its application in an industrial cracking furnace,* Can Chem Eng *94(2):262–272, 2016. Copyright 2016 John Wiley and Sons.*

Shekel function by JIN_free is −10.1546, which is closest to the true value, making JIN_free the best of the other surrogate models, followed by JIN_fixed, EC_free, and EC_fixed. Similar trends can be observed for other test functions as well (Jin et al., 2016).

Numerical experiments and engineering applications on surrogate modeling of an industrial ethene cracking model also validated the efficacy of the aforementioned methods (Jin et al., 2016). The results show that EC_free and JIN_free produce better surrogate models than EC_fixed and JIN_fixed. Moreover, the difference between the performance of EC_free and JIN_free is quite small, indicating that the steam cracking process does not have a highly nonlinear region in the considered operating conditions. Therefore, EC_free and JIN_free are recommended for surrogate modeling of complex chemical processes such as steam cracking.

2.4.2 Multiobjective Optimization of a Single Furnace

Steady-state optimization have been widely studied by many researchers (e.g., Gao et al., 2009a,b; Li et al., 2007; Tarafder et al., 2005); however, the effect of coke accumulation on the inner wall of the reactor coil on

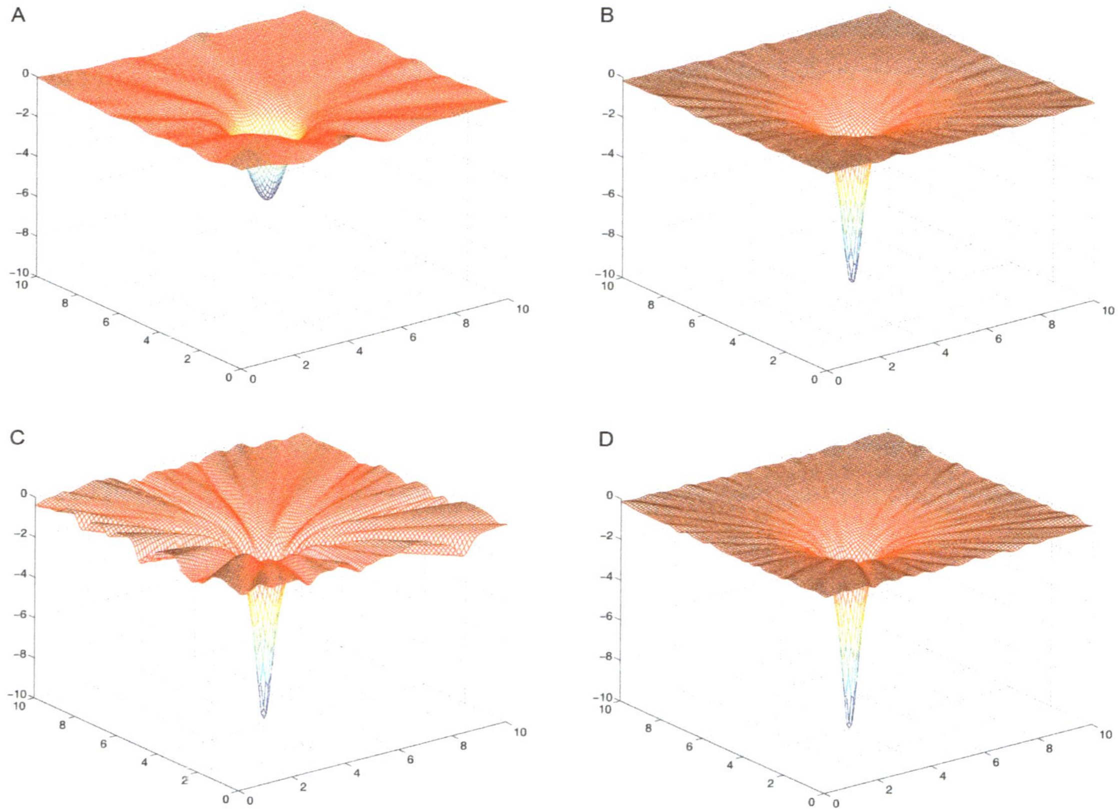

Fig. 35 The performance of the surrogate model obtained by (A) EC_fixed; (B) EC_free; (C) Jin_fixed; (D) Jin_free. *Adapted with permission from Jin Y, Li J, Du W, et al.: Adaptive sampling for surrogate modelling with artificial neural network and its application in an industrial cracking furnace,* Can J Chem Eng *94(2):262–272, 2016. Copyright 2016 John Wiley and Sons.*

the cracking performance and product yield were rarely considered, especially for research on multiobjective optimization. Since it has been reported that furnace optimization taking dynamic behavior of the coke formation can further improve the profits (Berreni and Wang, 2011; Edwin and Balchen, 2001; Gao et al., 2008; Masoumi et al., 2006), a surrogate model-based, multiobjective optimization of an industrial naphtha cracking furnace was done for the first time (Jin et al., 2015b).

As discussed in the previous section, ANN is a good candidate for surrogate model development. In this study, a feed-forward neural network-based state space equation model was used for optimization to reduce computational costs. The feedstock flow rate and COT were chosen as input variables, while the coke layer thickness was taken as the state variable. Due to the different timescales between the steam cracking and the coke formation, the surrogate model was considered as pseudo steady-state; i.e., the coke layer thickness was assumed to be constant during each discrete time step and was updated between the two time steps. The output variables are the TMT and the yield of valuable products such as ethene, propene, hydrogen, 1,3-butadiene, and benzene. The sample points were generated by the COILSIM1D software (Van Geem et al., 2007b) under the operating conditions covered by an industrial naphtha cracking furnace. The total number of the samples was 1666, in which 1204 samples were used for model training and the remainder were taken as test points. The performance of the developed surrogate model was evaluated by the predicted root mean square error (PRMSE) and the predicted relative root mean square error (PRRMSE), defined as follows:

$$\mathrm{PRMSE} = \sqrt{\frac{\sum_{i=1}^{n} \left(\tilde{z}_i - z_i\right)^2}{n}} \tag{1}$$

$$\mathrm{PRRMSE} = \sqrt{\frac{\sum_{i=1}^{n} \left(\tilde{z}_i - z_i\right)^2 / n}{\sum_{i=1}^{n} z_i / n}} \tag{2}$$

where z_i is the values of the test points and $\tilde{z}_i$ is the predicted value by the surrogate model. Table 3 lists the results of PRMSE and PRRMSE for every output variable, where good prediction accuracy of the surrogate model can be observed.

The multiobjective optimization was then carried out using the established surrogate model, taking the run-length behavior into account. The time interval was 1 h; i.e., the coking layer thickness was updated

Table 3 PRMSE and the PRRMSE During Test Time Series

Performance	H_2	C_2H_4	C_3H_6	C_4H_6	Benzene	TMT
PRMSE	0.0009	0.0132	0.0128	0.0025	0.0052	0.8606
PRRMSE	0.0898	0.0499	0.0866	0.0439	0.0733	0.0847

hourly. Two objectives were considered to be maximized, which are the average profits and the average total yield of the valuable products:

$$\text{Max obj}_p = \frac{\sum_{t=1}^{N}\sum_{j}\left[u_F(t)y_j(t)p_jT_s(t)\right]}{T_f + T_c} \tag{3}$$

$$\text{Max obj}_Y = \frac{\sum_{j=1}^{N_j}\sum_{t=1}^{N}\left[u_F(t)y_j(t)\right]}{\sum_{t=1}^{N}\left[u_F(t)\right]} \tag{4}$$

where $u_F(t)$ and $y_j(t)$ are the mass feedstock flow rate of hydrocarbon and the yield of the valuable product j at the tth time step, respectively; p_j represents the price of the product j; $T_s(t)$ is the time interval between the tth and the previous time steps; and T_f and T_c are the run length and the decoking time, respectively. For this industrial cracking furnace, the feedstock mass flow rate and the COT were in the range of 22,000–28,000 kg/h and 820–840°C. When the maximum TMT reaches 1090°C or the coke layer thickness is larger than 0.015 m, the furnace has to be shut down for decoking; therefore, the simulation should stop at that point. The optimization was investigated using General Algebraic Modeling System (GAMS), which is software for mathematical programming problems.

The Pareto optimal frontier obtained by the ε-constraint method in GAMS is depicted in Fig. 36. Five steady-state operation cases from the industrial data, denoted as c1–c5, are also shown for comparison. It is very clear that these points are all dominated by the solution of the frontier points, which means that the real process can still be improved significantly. For multiobjective decision making, five cases on the Pareto optimal frontier are selected and discussed. The loss of average yield by shifting from case 1 to case 2 is quite small, while the average profit improves a great deal. Similarly, the furnace will have more income, with little sacrifice of product yield from case 5 to case 4. It is worth noting here that in all five cases, the optimal feedstock flow rate is at its upper limit (i.e., 28,000 kg/h),

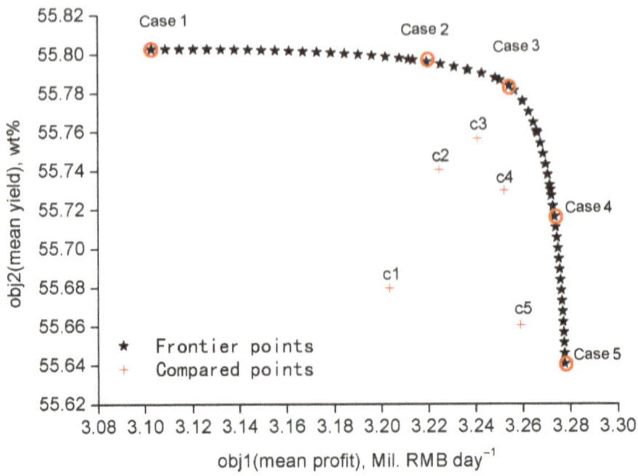

Fig. 36 The pareto-optimal frontier by maximizing objP and objY. *Reprinted with permission from Jin Y, Li J, Du W, et al.: Multi-objective optimization of pseudo-dynamic operation of naphtha pyrolysis by a surrogate model,* Chem Eng Technol *38(5):900–906, 2015b. Copyright 2015 John Wiley and Sons.*

indicating that the furnace has the potential to improve both objectives by increasing the feedstock flow rate. With the complete solution space obtained by the multiobjective optimization, the final decision can be easily made by taking other factors into account, such as the run length, by-product, and decoking cost. For example, if the plant is aiming to produce more valuable products, the corresponding operating condition for case 1 is preferred, whereas if profitability is the main concern, then the furnace should be operated under case 5.

2.4.3 Integrated Operation and Cyclic Scheduling Optimization of Furnace Systems

An ethene plant usually comprises many cracking furnaces running in parallel. They are connected with the upstream tanks containing variant types of feedstock, as shown in Fig. 37. In order to make the entire plant operated in the most economical way, research on scheduling of the furnace system has been carried out (And and Grossmann, 1998; Lim et al., 2006, 2009; Schulz et al., 2006; Wang et al., 2014; Zhao et al., 2010, 2011) to determine the allocation of different feedstocks to each of the furnaces, the on-stream time of the furnaces, and the corresponding operating conditions. Despite of the progress in the furnace system scheduling, the operating conditions were not optimized during the scheduling. However, even if the furnace is operated under the optimal conditions at the start of the run, this cannot guarantee a long-term optimal performance of the furnace due to the coke formation.

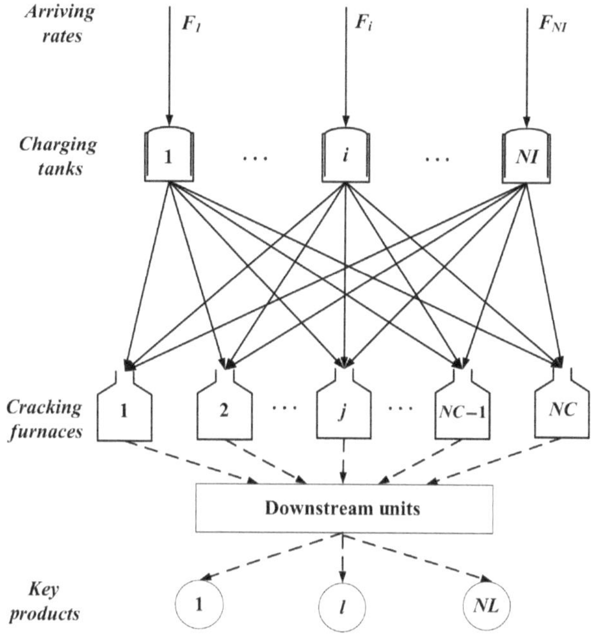

Fig. 37 A typical ethene cracking furnaces system. *Reprinted with permission from Jin Y, Li J, Du W, et al.: Integrated operation and cyclic scheduling optimization for an ethylene cracking furnaces system,* Ind Eng Chem Res *54(15):3844–3854, 2015a. Copyright 2015ACS Publications.*

Hence, the operating conditions are often changing over time in practice, making it difficult for the scheduling schemes proposed by the foregoing researchers to be applied in the actual plant. Therefore, we conducted a study on the scheduling of an industrial cracking furnace system with integrated pseudodynamic operation optimization (Jin et al., 2015a).

The surrogate models developed using ANNs were used to replace the rigorous ethene cracking furnace simulation. The dynamic behavior of the cracking process is approximated by a series of steady-state models under different time steps with different coke layer thicknesses. Since the timescale for coke formation is quite big compared to that of the cracking reaction, this approximation is generally valid so long as a proper time step is chosen. For the surrogate model training stage, it was set as 1 h, which was more than enough.

Because of the incorporation of the feedstock allocation between parallel cracking furnaces and time-varying operation variables. The integrated scheduling problem of a furnaces system is a complex mixed integer dynamic optimization (MIDO) problem. Fig. 38 shows the Gantt chart for this

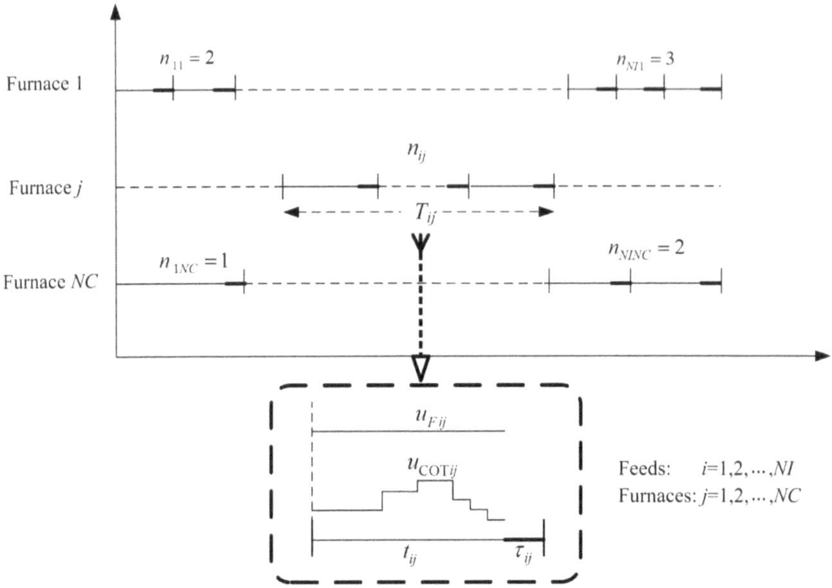

Fig. 38 Gantt chart for an example of integrating operation optimization into cyclic scheduling for an ethene cracking furnace system. *Reprinted with permission from Jin Y, Li J, Du W, et al.: Integrated operation and cyclic scheduling optimization for an ethylene cracking furnaces system,* Ind Eng Chem Res *54(15):3844–3854, 2015a. Copyright 2015 ACS Publications.*

MIDO problem for an ethene cracking furnace system. The cyclic scheduling scheme has to determine the processing time in one subcycle; i.e., the run length for feedstock i in furnace j, the number of subcycles for feedstock i in furnace j, the feedstock flow rate, and the total cyclic horizon.

The objective of the integrated scheduling is to maximize the economic criteria by simultaneously determining two types of decisions (cyclic scheduling decisions and operation optimization decisions):

$$\text{Max obj} = \frac{\sum_{ij}\left\{n_{ij}\cdot\left[\sum_{l}\left(\int_0^{t_{ij}} u_{Fij}(t)\cdot y_{Fijl}(t)\cdot P_l\cdot dt\right) - Cs_{ij}\right]\right\}}{H} \tag{5}$$

where $u_{Fij}(t)$ is the mass flow rate of the feedstock i in furnace j at the tth time step; $y_{Fijl}(t)$ is the corresponding yield of product l, with its price represented as P_l; n_{ij} is the subcycle number of the feedstock i in furnace j; Cs_{ij} stands for the corresponding decoking cost; and H is the total cyclic scheduling horizon. The constraints of the operating conditions include feedstock mass flow rate, maximum TMT, and COT operating range. In addition, material balance has to be fulfilled as the feedstocks consumed during the total cyclic scheduling horizon should equal those supplied from the upstream tanks.

The foregoing integrated cyclic scheduling problem was first converted to a large-scale mixed-integer nonlinear programming (MINLP) problem by discretization using the simultaneous approach of orthogonal collocation on finite elements (OCFE) (Biegler, 2007). The resultant MINLP problem was then modeled in the GAMS using standard branch and bound (SBB) as the MINLP solver with CONOPT, which is a solver for large-scale nonlinear optimization.

Two scheduling problems were studied in this work: a single furnace with multiple feedstocks and a multiple furnace with multiple feedstocks. In the first case, four different feedstocks—namely, light naphtha, naphtha, heavy vacuum gas oil (HVGO), and atmospheric gas oil (AGO)—were available for an industrial GK6 furnace. The scheduling scheme is shown in Fig. 39. For comparison purposes, the same problem was also solved using

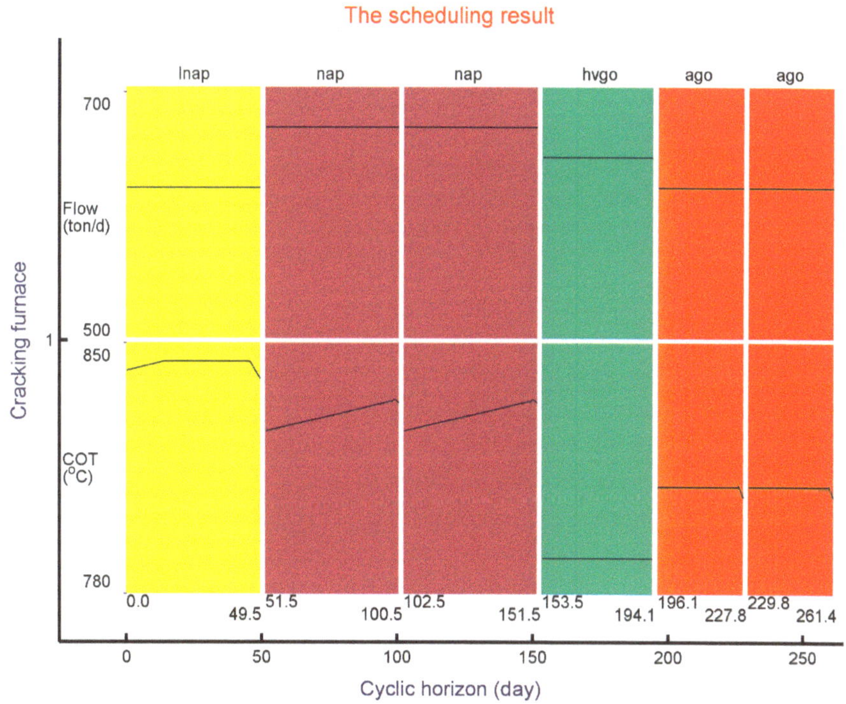

Fig. 39 Gantt chart of integrated cyclic scheduling with feedstock flow rate and COT profiles for the case of multiple feedstock in a single furnace. *Reprinted with permission from Jin Y, Li J, Du W, et al.: Integrated operation and cyclic scheduling optimization for an ethylene cracking furnaces system,* Ind Eng Chem Res *54(15):3844–3854, 2015a. Copyright 2015 ACS Publications.*

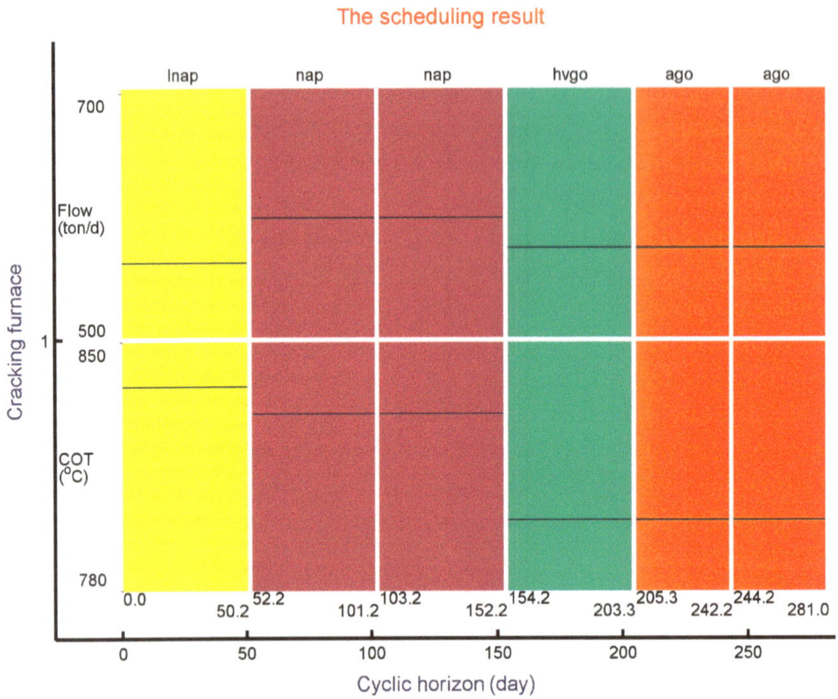

Fig. 40 Gantt chart of traditional cyclic scheduling with the feedstock flow rate and COT profiles for the case of multiple feedstock in a single furnace. *Reprinted with permission from Jin Y, Li J, Du W, et al.: Integrated operation and cyclic scheduling optimization for an ethylene cracking furnaces system,* Ind Eng Chem Res *54(15):3844–3854, 2015a. Copyright 2015 ACS Publications.*

the traditional scheduling approach, which did not take the dynamic optimization into account, as illustrated in Fig. 40. As can be seen from these two figures, both methods provide the same subcycle number; i.e., the cyclic horizon consists of six complete run lengths. The number and order for the four feedstocks are also identical in both methods. Light naphtha is used as feedstock throughout the first run length. Naphtha was cracked for the next two run lengths, followed by one run length with HVGO. The last two run lengths utilize AGO as cracking feedstock. Furthermore, the total cyclic scheduling horizon optimized by the two methods are quite close—about 281 and 261 days using the traditional and dynamic scheduling procedures, respectively. However, because of the integration of the dynamic optimization, the optimal COT changes with the on-stream time for all feedstocks except HVGO, while the COT remains unchanged during the run length in the traditional scheduling scheme. The calculated objectives are 2,841,296 RMB/d and 2,502,517 RMB/d for the integrated and the traditional scheduling schemes, respectively. This means that the new approach proposed by Jin et al. (2015a) help to increase the profit of the entire cracking furnace system to about 13.5%. Similar results are also observed in the multiple furnace and multiple feedstock case with 10.3% increase in the objective by the new approach. Because of the space limitation, the detailed results are not shown here; however, both cases suggest that the proposed integrated scheduling method is superior to the traditional one and can significantly improve the performance of a complex steam cracking furnace system.

3. EDC CRACKING FURNACE

3.1 Coupled Furnace-Reactor Simulation

Research with respect to the coupled furnace-reactor simulation of industrial EDC cracking furnaces is rather rare due to the enormous computational cost to do so, especially using the sophisticated models for detailed EDC cracking mechanisms, turbulent combustion, and radiative heat transfer. Park et al. (2005) performed coupled simulation to investigate the optimal operating condition of an EDC cracking furnace using a 2D CFD model and detailed kinetics developed by Choi et al. (2001). By defining an operation modification factor α, which is the ratio of the fuel flow rate in the upper burner to that in the lower burners, the effect of this parameter on furnace thermal efficiency and the heat flux imbalance in different reactor coils was studied. Schirmeister et al. (2009) investigated the relation between cracking selectivity and the purity of EDC feedstock. The influence of the cracking severity on the cost of vinyl chloride production was also evaluated by means of run length simulation. A simplified reaction model proposed by the authors and a zero-dimensional furnace model were used in their work.

The heat flux profile, as a link between the reactor side and the fire side in coupled simulation, is essential to cracking reaction modeling because it determines the cracking severity and the product distribution. Since it is strongly affected by the flow pattern and heat transfer on both sides, an accurate CFD-based furnace model is necessary, but extremely computationally expensive. Considering the purpose of the EDC cracking furnace modeling is to provide a better understanding of the effects of operating conditions on the furnace performance for the following optimization, a good compromise is to build up simplified models to find the optimal region in a wide range of operating conditions in the first step. Then the sophisticated models can be used to perform simulations within this region to further optimize the operating conditions.

To have a comprehensive understanding of the EDC cracking process and the subsequent operating optimization, we studied an industrial EDC cracking furnace used the aforementioned approach (Li et al., 2013a). The configuration of the radiant section of this furnace is depicted in Fig. 41. The reactor coil located in the center of the furnace has 20 passes from top to bottom, and is heated by four rows of radiant burners mounted on the side wall of the furnace. Combusted flue gas flows up to the top of the furnace and gradually transfers the heat taken by it to the reactor coil; hence, the temperature gradient mainly occurs in the vertical direction of the furnace. In light of this, a 1D Lobo-Evans method (Lobo, 1974) was adopted to describe the heat transfer process in the fire side and the furnace was divided into four sections in the vertical direction, as shown in Fig. 41. Since there is no floor burner that creates large flue gas recirculation zone in the furnace, the flow pattern of the fire side was simplified as a vertical plug flow without any velocity gradient in other directions. This assumption is usually acceptable for furnaces with only radiant burners and has been verified by the

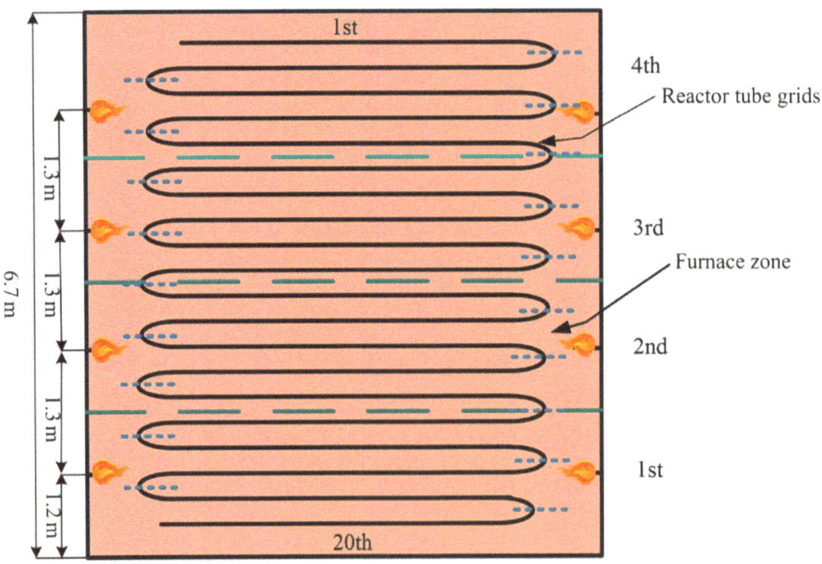

Fig. 41 Structure of the radiant section in the furnace. *Adapted with permission from Li C, Hu G, Zhong W, et al.: Coke deposition influence based on a run length simulation of a 1,2-dichloroethane cracker,* Ind Eng Chem Res *52(49):17501–17516, 2013b. Copyright 2013 ACS Publications.*

velocity profiles calculated by Stefanidis et al. (2007). For the reactor coil, a 1D plug flow model is generally acceptable because of the highly turbulent feedstock flow and the large length-to-diameter ratio of the reactor coil. A first-order series molecular reaction mechanism was considered, where EDC is cracked into vinyl chloride in the first step and then converted to acetylene as a by-product due to secondary reaction.

For this specific EDC furnace, the fuel gas flow rate can be adjusted by two valves. One valve controls the flow rate through the burners in the three upper rows, and the other controls the burners in the bottom row. Therefore, a fuel gas allocation factor α was defined as the ratio of the fuel flow rate in the bottom row to the total fuel flow rate. The influence of this factor on the furnace performance was studied numerically.

As shown in Fig. 42, the average flue gas temperature in the furnace changes dramatically with the fuel gas allocation factor α. The flue gas temperature in the bottom section (i.e., the first section) increases gradually with α, which is reasonable since a larger gas allocation factor means more fuel will be distributed to the burners located in the bottom section. It is also interesting to see that all curves in Fig. 42 nearly intersect at the second

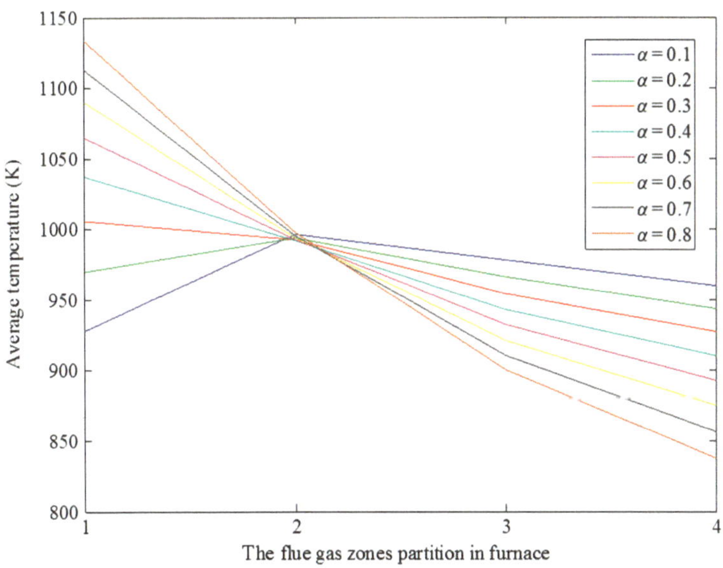

Fig. 42 Fuel gas temperature profiles along the furnace sections. *Reprinted with permission from Li C, Hu G, Zhong W, et al.: Comprehensive simulation and optimization of an ethylene dichloride cracker based on the one-dimensional Lobo–Evans method and computational fluid dynamics,* Ind Eng Chem Res *52(2):645–657, 2013a. Copyright 2013 ACS Publications.*

section of the furnace; however, this is believed to be caused by the 1D assumption and the inadequate number of furnace zones used in this study, which was only four.

Fig. 43 shows the effect of the fuel gas allocation factor on EDC conversion, VCM selectivity, and fuel gas consumption per unit of VCM production. The trend of the EDC conversion and the VCM selectivity with fuel gas allocation factor α is mainly due to the corresponding COT of the cracking reactor. As the coil outlet is located in the bottom section of the furnace, increasing α will result in higher COT and consequently higher cracking severity, leading to increased EDC conversion. On the other hand, the VCM selectivity decreases as a result of the secondary reaction promoted by the increased COT. Therefore, the fuel gas consumption per unit of VCM production is a combination of the EDC conversion and VCM selectivity. For this EDC cracking furnace, the decrease of the VCM selectivity is more pronounced and cannot be compensated by the increase of EDC conversion. Hence, the fuel gas consumption per unit of VCM production decreases with the fuel gas allocation factor, as shown in Fig. 43C.

Apart from the fuel gas allocation factor, the influence of COT on furnace performance was investigated. In this study, the total fuel gas flow rate was altered to obtain different COTs ranging from 710K to 796K, while the fuel gas allocation factor was set to the value of 0.365 and remained unchanged. The results are depicted in Fig. 44, where the trend of the EDC conversion and the VCM selectivity agree well with the foregoing discussion. From Fig. 44A, it can be seen that the unconverted EDC at the reactor outlet decreases quickly with COT due to the increased conversion. While higher COT also leads to accelerated increase of the by-product as a consequence of the secondary reaction, resulting in decreased VCM selectivity. Although the VCM production still increases with COT, the rate of increase is gradually slowing down; therefore, it can be expected that the VCM production will be at its maximum with COT. As more fuel is required to achieve higher COT, the fuel gas consumption per unit of VCM production achieves a minimum value with COT. This should be the optimal operating condition if the cost for separation of the products is not taken into account.

From the previous discussion, the optimal value of the fuel gas allocation factor and the COT should be 0.2K and 742K. However, the fuel gas allocation factor can only be adjusted within a small range of 0.3–0.4 to avoid flameout or damage to the firebricks, and so the value was set to 0.365. Moreover, the optimal COT was determined to be 756K to satisfy the conversion constraint for the purpose of reducing separation costs.

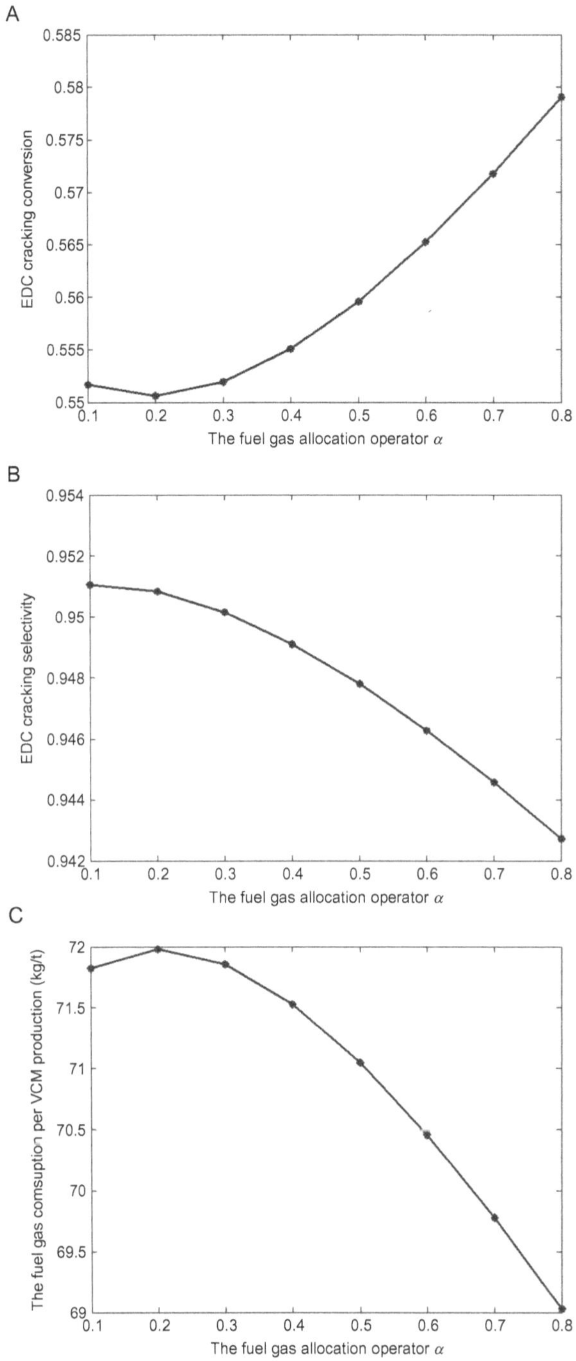

Fig. 43 (A) EDC cracking conversion; (B) EDC cracking severity; (C) fuel gas consumption per unit VCM production as a function of the fuel gas allocation factor. *Reprinted with permission from Li C, Hu G, Zhong W, et al.: Comprehensive simulation and optimization of an ethylene dichloride cracker based on the one-dimensional Lobo–Evans method and computational fluid dynamics,* Ind Eng Chem Res *52(2):645–657, 2013a. Copyright 2013 ACS Publications.*

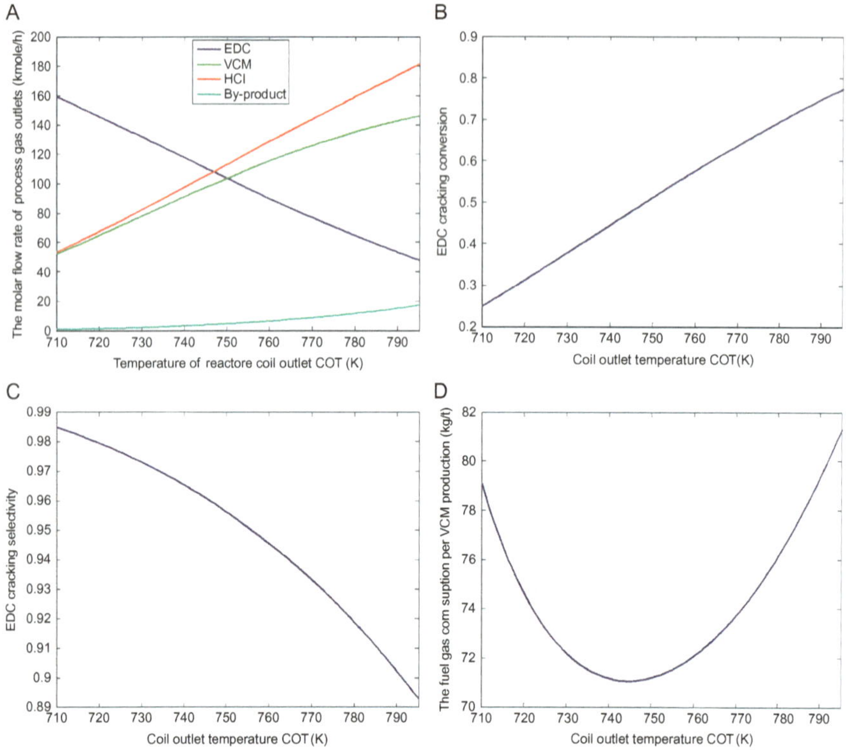

Fig. 44 (A) Process gas outlet composition; (B) EDC cracking conversion; (C) EDC cracking severity; (D) fuel gas consumption per unit VCM production as a function of the COT. *Reprinted with permission from Li C, Hu G, Zhong W, et al.: Comprehensive simulation and optimization of an ethylene dichloride cracker based on the one-dimensional Lobo–Evans method and computational fluid dynamics,* Ind Eng Chem Res *52(2):645–657, 2013a. Copyright 2013 ACS Publications.*

To validate the results obtained from the Lovo–Evans model and to provide a detailed temperature field for the furnace, the optimized results were also studied by coupled CFD simulations of the furnace and the reactor coil. The same cracking mechanism was used in the CFD simulation to be consistent with that of the 1D simulation. The two-step combustion mechanism proposed by Westbrook and Dryer (1981) was adopted for the furnace modeling, with the finite-rate eddy-dissipation model used to account for the turbulence-reaction interaction. The radiative transfer equation was solved by the DO model, taking CO_2 and H_2O as the participating gases.

The furnace temperature contour at different cross sections can be found in Fig. 45A–E, which shows a clear temperature gradient in the vertical direction of the furnace. With increased furnace elevation, the average flue gas

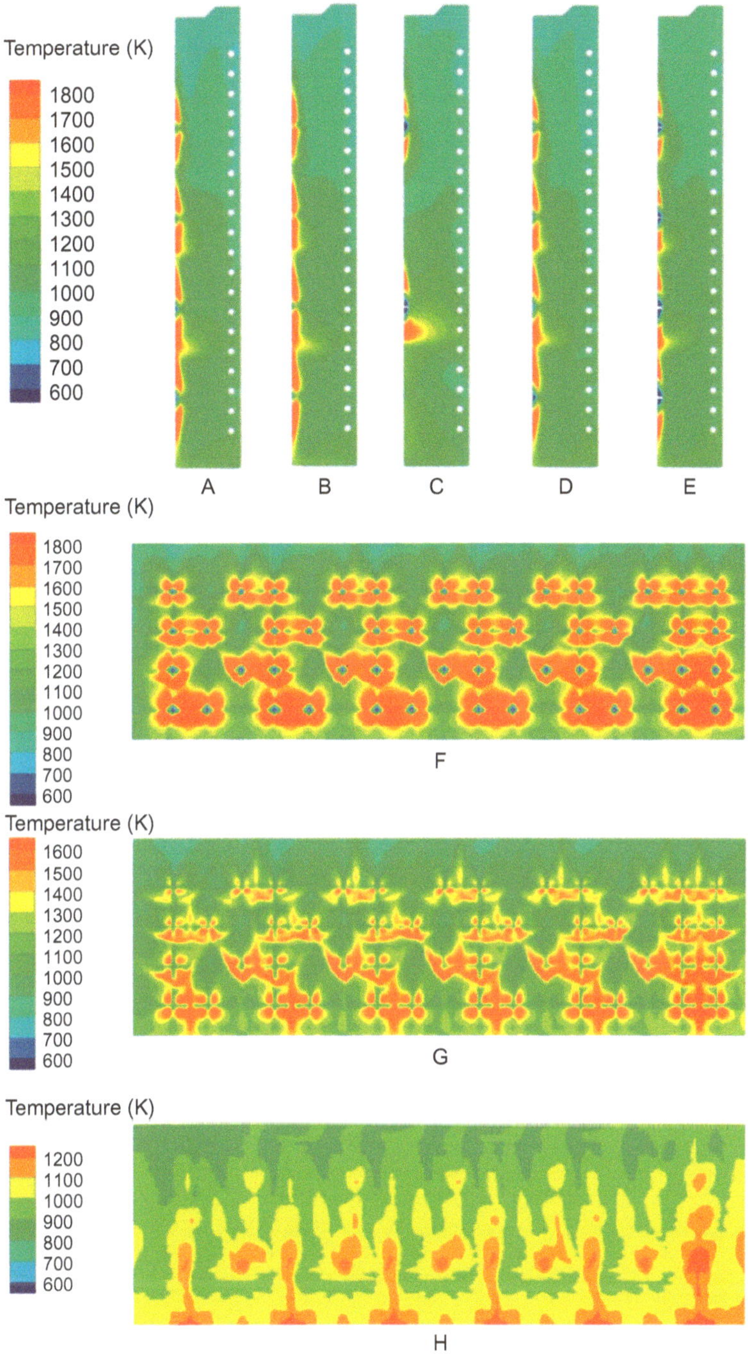

Fig. 45 Flue gas temperature contour at different planes in the furnace: (A) $x=1.5$ m; (B) $x=5$ m; (C) $x=10.5$ m; (D) $x=15$ m; (E) $x=18.5$ m; (F) $y=0.0485$ m; (G) $y=0.097$ m; (H) $y=0.71285$ m. *Reprinted with permission from Li C, Hu G, Zhong W, et al.: Comprehensive simulation and optimization of an ethylene dichloride cracker based on the one-dimensional Lobo–Evans method and computational fluid dynamics,* Ind Eng Chem Res *52(2):645–657, 2013a. Copyright 2013 ACS Publications.*

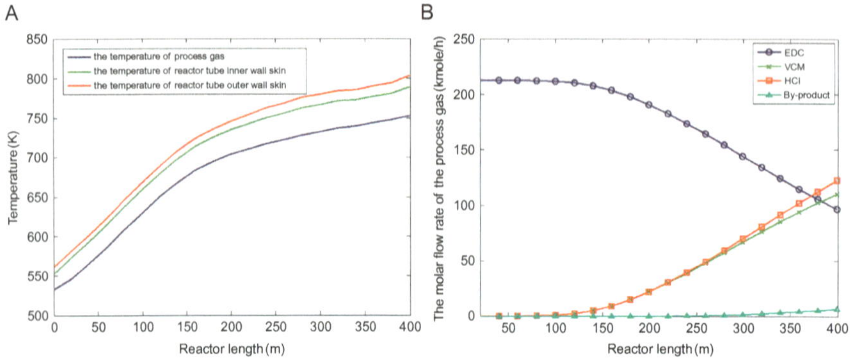

Fig. 46 (A) Temperature profiles; (B) process gas composition profiles along the reactor length. *Adapted with permission from Li C, Hu G, Zhong W, et al.: Comprehensive simulation and optimization of an ethylene dichloride cracker based on the one-dimensional Lobo–Evans method and computational fluid dynamics,* Ind Eng Chem Res *52(2):645–657, 2013a. Copyright 2013 ACS Publications.*

temperature gradually decreases. It can be observed from Fig. 45F that the premixed combustion flames of the burners located in the fourth row are larger than those of the other three rows. This can be explained by the unevenly distributed fuel flow rate because the fuel gas allocation factor was set as 0.365. As a consequence, the flue gas temperature is much higher in the bottom of the furnace, even at the regions away from the burners (see Fig. 45E).

Simulation results of the reactor coil are shown in Fig. 46. The process gas temperature increases rapidly in the first 150 m of the reactor coil because the absorbed heat is mainly used to heat the process gas. This is confirmed by Fig. 46B, as the EDC is not consumed until the temperature reaches about 670K at the position of 150 m. When the cracking starts, the slope of the process temperature curve becomes smaller and the VCM production increases. Close to the end of the reactor coil, secondary reaction is accelerated because of the rising process gas temperature, leading to significant increases of the by-product flow rate.

Finally, the 1D Lobo–Evans method and the CFD method are compared in Table 4. It can be seen that the results obtained by both methods agree well with the industrial data. Therefore, the validity of the optimized case using the Lobo–Evans method is confirmed by the CFD simulation.

3.2 Effect of Coke Deposition on the Run Length of EDC Cracker

The EDC thermal cracking process is also accompanied with coke deposition, which in turn affects the cracking performance and the run length of

Table 4 Comparison of Simulation Results and Industrial Data

Item	Industrial Data	Simulation Results by the Lobo–Evans Method	Simulation Results by CFD
Flue gas outlet temperature (K)	900.0	913.0	894.5
Process gas outlet temperature (K)	756.0	756.0	753.6
Coil inlet pressure (kPa)	2904	2953	2904
Coil pressure drop (kPa)	549	598	491
Conversion	0.550	0.554	0.547
Selectivity	0.950	0.949	0.945
Fuel gas consumption (per unit VCM) (kg/t)	72.0	71.7	72.5

Adapted with permission from Li C, Hu G, Zhong W, et al.: Comprehensive simulation and optimization of an ethylene dichloride cracker based on the one-dimensional Lobo–Evans method and computational fluid dynamics, *Ind Eng Chem Res* 52(2):645–657, 2013a. Copyright 2013 ACS Publications.

EDC cracking furnaces. Thus, it is necessary to take coke formation into account in EDC cracking furnace modeling and optimization. In addition, CCl_4 was considered as a promoter to accelerate the cracking rate of EDC and hence increase the conversion of EDC (Choi et al., 2001; Ranzi et al., 1992); however, adding CCl_4 will aggravate the undesired coke formation as well. Therefore, a trade-off between the on-stream time and the EDC conversion has to be made in order to improve the furnace performance and reduce energy consumption.

To this end, we carried out coupled furnace-reactor simulations for an industrial EDC cracking furnace (Li et al., 2013b). In this study, the furnace and the reactor models were same as discussed in the previous section. A simplified radical mechanism proposed by Schirmeister et al. (2009), which consists of 24 species and 31 reactions, was used to account for the EDC cracking reactions. Coke formation was also taken into account in this mechanism, in which the acetylene was considered the only coking precursor. As the timescale for coke formation is quite large compared to that of the cracking process, pseudo-steady-state run length simulations were performed on the assumption that the coking rate is constant during one time step. Therefore, the coke layer thickness was updated in each time step by multiplying the coking rate by the time interval. The coupling procedure is as follows: In the first step, estimated flue gas temperature profiles were used as boundary conditions to perform the reactor simulations. Then the heat flux profiles obtained from the reactor simulations were exported to the furnace model to calculate the flue gas temperature profiles, which were again used for the reactor simulations. The coupling procedure was repeated until the difference in the flue gas temperature between two iterations is smaller than a threshold, which was set as 0.1K. For the run length simulation, the coupling procedure was applied in each time step until the maximum TMT reaches the operational limit for decoking.

The simulated coke layer thickness as a function of the run length and the reactor length is shown in Fig. 47A. It can be clearly seen that the distribution of the coke layer is highly nonuniform, with most of the coke deposited near the outlet of the reactor coil. This is because the process gas temperature is not high enough for the coking precursor to be produced substantially, and for the coking rate to reach a certain level until the end of the reactor coil. The maximum coke layer thickness is at the outlet of the reactor coil, which is consistent with the position of the highest TMT, as shown in Fig. 47B. Therefore, the run length of the entire furnace is limited by the coke formation at the reactor outlet.

Fig. 48 illustrates changes of the species molar flow rate, the EDC conversion, and selectivity during the run length. As a result of the coke formation, the pressure drop increases with on-stream time, and so the inlet pressure of the reactor coil also has to be increased to maintain the same outlet pressure. This leads to a decrease in the EDC conversion because cracking reactions prefer lower pressure, giving rise to more unreacted EDC with on-stream time. Despite the slight increase in VCM selectivity, the production of the VCM still decreases during the run length since less EDC is converted. Meanwhile, on the furnace side, the thermal efficiency becomes less at the end of the run due to coke formation, which results in higher fuel gas consumption per unit of VCM production, as depicted in Fig. 48D.

The effect of the CCl_4 concentration on the run length of the EDC cracking furnace was also studied (see Fig. 49). It is very clear that adding CCl_4 helps to increase the EDC conversion, but the downside of this is the reduction of the run length. Due to the increased EDC conversion, more VCM can be produced under the same operating conditions, therefore improving the furnace performance in terms of fuel gas consumption per unit of VCM production. Although gradually increasing with on-stream time, the gas consumption per unit VCM for the EDC cracking case with 150 ppm CCl_4 is less than that of the CCl_4 free case during the whole

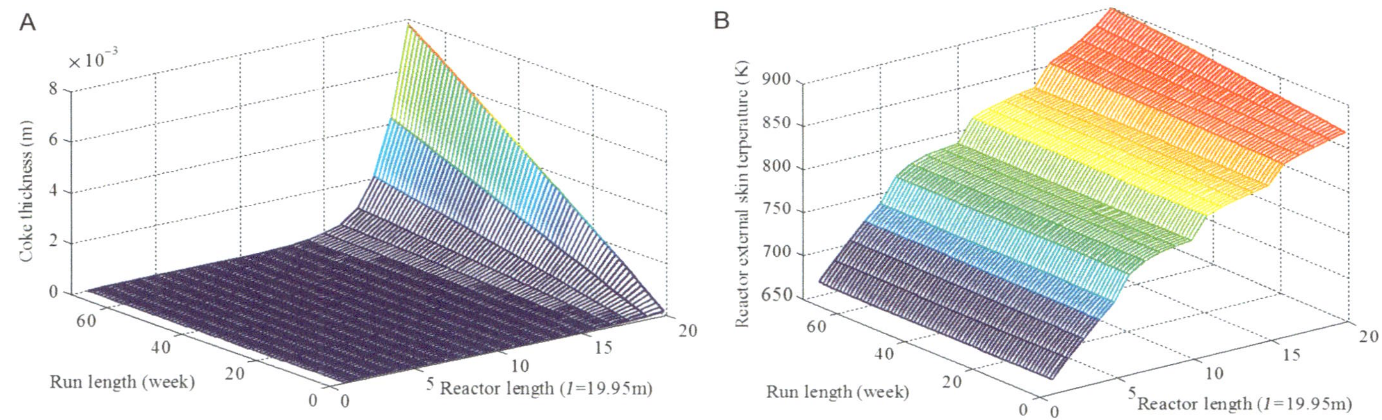

Fig. 47 (A) Coke thickness; (B) external tube temperature along the reactor tube length in an operation cycle. *Adapted with permission from Li C, Hu G, Zhong W, et al.: Coke deposition influence based on a run length simulation of a 1,2-dichloroethane cracker,* Ind Eng Chem Res *52(49):17501–17516, 2013b. Copyright 2013 ACS Publications.*

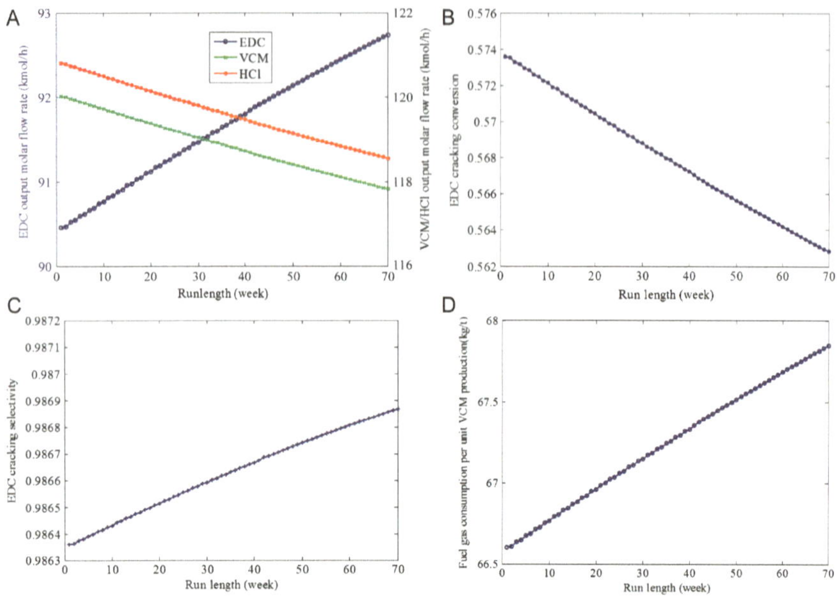

Fig. 48 (A) EDC/VCM/HCl molar flow rates in the reactor outlet; (B) EDC cracking conversion; (C) EDC cracking selectivity; (D) fuel gas consumption per unit VCM production during an operation cycle. *Adapted with permission from Li C, Hu G, Zhong W, et al.: Coke deposition influence based on a run length simulation of a 1,2-dichloroethane cracker,* Ind Eng Chem Res *52(49):17501–17516, 2013b. Copyright 2013 ACS Publications.*

operation cycle. On the other hand, the run length is not significantly reduced, which shows the potential of CCl_4 in furnace performance improvement.

As an important operating parameter of this furnace, the effect of the fuel gas allocation factor α was also investigated by run length simulations. From Fig. 50A, we can see that the predicted run length is very sensitive to the fuel gas allocation factor. The reason is that the fuel gas allocation factor mainly determines the reactor external skin temperature close to the outlet, where the highest coke layer thickness is observed. As a result of the higher heat input to the last pass of the reactor by increasing α, COT also increases, which leads to improved EDC conversion and consequently lower fuel gas consumption per unit of VCM production. However, the main disadvantage is the significant reduction in the run length. Therefore, a trade-off between the run length and the EDC conversion has to be made to achieve the best performance for the long-term profitability of the EDC cracking furnace.

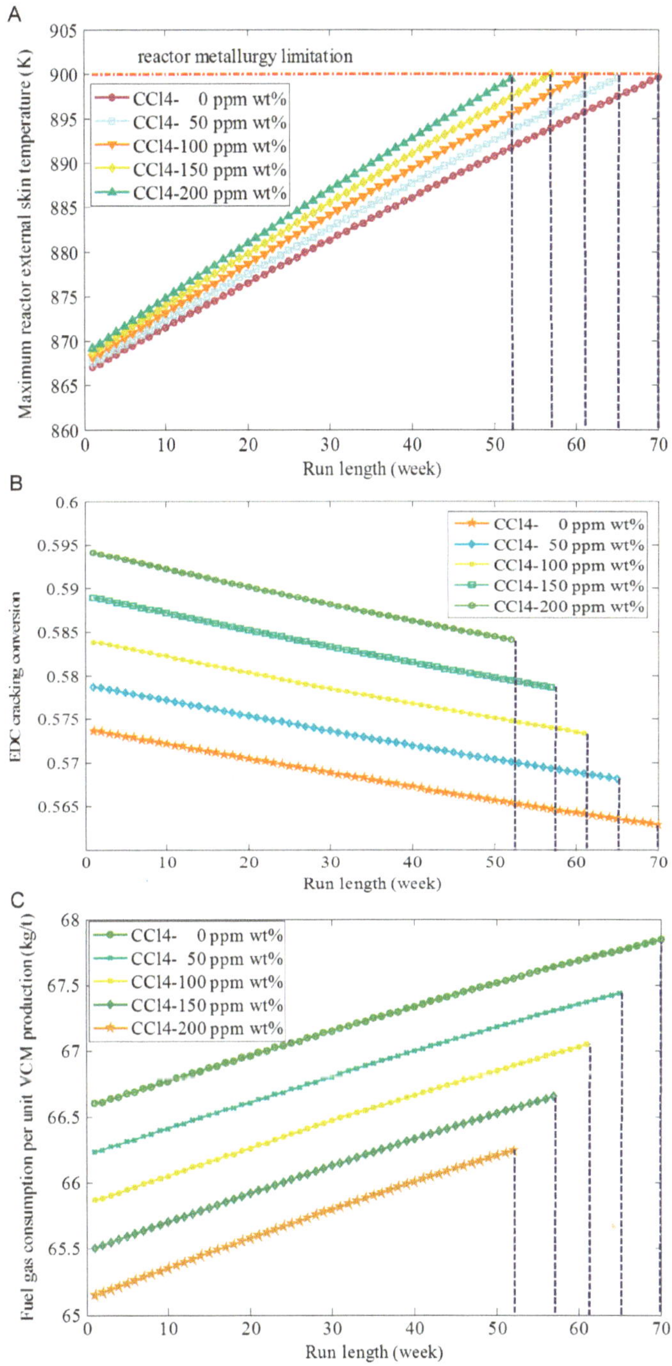

Fig. 49 (A) Maximum reactor external skin temperature; (B) EDC cracking conversion; (C) fuel gas consumption per unit VCM production as a function of the run length. *Adapted with permission from Li C, Hu G, Zhong W, et al.: Coke deposition influence based on a run length simulation of a 1,2-dichloroethane cracker,* Ind Eng Chem Res *52(49):17501–17516, 2013b. Copyright 2013 ACS Publications.*

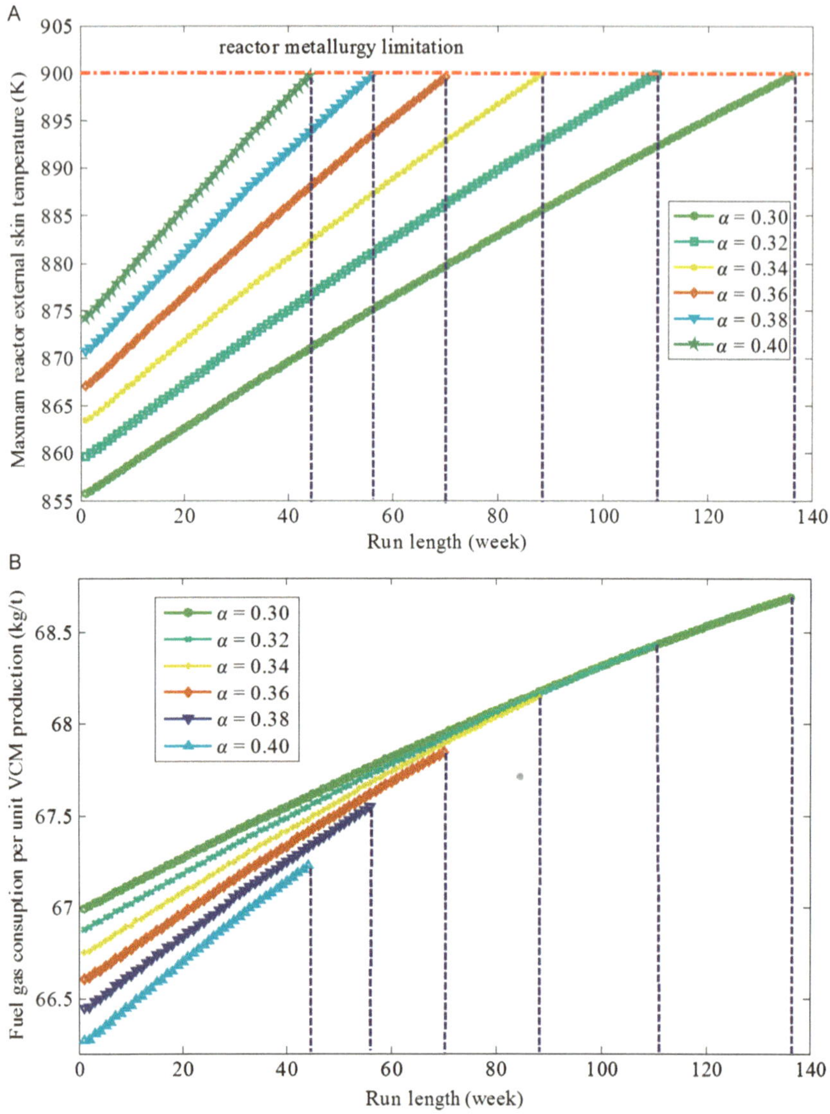

Fig. 50 (A) Maximum reactor external skin temperature; (B) fuel gas consumption per unit VCM production as a function of the run length. *Adapted with permission from Li C, Hu G, Zhong W, et al.: Coke deposition influence based on a run length simulation of a 1,2-dichloroethane cracker,* Ind Eng Chem Res *52(49):17501–17516, 2013b. Copyright 2013 ACS Publications.*

4. SUMMARY AND FUTURE OUTLOOK

The most recent progress in the numerical modeling of ethene and propene cracking furnaces was presented in this chapter. For the steam cracking furnace, not only the radiant section, which is most widely investigated, was discussed, but also the convection section and the optimization and scheduling of furnace systems. Detailed information on both the fire side and the process side obtained from coupled furnace-reactor simulations was presented. The DO model was proven to be the most suitable radiation model for large-scale furnace simulation. It was also shown that using the nongray gas model is essential to get accurate reactor simulation results. The results of the convection section simulation, in which all tube banks were completely coupled with the flue gas side for the first time, were also discussed. Based on the established model, integrated dynamic scheduling optimization of a cracking plant consisting of multifurnace and multi-feedstock showed the remarkable increase in economic performance. In addition, for the EDC cracking furnace, the operating optimization results using coupled furnace-reactor simulations were presented. By performing run length simulations, the quantitative effect of CCl_4 concentration on EDC conversion and the furnace run length was investigated for the first time in an industrial furnace.

Nowadays, it has become more and more difficult for the traditional steam cracking furnace design and operation to fulfill the new requirements, such as the extremely low NOx emissions due to environmental concerns. Therefore, the burners have to be designed to be able to generate flames with lower temperature peaks to reduce the production of thermal NO_x while maintaining the same firing capacity. Another aspect is that flue gas produced by the traditional air combustion is not a concentrated CO_2 stream; thus, it cannot be easily captured, stored, or used for other applications. The oxy-fuel combustion process, which burns the fuel in pure oxygen instead of air (Giménez-López et al., 2015), may be a good solution to this problem because it produces a highly concentrated stream of CO_2, ready for carbon capture and storage. Besides carbon capture and storage, oxy-fuel combustion has other advantages, such as the possibility to significantly reduce the size of the combustor unit compared to that used in the corresponding air-fired combustion (Tan et al., 2002). In addition, oxy-fuel combustion has a higher potential for NOx reduction compared to air combustion processes due to the lack of nitrogen inside the oxidizer stream. However, as the behavior of oxy-fuel combustion differs greatly from the air combustion in flame propagation, burning velocity, and so on (Oh and Noh, 2014), it is not straightforward to apply this process to cracking furnaces. Moreover, the radiative heat transfer will also be affected as a result of the different flue gas composition in oxy-fuel combustion. Therefore, it is interesting to study the combustion and radiative heat transfer of steam cracking furnaces under oxy-fuel combustion conditions to evaluate their potential in improving the performance of industrial cracking furnaces.

REFERENCES

And VJ, Grossmann IE: Cyclic scheduling of continuous parallel-process units with decaying performance, *AIChE J* 44(7):1623–1636, 1998.

Ashmore PG, Gardner JW, Owen AJ, et al: Chlorine-catalysed pyrolysis of 1,2-dichloroethane. Part 1. Experimental results and proposed mechanism, *J Chem Soc Faraday Trans 1* 78(3):657–676, 1982a.

Ashmore PG, Owen AJ, Robinson PJ: Chlorine-catalysed pyrolysis of 1,2-dichloroethane. Part 2. Unimolecular decomposition of the 1,2-dichloroethyl radical and its reverse reaction, *J Chem Soc, Faraday Trans 1* 78(3):677–693, 1982b.

Barton DHR: The kinetics of the dehydrochlorination of substituted hydrocarbons. Part I. Induced dehydrochlorination, *J Chem Soc* 148–155, 1949. http://dx.doi.org/10.1039/JR9490000148.

Barton DHR, Howlett KE: The kinetics of the dehydrochlorination of substituted hydrocarbons. Part II. The mechanism of the thermal decomposition of 1:2-dichloroethane, *J Chem Soc* 155–164, 1949. http://dx.doi.org/10.1039/JR9490000155.

Berreni M, Wang M: Modelling and dynamic optimization of thermal cracking of propane for ethylene manufacturing, *Comput Chem Eng* 35(12):2876–2885, 2011.

Biegler LT: An overview of simultaneous strategies for dynamic optimization, *Chem Eng Process Process Intensif* 46(11):1043–1053, 2007.

Charon-Revellin N, Dulot H, López-García C, et al: Kinetic modeling of vacuum gas oil hydrotreatment using a molecular reconstruction approach, *Oil Gas Sci Technol Rev IFP Energies nouvelle* 66:479–490, 2010.

Choi B-S, Oh JS, Lee S-W, et al: Simulation of the effects of CCl4 on the ethylene dichloride pyrolysis process, *Ind Eng Chem Res* 40(19):4040–4049, 2001.

De Schepper SCK, Heynderickx GJ, Marin GB: Coupled simulation of the flue gas and process gas side of a steam cracker convection section, *AIChE J* 55(11):2773–2787, 2009a.

De Schepper SCK, Heynderickx GJ, Marin GB: Modeling the evaporation of a hydrocarbon feedstock in the convection section of a steam cracker, *Comput Chem Eng* 33(1):122–132, 2009b.

De Schepper SCK, Heynderickx GJ, Marin GB: Modeling the coke formation in the convection section tubes of a steam cracker, *Ind Eng Chem Res* 49(12):5752–5764, 2010.

Detemmerman T, Froment GF: Three dimensional coupled simulation of furnaces and reactor tubes for the thermal cracking of hydrocarbons, *Oil Gas Sci Technol Rev IFP* 53(2):181–194, 1998.

Eason J, Cremaschi S: Adaptive sequential sampling for surrogate model generation with artificial neural networks, *Comput Chem Eng* 68:220–232, 2014.

Edwards DK: Molecular gas band radiation. In Thomas FI, James PH, editors: *Advances in heat transfer*, vol. 12, New York, 1976, Academic Press, pp 115–193.

Edwin EH, Balchen JG: Dynamic optimization and production planning of thermal cracking operation, *Chem Eng Sci* 56(3):989–997, 2001.
Gao X, Chen B, He X, et al: Multi-objective optimization for the periodic operation of the naphtha pyrolysis process using a new parallel hybrid algorithm combining NSGA-II with SQP, *Comput Chem Eng* 32(11):2801–2811, 2008.
Gao GY, Wang M, Pantelides CC, et al: Mathematical modeling and optimal operation of industrial tubular reactor for naphtha cracking. In Rita Maria de Brito Alves CAOdN, Evaristo Chalbaud B, editors: *Computer aided chemical engineering*, vol. 27, Salvador-Bahia, Brazil, 2009a, Elsevier, pp 501–506.
Gao GY, Wang M, Ramshaw C, et al: Optimal operation of tubular reactors for naphtha cracking by numerical simulation, *Asia Pac J Chem Eng* 4(6):885–892, 2009b.
Ghassabzadeh H, Darian JT, Zaheri P: Experimental study and kinetic modeling of kerosene thermal cracking, *J Anal Appl Pyrolysis* 86(1):221–232, 2009.
Giménez-López J, Millera A, Bilbao R, et al: Experimental and kinetic modeling study of the oxy-fuel oxidation of natural gas, CH4 and C2H6, *Fuel* 160:404–412, 2015.
Gyorffy M, Hineno M, Hashimoto K, et al: *Mert performance and Technology Update,* Paper presented at AIChE Spring meeting: ethylene producers conference, Tampa Bay, USA, 2009.
Habibi A, Merci B, Heynderickx GJ: Impact of radiation models in CFD simulations of steam cracking furnaces, *Comput Chem Eng* 31(11):1389–1406, 2007a.
Habibi A, Merci B, Heynderickx GJ: Multiscale modeling of turbulent combustion and NOx emission in steam crackers, *AIChE J* 53(9):2384–2398, 2007b.
Hagan MT, Demuth HB, Beale MH, et al: *Neural Network Design,* ed 2, 2014. Martin Hagan. http://hagan.okstate.edu/nnd.html.
Heynderickx GJ, Froment GF: Simulation and comparison of the run length of an ethane cracking furnace with reactor tubes of circular and elliptical cross sections, *Ind Eng Chem Res* 37(3):914–922, 1998.
Heynderickx GJ, Nozawa M: High-emissivity coatings on reactor tubes and furnace walls in steam cracking furnaces, *Chem Eng Sci* 59(22–23):5657–5662, 2004.
Heynderickx GJ, Nozawa M: Banded gas and nongray surface radiation models for high-emissivity coatings, *AIChE J* 51(10):2721–2736, 2005.
Heynderickx GJ, Cornelis GG, Froment GF: Circumferential tube skin temperature profiles in thermal cracking coils, *AIChE J* 38(12):1905–1912, 1992.
Heynderickx GJ, Oprins AJM, Marin GB, et al: Three-dimensional flow patterns in cracking furnaces with long-flame burners, *AIChE J* 47(2):388–400, 2001.
Hottel HC, Sarofim AF: Models of radiative transfer in furnaces, *J Eng Phys Thermophys* 19(3):1102–1114, 1970.
Hu G, Wang H, Qian F, et al: Comprehensive CFD simulation of product yields and coking rates for a floor- and wall-fired naphtha cracking furnace, *Ind Eng Chem Res* 50(24):13672–13685, 2011.
Hu G, Wang H, Qian F, et al: Coupled simulation of an industrial naphtha cracking furnace equipped with long-flame and radiation burners, *Comput Chem Eng* 38:24–34, 2012.
Hudebine D, Verstraete JJ: Molecular reconstruction of LCO gasoils from overall petroleum analyses, *Chem Eng Sci* 59(22–23):4755–4763, 2004.
Huybrechts G, Wouters G: Mechanism of the pyrolysis of 1,2-dichloroethane in the absence and presence of added chlorine, *In J Chem Kinet* 34(5):316–321, 2002.
Jegla Z: The conceptual design of a radiant chamber and preliminary optimization of a process tubular furnace, *Heat Transfer Eng* 27(6):50–57, 2006.
Jin Y, Li J, Du W, et al: Integrated operation and cyclic scheduling optimization for an ethylene cracking furnaces system, *Ind Eng Chem Res* 54(15):3844–3854, 2015a.
Jin Y, Li J, Du W, et al: Multi-objective optimization of pseudo-dynamic operation of naphtha pyrolysis by a surrogate model, *Chem Eng Technol* 38(5):900–906, 2015b.

Jin Y, Li J, Du W, et al: Adaptive sampling for surrogate modelling with artificial neural network and its application in an industrial cracking furnace, *Can J Chem Eng* 94(2):262–272, 2016.

Kattan N, Thome JR, Favrat D: Flow boiling in horizontal tubes: part 1—development of a diabatic Two-phase flow pattern Map, *J Heat Transfer* 120(1):140–147, 1998.

Kumar P, Kunzru D: Modeling of naphtha pyrolysis, *Ind Eng Chem Process Des Dev* 24(3):774–782, 1985.

Lan X, Gao J, Xu C, et al: Numerical simulation of transfer and reaction processes in ethylene furnaces, *Chem Eng Res Des* 85(12):1565–1579, 2007.

Leckner B: Spectral and total emissivity of water vapor and carbon dioxide, *Combust Flame* 19(1):33–48, 1972.

Li C, Zhu Q, Geng Z: Multi-objective particle swarm optimization hybrid algorithm: an application on industrial cracking furnace, *Ind Eng Chem Res* 46(11):3602–3609, 2007.

Li C, Hu G, Zhong W, et al: Comprehensive simulation and optimization of an ethylene dichloride cracker based on the one-dimensional Lobo–Evans method and computational fluid dynamics, *Ind Eng Chem Res* 52(2):645–657, 2013a.

Li C, Hu G, Zhong W, et al: Coke deposition influence based on a run length simulation of a 1,2-dichloroethane cracker, *Ind Eng Chem Res* 52(49):17501–17516, 2013b.

Lim H, Choi J, Realff M, et al: Development of optimal decoking scheduling strategies for an industrial naphtha cracking furnace system, *Ind Eng Chem Res* 45(16):5738–5747, 2006.

Lim H, Choi J, Realff M, et al: Proactive scheduling strategy applied to decoking operations of an industrial naphtha cracking furnace system, *Ind Eng Chem Res* 48(6):3024–3032, 2009.

Lobo W: Design of furnaces with flue gas temperature-gradients, *Chem Eng Prog* 70(1):65–71, 1974.

Masoumi ME, Sadrameli SM, Towfighi J, et al: Simulation, optimization and control of a thermal cracking furnace, *Energy* 31(4):516–527, 2006.

Modest MF: The treatment of nongray properties in radiative heat transfer: from past to present, *J Heat Transfer* 135(6):061801, 2013.

Nuchitprasittichai A, Cremaschi S: An algorithm to determine sample sizes for optimization with artificial neural networks, *AIChE J* 59(3):805–812, 2013.

Oh J, Noh D: The effect of CO_2 addition on the flame behavior of a non-premixed oxy-methane jet in a lab-scale furnace, *Fuel* 117(Pt. A):79–86, 2014.

Oprins AJM, Heynderickx GJ: Calculation of three-dimensional flow and pressure fields in cracking furnaces, *Chem Eng Sci* 58(21):4883–4893, 2003.

Oprins AJM, Heynderickx GJ, Marin GB: Three-dimensional asymmetric flow and temperature fields in cracking furnaces, *Ind Eng Chem Res* 40(23):5087–5094, 2001.

Park Y, Seok Choi B, Yi J: Simulation of imbalance reduction between two reactors in an ethylene dichloride cracker, *Chem Eng Sci* 60(5):1237–1249, 2005.

Plehiers PM, Reyniers GC, Froment GF: Simulation of the run length of an ethane cracking furnace, *Ind Eng Chem Res* 29(4):636–641, 1990.

Pyl SP, Van Geem KM, Reyniers M-F, et al: Molecular reconstruction of complex hydrocarbon mixtures: an application of principal component analysis, *AIChE J* 56(12):3174–3188, 2010.

Ramana Rao MV, Plehiers PM, Froment GF: The coupled simulation of heat transfer and reaction in a pyrolysis furnace, *Chem Eng Sci* 43(6):1223–1229, 1988.

Ranzi E, Dente M, Rovaglio M, et al: Pyrolysis and chlorination of small hydrocarbons, *Chem Eng Commun* 117(1):17–39, 1992.

Ranzi E, Frassoldati A, Granata S, et al: Wide-range kinetic modeling study of the pyrolysis, partial oxidation, and combustion of heavy n-alkanes, *Ind Eng Chem Res* 44(14): 5170–5183, 2004.

Ren T, Patel M, Blok K: Olefins from conventional and heavy feedstocks: energy use in steam cracking and alternative processes, *Energy* 31(4):425–451, 2006.

Reyniers PA, Schietekat CM, Van Cauwenberge DJ, et al: Necessity and feasibility of 3D simulations of steam cracking reactors, *Ind Eng Chem Res* 54(49):12270–12282, 2015.

Rice FO: The thermal decomposition of organic compounds from the standpoint of free radicals. I. Saturated hydrocarbons, *J Am Chem Soc* 53(5):1959–1972, 1931.

Sabbe MK, Van Geem KM, Reyniers M-F, et al: First principle-based simulation of ethane steam cracking, *AIChE J* 57(2):482–496, 2011.

Schietekat CM, Van Cauwenberge DJ, Van Geem KM, et al: Computational fluid dynamics-based design of finned steam cracking reactors, *AIChE J* 60(2):794–808, 2014.

Schirmeister R, Kahsnitz J, Träger M: Influence of EDC cracking severity on the marginal costs of vinyl chloride production, *Ind Eng Chem Res* 48(6):2801–2809, 2009.

Schulz EP, Bandoni JA, Diaz MS: Optimal shutdown policy for maintenance of cracking furnaces in ethylene plants, *Ind Eng Chem Res* 45(8):2748–2757, 2006.

Singh P, Deschrijver D, Dhaene T: *A balanced sequential design strategy for global surrogate modeling,* Paper presented at: 2013 Winter simulations conference (WSC), 2013.

Smith TF, Shen ZF, Friedman JN: Evaluation of coefficients for the weighted sum of gray gases mode, *J Heat Transf* 104(4):602–608, 1982.

Stefanidis GD, Merci B, Heynderickx GJ, et al: CFD simulations of steam cracking furnaces using detailed combustion mechanisms, *Comput Chem Eng* 30(4):635–649, 2006.

Stefanidis GD, Merci B, Heynderickx GJ, et al: Gray/nongray gas radiation modeling in steam cracker CFD calculations, *AIChE J* 53(7):1658–1669, 2007.

Stefanidis GD, Van Geem KM, Heynderickx GJ, et al: Evaluation of high-emissivity coatings in steam cracking furnaces using a non-grey gas radiation model, *Chem Eng J* 137(2):411–421, 2008.

Sun W, Saeys M: Construction of an ab initio kinetic model for industrial ethane pyrolysis, *AIChE J* 57(9):2458–2471, 2011.

Tan Y, Douglas MA, Thambimuthu KV: CO2 capture using oxygen enhanced combustion strategies for natural gas power plants, *Fuel* 81(8):1007–1016, 2002.

Tarafder A, Lee BCS, Ray AK, et al: Multiobjective optimization of an industrial ethylene reactor using a nondominated sorting genetic algorithm, *Ind Eng Chem Res* 44(1):124–141, 2005.

Van Cauwenberge DJ, Schietekat CM, Floré J, et al: CFD-based design of 3D pyrolysis reactors: RANS vs. LES, *Chem Eng J* 282:66–76, 2015.

Van Geem KM, Heynderickx GJ, Marin GB: Effect of radial temperature profiles on yields in steam cracking, *AIChE J* 50(1):173–183, 2004.

Van Geem KM, Reyniers M-F, Marin GB, et al: Automatic reaction network generation using RMG for steam cracking of n-hexane, *AIChE J* 52(2):718–730, 2006.

Van Geem KM, Hudébine D, Reyniers MF, et al: Molecular reconstruction of naphtha steam cracking feedstocks based on commercial indices, *Comput Chem Eng* 31(9):1020–1034, 2007a.

Van Geem KM, Žajdlík R, Reyniers M-F, et al: Dimensional analysis for scaling up and down steam cracking coils, *Chem Eng J* 134(1–3):3–10, 2007b.

Van Geem KM, Reyniers MF, Marin GB: Challenges of modeling steam cracking of heavy feedstocks, *Oil Gas Sci Technol—Rev IFP* 63(1):79–94, 2008.

van Goethem MWM, Kleinendorst FI, van Leeuwen C, et al: Equation-based SPYRO® model and solver for the simulation of the steam cracking process, *Comput Chem Eng* 25(4–6):905–911, 2001.

Verhees P, Amghizar I, Goemare J, et al: 1D model for coupled simulation of steam cracker convection section with improved evaporation model, *Chem Ingenieur Tech* 88(11):1650–1664, 2016.

Verstraete JJ, Dulot H, Revellin N, et al: Molecular reconstruction of vacuum gasoils, *Prepr Pap Am Chem Soc Div Fuel Chem* 49(1):20–21, 2004.

Verstraete JJ, Schnongs P, Dulot H, et al: Molecular reconstruction of heavy petroleum residue fractions, *Chem Eng Sci* 65(1):304–312, 2010.

Wang Z, Feng Y, Rong G: Synchronized scheduling approach of ethylene plant production and naphtha oil inventory management, *Ind Eng Chem Res* 53(15): 6477–6499, 2014.

Westbrook CK, Dryer FL: Simplified reaction mechanisms for the oxidation of hydrocarbon fuels in flames, *Combust Sci Technol* 27(1–2):31–43, 1981.

Zendehboudi S, Zahedi G, Bahadori A, et al: A dual approach for modelling and optimisation of industrial urea reactor: smart technique and grey box model, *Can J Chem Eng* 92(3):469–485, 2014.

Zhang Y, Qian F, Schietekat CM, et al: Impact of flue gas radiative properties and burner geometry in furnace simulations, *AIChE J* 61(3):936–954, 2015.

Zhao C, Liu C, Xu Q: Cyclic scheduling for ethylene cracking furnace system with consideration of secondary ethane cracking, *Ind Eng Chem Res* 49(12):5765–5774, 2010.

Zhao C, Liu C, Xu Q: Dynamic scheduling for ethylene cracking furnace system, *Ind Eng Chem Res* 50(21):12026–12040, 2011.

CHAPTER FOUR

Gas Turbines and Engine Simulations

B. Cuenot[1]
CFD Combustion Team, CERFACS, Toulouse cedex, France
[1]Corresponding author: e-mail address: benedicte.cuenot@cerfacs.fr

Contents

Advances in Chemical Engineering, Volume 49
ISSN 0065-2377
http://dx.doi.org/10.1016/bs.ache.2016.09.004

Abstract

Turbulent combustion is a complex phenomenon combining the random nature of turbulence with the nonlinearity of chemistry. As a major process for energy production, in particular in propulsion systems, turbulent combustion must be perfectly controlled to ensure maximum efficiency with minimum environmental impact in terms of both fuel consumption and pollutant emissions. Today, the design and development of new concepts cannot be efficiently performed with experiment only and numerical simulation is required. An important step has been made in the last decades with the help of high-performance computing, which has allowed to go beyond the limited description of turbulence by the mean, as is done in the Reynolds-averaged approach. Thanks to large eddy simulation (LES), numerical simulation has recently emerged as a predictive tool, complementary to experiment and essential to understand the subtle interactions between fluid mechanics, thermochemistry, and heat transfer. In this chapter the main elements and current research trends of turbulent combustion modeling in the context of LES are described. First a brief introductory overview of technical challenges faced by engine manufacturers is given in the fields of automotive, aeronautical, and spatial propulsion. Combustion basics including thermochemistry and laminar flames are then presented to introduce the various concepts of turbulent combustion modeling. To illustrate the capacities of LES, recent examples of simulations are presented for various propulsion systems, in steady or transient operation phases where LES is undoubtedly the best adapted approach. Finally, recent first steps toward multiphysics computations, including two-phase flows and heat transfer, are reported.

1. INTRODUCTION

Most propulsion systems today still rely on combustion to convert chemical energy into mechanical energy. The main alternative is electric propulsion, which has been used for long in railway transportation and, more recently, in car engines. The use of electricity raises the practical issue of energy supply. In the case of trains, this was solved by using electrified rails or lines, which allowed an early development of electric propulsion and avoided other problems of thermal propulsion such as fuel supply for long-distance journeys or smoke accumulation in tunnels. The situation is different for electric cars, which require high-performance batteries or fuel cells, and a network of recharging stations. However the positive impact of no-emission cars on the local air quality is a strong motivation to develop such technology. Electric propulsion is also found in the spatial sector, for on-orbit altitude control of satellites. Systems such as Hall-effect thrusters have recently regained interest, thanks to their high specific impulse that allows significant saving in propellant mass. These electricity-based technologies however still represent only a small fraction of the global transport and propulsion market, and they are often combined with combustion, as in the hybrid engines developed by the automotive industry. More than that, no realistic alternative for combustion exists today for some transport means, such as aviation. Another possible alternative for the future is hydrogen-based fuel cell technology. Although its potential for low-carbon energy supply is promising, its cost and the necessity to produce hydrogen are current important obstacles to its deployment at a large scale. Fig. 1 confirms the dominant use of combustion in the future, showing the expected evolution of fuel consumption by transport according to different scenarios. In the best case, electric propulsion reaches up to 25% of the energy source, but in all other cases, it never exceeds 10% and is often lower. The distribution of fuel consumption among the different transport means is shown in Fig. 2, where it can be seen that ground transportation is by far the major contributor. This explains why regulations and efforts toward more efficient, environmentally friendly engines have first focused on automotive propulsion. This is now becoming also a concern for air transport, in particular for the preservation of the environment around airports, which are more and more embedded in urban areas.

The still dominant place of combustion is also valid for the general energy sector. Almost 90% of energy on Earth is produced today by burning fossil or renewable fuel. Even the exploitation of other energy sources, such as wind or sun, partly relies on combustion to compensate for their natural variability.

As a consequence, although combustion has become a mature technological field, which is well mastered by engineers to design efficient and reliable engines, new issues raised by sustainable development and climate change call for further optimization of combustors. Indeed, one major

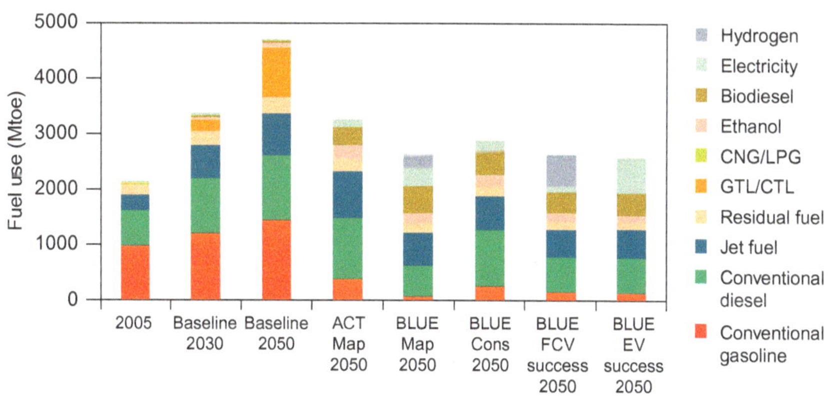

Fig. 1 Transport energy use for different technology scenarios. *Source: © OECD/IEA 2008 Energy Technology Perspectives: Scenarios & Strategies to 2050, IEA Publishing. Licence: www.iea.org/t&c.*

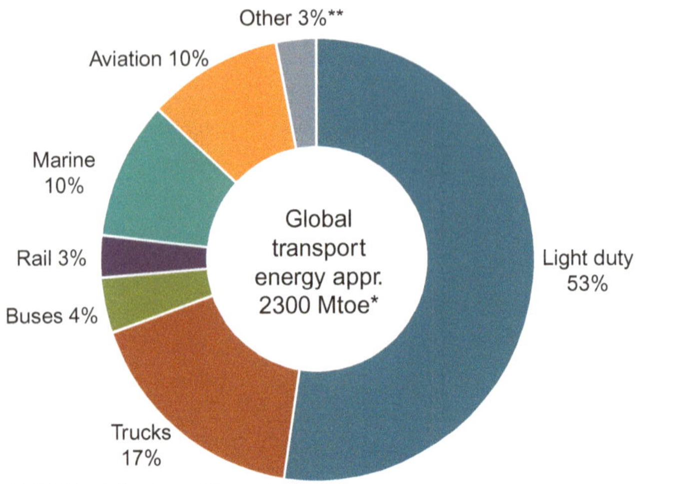

Fig. 2 Global view on transport modes in 2009 (AMF 2011: IEA Advanced Motor Fuels Annual Report 2011).

drawback of combustion-based propulsion is the emission of pollutant species, which are subjected to more and more stringent regulations. Pollutants may be categorized in three types. The first one corresponds to products of combustion such as carbon dioxide (CO_2) and is directly linked to fuel consumption and combustion efficiency. It contributes to the greenhouse gas effect, strongly impacting the climate as demonstrated in the Fifth Assessment Report on climate change published by IPCC (www.ipcc.ch). Another category of pollutants is linked to combustion chemistry which produces carbon monoxide (CO) and nitric oxides (NO_x). These chemical species represent serious dangers for human health and are also targeted by environmental regulations. Fig. 3 represents the CO and NO_x production as a function of the burnt gas temperature for a hypothetical burner (Lefebvre, 1999). The emission of CO is the direct consequence of incomplete oxidation into CO_2, either due to lack of oxidizer or due to too low temperature which slows down the oxidation process, leading to a peak production at low burnt gas temperature. On the other hand, nitric oxides are produced at high temperature and favored by long residence time of the hot gas volume. As shown in Fig. 3, it is therefore difficult to reach minimum levels for both CO and NO_x production, and a compromise must be found. The intermediate zone between 1700K and 1900K where both pollutants are emitted at reasonably low levels is the target zone of most industrial burners.

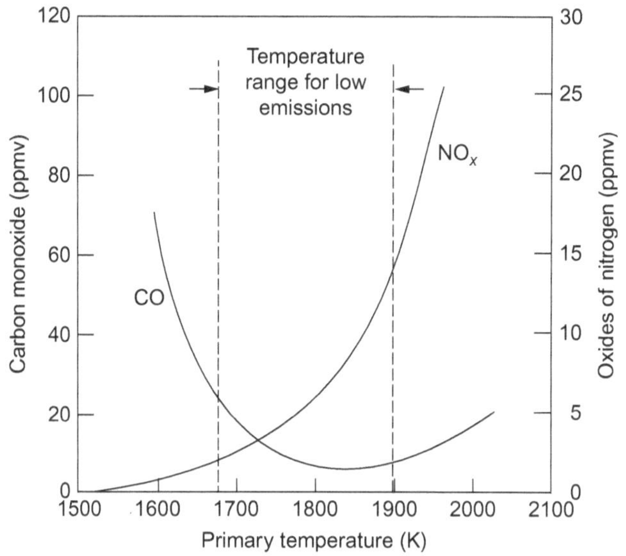

Fig. 3 Influence of primary-zone temperature on CO and NO_x emissions (Lefebvre, 1999).

Unburnt hydrocarbons are also very harmful and may be emitted if combustion is not complete. This happens if fuel is trapped in zones that are not reached by the flame, for example, in corner recirculation zones (CRZs) or in liquid fuel films that may build on the combustion chamber walls. Finally, soot is highly undesirable and a major concern in urban areas as it is a source of breathing difficulties or lung disease. Soot emission is the result of complex heterogeneous chemistry processes and is difficult to avoid when burning heavy hydrocarbons. However, soot can be reoxidized if the burnt gas contains sufficient remaining air after burning.

The current challenge of combustion science is therefore to allow burning fossil and renewable fuels without wasting them and with limited pollutant emissions for good air quality and minimum impact on the global climate. To reach these goals, all the physics involved in the combustion process and engine operation must be taken into account. This includes thermochemistry, turbulence, multiphase flows, acoustics, and heat transfer. All these multiscale, nonlinear phenomena occur simultaneously in the combustion chamber and deeply interact with each other, possibly leading to very negative effects such as combustion instabilities, which can destroy the engine in extreme cases. They also drive key processes such as ignition and blow-off, combustion efficiency, and noise or pollutant emissions, which makes the design of new systems a complex and delicate process.

Such complexity requires a detailed and deep understanding of all phenomena and their interaction and cannot be handled with simple engineering tools or experiments only. In this context, numerical simulation plays a key role and is essential to make significant steps in the development of the new generation of combustion systems.

The accurate simulation of turbulent combustion remains one of the greatest challenges of modern sciences. It has made great progress in the recent years thanks to high-performance computing (HPC). The access to more and more powerful massively parallel computers not only has renewed combustion science by allowing a detailed description of complex phenomena but also has started the "numerical revolution" in the industrial sector of aeronautical and car engines, by making it possible to handle complex, real burner geometries in real operating conditions. This implies that simulation tools are able to give accurate and reliable results, leading to a correct understanding and system behavior, in a reasonable time. Beyond their scientific interest, such tools give to the industrial user an important advantage over its competitors in terms of cost of design and development of new products: instead of manufacturing a number of prototypes, simulation and optimization may be used to select only a few for testing. For this reason, it is foreseen that in the next decade, numerical simulation will take more and more place in industry and will become a major engineering tool.

This chapter gives the main, state-of-the-art elements of high-fidelity combustion simulation and its application to propulsion systems.

2. PROPULSION SYSTEMS

We will focus in this chapter on three types of engines, which represent the vast majority of propulsion systems, namely internal combustion engines (ICEs) used for automotive propulsion, gas turbines (GTs) encountered in civil aviation, and rocket engines. Other combustion-based engines such as solid propulsion boosters which produce the thrust for space launchers, or ramjets and scramjets used for very high speed vehicles, are very specific and not described here. Engines are designed to guarantee target performances in terms of combustion efficiency or ignition capabilities, while minimizing undesired effects such as thermal and mechanical fatigue, noise, or pollutant emission. Various concepts have been developed to reach these objectives, whose main principles are described below. Based on these basic concepts, multiple variations exist that aim at improving the overall behavior of the system, in an attempt to reach the best trade-off between positive and adverse effects. Describing them all is out of scope of this chapter and the reader is referred to Lumley (1999), Boyce (2011), MacIsaac and Langton (2011), and Sutton and Biblarz (2010), which provide complete reviews of existing technologies.

In the following sections, the thermodynamic principles and the design challenges are detailed for the three types of engines addressed here.

2.1 Internal Combustion Engines

ICEs are found in many applications daily used in our society. They still equip the vast majority of cars, but also power pumps, portable generators, or agricultural machines. They are also used in aircraft propellers and ships. About 70% of crude oil consumed daily world wide is used in ICE for transportation.

The main principle of ICE is burning at constant volume. This is achieved thanks to moving parts that open for injection and convert the chemical energy into mechanical energy. The geometrical arrangement of these moving parts leads to the various types of ICE, having either reciprocating piston or pistonless rotary motion; poppet valves, sleeve valves, side valves, or overhead valves; and various camshaft placements. In this chapter, we will focus on reciprocating ICE, also known as piston engines. In such engines, one or more reciprocating pistons convert pressure (produced by the hot, burnt gas expansion) into a rotating motion. Ignition at each cycle may be achieved either with a spark in spark-ignition (SI), gasoline engines or via compression in compression-ignition (CI), diesel engines.

Piston engines are characterized by their number n_c and volume V_c of cylinders, compression ratio r_c, and stroke (travel distance of the piston), which all together drive the engine power. Fig. 4 illustrates the different components of a piston engine. The piston moves between the Top Dead Center (TDC) and the Bottom Dead Center (BDC), determining the compression ratio which is the volume ratio between the two positions. The piston motion is then transmitted to the crank, which makes one rotation at each up and down displacement of the piston.

The mean effective pressure (MEP) is the mean pressure corresponding to the total work W: $\text{MEP} = W/n_cV_c$. It allows to determine the power P of the engine as $P = \text{MEP}n_cV_cNn_r$, where N is the rotation speed and n_r is the number of rotations in a cycle (equal to 2 in a four-stroke engine). This simple relation allows to give a first estimate of the engine power from the cylinder capacity n_cV_c and the mean pressure. Cylinder capacity may vary

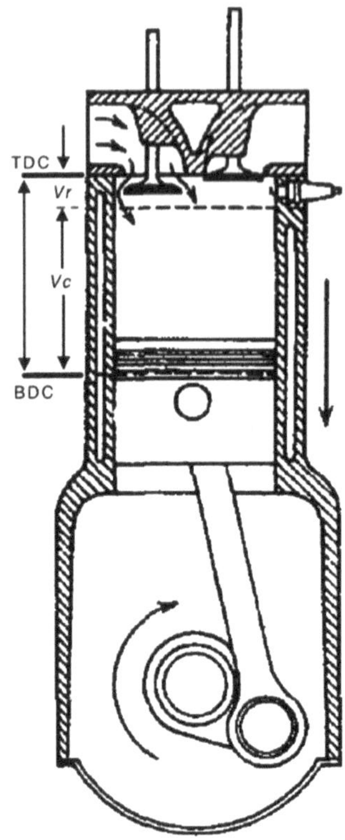

Fig. 4 Components of a piston engine cylinder (Heywood, 1988).

from few tens of cm^3 for pumps or mopeds to few m^3 for diesel engines propulsing cargo boats (Fig. 5). This leads to engine power ranging from 1 kW up to 5 MW (i.e., approximately from 1 to 10,000 hp).

2.1.1 Thermodynamic Cycle

The ideal thermodynamic cycle of a four-stroke piston engine is the Otto/Beau de Rochas cycle, illustrated in Fig. 6 in a *P–V* diagram. After injection (A to B), valves are closed and the piston moves up for isentropic compression (B to C). Then combustion occurs at constant volume (C to D), raising again the pressure. The hot gas pushes the piston down in an isentropic expansion (D to E) and the cycle ends with the hot gas exhaust (E to B).

The work W corresponds to the pressure work in the compression and expansion phases. From the first principle of thermodynamics, applied to a closed, adiabatic system, the work is related to the combustion energy source

Fig. 5 View of the 14-cylinder RT-flex96C cargo diesel engine. The 430 ton crankshaft spins at 102 rpm for maximum power of 80,080 kW. *Source: Photo copyright @ Winterthur Gas & Diesel Ltd.*

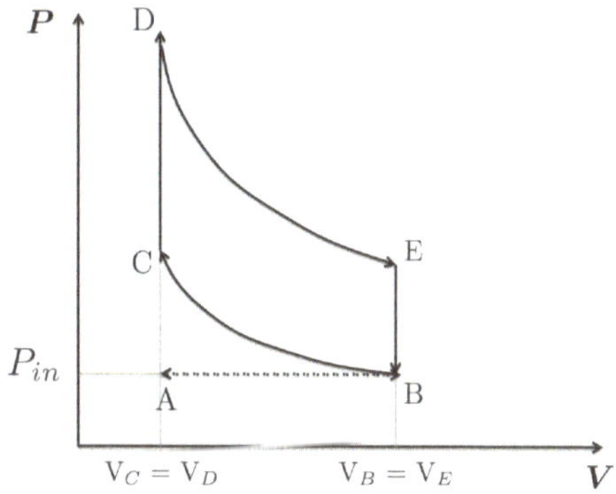

Fig. 6 The Otto/Beau de Rochas cycle.

Q and the variation of internal energy as $W + Q = \Delta E$, where ΔE is here equal to $E_E - E_B$. The energy released by combustion Q is expressed as the thermal energy produced by the combustion of 1 kg of fuel multiplied by the fuel consumption $\dot{m}_f$. Applying thermodynamic relations to compute the pressure and temperature at all states of the cycle, one gets the energy difference $E_E - E_B = r_c^{1-\gamma} Q$, where γ is the heat capacity ratio, of the order of 1.4. This finally gives $W = (r_c^{1-\gamma} - 1)Q$ and the cycle efficiency $\eta = -W/Q = 1 - r_c^{1-\gamma}$. The compression ratio is therefore a critical parameter and must be as high as possible to maximize the efficiency.

The Otto/Beau de Rochas cycle leads generally to very high maximum pressure P_D, which is not desirable for the durability of the piston and the cylinder. To reduce the maximum pressure, the dual cycle is introduced in which combustion starts at constant volume, as in the Otto/Beau de Rochas cycle, but ends at constant pressure (Fig. 7). This introduces two additional parameters: $\alpha = P_D/P_C$ measures the pressure increase during the first stage of combustion, while $\beta = V_{D'}/V_D$ measures the volume increase in the second stage of combustion. This leads to a more complex expression for the thermodynamic efficiency:

$$\eta = 1 - r_c^{1-\gamma} \frac{\alpha\beta^\gamma - 1}{\alpha - 1 + \gamma\alpha(\beta - 1)} \tag{1}$$

The reduction of the maximum pressure clearly lowers the efficiency of the dual cycle, but allows to decrease the mechanical efforts in the engine.

In practice, heat and pressure losses occur in the engine and decrease the cycle efficiency. In addition, other technical constraints may significantly modify the cycle: combustion is not instantaneous and may be incomplete, possible endothermic dissociation of products may decrease the temperature, the synchronization of the valves is sometimes modified, and ignition advance is often used to avoid engine knocking. To distinguish all these effects, three different efficiencies are introduced: the combustion efficiency η_{comb} includes incomplete burning and dissociation effects, the cycle efficiency η_{cycle} takes into account heat losses and valves synchronization, and the mechanical efficiency η_{mech} introduces friction and pressure losses.

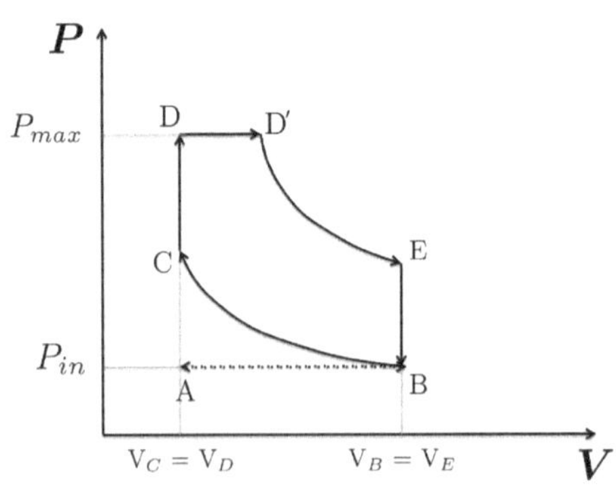

Fig. 7 The dual cycle.

The values of the efficiencies η_{comb}, η_{cycle}, and η_{mech} vary depending on the engine. Typically, the global efficiency of engines of passenger cars is of the order of 30–33%.

2.1.2 Flow Dynamics and Mixing

The performance of ICE is primarily linked to the quality of mixing, which allows a fast and complete combustion at each cycle. Mixing is the result of turbulence, generated in the cylinder by the fuel injection, the piston motion, and the interaction with the wall.

Typical turbulent Reynolds numbers $Re_t = u'L/\nu$, where u' is the flow velocity fluctuation, L is a characteristic length scale of the system, and ν is the kinematic viscosity, are between 100 and 10,000, with u' being in the range of 1–10 m/s and L between 0.1 and 1 cm. Turbulence is a key mechanism for ICE. Indeed, laminar burning, with a flame speed of a few tens of cm/s, does not allow to exceed rotation speeds of the order of 100 rpm. Only turbulent mixing and combustion may lead to sufficient power. A particular flow feature specific to ICE is the tumble, illustrated in Fig. 8, whose strength has been shown to be strongly correlated to the turbulence level at the time of ignition (Arcoumanis et al., 1998). This large-scale vortex appears during the admission phase and is highly

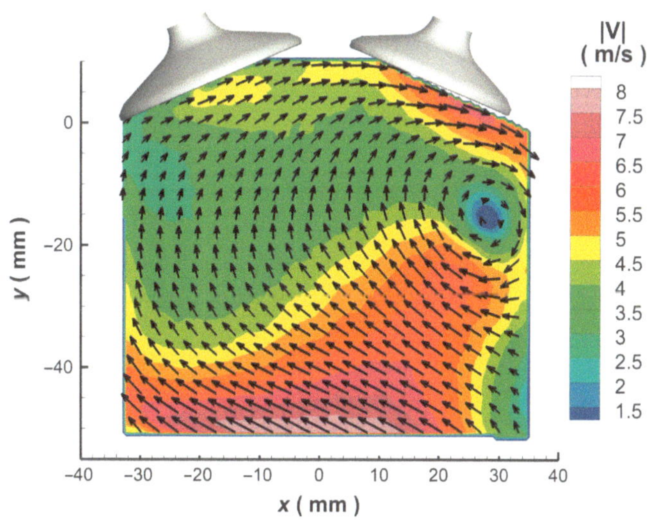

Fig. 8 Visualization of the tumble in ICE: measured velocity field (averaged) (Baum et al., 2014). *From Baum E, Peterson B, Böhm B, Dreizler A: On the validation of LES applied to internal combustion flows: Part 1: Comprehensive experimental database.* Flow Turbul Combust *92, 269–297, 2014.*

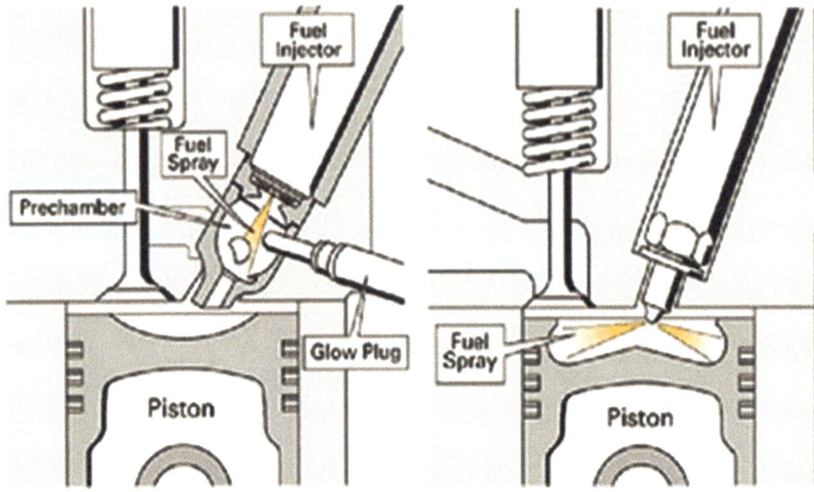

Fig. 9 Indirect vs direct injection in ICE. *Source: www.car-engineer.com.*

dependent on the injection system (Fig. 9). Indirect injection consists in injecting the fuel in the admission pipe, upstream of the admission valve, i.e., before entering in the cylinder. The fuel has therefore sufficient time to preevaporate and mix with the air in the pipe, and when the valve opens, a fuel–air mixture is injected which is further homogenized during the compression phase. On the contrary, in the direct injection method, the fuel is sprayed directly in the cylinder. Direct injection was always used in CI engines, but only recently applied to SI engines for which indirect injection has been preferred for a long time. Apart from its lower price, the main reason to prefer indirect injection was the better fuel–air mixing and the absence of liquid in the cylinder. However, new engine concepts, and in particular downsizing, have pushed toward direct injection for a better control of the injected fuel, the absence of liquid in the admission pipe, and most importantly the possibility to burn in a stratified mode which, if well controlled, gives more flexibility to reach high efficiency while reducing pollutants.

2.1.3 Injection and Ignition

In CI engines (diesel engines), the flame auto-ignites during the compression phase. This process is controlled by the auto-ignition delay, which is highly dependent on the fuel composition and the operating conditions. In particular, some fuels exhibit a negative temperature coefficient behavior, in which the auto-ignition delay increases with the temperature. Auto-ignition combined with direct injection as in diesel engines leads to nonpremixed

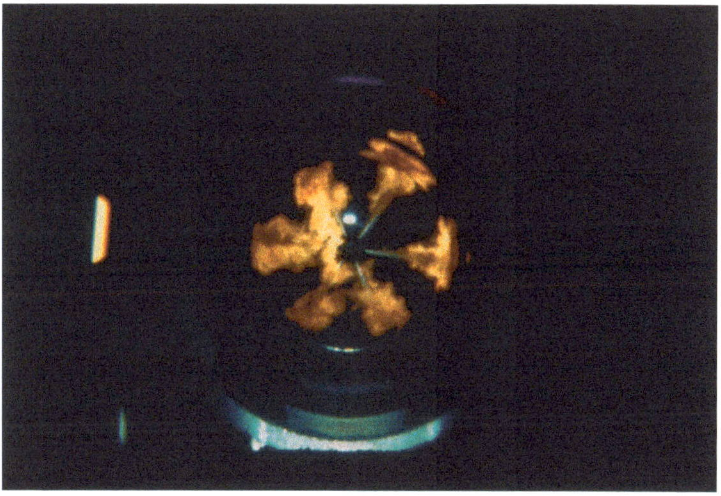

Fig. 10 Diesel combustion: view of direct injection and diffusion flame. *From Mercedes-Benz Classic.*

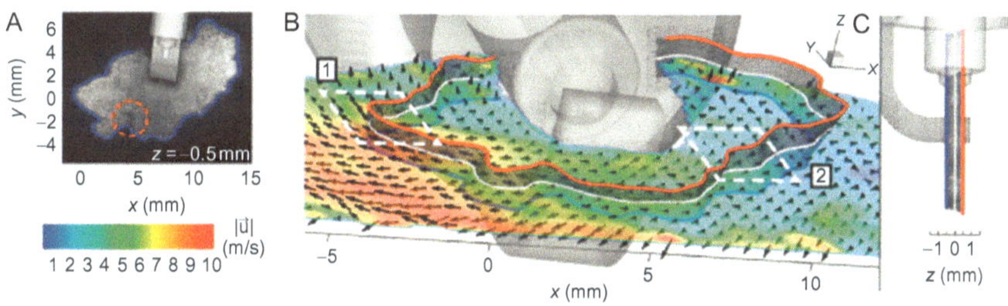

Fig. 11 Spark ignition engine. (A) Raw LIF-image and reaction zone (*blue line*). (B) 3D flame surface and flow field. (C) Side view of 3D flame surface. *From Peterson B, Baum E, Böhm B, Dreizler A: Early flame propagation in a spark-ignition engine measured with quasi-4D diagnostics.* Proc Combust Inst *35, 3829–3837, 2015.*

turbulent combustion, as illustrated in Fig. 10. In SI engines, the combustion is started by the breakdown of the spark plug at each cycle. The timing of the sparking depends on the engine speed and load and is expressed in crankshaft degrees prior to TDC. As already mentioned, ignition advance allows to increase combustion efficiency and is between 10 and 40 crankshaft degrees prior to TDC. The spark allows to create a small flame kernel that grows rapidly before transitioning to a turbulent propagating flame (Fig. 11).

2.1.4 Cyclic Variation and Abnormal Combustion

In normal combustion, pressure increases smoothly to a peak value optimally reached when combustion is complete, and then decreases as the piston

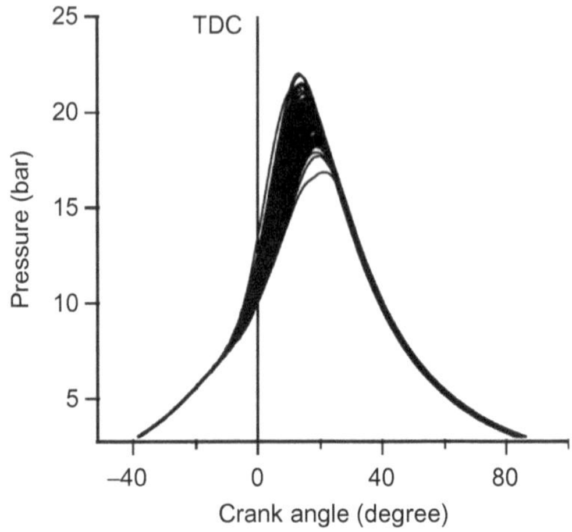

Fig. 12 Temporal evolution of the in-cylinder pressure: example of cyclic variation (Granet et al., 2012).

moves down to the BDC, transferring mechanical work to the system. Typical pressure curves are shown in Fig. 12, which notably vary around a mean cycle due to the important role of turbulence. Indeed the random nature of turbulence leads to cycle-to-cycle variations, i.e., differences in the peak pressure from one cycle to the other. This has an important impact on the engine efficiency and makes it difficult to optimize the engine operating conditions. However due to the random and uncontrolled nature of this cyclic variation, it is very difficult to include it in the design phase of an engine and requires sophisticated numerical approaches to predict it with good accuracy.

When a fuel/air mixture pocket forms outside the turbulent flame, and if local pressure and temperature conditions allow it, this pocket may auto-ignite and create a sharp pressure increase. In extreme cases, detonation can even occur, which is very destructive and highly undesirable. This phenomenon, known as knock, has been identified very early, but the exact conditions leading to knock and controlling its intensity are not yet fully understood. Among the influencing parameter is the fuel octane rating, which prevents detonation and knock if sufficiently high. Knock is a different phenomenon from pre-ignition, which occurs when the fuel/air mixture pre-ignites before the spark plug fires. This typically occurs when residual hot spots are present in the combustion chamber in the admission phase, and can also cause severe damages to the engine. Knock and

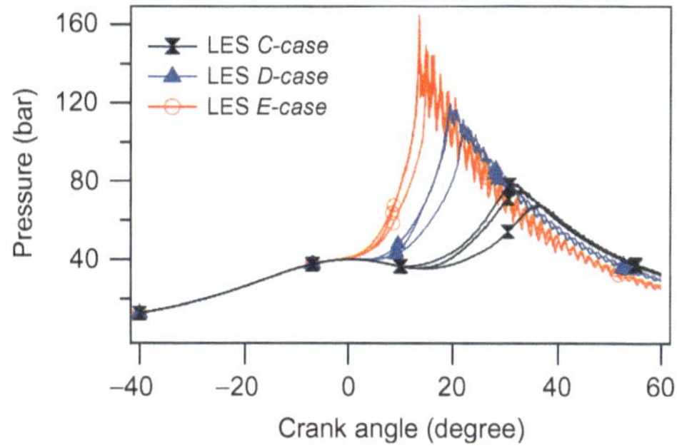

Fig. 13 Temporal evolution of the in-cylinder pressure: example of knock for three cases with variable spark timing (Poinsot et al., 2015).

pre-ignition phenomena can be identified on the cycle pressure curve, as shown in Fig. 13.

2.1.5 Pollutant Emissions

Passenger cars still being the most used transportation mean, manufacturers are facing stringent regulations for the emission of pollutant species and soot, set, for example, in Europe by European emission standards (Euro 6 norm). However, their reduction is often at the price of lower engine performance. As NO_x are very sensitive to high temperature, low-NO_x engines operate at lower temperature. This is reached either by burning lean mixtures, i.e., with excess air, or by reinjecting a portion of the hot combustion products in the combustion chamber to dilute the fresh gas mixture. This last technique is known as Exhaust Gas Recirculation (EGR) and is efficient in reducing the NO_x emission, but it also lowers the engine efficiency. Soot particles are filtered, captured, or reoxidized in after-burners, but this implies more complex systems with a decrease of mechanical efficiency.

2.1.6 New Trends in ICE

In the sector of automotive propulsion, the development of new engine concepts is driven by the necessity to improve combustion efficiency (to reduce fuel consumption), while minimizing pollutant emissions and still guaranteeing good engine behavior with regard to ignition and knock. One current tendency is engine downsizing, i.e., increase of the power to size ratio, and down-speeding, i.e., decrease of the rotation speed at given highway speed. This requires boosting devices such as turbochargers and superchargers, and direct injection technology, either wall guided, air guided, or spray guided, for a better control of the flame. However, this also leads to an increased propensity to knock and abnormal combustion and increases notably wall heat transfers.

Another recent concept is homogeneous charge compression ignition (HCCI), where the fuel and oxidizer are fully premixed and burn at lower temperature, reducing pollutant emissions and improving combustion efficiency. The difficulty of HCCI is however to guarantee perfect premixing, which is obtained with turbulence and therefore difficult to control.

Finally, new fuels and fuel blends are now available and may be used to modify the engine behavior. It has been shown, for example, that optimum fuel characteristics change according to the load (Inagaki et al., 2006), with high-cetane fuels being more adapted at light loads and, conversely, low-cetane fuels at high loads (Kokjohn et al., 2009). This led to the reactivity controlled compression ignition (RCCI) concept (Reitz and Duraisamy, 2015), using two fuels of different reactivity and multiple injection to control combustion. However, it should be kept in mind that the use of alternative fuels raises the issue of their supply, contrary to fossil fuels which are naturally available in large amounts.

2.2 Gas Turbines

As ICE, GTs are used in various applications and are of different sizes. Aeronautical engines fit in engine nacelles of 1–2 m long, and their power typically ranges between 500 shaft-hp and 3000 shaft-hp (350–2500 kW) for helicopters, and up to about 80 MW for civil aviation. Jet engines are usually rather characterized by their thrust, which is the force pushing the aircraft forward and varies between 50 kN for business jets up to more than 400 kN for wide-body jets. For example, the famous CFM engines (designed and manufactured by a joint venture GE-Safran), which equip most single-aisle commercial jets, can reach a thrust of approximately 150 kN. Aeronautical engines are much smaller than ground-based energy production GT, which reach extremely high powers of several hundreds of MW and are more than 10 m long (Fig. 14). On the other side, micro-GTs are developed for hybrid electric vehicles or as a portable energy supply for military usage and may deliver less than 1 kW.

Contrary to ICE, a GT is an open thermodynamic system with in- and outflow, which means that pressure does not increase much during combustion. As the engine power is directly linked to its capacity to burn a maximum of fuel in a minimum volume (or time), it is necessary to feed the combustion chamber with a pressurized flow which leads to a higher mass

Fig. 14 The SGT5-8000H gas turbine, developed by Siemens, is rated for a power output of 375 MW. The picture shows the gas turbine at Berlin facility ready for shipment.

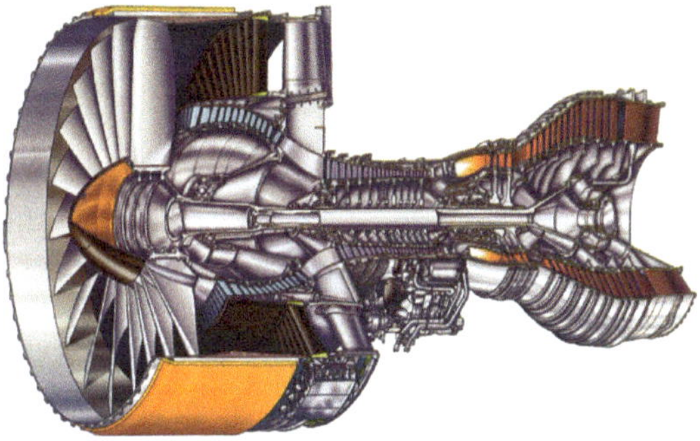

Fig. 15 View of a GT engine.

flow rate and enhances the fuel oxidation process. GTs are therefore composed of three main elements: a compressor, a combustion chamber, and a turbine (Figs. 15 and 16). The combustion chamber has a toroidal shape (Figs. 17 and 18) and contains a series of equidistant injectors on which flames stabilize and possibly interact with each other if sufficiently close. All injectors are similar, except in some configurations where so-called ignition injectors, placed close to the spark-plugs, are optimized to enhance ignition.

2.2.1 Thermodynamic Cycle

The conversion of energy in a GT follows the ideal thermodynamic Joule–Brayton cycle (Fig. 19). After the isentropic compression through the compressor (A to B), air is injected in the combustion chamber where it mixes with fuel vapor to produce heat and hot combustion products (B to C). The high pressure, high temperature exhaust fluid is then expanded in the turbine which finally produces thrust or entrains the motor shaft (C to D). To take into account the work brought to the system during

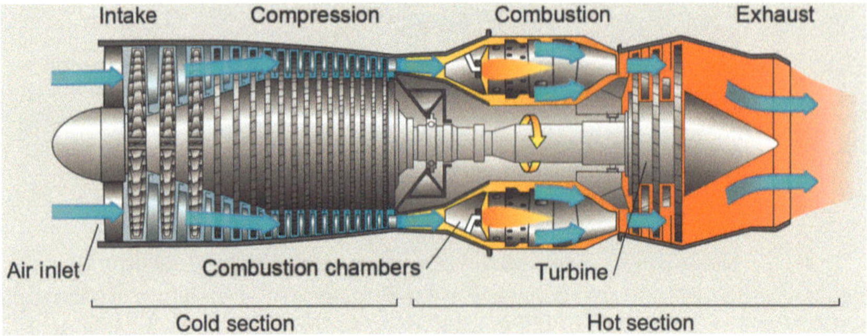

Fig. 16 Sketch of a GT engine.

Fig. 17 RAF Panavia Tornado Aircraft Rolls Royce RB199 Jet Engine Combustion Chamber. *Source: Rolls-Royce.*

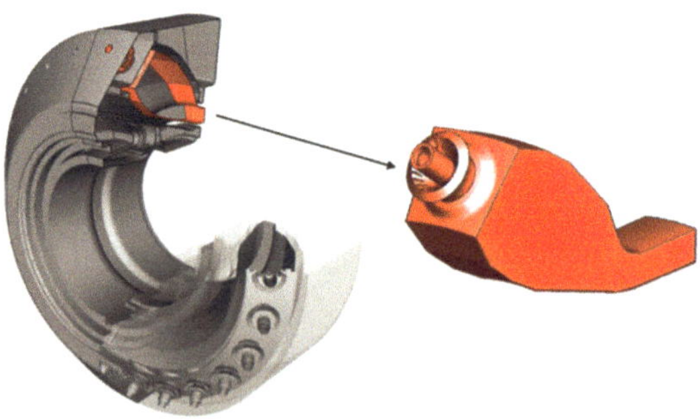

Fig. 18 View of an annular ground GT combustion chamber and detail of one sector.

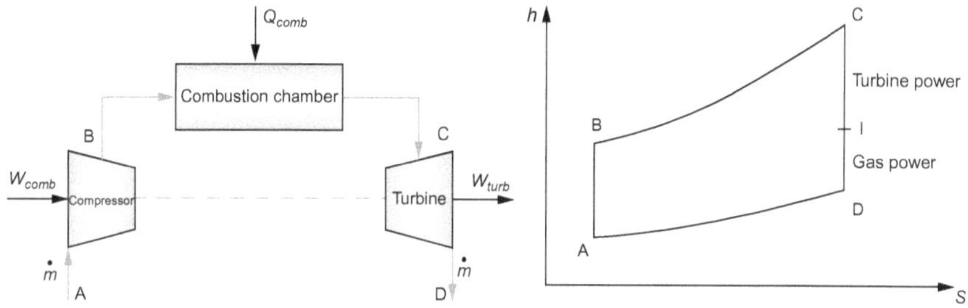

Fig. 19 The Joule–Brayton cycle.

the compression phase in the calculation of the efficiency, this last step is split into two parts corresponding to the power needed by the compressor (C to I), called the turbine power, and the remaining, useful power (I to D) called the gas power. Using isentropic relations and the first thermodynamic principle, the efficiency of the cycle is found to be:

$$\eta = 1 - \epsilon^{-(\gamma-1)/\gamma} \tag{2}$$

where $\epsilon = P_B/P_A$ is the compression ratio and appears as the main controlling parameter for thermodynamic efficiency. The useful, specific gas power $W_{s,\ ID}$ is given by:

$$W_{s,ID} = C_p T_C\left(1 - \epsilon^{(1-\gamma)/\gamma}\right) - C_p T_A\left(\epsilon^{(\gamma-1)/\gamma}\right) \tag{3}$$

This allows to determine the optimum pressure ratio ϵ_{opt} leading to the maximum gas power, as the one corresponding to the $T_B = T_D = \sqrt{T_A T_C}$. In the above analysis, pressure losses, mechanical losses, and other kinds of losses have been ignored and should be included to establish the real cycle.

2.2.2 Engine Components

Turbomachinery (compressor or turbine) is a complex system which requires dedicated tools for analysis and prediction, and is not detailed in this chapter which focuses on burners. Although the combustor has little effect on the cycle efficiency, it has a significant impact on engine operability and other behaviors. The typical flow velocity at the compressor outlet of about 150 m/s being much higher than the turbulent flame velocity, the stabilization, and completion of combustion in a small volume are difficult to obtain. To create low-velocity regions where the flame can survive, obstacles were first used to maintain the flame in the wake of the so-called flame holders. However, this raised mechanical and thermal issues, with pressure and heat losses and overheating of the obstacles. A far better solution, now applied in most combustors, is to establish a swirled flow which induces low velocity or even recirculating velocity regions for sufficiently high swirl numbers (Syred and Beer, 1974). Such flow is obtained thanks to the injection systems in which axial and/or radial swirlers generate a rotational motion. This has a positive side effect by increasing turbulence intensity and mixing efficiency.

Fig. 20 shows the architecture of a sector (one injector) of a conventional GT combustion chamber. A diffuser smoothly reduces the incoming gas velocity with a limited pressure drop upstream of injection and feeds a first cavity called the plenum which distributes the air between the injection system and the bypass flow around the chamber. Bypass provides a lower thrust-specific fuel consumption and reduces jet noise. It is also used for cooling the combustion chamber wall. In high bypass-ratio engines, most of the thrust is produced by the bypass flow. A liquid fuel injection system is placed centrally to allow atomization and evaporation by the injected air flow. The ignition system is usually installed on a side wall for practical reason (electric supply) in the primary zone where combustion occurs. Finally, cooling devices allow to minimize heat transfers and preserve the combustor integrity by establishing a cold air film on the walls, either through tangential injection (as in Fig. 20) or normal injection via multi-perforated plates.

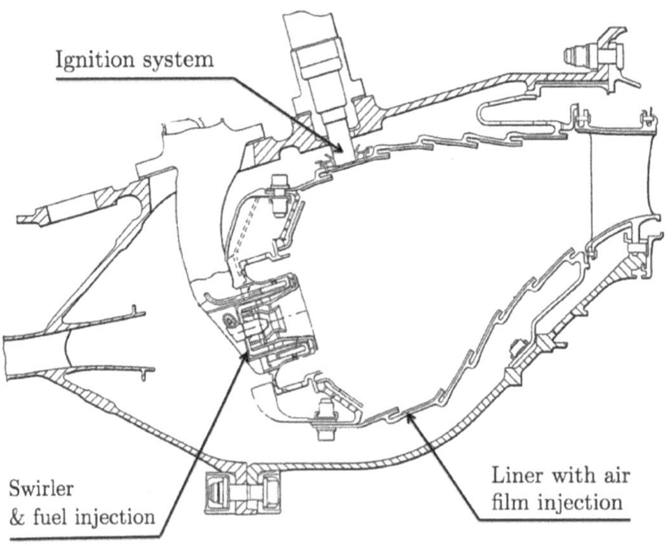

Fig. 20 Conventional combustion design.

2.2.3 Operability

Operability of aeronautical GT describes the engine behavior in the desired flight conditions, in terms of ignition and blow-out, stability, noise, and pollutant emissions. Due to the combustion chamber architecture, ignition is a two-step process: first the injector closest to the spark plug must be ignited, and then the whole combustor must ignite via a neighbor-to-neighbor flame propagation process. The size of the combustor is an important design parameter for ignition, as the residence time must be higher than the ignition delay. This is even more critical for high-altitude relight capability (Nicholas, 2005), which is a certification constraint and corresponds to the least favorable conditions (low temperature and pressure) to ignite. Fig. 21 shows the flight envelope in a Mach-altitude graph along with iso-levels of temperature. Dashed lines show how changes of the combustor volume modify the ignition performance: from the baseline design (thick dashed line), a reduction of the combustor volume reduces the residence time and directly lowers the relight capability limit (Doerr, 2012). The blow-out limit is also an important design parameter and is characterized by the fuel-to-air ratio (FAR) at which extinction occurs. This is of particular importance for helicopter engines that may be subjected to sudden and fast fuel injection variations under the action of the pilot. The extinction

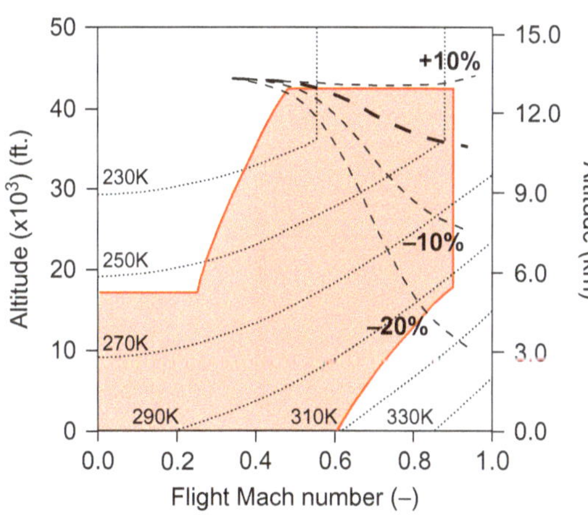

Fig. 21 Effect of combustor size on the relight capability in an altitude–Mach number graph: *thin dashed lines* indicate ignition performances for various combustor volumes, compared to the initial design (*thick dashed lines*). Typical flight envelope (*red*) and iso-line of temperature (*dotted lines*) are also reported. *Source: Adapted from Doerr, T., 2012. Introduction to aero-engine gas turbine combustion. In: Von Karman Institute for Fluid Dynamics (Ed.), Lecture Series 2012-04. Von Karman Institute for Fluid Dynamics.*

FAR depends on the quality of fuel evaporation and mixing, on the turbulent flame structure, and on the geometrical arrangement of burners in the annular combustor geometry.

The combustion chamber outflow is composed of hot gases which feed the turbine placed downstream and therefore have a strong impact on its efficiency. But more important than that, the hot combustion gases induce a very high thermal load on the turbine blades. To limit the thermal fatigue and ensure sufficient durability, the turbine blades are cooled thanks to an internal cold flow, and the combustion products are diluted with fresh air prior to exiting the combustor to decrease their temperature from around 1500K or more to less than 700K. This dilution must be as homogeneous as possible, to avoid the presence of hot pockets that could severely damage the turbine blades. This is achieved thanks to a smart arrangement of cooling air injections and intense turbulent mixing and is very delicate to guarantee when designing a new chamber. The temperature fluctuations of the flow at the exit of the combustion chamber are evaluated with the radial temperature distribution function, which corresponds to the radial profile of the mean azimuthal variations of the temperature elevation relative to a reference value (Fig. 22).

2.2.4 Pollutant Emissions and Noise

As illustrated in Fig. 3, the design of modern GT combustors requires an optimum compromise between NO_x and CO emissions at the different engine regimes. Various concepts have been proposed to reduce pollutant

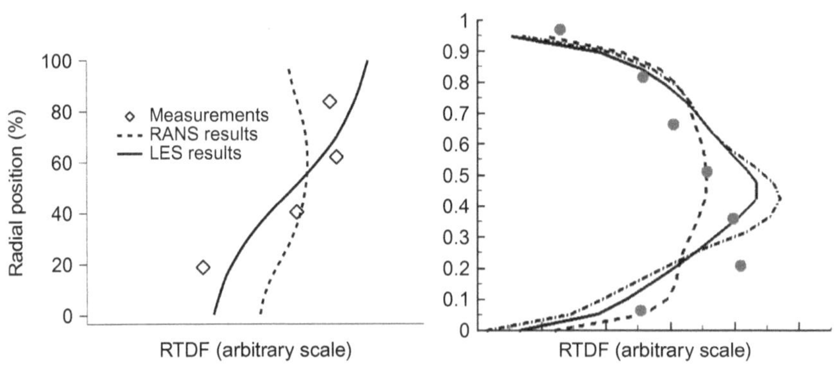

Fig. 22 Outlet normalized temperature profiles (RDTF). *Left*: Turbomeca engine: Experiment (*symbols*), LES (*solid line*), and RANS (*dashed line*) (Boudier et al., 2007). *Right*: Rolls-Royce engine: Experiment (*symbols*), LES with and without the casing (*dashed lines*), and RANS (*solid line*) (James et al., 2006).

emissions while keeping the engine performances (Tacina, 1990) and are all based on the control of the hot gas temperature and the residence time. One way to decrease the production of NO_x is to burn in lean conditions, which leads to low flame temperatures. However lean combustion, being less intense, is also less stable. In the RQL concept (Rich burn, Quick quench, Lean burn (Novick and Troth, 1981)) a small rich flame is established and rapidly diluted to stabilize lean combustion. The LPP (lean premixed prevaporized) technology preevaporates the liquid fuel which then mixes with air before burning to enhance combustion. A recent approach is to use multiple staged injections, at various equivalence ratios, to stabilize a globally lean flame. In all these systems, the dilution and fuel split between the different injections have a strong impact on the final emissions and usually result in fuel stratification which also impacts combustion efficiency. They require fine tuning and are therefore difficult to optimize.

Combustion is a compressible phenomenon, i.e., it is associated to pressure variation. This results in noise, which is also considered today as an environmental pollution and must be decreased. This is also the source of thermoacoustic instabilities, which are the result of the coupling between the acoustic resonance of the combustion chamber and the flame acoustic source. Thermoacoustic or combustion instabilities may be amplified to extremely high-pressure levels that reduce the engine efficiency, induce mechanical vibration, and may damage or in some extreme cases destroy the burner. The amplification occurs if the combustion heat release and the pressure oscillate in phase: this is known as the Rayleigh criterion that simply writes:

$$\int_V Q' P' dV > 0 \tag{4}$$

where Q' and P' are the heat release and pressure fluctuations, integrated over the burner volume V. Acoustic analysis may include the flame acoustic transfer function to derive the possible modes of combustion instabilities in a burner, as illustrated in Fig. 23. However, as the triggering mechanism is usually unknown, the occurrence of combustion instabilities is very difficult to predict and require sophisticated numerical tools to be studied.

2.2.5 New Concepts in GTs

In 2016, SAFRAN Helicopter engines proposed the rotating combustion engine, which allows to significantly increase combustion efficiency by orienting the flame propagation in the azimuthal direction of the annular

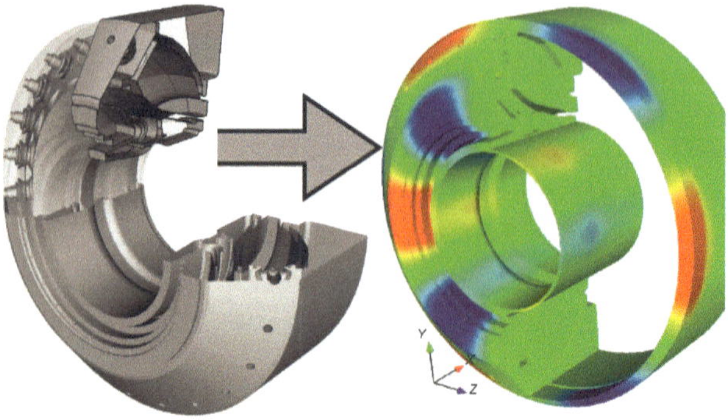

Fig. 23 Example of azimuthal acoustic mode in an annular chamber calculated with a Helmholtz solver.

combustion chamber. This requires a perfect control of fuel atomization and burning, and efficient cooling devices.

Another recent concept is based on constant volume combustion, to benefit from the better efficiency of closed systems. This raises difficult technical issues, such as opening and closing devices, but has recently gained high interest as a highly innovative technique (Labarrere, 2016).

As for ICE, the question of an alternative to fossil fuel is now raised for aeronautical engines. As demonstrated by the first flight with green diesel (15% green diesel and 85% petroleum jet fuel in the left engine) performed by Boeing in 2014, using biofuels is a real possibility for air transport. It is now important to characterize the impact on engine operability and efficiency as well as on pollutant emissions. Another issue is the production, supply, and quality of the large amounts of biofuel that would be demanded by aviation.

2.3 Rocket Engines

The thrust that generates the lift-off and acceleration of a spatial launcher is produced first by solid-fuel rockets (or boosters), which provide about 90% of the power at take-off, but also by liquid rocket engines (LREs), which are also used for the acceleration of the upper stage. These two different types of engines are complementary: the boosters generate a very large thrust at take-off (540 tons for Ariane 5) but are depleted rapidly, whereas the main LRE generates a smaller thrust (115 tons for Vulcain, Fig. 24) but for a longer period of time. Upper-stage LREs are smaller, as, for example, the Vinci engine that has a thrust of 18 tons.

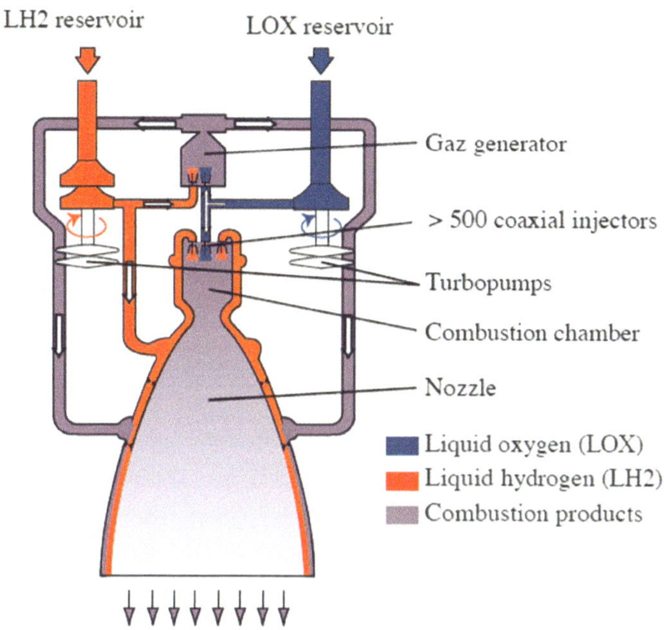

Fig. 24 Operating principle of the Vulcain 2 engine (Snecma, 2011).

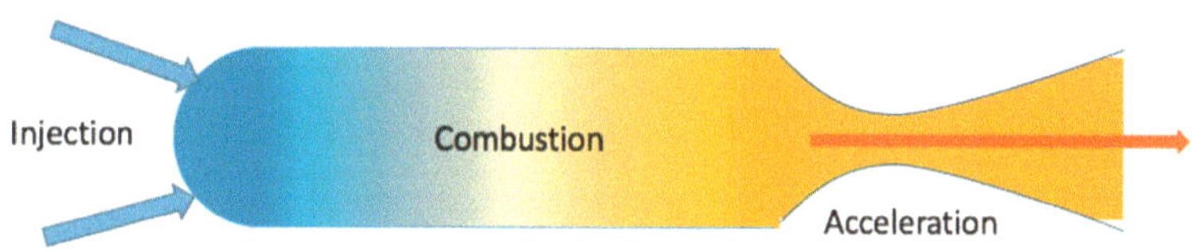

Fig. 25 Schematic layout of an LRE (Snecma, 2011). *From Safran.*

Propulsion in boosters and LREs is obtained through the action–reaction principle: thrust is generated by the ejection of fluid momentum from the nozzle of the engine. In this system (Fig. 25), the combustion chamber plays a central role, since it converts chemical energy, stored in the reactants (also called propellants), into mechanical energy, subsequently converted into kinetic energy through the nozzle. The momentum theorem applied to the system of Fig. 25 leads to:

$$m\frac{\partial V}{\partial t} = T \tag{5}$$

where m is the launcher mass, V is its velocity, and T is the thrust:

$$T = \dot{m}V_e + A_e(P_e - P_a) \tag{6}$$

where $\dot{m}$ is the ejected mass rate, V_e the ejection velocity of the burnt gas in the launcher referential, A_e the ejection area, and $P_e - P_a$ is the pressure difference between the nozzle exit and the ambient pressure. If $P_e = P_a$, the nozzle is said adapted and the thrust is reduced to the gas ejection. If the thrust is maintained at a constant value, the acceleration of the rocket increases during the launch as m decreases due to gas ejection. The specific impulse is defined as the time (in seconds) during which 1 kg of propellant delivers a thrust of 1 kgf = 9.81 N. It can be calculated as

$$I_s = \frac{T}{\dot{m}g} \tag{7}$$

where g is the gravity. It does not depend on the engine but is only characteristic of the propellant mixture. The H_2–O_2 mixture has the highest specific impulse, it is often preferred to kerosene–O_2 or CH_4–O_2 mixtures, although they are cheaper and require simpler technology.

To reduce the dimension of the reservoirs and increase the possible payload, reactants are stored at a very low temperature and high pressure in the launcher reservoirs, typically 100K and 100 bars for O_2 and H_2, which significantly increases their density. In these conditions, the fluids are in the supercritical thermodynamic state; i.e., they cannot be considered anymore as perfect gases. The supercritical state of a chemical component is reached above its critical pressure (50 bars for O_2 and 13 bars for H_2). In the (P–T) diagram, the boiling line vanishes at the critical point (critical pressure and temperature) and the liquid–gas equilibrium does not exist anymore above this point (Fig. 26). The fluid has instead the capacity to evolve continuously from a light gas-like density to a high liquid-like density, without

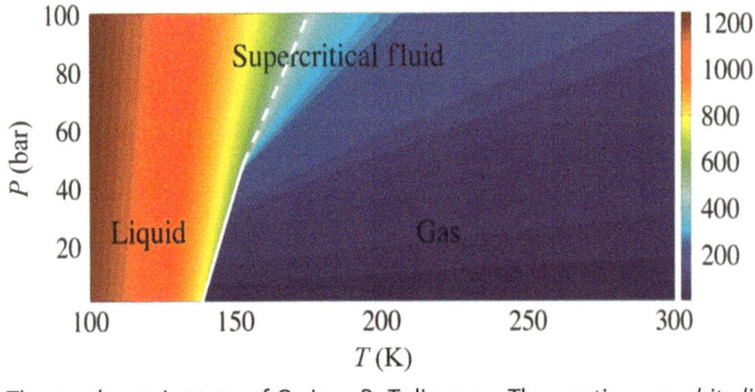

Fig. 26 Thermodynamic state of O_2 in a P–T diagram. The *continuous white line* is the boiling line, while the *dashed white line* is the pseudo-boiling line.

experiencing phase change nor surface tension. However, just after injection, the cryogenic, supercritical fluids burn and rapidly reach very high temperatures inside the combustion chamber. The adiabatic flame temperature of a pure H_2/O_2 flame at 10 MPa is 3800K: at such high temperature, the fluid density goes back to low values and the fluids behave again as perfect gases. Another transient situation corresponds to the start of the engine at ground level, i.e., at atmospheric pressure inside the combustion chamber: at this pressure the subcritical, cold oxygen is liquid. The two-phase combustion continues until the pressure has increased to a supercritical value inside the chamber. In both situations, the fluid crosses the critical pressure or the pseudo-boiling line and the process is called transcritical.

Injectors are usually coaxial tubes, with or without recess, arranged in a purposefully nonregular pattern on an injection plate, in an attempt to attenuate possible acoustic modes. To obtain short flames, the number of injectors is as high as possible and reaches, for example, 566 in Vulcain 2.

2.3.1 Technical Challenges

The important market of satellite launching has motivated programs for spatial launchers in about 10 countries and Europe. The recent announcement of a low-cost, reusable launcher by Space-X in the United States has definitively put the development of spatial launchers in a commercial environment, with important consequences on technical choices. Still, the high cost of a satellite keeps reliability as a main driver, which demands further work on some technical challenges that are not yet fully mastered.

Due to the injection technique and the very fast combustion of H_2–O_2, the flame is attached to the injectors despite the very high shear and no premixing occurs between the two reactants. As a consequence the combustion occurs in a pure nonpremixed, diffusion-controlled regime which, as will be seen in Section 3.3.3, leads to locally high temperatures. The current tendency to replace H_2 with CH_4 may however lead to a different flame structure as slower chemistry favors flame detachment and possible premixing.

If pollutant emissions are not a major concern for LREs, all other issues already pointed out for automotive or aeronautical engines are also relevant. Ignition is a very delicate process, which must be achieved without too high over-pressure peaks that could damage the rocket or the satellite. To obtain such smooth ignition, it is crucial to limit the premixing of the reactants. On the other hand, ignition success must be guaranteed which means that there must be sufficient premixing. The compromise between these two contradictory constraints is very difficult and requires sophisticated models.

LREs are also subjected to combustion instabilities, which have been the cause of numerous failures in the past. Considering the mean pressure level in the chamber of 100 bars, and the power of the engine, these thermo-acoustic instabilities may have a much more destructive nature. Many rocket engines are therefore equipped with Helmholtz resonators, baffles, or acoustic damping liners.

Finally, one major issue is thermal fatigue. Until recently, the objective was to cool the engine to guarantee its correct operation during the launch duration time. However, with the increasing commercial pressure in the market of satellite launching, the idea of reusable systems has been introduced. This requires to preserve the system integrity and performance even in the extreme operating conditions of rocket engines and has motivated the development of new concepts, with special effort on the reduction of the thermal fatigue.

3. COMBUSTION

Combustion in propulsion systems is a complex phenomenon that must be fully controlled to guarantee reliability, security, and performance. Essentially, combustion is a thermochemical process, resulting from the oxidation of a fuel or a blend of fuels with either air or oxygen, through a set of gaseous, exothermic chemical reactions. This chemical conversion can be studied in homogeneous or perfectly stirred reactors (PSRs), allowing to characterize basic features such as equilibrium state, auto-ignition delay, chemical time scales and any chemical path of interest. To obtain a flame however, it is necessary to introduce transport phenomena which bring together the reactants and heat. Flames are therefore described with the standard fluid equations of motion and energy, in which chemical source terms are introduced.

3.1 Thermochemistry

Oxidation is the conversion of reactants into products having a lower chemical energy, the difference being transferred to the fluid mixture in the form of heat. This energy release is characteristic of each fuel and depends on its composition. For comparison purposes a reference value of the energy release is defined as the heat produced by the complete combustion at atmospheric pressure of 1 kg of fuel initially taken at 0°C, the combustion products being also cooled down to 0°C (Fig. 27). In other words, this corresponds to the difference in chemical energy between reactants and products taken at the same temperature of 0°C. This quantity is named the higher heating value (HHV) and can be found in thermodynamic tables for any fuel. Typical values are given in Fig. 28 for usual fuels. Note that alkanes and alkenes have similar values, as the number of energetic chemical bonds is close to proportional to the size of the molecule. Values are however different for methanol and hydrogen which have a different molecular structure. Taking the fluids at 0°C raises however an issue

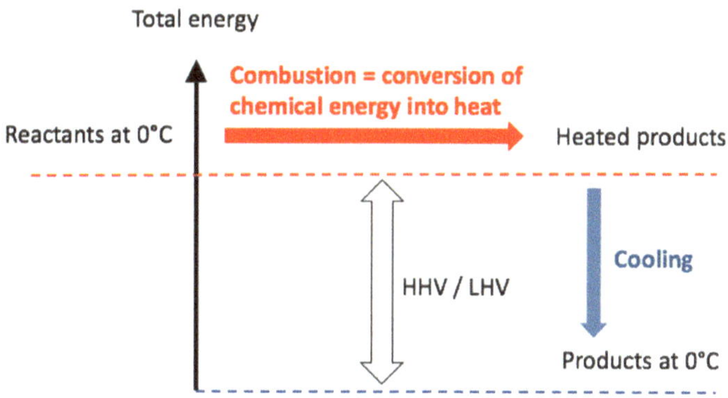

Fig. 27 Definition of HHV/LHV.

	LHV (MJ/kg)	HHV (MJ/kg)
Methane	50.01	55.50
Propane	47.79	51.78
Octane	44.43	47.89
Ethylene	47.19	50.31
Benzene	40.17	41.80
Methanol	19.94	22.70
Hydrogen	120.97	141.83

Fig. 28 Values of HHV and LHV for usual fuels.

because at this temperature and atmospheric pressure, one main combustion product, water, is liquid. This means that water has condensed and therefore has also released energy to the system. This additional energy release depends on the quantity of water produced and differs from one fuel to the other. In order to eliminate this effect from the HHV, the low heat value (LHV) is defined as the HHV but considering that water stays in its vapor state, leading to a lower value: LHV = HHV $-L_v$, where L_v is the latent heat of water evaporation. Fig. 28 shows the difference between HHV and LHV for usual fuels, which is of the order of 10% of the HHV.

The gaseous mixture composed of fuel, oxidizer, and products is considered as a thermodynamic system described by state variables (P, T, n_k), where n_k is the mole number of the species k and $n=\sum_k n_k$ is the total number of moles. Note that chemical reactions do not necessarily conserve the mole number n, but only conserve mass. Assuming a perfect gas, the state equation writes:

$$P=\frac{n}{\mathcal{V}}RT=\rho rT \tag{8}$$

where $\mathcal{V}$ is the volume of the system, $\rho=nW/\mathcal{V}$ is the density with W the mean molar weight of the mixture, and $r=R/W$. The mixture composition is usually described with species mole fractions $X_k = n_k/n$ or mass fractions $Y_k = m_k/m$ where $m_k = W_k n_k$ and $m = nW$ are, respectively, the mass of species k and the mass of mixture in the volume V, with W_k the molar mass of species k. By definition, $\sum_k X_k=\sum_k Y_k=1$. Chemical reaction rates are expressed in molar concentration $[C_k]=n_k/\mathcal{V}$ or mass concentration $[C_k]^m=m_k/\mathcal{V}$. They are linked to the mole and mass fractions through the density as

$$[C_k]=\rho\frac{Y_k}{W_k}=\rho\frac{X_k}{W} \quad \text{and} \quad [C_k]^m=\rho Y_k=\rho X_k\frac{W_k}{W} \tag{9}$$

Other useful state functions are the molar internal energy of the system e and its molar enthalpy $h = e + RT$. These functions are built from the species state functions of all species e_k and $h_k = e_k + RT$, following standard mixing rules $e=\sum_k e_k X_k$ and $h=\sum_k h_k X_k$. For perfect gas, the state functions e_k and h_k are functions of the temperature only and write:

$$e_k(T) = e_k(T^0) + \int_{T^0}^{T} C_{v,k}(T)dT \quad \text{and}$$
$$h_k(T) = h_k(T^0) + \int_{T^0}^{T} C_{p,k}(T)dT \tag{10}$$

where $e_k(T^0)$ and $h_k(T^0)$ are the energy and enthalpy of formation at the reference temperature T^0 of the species k, and the quantities $C_{v,\ k}$ and $C_{p,\ k} = C_{v,\ k} + R$ are its heat capacities at constant volume and constant pressure, respectively. It is recalled that the enthalpy of formation of a pure body in its standard state is 0. Mixing rules also apply to the reference values and the heat capacities to build mean mixture values $e(T^0) = \sum_k e_k(T^0)X_k$, $h(T^0) = \sum_k h_k(T^0)X_k$, $C_v = \sum_k C_{v,k}X_k$, and $C_p = \sum_k C_{p,k}X_k$, allowing to write:

$$e(T) = e(T^0)X_k + \int_{T^0}^{T} C_v(T)dT \quad \text{and} \quad h(T) = h(T^0)X_k + \int_{T^0}^{T} C_p(T)dT \tag{11}$$

Note that all the above quantities and relationships also hold if expressed in mass units, which are obtained from molar quantities by simply dividing by the molar mass.

3.1.1 Equilibrium State

According to the second thermodynamic law, the system will naturally evolve toward its equilibrium, which corresponds to the maximum of entropy s. To account for the energy released by chemical reactions, the Gibbs energy $g = h - Ts$ is introduced. The equilibrium state then corresponds to the minimum of g at constant pressure and temperature. This leads to the law of mass action:

$$\ln(K_p) = \ln\left[\prod_k \left(\frac{P}{P^0}X_k\right)^{\nu_k}\right] = \frac{1}{RT}\sum_k g_k(T^0)\nu_k \tag{12}$$

where K_p is the equilibrium constant. In a system with N species composed of e elements, the equilibrium composition is the solution of $N - e$ equations of the type of Eq. (12), written for $N - e$ chemical reactions.

The equilibrium temperature corresponds to the temperature of the mixture having the equilibrium composition and the same energy (more specifically, the same internal energy in closed systems, or the same enthalpy in open systems) as the initial mixture. It does not depend on the process that has led to equilibrium, but only on the species that are present or may appear in the mixture.

3.1.2 Complete Combustion

Complete combustion is an interesting limit case which introduces other reference values of interest. Consider, for example, a hydrocarbon of formula C_nH_m. Its complete combustion in air can be represented by the following global chemical balance between the final and initial states:

$$C_nH_m + \left(n + \frac{m}{4}\right)(O_2 + \alpha N_2) \rightarrow nCO_2 + \frac{m}{2}H_2O + \left(n + \frac{m}{4}\right)\alpha N_2 \quad (13)$$

where $\alpha = 3.76$ is the number of moles of N_2 for 1 mole of O_2 in air. In the above expression, the stoichiometric coefficients have been adjusted so as to conserve atoms, in the case where the reactants are in stoichiometric conditions, indicated by the fact that there is no fuel or oxygen left in the products (right side). Note that, although it is an inert species with no contribution to the chemical conversion, it is kept in the balance equation to include it in the final state calculations. It is important to insist here on the fact that Eq. (13) does not represent a chemical reaction, but only a chemical balance between an initial state (reactants) and a final state (products). It does not tell anything about the true chemistry of the combustion.

From Eq. (13), the stoichiometric ratio α_s of the fuel can be determined, which is the ratio of the mass of fuel and of oxidizer when they are in stoichiometric proportions. In the present example, it writes:

$$\alpha_{st.} = \left(\frac{m_{Fu}}{m_{Ox}}\right)_{st.} = \frac{nW_C + mW_H}{2(n + m/4)W_O} \quad (14)$$

With this value, one can now determine the equivalence ratio Φ of any mixture, defined as the fuel-to-oxidizer ratio of the mixture divided by the fuel-to-oxidizer ratio at stoichiometry:

$$\Phi = \frac{m_{Fu}/m_{Ox}}{(m_{Fu}/m_{Ox})_{st.}} = \frac{m_{Fu}}{\alpha_{st.} m_{Ox}} \quad (15)$$

A rich mixture is obtained for $\Phi > 1$ (fuel in excess) and a lean mixture for $\Phi < 1$ (oxidizer in excess). Note that both $\alpha_{st.}$ and Φ can be defined with air as oxidizer instead of oxygen, which changes $\alpha_{st.}$ but leads to exactly the same value of Φ. The two quantities can also be calculated with moles or mass fractions, again giving the same values for both. However, mole fractions should not be used in general, as the number of moles is not always conserved in the balance of Eq. (13).

Following the first principle of thermodynamics, the energy balance can then be expressed in either enthalpy (open system, constant pressure) or internal energy (closed system, constant volume). Denoting T_i and T_f, respectively, the initial and final temperatures, the enthalpy balance corresponding to the mass balance of Eq. (13) writes:

$$\sum_k \nu'_k h_k(T_i) = \sum_k \nu''_k h_k(T_f) \tag{16}$$

where ν'_k and ν''_k are the stoichimetric coefficients of species k, respectively, in the reactants (initial mixture) and in the products (final mixture). Expressing the species enthalpy as in Eq. (10), the enthalpy balance can be rewritten as:

$$\sum_k (\nu'_k - \nu''_k) h_k(T^0) = \int_{T^0}^{T_f} \sum_k \nu''_k C_{p,k}(T)dT - \int_{T^0}^{T_i} \sum_k \nu'_k C_{p,k}(T)dT \tag{17}$$

The quantity $\Delta H_r = \sum_k (\nu'_k - \nu''_k) h_k(T^0)$ is the heat of reaction, i.e., the heat produced by the combustion of ν'_{Fu} moles of fuel. Usually the chemical balance is written with $\nu'_{Fu} = 1$ (as in Eq. 13), so that ΔH_r corresponds to the combustion of 1 mole of fuel. It is then related to the LHV by $\Delta H_r = \text{LHV}.W_{Fu}$.

In the example of C_nH_m, recalling that $h_{O_2}(T^0) = 0$ it is easily found that:

$$\Delta H_r = n h_{CO_2}(T^0) + \frac{m}{2} h_{H_2O}(T^0) - h_{C_nH_m}(T^0) \tag{18}$$

In case of a lean mixture, the chemical balance writes for the C_nH_m example:

$$\begin{aligned} \Phi C_nH_m + \left(n + \frac{m}{4}\right)(O_2 + \alpha N_2) &\rightarrow \Phi n CO_2 \\ &+ \Phi \frac{m}{2} H_2O + \left(n + \frac{m}{4}\right)(1 - \Phi)O_2 + \left(n + \frac{m}{4}\right)\alpha N_2 \end{aligned} \tag{19}$$

and the heat of reaction is then:

$$\Delta H_r|_{\Phi<1} = \Phi\{nh_{CO_2}(T^0) + \frac{m}{2}h_{H_2O}(T^0) - h_{C_nH_m}(T^0)\} = \Phi\Delta H_r|_{\Phi=1} \tag{20}$$

The balance is different for a rich mixture:

$$\begin{aligned}\Phi C_nH_m + \left(n+\frac{m}{4}\right)(O_2+\alpha N_2) &\to nCO_2 \\ &+ \frac{m}{2}H_2O + (\Phi-1)C_nH_m + \left(n+\frac{m}{4}\right)\alpha N_2\end{aligned} \tag{21}$$

which leads to the heat of reaction:

$$\Delta H_r(|_{\Phi>1} = nh_{CO_2}(T^0) + \frac{m}{2}h_{H_2O}(T^0) - h_{C_nH_m}(T^0) = \Delta H_r|_{\Phi=1} \tag{22}$$

The final temperature T_f is obtained by solving Eq. (17), rewritten as:

$$\int_{T^0}^{T_f} \sum_k \nu_k'' C_{p,k}(T)dT = \Delta H_r + \int_{T^0}^{T_i} \sum_k \nu_k' C_{p,k}(T)dT \tag{23}$$

If the initial mixture is taken at ambient conditions, i.e., $T_i = T^0$, and assuming that heat capacities do not depend on temperature, a very simple equation is obtained for the final temperature:

$$T_f = T_i + \frac{\Delta H_r}{\sum_k \nu_k'' C_{p,k}} \tag{24}$$

Although it is inert, N_2 must be included in this last expression because it is also heated as any mixture component. This explains why combustion in air always leads to smaller final temperatures compared to oxy-combustion at the same equivalence ratio, in which there is no inert component that is heated without contributing to heat production. This is also the reason why rich combustion also leads to lower final temperatures compared to stoichiometry, even though the produced amount of heat is the same: the excess fuel is heated without contributing to heat production. In lean combustion, the final temperature is even lower due to the reduced heat of reaction. As a consequence the final temperature is maximum at $\Phi = 1$, as illustrated in Fig. 29.

Because it corresponds to the limit case of complete burning, the final temperature is a maximum that can never be exceeded, or only slightly if there exists exothermic reverse chemical reactions. In particular, it is different, and usually

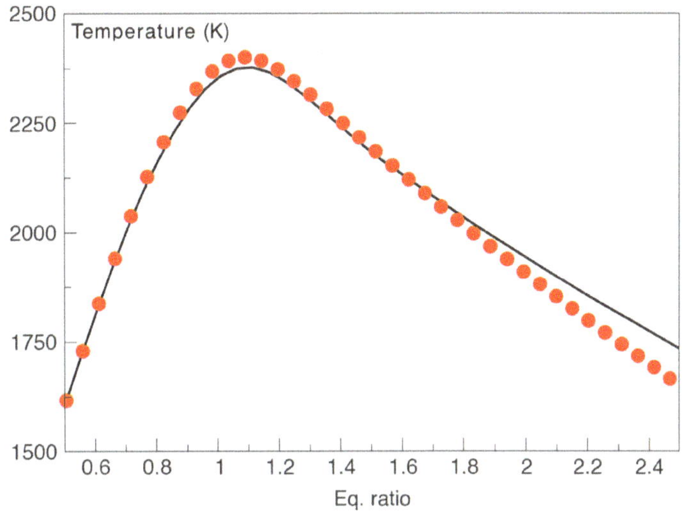

Fig. 29 Final temperature (*solid line*) and equilibrium temperature (*red circles*) as a function of the equivalence ratio for an ethylene–air flame at initial ambient conditions.

Fig. 30 The perfectly stirred reactor (PSR).

higher, than the equilibrium temperature which is the true temperature effectively reached by burnt gases (Fig. 29). Note that both temperatures are fully independent of the details of the conversion process and only represent a final state based on the thermodynamic and conservation principles. As such, they should be calculated with the same set of chemical species if to be compared.

3.2 Introducing Chemical Kinetics: The PSR

To study the evolution from the initial to the final state, and in particular to introduce chemical time scales, the configuration of PSR is used. It is an open system, as illustrated in Fig. 30, in which a mixture flows through at a mass flow rate $\dot{m}$ and with a residence time τ, which is related to the reactor volume $\mathcal{V}$ by $1/\tau = \dot{m}/\rho\mathcal{V}$. The mixture spontaneously evolves in the reactor, where all species, i.e., reactants, products, and intermediate species, are supposed to mix instantaneously.

The PSR reproduces the time evolution of combustion chemistry, illustrated in Fig. 31. Although chemical reactions start at the very beginning,

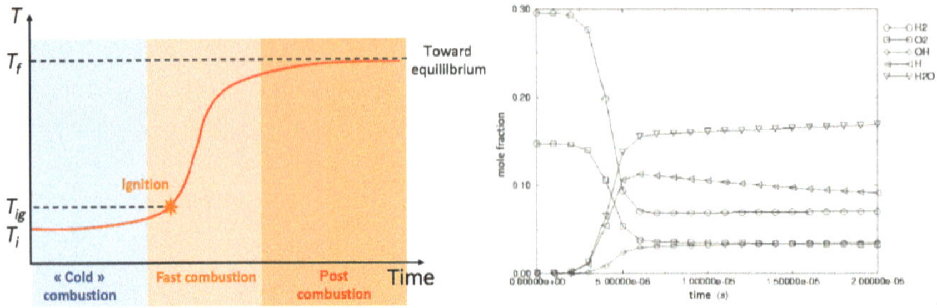

Fig. 31 *Left*: Time evolution of temperature in a PSR. *Right*: Evolution of a stoichiometric mixture of H_2/O_2 at ambient conditions in a PSR.

because the initial mixture is not in equilibrium, they stay however very slow as long as the temperature remains below a minimum value T_{ig}, and the temperature increases very slowly. This is called "cold" combustion and may last very long if no additional energy is brought to the system. Indeed, chemical reactions of combustion require a minimum level of energy to dissociate reactant molecules and initiate the fast combustion process. This can be achieved either by using an ignition system (spark plug, laser beam, or other) or by heating the mixture to the ignition temperature T_{ig}, typically around 700–800K, allowing auto-ignition. After this initiation phase, fast combustion occurs where exothermic chain-branching reactions consume and produce highly reactive radicals and lead to a thermal runaway. Finally, recombination reactions produce more stable molecules and the mixture progressively evolves toward equilibrium. Note that in the case of very high burnt gas temperature, some endothermic redissociations may occur. In the PSR configuration, the output mixture is the result of chemical conversion after a time duration of τ, which can be anywhere in the plot of Fig. 31. Infinite residence time gives therefore a mixture in equilibrium, whereas short residence times will not show significant evolutions if the initial temperature is below T_{ig}. PSR calculations can then be used to characterize the auto-ignition delay of any mixture as a function of the initial temperature. By decreasing the residence time of a burning case, blow-out limits can also be determined.

PSR calculations therefore require to describe the details of combustion kinetics. These are represented by a set of M elementary reactions, written as:

$$\nu_A A + \nu_B B \leftrightarrow \nu_C C + \nu_D D \tag{25}$$

The reaction rate of the forward reaction is written using the Arrhenius law:

$$\dot{\omega}_f = A_f [C_A]^{\nu_A} [C_B]^{\nu_B} exp(-T_a/T) \tag{26}$$

where A_f is the preexponential factor, driving the chemical time scale, and T_a is the activation temperature, responsible for the "cold" combustion behavior and the thermal runaway phenomenon. Note that A_f has no predefined unit, as it depends on the order of the reaction. The reverse reaction rate is written similarly with the Arrhenius law, but the preexponential factor A_b is calculated from the equilibrium constant K_p:

$$A_b = \frac{A_f}{(P^0/RT)^{\nu_{tot}} K_p} \tag{27}$$

where $\nu_{tot} = \nu_C + \nu_D - \nu_A - \nu_B$.

The number of reactions M necessary to describe the combustion of hydrocarbons stays reasonable for methane (about 500), but grows rapidly with the size of the molecule. It increases even more when pollutant chemistry is included, such as for the prediction of NO_x. This motivates the use of reduced kinetic schemes, either analytically derived (Lu and Law, 2008) or globally fitted (Franzelli et al., 2010).

3.3 Flames

Flames are reaction zones that occur in nonhomogeneous mixtures and that are sustained by transport phenomena. Depending on the way the reactants are brought to the flame, two generic flame structures are distinguished. In the premixed flame, the reactants are fully premixed and burn when sufficiently heated by the flame, which separates them from the hot combustion products. In the nonpremixed, or diffusion flame, the reactants are injected separately and burn when they meet at the flame location (Fig. 32).

To describe transport phenomena in fluids, the standard conservation equations for density, momentum, and energy plus the k species mass fractions are used, with additional source terms due to combustion. As the density varies strongly, the incompressible formulation is not adapted and either low-Mach or fully compressible equations must be used (Kuo, 2005).

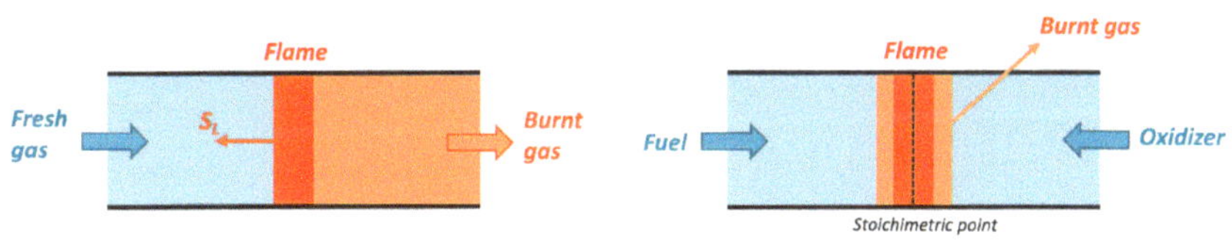

Fig. 32 Generic flame structures: premixed (*left*) and nonpremixed (*right*) flames.

3.3.1 Governing Equations

The Navier–Stokes equations for multispecies reacting mixtures using Einstein index notation write:

$$\frac{\partial \rho}{\partial t} + \frac{\partial \rho u_i}{\partial x_i} = 0 \tag{28}$$

$$\frac{\partial \rho u_j}{\partial t} + \frac{\partial \rho u_i u_j}{\partial x_i} + \frac{\partial P \delta_{ij}}{\partial x_i} = -\frac{\partial \tau_{ij}}{\partial x_i} \quad \text{for } j = 1, 3 \tag{29}$$

$$\frac{\partial \rho Y_k}{\partial t} + \frac{\partial \rho u_i Y_k}{\partial x_i} = -\frac{\partial J_{i,k}}{\partial x_i} + \dot{\omega}_k \quad \text{for } k = 1, N \tag{30}$$

$$\frac{\partial \rho E}{\partial t} + \frac{\partial \rho u_i E}{\partial x_i} + \frac{\partial u_j P \delta_{ij}}{\partial x_i} = -\frac{\partial q_i}{\partial x_i} - \frac{\partial u_j \tau_{ij}}{\partial x_i} + \dot{\omega}_T \tag{31}$$

with u_j the jth component of the velocity, P the static pressure, τ_{ij} the viscous tensor, $J_{i,\,k}$ the ith component of the diffusive flux of species k, $\dot{\omega}_k$, and $\dot{\omega}_T$ the species k and energy chemical source term, E the total energy, and q_i the ith component of the energy flux. δ_{ij} is the Kronecker symbol equal to 1 if $i = j$ and 0 otherwise. Note that the summation of the N species equations leads to the continuity equation, as the conservation of atoms in the kinetic scheme implies that $\sum_k \dot{\omega}_k = 0$. This means that either the continuity equation or the Nth species conservation equation can be omitted.

The viscous tensor is proportional to the symmetric part of the deformation tensor (Newtonian fluid):

$$\tau_{ij} = 2\mu\left(S_{ij} - \frac{1}{3}\delta_{ij} S_{kk}\right) \tag{32}$$

with μ the dynamic viscosity and S_{ij} the deformation tensor:

$$S_{ij} = \frac{1}{2}\left(\frac{\partial u_j}{\partial x_i} + \frac{\partial u_i}{\partial x_j}\right) \tag{33}$$

The computation of the exact formulation of the diffusive flux of species k in the mixture is a very complex task and requires to calculate at each time-step the binary diffusion coefficients D_{jk} between species j and k. Simplified multicomponent transport may be introduced following the Hirschfelder and Curtiss approximation (1969):

$$D_k = \frac{1 - Y_k}{\sum_{j \neq k} X_j / D_{jk}} \tag{34}$$

where D_k is the equivalent diffusion coefficient of species k in the rest of the mixture. From Eq. (34), it can be seen that the diffusion coefficient still varies with the mixture composition. However the variations stay small and are ignored for further simplification. The diffusive flux of species k then writes:

$$J_{i,k} = \rho Y_k V_{k,i} = -\rho\left(D_k \frac{W_k}{W}\frac{\partial X_k}{\partial x_i} - Y_k V_i^c\right) \tag{35}$$

where V_i^c is a correction velocity ensuring global mass conservation:

$$V_i^c = \sum_{k=1}^{N} D_k \frac{W_k}{W}\frac{\partial X_k}{\partial x_i} \tag{36}$$

The energy flux is the combination of the Fourier flux and the enthalpy flux induced by species diffusion:

$$q_i = -\lambda \frac{\partial T}{\partial x_i} + \sum_{k=1}^{N} J_{i,k} h_{s,k} \tag{37}$$

where λ is the heat conduction coefficient of the mixture and $h_{s,k}$ is the sensible enthalpy of species k. Finally the reaction energy source term is linked to the species source terms by:

$$\dot{\omega}_T = -\sum_{k=1}^{N} \Delta H_{f,k}^0 \dot{\omega}_k \tag{38}$$

3.3.2 Transport Properties

Similar to the species diffusion coefficients, the dynamic viscosity μ of the mixture can be assumed independent of the mixture composition and only depend on the temperature, for example, through a standard power law:

$$\mu = \mu_0 \left(\frac{T}{T_0}\right)^b \tag{39}$$

where the exponent b depends on the mixture and typically lies between 0.6 and 1.0. The species diffusion coefficients can be then expressed with Schmidt numbers Sc_k as:

$$D_k = \frac{\mu}{\rho Sc_k} \tag{40}$$

The Schmidt number is a dimensionless number that compares the viscous and species diffusion rates. Similarly, the thermal conductivity λ is calculated from the Prandtl number Pr of the mixture:

$$\lambda = \frac{\mu C_p}{Pr} = \rho C_p D_{th} \quad \text{with } D_{th} = \frac{\nu}{Pr} \tag{41}$$

which compares the viscous and thermal diffusion rates. Finally, the Lewis number Le_k that compares the thermal and species diffusion rates is introduced as:

$$Le_k = \frac{D_{th}}{D_k} = \frac{Sc_k}{Pr} \tag{42}$$

In most cases, it is observed that the Schmidt and Prandtl numbers vary little with the flow properties and can be assumed constant.

3.3.3 Laminar Flames

The above set of equations may be applied to the generic one-dimensional, laminar flame structures introduced in Fig. 32, which leads to:

$$\frac{\partial \rho}{\partial t} + \frac{\partial \rho u}{\partial x} = 0 \tag{43}$$

$$\frac{\partial \rho u}{\partial t} + \frac{\partial \rho u^2}{\partial x} + \frac{\partial P}{\partial x} = 0 \tag{44}$$

$$\frac{\partial \rho Y_k}{\partial t} + \frac{\partial \rho u Y_k}{\partial x} = \rho D_k \frac{\partial^2 Y_k}{\partial x^2} + \dot{\omega}_k \text{ for } \quad k = 1, N \tag{45}$$

$$\frac{\partial \rho C_p T}{\partial t} + \frac{\partial \rho u C_p T}{\partial x} = \lambda \frac{\partial^2 T}{\partial x^2} + \dot{\omega}_T \tag{46}$$

where viscous terms have been neglected as well as the contribution of pressure in the temperature equation, and where diffusion fluxes for species and temperature have been simplified. Premixed and nonpremixed flames are both solutions of the above system of equations, but with different boundary conditions.

3.3.3.1 Premixed Flames

In premixed flames, the mixture of fresh reactants is converted into hot products through the flame, as illustrated in Fig. 33. The initiation chemical reactions are triggered by the preheating of the reactants by the hot products through heat conduction and start the high-activation, exothermic chain-branching reactions, which develop until one of the reactant has been totally

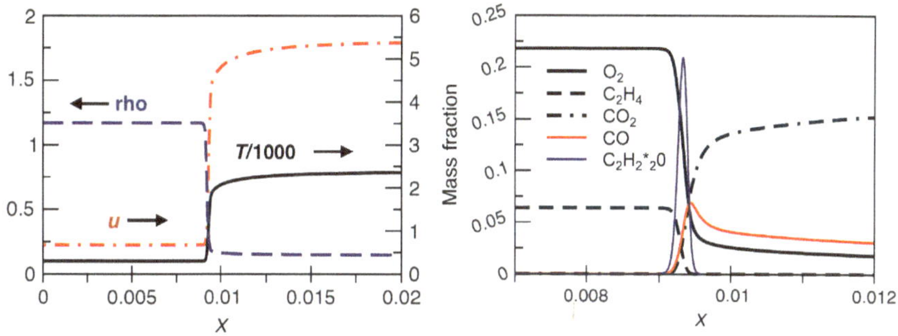

Fig. 33 Structure of a laminar, stoichiometric ethylene–air premixed flame at ambient conditions. *Left*: Temperature, density, and velocity profiles. *Right*: Species profiles.

consumed. The flame is then followed by a postflame zone where the mixture slowly relaxes to equilibrium.

In the case of premixed flames, steady solutions exist if the injection velocity of the reactant mixture is equal to the propagating flame velocity S_L, which results from the balance between diffusion and reaction and scales as $\sqrt{AD_{Fu}}$. The continuity equation in steady state simplifies to:

$$\rho u = const. = \rho_u S_L \tag{47}$$

where the subscript$_u$ refers to the fresh, unburnt reactants. The laminar flame speed S_L is a parameter of premixed flames that can be measured. It only depends on the reactant mixture state and reflects the speed of combustion. As chemistry is faster at higher temperature, the flame speed changes with the equivalence ratio similar to the burnt gas temperature and is maximum at or near stoichiometry (Fig. 34).

The velocity profile through the flame can be then expressed as:

$$u = \frac{\rho_u S_L}{\rho} \tag{48}$$

and the momentum equation is only useful to determine the pressure profile.

The steady fuel mass fraction and temperature equations can now be written as:

$$\rho_u S_L \frac{\partial Y_{Fu}}{\partial x} = \rho D_{Fu} \frac{\partial^2 Y_{Fu}}{\partial x^2} + \dot{\omega}_{Fu} \tag{49}$$

$$\rho_u S_L \frac{\partial C_p T}{\partial x} = \lambda \frac{\partial^2 T}{\partial x^2} - Q\dot{\omega}_{Fu} \tag{50}$$

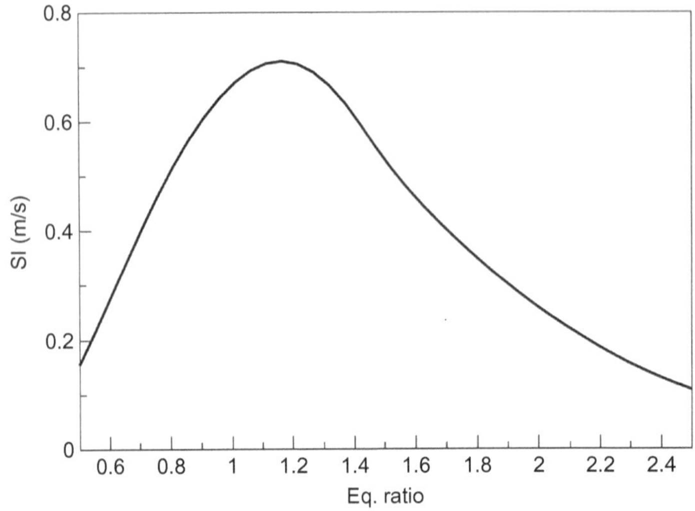

Fig. 34 Laminar flame speed of ethylene–air combustion at ambient conditions, as a function of the equivalence ratio.

In the above equations, the heat release $\dot{\omega}_T$ has been expressed with the heat of reaction and the burning rate of fuel as $\dot{\omega}_T = -Q\dot{\omega}_{Fu}$, where $Q = \Delta H_r / W_{Fu}$ is the heat produced by the burning of one mass unit of fuel. Assuming further that C_p is constant and that both the fuel and temperature diffuse at the same rate, i.e., that the Lewis number $Le = \lambda/\rho C_p D_{Fu} = 1$, the above equations can be combined to write the transport equation for the quantity $Y_{Fu} + C_p T/Q$:

$$\rho_u S_L \frac{\partial (Y_{Fu} + C_p T/Q)}{\partial x} = \rho D_{Fu} \frac{\partial^2 (Y_{Fu} + C_p T/Q)}{\partial x^2} \tag{51}$$

One possible solution is the function $exp(D_{Fu}/S_L x)$, which tends to infinity with increasing x and is therefore not acceptable. The only other possible solution is therefore a constant function, evaluated either in the reactants or in the products side:

$$Y_{Fu} + C_p T/Q = Y_{Fu,u} + C_p T_u/Q = Y_{Fu,b} + C_p T_b/Q \tag{52}$$

where the subscripts $_u$ and $_b$ refer to the unburnt and burnt sides, respectively. This means that Y_{Fu} and T are proportional and can be therefore described by one single variable, named the progress variable c. Note that this allows to find also:

$$T_b - T_u = \frac{Q}{C_p}(Y_{Fu,u} - Y_{Fu,b}) \quad (53)$$

which is similar to Eq. (24), but expressed here with mass quantities. The same derivation can be done to show that, similarly, the quantity $Y_{Fu} - \alpha_{st.} Y_{ox}$ and, consequently, the equivalence ratios are constants through the flame, so that Y_{ox} is also a function of the progress variable.

The progress variable can be indifferently built from the temperature or species mass fractions, normalized to vary between 0 and 1, as:

$$c = \frac{T - T_u}{T_b - T_u} = \frac{Y_{Fu,u} - Y_{Fu}}{Y_{Fu,u} - Y_{Fu,b}} = \frac{Y_{Ox,u} - Y_{Ox}}{Y_{Ox,u} - Y_{Ox,b}} \quad (54)$$

Solving the transport equation for the progress variable only is then sufficient to get the full premixed flame solution, illustrated in Fig. 35.

The gradient of the variable profiles is driven by the competition between diffusion and reaction and may be used to define a flame thickness that can be estimated, for example, as:

$$\delta_{th} = \frac{T_b - T_u}{(\partial T/\partial x)_{max}} = \frac{1}{(\partial c/\partial x)_{max}} \quad (55)$$

As it is calculated from the temperature profile, δ_{th} is named the thermal thickness. With the assumptions made here, it can be equivalently calculated from any variable with the same result. This is not true in the general case, where the molecular and thermal diffusion coefficients differ. Another flame thickness δ_r can be defined as the reaction zone thickness, measuring, for example, the width of the peak of heat release (Fig. 33). Blint (1986) proposed an expression for this flame thickness:

$$\delta_r = 2\delta_L \left(\frac{T_b}{T_u}\right)^{0.7} \quad (56)$$

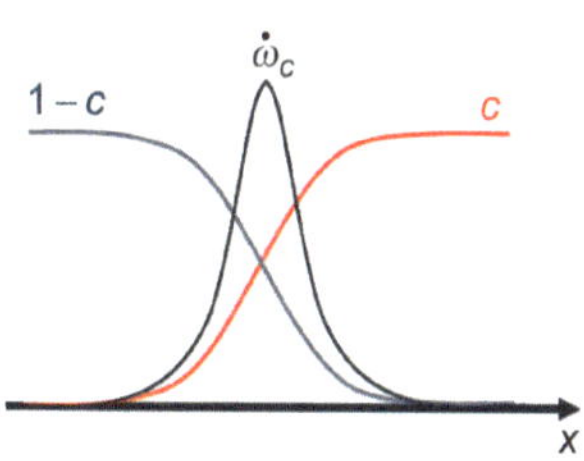

Fig. 35 Description of a premixed flame with the progress variable.

where $\delta_L = D_{Fu}/S_L$ is a flame thickness obtained from dimensional analysis, which scales as $\sqrt{D_{Fu}/A}$. Although both flames' thicknesses δ_{th} and δ_r are driven by the same mechanisms, they differ slightly.

Balance equations for mass and energy can be obtained by integrating Eqs. (49) and (50) :

$$\int_{-\infty}^{+\infty} \rho_u S_L \frac{\partial Y_{Fu}}{\partial x} dx = \int_{-\infty}^{+\infty} \rho D_{Fu} \frac{\partial^2 Y_{Fu}}{\partial x^2} + \int_{-\infty}^{+\infty} \dot{\omega}_{Fu} dx \quad (57)$$

which gives:

$$\rho S_L (Y_{Fu,b} - Y_{Fu,u}) = \int_{-\infty}^{+\infty} \dot{\omega}_{Fu} dx \quad (58)$$

as diffusion fluxes are zero at ∞. Similarly, one obtains for the temperature:

$$\rho S_L C_p (T_b - T_u) = -Q \int_{-\infty}^{+\infty} \dot{\omega}_{Fu} dx \quad (59)$$

This means that the total fuel consumption rate and the total heat release of a one-dimensional premixed flame are proportional to the laminar flame speed S_L, which appears again as a characteristic parameter of the flame and the combustion chemistry.

3.3.3.2 Nonpremixed Flames

The structure of nonpremixed flames is illustrated in Fig. 36. The reactants are now injected separately on both sides of the flame and the temperature profile has a peak shape. Similar to premixed flames, the hot products preheat the reactants on each side, and combustion occurs when they mix, with a maximum reactivity at the stoichiometric point.

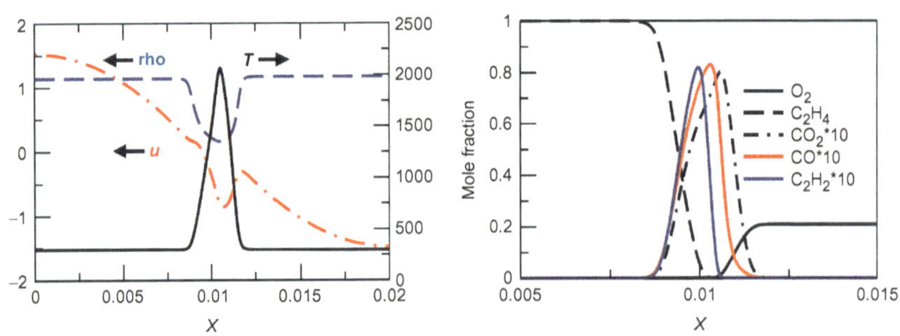

Fig. 36 Structure of a laminar, stoichiometric ethylene–air diffusion flame at ambient conditions. *Left*: Temperature, density, and velocity profiles. *Right*: Species profiles.

Contrary to premixed flames, nonpremixed flames in one-dimensional flows are not steady, and the time derivative in Eqs. (43)–(46) must be kept. A first consequence is that, from the unsteady continuity equation, momentum is not anymore uniform and constant, so that the quantity ρu cannot be simplified. The nonpremixed flame equations then write:

$$\frac{\partial \rho Y_{Fu}}{\partial t}+\frac{\partial \rho u Y_{Fu}}{\partial x}=\rho D_{Fu}\frac{\partial^2 Y_{Fu}}{\partial x^2}+\dot{\omega}_{Fu} \tag{60}$$

$$\frac{\partial \rho C_p T}{\partial t}+\frac{\partial \rho u C_p T}{\partial x}=\lambda\frac{\partial^2 T}{\partial x^2}-Q\dot{\omega}_{Fu} \tag{61}$$

As in the case of premixed flames, the two above equations can be combined to form a transport equation for the same quantity $Y_{Fu}+C_pT/Q$, that is, similar to Eq. (51) where the time derivative is now kept:

$$\frac{(\partial Y_{Fu}+C_pT/Q)}{\partial t}+\frac{\partial \rho u(Y_{Fu}+C_pT/Q)}{\partial x}=\rho D_{Fu}\frac{(\partial^2 Y_{Fu}+C_pT/Q)}{\partial x^2} \tag{62}$$

This is the equation of a conserved (or passive) scalar noted Z, which does not have simple solution as in the case of premixed flame, due to the unsteady term and the different boundary conditions (Fig. 32). Eq. (62) has no chemical source term, so that the solution only depends on the velocity field and can be solved independently of the flame. Similar to the progress variable, the passive scalar Z can be built from various combinations of variables, such as, for example, $Y_{Fu}-\alpha_s Y_{Ox}$ or $\alpha_s Y_{Ox}+C_pT/Q$. When normalized by the boundary values, as:

$$z=\frac{Z-Z_f}{Z_o-Z_f} \tag{63}$$

where the subscripts $_f$ and $_o$ refer to the fuel and oxidizer injection sides, respectively, all definitions of Z lead to the same z that goes from 0 in the fuel side to 1 in the oxidizer side through the flame. When computed from the fuel and oxidizer mass fractions, z is also called the mixture fraction and can be rewritten as:

$$z=\frac{\Phi}{\Phi+1}-\frac{Y_{Ox}}{Y_{Ox,o}}\frac{\varphi-1}{\Phi+1} \tag{64}$$

where $\Phi=Y_{Fu,\,f}/\alpha_{st.}Y_{Ox,\,o}$ is computed from the mass fractions of fuel and oxidizer, respectively, in the fuel and oxidizer injections, and

$\varphi = Y_{Fu}/\alpha_s Y_{Ox}$ is the local equivalence ratio that, contrary to premixed combustion, changes through the flame. At stoichiometry, $\varphi = 1$ and the mixture fraction $z_{st.}$ is:

$$z_{st.} = \frac{\Phi}{1+\Phi} \tag{65}$$

By describing the effect of the flow on the mixing of both reactants, the mixture fraction allows to decouple fluid transport and chemical phenomena in the flame. This can be translated mathematically by the change of variable $(x,t) \rightarrow (z,t)$, leading to:

$$\left[\frac{\partial \rho Y_{Fu}}{\partial t}\right]_{z_{const.}} = \rho D_{Fu}\left(\frac{\partial z}{\partial x}\right)^2 \frac{\partial^2 Y_{Fu}}{\partial z^2} + \dot{\omega}_{Fu} \tag{66}$$

$$\left[\frac{\partial \rho T}{\partial t}\right]_{z_{const.}} = \rho D_{Fu}\left(\frac{\partial z}{\partial x}\right)^2 \frac{\partial^2 T}{\partial z^2} - \frac{\Delta H_r}{C_p}\dot{\omega}_{Fu} \tag{67}$$

In the above equations, the quantity $\chi = \rho D_{Fu}(\partial z/\partial x)^2$ is the scalar dissipation of z. It is homogeneous to an inverse of time and representative of the flow time scale. Assuming that the flame variable unsteadiness is the same as the unsteadiness of z, i.e., $(\partial/\partial t)_{z_{const.}} = 0$, one finally obtains:

$$\chi \frac{\partial^2 Y_{Fu}}{\partial z^2} = -\dot{\omega}_{Fu} \tag{68}$$

$$\chi \frac{\partial^2 T}{\partial z^2} = \frac{Q}{C_p}\dot{\omega}_{Fu} \tag{69}$$

In nonpremixed flames, the reaction rate is maximum at stoichiometric conditions where, as was demonstrated in premixed combustion, the maximum of heat is produced. This is where the flame is located, separating the fuel and oxidizer flows. If combustion is fast enough, the flame is thin and χ does not deviate much from its stoichiometric value in the reaction zone. It is then possible to assume $\chi = \chi_{st.}$ in the above equation, making the scalar dissipation a solution parameter. For a given chemistry, all flame variables can then be expressed as functions of $(z, \chi_{st.})$.

Eqs. (68) and (69) have a simple solution in the limit case of infinitely fast chemistry, i.e., $\dot{\omega}_{Fu} = \delta(x - x_{st.})$, where $\delta(x)$ is the Dirac function. In that case, the zero chemical source term leads to linear solutions in both sides

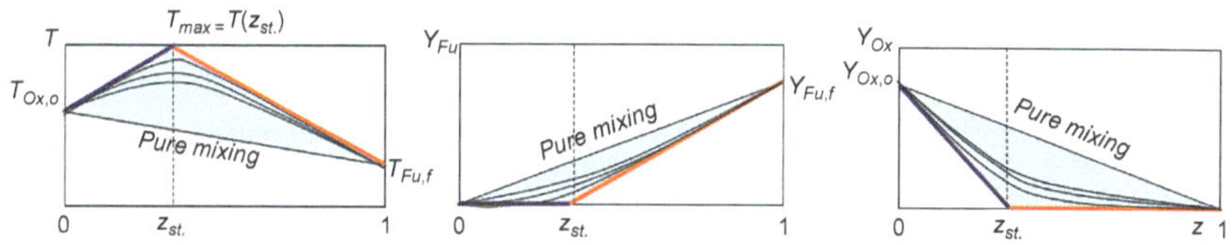

Fig. 37 Structure of a laminar diffusion flame in the *z*-space. The infinitely fast chemistry solution is represented by the *blue* and *red lines*. The finite rate chemistry solutions are represented with *black lines* for various values of $\chi_{st.}$: the difference with the infinitely fast chemistry increases with $\chi_{st.}$.

of the flame, with the matching conditions $Y_{Fu} = Y_{Ox} = 0$ and $T = T_{st.}$ at the stoichiometric point (Fig. 37):

$$\text{for} \quad 0 < z < z_{st.} \quad Y_{Fu} = 0, \quad Y_{Ox} = Y_{Ox,o}\frac{z_{st.} - z}{z_{st.}} \text{ and } T = T_{Ox,o} + (T_{st.} - T_{Ox,o})\frac{z}{z_{st.}} \tag{70}$$

$$\text{for} \quad z_{st.} < z < 1 \quad Y_{Fu} = Y_{Fu,f}\frac{z - z_{st.}}{1 - z_{st.}}, \; Y_{Ox} = 0 \text{ and } T = T_{Fu,f} + (T_{st.} - T_{Fu,f})\frac{1 - z}{1 - z_{st.}} \tag{71}$$

To determine $T_{st.}$, one can notice that the quantity $Y_{Fu} + C_pT/Q$ is a linear function of z in the whole domain as it has a zero source term. This implies that the slope is the same on both sides:

$$\frac{C_p}{Q}\frac{T_{st.} - T_{Ox,o}}{z_{st.}} = \frac{Y_{Fu,f}}{1 - z_{st.}} - \frac{C_p}{Q}\frac{T_{st.} - T_{Fu,f}}{1 - z_{st.}} \tag{72}$$

which leads to:

$$T_{st.} = z_{st.}T_{Fu,f} + (1 - z_{st.})T_{Ox,o} + z_{st.}\frac{Q}{C_p}Y_{Fu,f} \tag{73}$$

If $T_{Fu,f} = T_{Ox,o} = T_u$, the above expression reduces to:

$$T_{st.} = T_u + z_{st.}\frac{Q}{C_p}Y_{Fu,f} \tag{74}$$

The quantity $z_{st.}Y_{Fu,f}$ is the quantity of fuel at stoichiometry in a frozen flow: the flame temperature $T_{st.}$ corresponds then to the burning of the quantity of fuel available after mixing. Note that assuming infinitely fast chemistry implies no sensitivity to the flow, i.e., to the scalar dissipation that has disappeared in the above solution.

In the real case of finite rate chemistry, Eqs. (68) and (69) can be solved numerically, where $\chi_{st.}$ acts now as a parameter, representative of the flow time scale. Solutions always lay inside the triangle formed by the mixing line, the nonreacting line, and the infinitely fast chemistry lines on the fuel and oxidizer sides (Fig. 36). The limit $\chi_{st.} \to 0$ corresponds to the infinitely fast chemistry, and the solutions depart from this limit solution when $\chi_{st.}$ increases.

In the physical space, the flame thickness, evaluated from the variable gradients, is therefore proportional to the gradient of z at stoichiometry:

$$\delta_{th} \propto \left(\frac{\partial z}{\partial x}\right)_{st.}^{-1} \tag{75}$$

As for premixed flames, it is possible to write mass and energy balances by integrating Eqs. (68) and (69) in the x-space:

$$\begin{aligned} \dot{\Omega}_{Fu} &= \int_{-\infty}^{+\infty} \dot{\omega}_{Fu} dx = \left(\frac{\partial z}{\partial x}\right)_{st.}^{-1} \int_0^1 \dot{\omega}_{Fu} dz = \left(\frac{\partial z}{\partial x}\right)_{st.}^{-1} \chi_{st.} \frac{\partial Y_{Fu}}{\partial z}(z=1) \\ &= \left(\frac{\partial z}{\partial x}\right)_{st.}^{-1} \chi_{st.} \frac{Y_{Fu,f}}{1-z_{st.}} \end{aligned} \tag{76}$$

where the derivative $\partial z / \partial x$ has been taken constant at the stoichiometric value. Finally the flame total consumption can be written as:

$$\dot{\Omega}_{Fu} = \sqrt{\rho D_{Fu} \chi_{st.}} \frac{Y_{Fu,f}}{1-z_{st.}} \tag{77}$$

Similarly:

$$\begin{aligned} \dot{\Omega}_T &= \int_{-\infty}^{+\infty} \frac{Q}{C_p} \dot{\omega}_{Fu} dx = -\left(\frac{\partial z}{\partial x}\right)_{st.}^{-1} \chi_{st.} \left(\frac{\partial T}{\partial z}(z=1) - \frac{\partial T}{\partial z}(z=0)\right) \\ &= -\left(\frac{\partial z}{\partial x}\right)_{st.}^{-1} \chi_{st.} \left(-\frac{T_{st.} - T_{Fu,f}}{1-z_{st.}} - \frac{T_{st.} - T_{Ox,o}}{z_{st.}}\right) \end{aligned} \tag{78}$$

$$\dot{\Omega}_T = \frac{Q}{C_p} \sqrt{\rho D_{Fu} \chi_{st.}} \frac{Y_{Fu,f}}{1-z_{st.}} \tag{79}$$

This shows that in the infinitely fast chemistry limit, the flame consumption rate is independent of chemistry and only controlled by the diffusion of reactants toward the flame zone. This is the reason why nonpremixed flames are also called diffusion flames. This conclusion also holds for fast chemistry, as long as $\tau_c \ll 1/\chi_{st.}$, where τ_c is a chemical time scale.

Solutions for z may be found in simple one-dimensional configurations. Due to the diffusion-controlled nature of the flame, there is no steady solution without a nonzero velocity that counteracts diffusion fluxes and brings reactants toward the flame where they are consumed.

Starting again from Eq. (62), the transport equation of z writes:

$$\frac{\partial z}{\partial t} + \frac{\partial \rho u z}{\partial x} = \rho D_{Fu} \frac{\partial^2 z}{\partial x^2} \tag{80}$$

with the boundary conditions:

$$z = 0 \text{ for } x \rightarrow -\infty \quad \text{and} \quad z = 1 \text{ for } x \rightarrow +\infty \tag{81}$$

To account for density variation the Howarth–Dorodnitsyn transformation is used:

$$\xi = \int_{x_{st.}}^{x} \frac{\rho(X)}{\rho_{st.}} dX \tag{82}$$

Applying to Eq. (80) leads to:

$$\frac{\partial z}{\partial t} = D_{Fu} \frac{\partial^2 z}{\partial \xi^2} \tag{83}$$

for which the solution writes:

$$z(x,t) = erf\left(\frac{\xi}{2\sqrt{D_{Fu} t}}\right) \tag{84}$$

where *erf* is the error function. It can be seen from Eq. (84) that the gradient of z at stoichiometry decreases with time as $1/2\sqrt{D_{Fu} t}$, which means that the flame thickness increases to infinity, while the consumption rate goes to 0.

Stationary diffusion flames can be obtained by adding a velocity field. In the two-dimensional counterflow configuration (Fig. 38) the velocity field derives from a potential flow and can be written as:

$$\rho u(x,y) = -\rho_{st.} a \xi \tag{85}$$

$$\rho v(x,y) = \rho a y \tag{86}$$

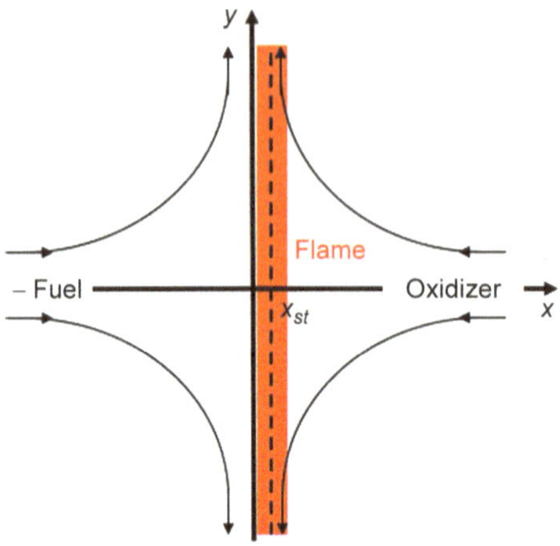

Fig. 38 Counterflow diffusion flame.

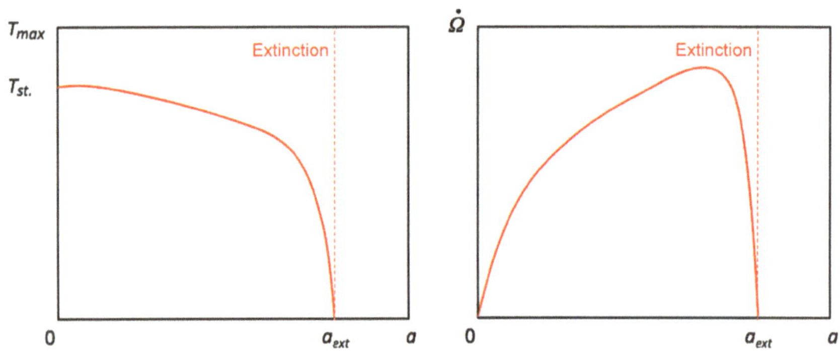

Fig. 39 Response to strain rate of a diffusion flame.

where a is the constant strain rate. Assuming no variation of ρ in the y-direction, the above expressions ensure a divergence-free flow field. Introducing this velocity field in Eq. (80), the solution writes:

$$z(x,t) = erf\left(\frac{\xi}{\sqrt{2D_{Fu}/a}}\right) \tag{87}$$

In this case the nonpremixed, diffusion flame has a steady solution, with a thickness proportional to $\sqrt{2D_{Fu}/a}$, and a consumption rate proportional to $a/2D_{Fu}$, i.e., imposed by the velocity strain rate. The typical response of such flame to strain rate is illustrated in Fig. 39. At a zero strain rate, the infinitely fast chemistry limit is obtained, with the singularity for the reaction rate which tends to a Dirac function. When the strain rate increases, diffusion occurs and the maximum temperature decreases. As the reactant mass fluxes increase, the flame consumption rate also increases, up to a maximum value at which chemistry is not anymore fast enough to burn all reactants brought by the flow. At this stage, the temperature decreases faster and makes the consumption rate decrease also, down to the final extinction which occurs abruptly at a limit strain rate a_{ext}.

Note that comparing the solutions of Eqs. (84) and (87), it appears that the unsteady flame has the same flame structure than the strained flame at $t = 1/2a$.

3.3.3.3 Summary

Characteristics of premixed and nonpremixed flames are summarized in Table 1. The two generic flames have different but clear structures, which can be used as reference flames and for modeling as will be explained in the next Section. However in real systems, flames rarely have purely premixed or nonpremixed structures, but are usually partially premixed with mixed local structures.

3.4 Turbulent Combustion

Fig. 40 features a typical turbulent flame as can be observed in a lab-scale experiment. It is also representative of the turbulent combustion that occurs in engines. The flame appears much wrinkled and highly intermittent. Indeed, the important role of transport has been demonstrated in the previous sections: when this transport is turbulent, the random nature of the flow reflects directly on the flame.

Quantitatively, turbulence has two major effects on the flame. The first effect is a global effect: under the action of vortices, the flame front wrinkles

Table 1 Summary of Laminar Flame Characteristics

	Premixed Flame	Nonpremixed Flame
Controlled by	Chemistry	Diffusion
Described by	Progress variable c	Mixture fraction z
Flame thickness	$\delta_L sqrt D_{Fu}/A$	$\delta \ \sqrt{D_{Fu}/\chi_{st.}}$
Consumption rate	$\rho_u S_L(Y_{Fu,\ b} - Y_{Fu,\ u})$	$\sqrt{\rho D_{Fu} \chi_{st.}} \dfrac{Y_{Fu,f}}{1 - z_{st.}}$

Fig. 40 Photograph of a nonpremixed turbulent jet flame of ethylene burning in quiescent air. *Source: Skeen SA, Axelbaum RL:* Laboratory for Advanced Combustion and Energy Research (LACER), *Washington University in St. Louis.*

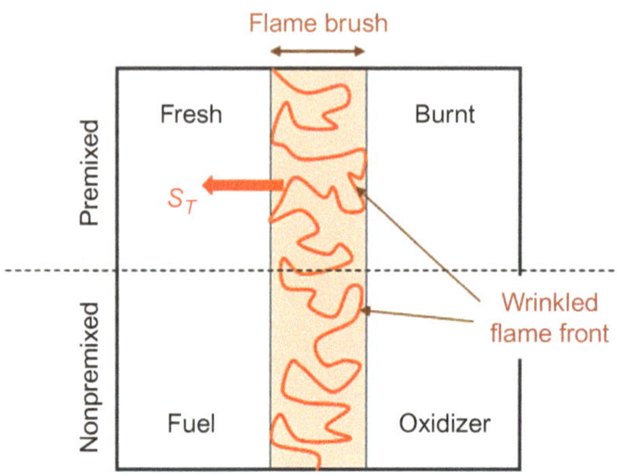

Fig. 41 Turbulent flame wrinkling.

and the flame surface area increases (Fig. 41). This means that in a given volume, turbulent flames lead to more reacting zones, i.e., more fuel consumption. In other words, a turbulent flame globally burns faster than a laminar flame. This is very useful for practical applications, as it allows to burn important fuel mass flow rates in a short time or volume. When the turbulence intensity increases, the strong wrinkling first leads to the formation of pockets, then the integrity of the flame surface disappears, and combustion tends progressively toward the perfectly mixed regime (Fig. 41).

The second effect is local: turbulence induces strong, unsteady velocity gradients, which modify the local transport of reactants and lead to local flame structures that depart significantly from the generic one-dimensional flame structures described in the previous section. This is quantified with the local stretch rate κ defined as:

$$\kappa = (\delta_{ij} - n_i n_j)\frac{\partial w_i}{\partial x_j} \tag{88}$$

where $\mathbf{n}$ is the unit vector normal to the flame surface and $\mathbf{w}$ is the velocity of the flame surface. Writing $\mathbf{w} = \mathbf{u} + S_d\mathbf{n}$, with $\mathbf{u}$ the velocity field and S_d the flame displacement speed, the stretch rate is obtained as:

$$\kappa = (\delta_{ij} - n_i n_j)\frac{\partial u_i}{\partial x_j} + S_d\frac{\partial n_i}{\partial x_i} \tag{89}$$

In the above expression, $\partial n_i/\partial x_i$ is simply the curvature of the flame surface, while the first term is the velocity gradient in the direction tangential to the flame surface, i.e., the strain rate.

Stretch can increase the local consumption rate by increasing mass fluxes toward the reaction zone. However, if the flame chemistry is not fast enough to fully burn these enhanced reactant fluxes, the local temperature decreases and the flame may eventually quench.

The flame–turbulence interaction is two way, and the flame also has an impact on the turbulent flow, due to the density and temperature variations. Although this may induce additional instabilities at the flame surface, the most observable effect is the reduction of turbulence intensity in the burnt gases due to their higher viscosity.

3.4.1 Turbulent Premixed Flames

The wrinkled flame appears as a flame brush, with an increased thickness and propagation speed S_T, called the turbulent flame speed. Essentially, S_T is the result of wrinkling:

$$\frac{S_T}{S_L} = \frac{A_T}{A} \tag{90}$$

where A_T/A is the wrinkling, defined as the ratio of the turbulent to laminar flame surface areas. Contrary to S_L, the turbulent flame speed S_T is not an intrinsic flame characteristic and depends on the flow. This explains why many correlations can be found in the literature, usually linked to the turbulent velocity statistics, as in Abdel-Gayed and Bradley (1985):

$$\frac{S_T}{S_L} = 1 + \alpha\left(\frac{u'}{S_L}\right)^n \tag{91}$$

where α and n are two model parameters close to unity and u' is the velocity standard deviation.

As for any coupled phenomena, flame–turbulence interaction is characterized by the ratio of characteristic space and time scales. This leads to the

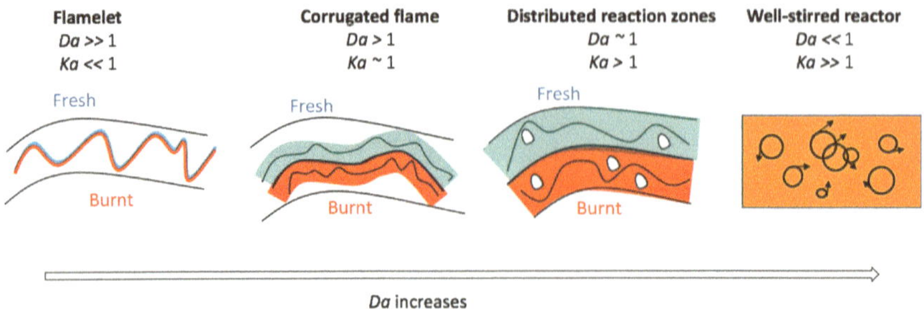

Fig. 42 Turbulent premixed flame regimes.

nondimensional Damköhler number Da, defined as the ratio of the chemical time scale $\tau_c = \delta_L / S_L$ and the flow time scale $\tau_f = l_t / u'$, where l_t is the integral scale of turbulence:

$$Da = \frac{\tau_f}{\tau_c} = \frac{l_t}{\delta_L} \frac{S_L}{u'} \tag{92}$$

When $Da \gg 1$, combustion is much faster than turbulence and the flame structure is little affected: combustion occurs in wrinkled, thin reaction zones which keep locally the same structure as a one-dimensional premixed flame burning in the same conditions (Fig. 42). This is the so-called flamelet regime and is the one most encountered in combustion chambers of engines. On the contrary when $Da \ll 1$, combustion is much slower than turbulence, which fully mixes reactants and products before burning, leading to a "well-stirred-reactor" regime. Between these two limit cases, a variety of combustion regimes occur, from pockets to corrugated flame front.

Recognizing however that turbulence is a multiscale phenomenon, with a wide range of scales, it is interesting to evaluate also the interaction of the flame with the smallest scale of turbulence, i.e., the Kolmogorov scale $\tau_\eta = \eta / u_\eta$, leading to the Karlovitz number:

$$Ka = \frac{\tau_c}{\tau_\eta} = \frac{\delta_L}{\eta} \frac{u_\eta}{S_L} \tag{93}$$

The same interpretation as for the Da number holds, referring now to the small scale of turbulence.

The various combustion regimes are represented in a $(l_t/\delta_L) - (u'/S_L)$ diagram, shown in Fig. 43, in which the Reynolds number, Damköhler number, and Karlovitz number limits appear as straight lines.

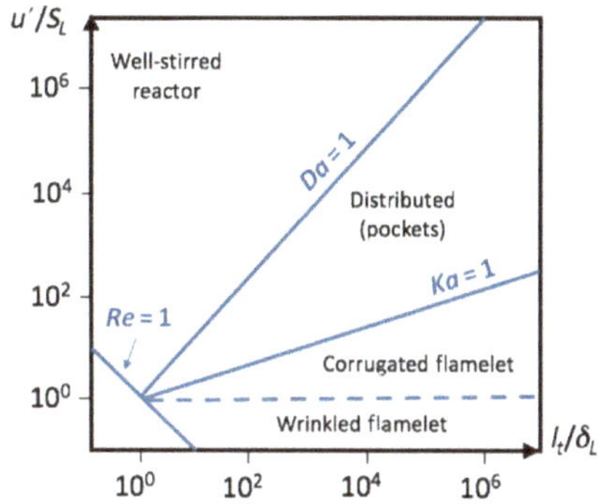

Fig. 43 Turbulent premixed combustion diagram. *Source: Adapted from Peters N: Laminar flamelet concepts in turbulent combustion. In:* 21st symposium (international) on combustion. *The Combustion Institute, Pittsburgh, pp 1231–1250, 1986.*

3.4.2 Turbulent Diffusion Flames

As nonpremixed combustion is controlled by the flow, turbulent diffusion flames are very sensitive to turbulence. If the flame wrinkling process is similar to the one observed in premixed flames, the main difference is that the local flame thickness and the consumption rate of a turbulent diffusion flame are driven by the turbulence through the local strain rate. This is reflected in the calculation of the Da number, where the flow time scale is written as the inverse of the scalar dissipation:

$$Da = \frac{1}{\tau_c \chi} \tag{94}$$

In the above expression, the scalar dissipation is evaluated in the whole range of turbulent scales. If $Da \gg 1$, combustion is very fast, the scalar dissipation is low, and the local flame structure is close to the infinitely fast chemistry limit (Fig. 39). This corresponds to the "flamelet" regime where the local flame structure is similar to a one-dimensional, strained, diffusion flame burning in the same conditions and subjected to the same strain rate. When $Da \ll 1$, the slow chemistry and the high strain rate lead to local quenching and the flame burns in a distributed regime. Between these two limits, chemistry is not fast enough to adjust to any change of the scalar dissipation and unsteady effects appear (Cuenot and Poinsot, 1994).

Contrary to premixed flames, it is not possible to derive a combustion regime diagram in a phase space defined by a flame thickness and a flame speed, as such parameters do not exist for diffusion flames. Cuenot and Poinsot (1994) proposed to use the flame thickness and time of the laminar

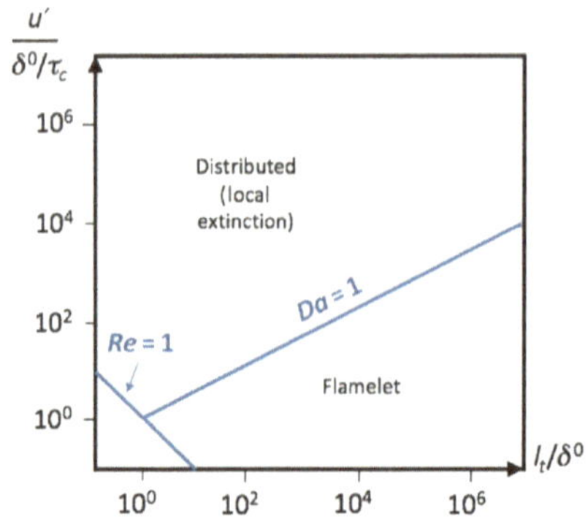

Fig. 44 Turbulent nonpremixed combustion diagram. *Source: Adapted from Cuenot B, Poinsot T: Effects of curvature and unsteadiness in diffusion flames. implications for turbulent diffusion flames,* Proc Combust Inst *25:1383–1390, 1994.*

flame that would exist without turbulence that they evaluated at the start of the flame–turbulence interaction (Fig. 44). Another possibility is to use the mean scalar dissipation rate as a reference.

4. MODELS FOR TURBULENT COMBUSTION SIMULATION

The simulation of turbulent flows is a complex topic due to the combination of two factors: the wide range of vortical structure scales and the nonlinearity of the convection term, which is the source of transition to turbulence and of the energy cascade from the largest to the smallest scales. Although an extensive literature already exists in the field, including textbooks such as Echeckki and Mastorakos (2011), research in turbulent flow modeling is still very active, motivated in particular by the constantly increasing computing power which continues to open new perspectives.

Introducing combustion in such simulations adds another level of complexity, in terms of number of variables, space and time scales, and nonlinearity of the chemical source term. This requires specific models as described below.

4.1 Turbulence Simulation and Modeling

The random nature of turbulence implies a statistical description. Usually the first two statistical moments (mean and root mean square) are sufficient to analyze the flow, but in some fundamental studies, higher order moments may be needed. To calculate these statistics, three classes of approach exist. In the Reynolds-averaged Navier–Stokes (RANS) approach, the statistical moments are directly computed as the solutions of transport equations derived from the flow equations (Eqs. 29–31). It is very different from the direct numerical simulation (DNS) approach where statistics are built from a series of different realizations of the same turbulent flow calculated with the flow equations (Eqs. 29–31). In the first case, the equations for the statistical moments introduce closure models which may not reflect the complexity of the turbulent flow, limiting the accuracy of the simulation. On the other hand, computing series of realizations requires to introduce randomness in the deterministic calculations and is very demanding in terms of computing time to reach statistical convergence. Indeed, each flow realization must describe the whole range of turbulent scales, which has a direct impact on the grid size and resolution. To alleviate this constraint, the LES technique has been proposed as an intermediate approach. It is still based on the principle of computing turbulent flow realizations, but assumes that the smallest scales of the turbulence can be modeled and need not to be resolved.

If the turbulent flow is statistically steady, i.e., the statistical moments do not vary with time, the ergodic hypothesis may be used, which states that a system will be most of the time in its most probable state. This implies that, in the case of DNS and LES, different instantaneous flow solutions may be viewed as different flow realizations and used to build statistics. In other words, this means that it is not necessary to perform a large number of simulations of the same turbulent flow, but that statistics can be built from only one simulation, run for a sufficiently long time to obtain a good statistical sample. The ergodic hypothesis is also useful for RANS, as it allows to search for steady solutions only and apply efficient numerical algorithms.

One major difficulty for all approaches is the simulation of turbulent flows with nonstationary statistics. This situation often occurs in reacting flows, as combustion has many transient processes as seen in the previous section. In DNS and LES, this requires to perform series of simulations of the same flow, as done, for example, in Esclapez et al. (2015) to study the probability of ignition. In RANS, the concept of unsteady RANS has been introduced, where the moment equations are now solved in their unsteady formulation, but it is efficient only if the characteristic time of the statistical moments is far from the characteristic time scales of the turbulence itself, which is difficult to establish.

4.1.1 The Hierarchy of DNS, LES, and RANS Approaches

Consider the example of a homogeneous, isotropic turbulent flow, having the characteristic energy spectrum illustrated in Fig. 45. Statistical moments of such flow are spatially uniform, so that RANS simulations do not have any spatial scale to resolve. On the contrary, DNS must solve the whole range of scales, from the largest ones (l_t) to the smallest ones, down to the Kolmogorov scale η. As an intermediate approach, LES solves only the largest scales of turbulence, down to a cutoff frequency k_c which must be placed between the inertial zone and the dissipation zone. This of course has a direct impact on the computing time, which increases significantly from RANS to LES and DNS. The counterpart is the accuracy, limited in RANS by the use of closure models and, to a less extent also in LES for the smallest scales.

Typical results obtained with RANS, LES, and DNS are illustrated in Fig. 46, reporting the temporal velocity signal at a fixed location. The

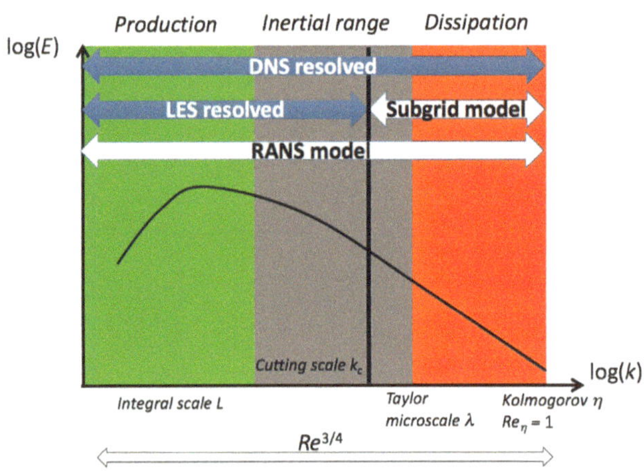

Fig. 45 The hierarchy of DNS, LES, and RANS illustrated in the energy spectrum of homogeneous, isotropic turbulence.

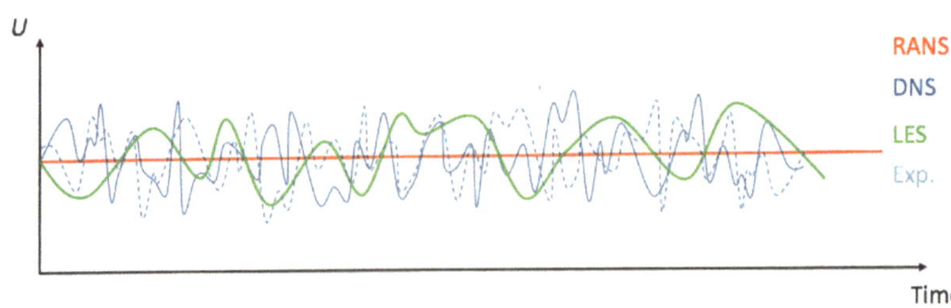

Fig. 46 Temporal velocity signal recorded at a fixed location in a statistically stationary turbulent flow.

RANS result is a constant as it corresponds to the constant mean velocity. The DNS signal is similar to the one obtained experimentally, in the sense that it has the same statistics, but it is a different realization: point-to-point values are different. This clearly illustrates why only statistics are meaningful when studying turbulence. Finally the LES result is intermediate between DNS and RANS and shows only the lowest frequencies of the signal.

The history of numerical simulation of turbulent flows is strongly related to the evolution of the power of computers, as illustrated in Fig. 47. It started of course with the fastest RANS approach, which was the only accessible approach on the first computers. Then, DNS emerged when the performances of the computers started to increase significantly. It may be surprising that DNS appeared first, although it is more demanding in terms of computing time than LES for the same flow. The reason is two-fold. First, the main objective of scientists was to understand the driving mechanisms of turbulence, and the idea to calculate if without modeling was very appealing. Second, LES aims at calculating realistic complex flows which implies large Reynolds numbers and geometry size, and at the end large computing time which was not available at the time when DNS started. LES therefore appeared later, and quite recently in the field of turbulent combustion.

Although RANS is still an interesting tool in the industrial context, as a parameterized meta-model based on a comprehensive calibration plan, it cannot be helpful for innovation or system understanding. On the other

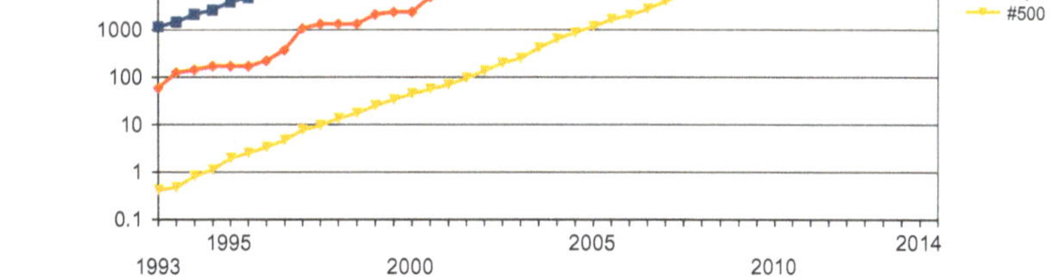

Fig. 47 Evolution of the power of computing architectures: performance in GFLOPS (log scale), combined performance of 500 largest supercomputers (*blue*), fastest supercomputer (*red*), and supercomputer in 500th place (*yellow*). *From https://en.wikipedia.org/wiki/History_of_supercomputing#/media/File:Supercomputers-history.svg.*

hand, DNS is too limited in terms of geometry size and complexity, and turbulence intensity, to be representative of real flows and is devoted to the detailed study of turbulent mechanisms in simplified academic configurations. For these reasons, CFD has massively moved from RANS to LES and entered in a new era as only LES has the capacity to describe with high fidelity and help understanding turbulence phenomena in real complex configurations. LES is therefore a tool suited for both academic studies and industrial applications, for which its predictive capacity can help reducing and optimizing real system testing. Therefore, after a brief description of RANS and DNS, the focus is made in this chapter on LES principles and methodology.

4.1.2 RANS Simulations

Considering any turbulent flow quantity $f(\mathbf{x}, t)$, as a random variable, obeying statistics which vary in space and time, the statistical moments of this quantity are defined from ensemble averages and write, in the discrete form:

$$\langle f(\mathbf{x},t)\rangle_n = \frac{1}{N}\sum_N f^n(\mathbf{x},t) \tag{95}$$

where $\langle . \rangle_n$ denotes the statistical moment of order n of the turbulent quantity, and N is the size of the statistical sample. The Reynolds decomposition then consists in writing $f(\mathbf{x}, t) = \langle f(\mathbf{x}, t)\rangle + f'(\mathbf{x}, t)$, where $\langle f(\mathbf{x}, t)\rangle$ is the mean value and $f'(\mathbf{x}, t)$ the fluctuation. Introducing this decomposition for the velocity in the flow equations (Eqs. 29–31) leads to an equation for the mean velocity, with an unclosed term, known as the Reynolds stress tensor, associated to the nonlinear convective term $\langle u_i' u_j' \rangle$. This term is a major mechanism of turbulent motion and is critical for the simulation of turbulent flow. It requires therefore modeling, which is the topic of a very large literature proposing numerous models having different levels of complexity and accuracy. The difficulty is that the Reynolds stress tensor has a very different behavior at large scales, where it produces turbulence, from small scales where it leads to mixing and dissipation. Therefore, building a model that is able to reproduce these two very different effects is still an open question. In addition, large-scale turbulent phenomena are mostly influenced by the configuration geometry and there is no universal model that can cover such diversity. A consequence of the Reynolds decomposition is that RANS does not describe a real flow, but only its statistics.

RANS turbulence models usually introduce a turbulent viscosity ν_t, which adds to the laminar viscosity ν so that the Reynolds number Re_{num} effectively "seen" by the RANS approach is:

$$Re_{num} = Re\frac{\nu}{\nu + \nu_t} \tag{96}$$

which is much smaller than the true Reynolds number Re. This is the reason why a RANS solution looks like a "laminar" flow, and does not contain any turbulent structure.

In combustion, the density variation introduces a supplementary difficulty, which is partially avoided by introducing the Favre decomposition:

$$\widetilde{\langle f \rangle} = \frac{\langle \rho f \rangle}{\langle \rho \rangle} \tag{97}$$

Undoubtedly the most interesting feature of RANS is that it is very fast. This explains why it has been, and is still widely used in industry to give global trends about the impact of a geometrical or operational evolution. Indeed, although RANS does not describe the turbulence mechanisms that are impacted by such evolution, it can still give some information on the mean flow provided that the model parameters have been previously calibrated on a sufficiently similar system. However, RANS does not have the capacity to predict a turbulent flow in blind test and does not give any insight on the underlying turbulent mechanisms of a complex flow.

4.1.3 Direct Numerical Simulation

The principle of DNS is to solve directly Eqs. (29)–(31), without any additional models. As already mentioned, this implies to solve the whole range of turbulent scales, which has a direct impact on the grid size. Taking again the example of the homogeneous, isotropic turbulent flow, the largest scale is of the order of $l_t \approx u'^3 \varepsilon$, while the Kolmogorov scale is $\eta \approx (\nu^3/\varepsilon)^{1/4}$, where ε is the turbulent dissipation. This leads to a large-to-small turbulent scale ratio $l_t/\eta \approx Re_t^{3/4}$, where the turbulent Reynolds number is defined as $Re_t = u' l_t/\nu$. Considering that the grid must contain $\approx$10 large scales for statistical convergence and must resolve the smallest scale with $\approx$3 cells, the number of grid points in each direction scales as $N = L/\Delta \approx 10 l_t/(\eta/3) \approx 3Re_t^{3/4}$. In three dimensions, the total number of grid points finally scales as $9Re_t^{9/4}$. This induces a strong limitation in Reynolds number, i.e., in turbulence intensity, that can be simulated with DNS. This limitation depends of course on the

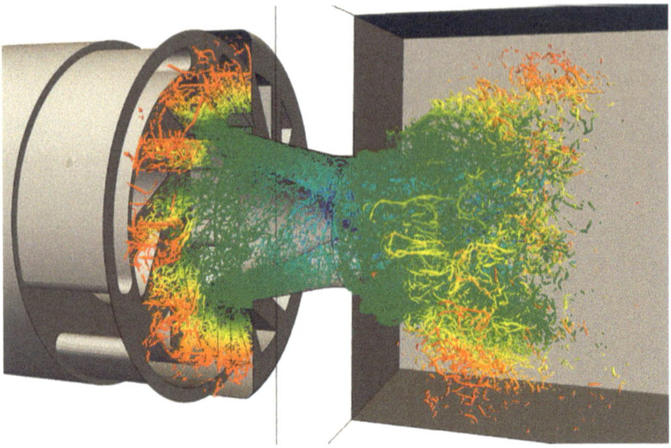

Fig. 48 Example of DNS study of a turbulent swirled flame: iso-contour of *Q*-criterion showing turbulent structures. The grid contains 2.6 billion tetrahedra (Moureau et al., 2011).

discretization scheme, and on the turbulent flow considered, but is still constraining today, even with the high-performance computers that are now available. Largest DNSs today use grids with several billion points and are run on several thousands of processors. As an example, B.J. Boersma (Delft University of Technology) performed three DNSs of turbulent pipe flow having Reynolds numbers between 25, 000 and 76, 000, using up to 7.6 billion grid points and 24,000 processors (Boersma, 2011). DNS is also applied to turbulent combustion, giving a unique insight on the flame–turbulence interaction as shown, for example, in Kim et al. (2015) and Moureau et al. (2011). Fig. 48 shows a typical DNS result obtained by Moureau et al. (2011) on a lab-scale swirled burner, using a grid composed of 2.6 billion tetrahedra. The simulation required 16,384 cores of an IBM Blue Gene/P machine during 80 h to achieve a physical time of 1.9 ms.

To take the best benefit of DNS, it is recommended to use a high-order numerical scheme for the spatial and temporal discretization, and to consider configurations with periodic directions, along which averaging can be performed to accelerate the statistical convergence. For these reasons, DNS is used mostly for fundamental studies on academic configurations, for which high accuracy and details of the mechanisms of turbulence are required.

4.1.4 Large Eddy Simulations

Recognizing that the smallest scales of turbulence mostly have a diffusive and dissipative behavior relatively independent of the flow configuration, they can be efficiently modeled, thus avoiding the very fine grids required by DNS. Following this idea, LES separates the large and small scales at a cutoff frequency, allowing to solve directly the large scales and their interaction with the configuration geometry, while modeling the diffusive character of the smallest scales. This is achieved thanks to a filtering operation, obtained by the convolution product of any scalar quantity with a filter $\mathcal{G}_\Delta$:

$$\bar{f}(\mathbf{x},t)=\int f(\mathbf{x},t)\mathcal{G}_\Delta(\mathbf{x}-\mathbf{x}')\,dx \tag{98}$$

The filtering operation can be applied explicitly to the conservation equations, or directly result from the discretization of the grid which is most commonly used for obvious reasons of computational cost.

As in RANS, any turbulent flow quantity can be decomposed into a filtered part $\bar{f}$ and an unresolved, or subgrid scale (SGS) contribution f', defined by:

$$f'(\mathbf{x},t)=f(\mathbf{x},t)-\bar{f}(\mathbf{x},t) \tag{99}$$

For combustion application, the Favre definition is preferred for the filtered quantity:

$$\tilde{f}=\frac{\overline{\rho f}}{\bar{\rho}} \tag{100}$$

Applying the filtering operation to Eqs. (29)–(31) yields:

$$\frac{\partial\bar{\rho}}{\partial t}+\frac{\partial\bar{\rho}\tilde{u}_i}{\partial x_i}=0 \tag{101}$$

$$\frac{\partial\bar{\rho}\tilde{u}_j}{\partial t}+\frac{\partial\bar{\rho}\tilde{u}_i\tilde{u}_j}{\partial x_i}+\frac{\partial\overline{P\delta_{ij}}}{\partial x_i}=\frac{\partial}{\partial x_i}\left[\bar{\tau}_{ij}-\bar{\rho}\left(\widetilde{u_iu_j}-\tilde{u}_i\tilde{u}_j\right)\right]\quad\text{for } j=1,3 \tag{102}$$

$$\frac{\partial\bar{\rho}\tilde{Y}_k}{\partial t}+\frac{\partial\bar{\rho}\tilde{u}_i\tilde{Y}_k}{\partial x_i}=\frac{\partial}{\partial x_i}\left[\bar{J}_{i,k}-\bar{\rho}\left(\widetilde{u_iY_k}-\tilde{u}_i\tilde{Y}_k\right)\right]+\bar{\dot{\omega}}_k\quad\text{for } k=1,N \tag{103}$$

$$\frac{\partial\bar{\rho}\tilde{E}}{\partial t}+\frac{\partial\bar{\rho}\tilde{u}_i\tilde{E}}{\partial x_i}+\frac{\overline{\partial u_iP\delta_{ij}}}{\partial x_i}=\frac{\partial}{\partial x_i}\left[\bar{q}_i-\bar{\rho}\left(\widetilde{u_iE}-\tilde{u}_i\tilde{E}\right)\right]+\overline{\tau_{ij}\frac{\partial u_i}{\partial x_j}}+\bar{\dot{\omega}}_T+\dot{\mathcal{Q}}_{sp} \tag{104}$$

In Eqs. (101)–(104), the left-hand side contains the filtered inviscid fluxes, while viscous and turbulent (or SGS) fluxes as well as source terms are in the right-hand side.

The filtered viscous fluxes appearing in the LES governing equations, $\overline{\tau_{ij}}$, $\bar{J}_{i,k}$, and $\overline{q_i}$, are directly computed from the filtered quantities as they are small compared to the turbulent transport and do not require high accuracy.

On the contrary, the unresolved SGS turbulent transport appearing in Eqs. (101)–(104) must be modeled. Generally, the influence of the SGS on the resolved motion is taken into account by an SGS model based on the introduction of a turbulent kinematic viscosity, ν_t (Boussinesq, 1877 hypothesis). Such an approach assumes that the effect of the SGS field on the resolved field is purely dissipative. This assumption is essentially valid within the cascade theory introduced by Kolmogorov (1941). SGS closures then write:

- **Unresolved Reynolds tensor**

  $$\overline{\tau}_{ij}^{sgs} = -\overline{\rho}(\widetilde{u_i u_j} - \tilde{u}_i \tilde{u}_j) = 2\overline{\rho}\nu_t\left(\widetilde{S}_{ij} - \frac{1}{3}\delta_{ij}\widetilde{S}_{kk}\right)$$

- **Unresolved species fluxes**

  $$\overline{J}_{i,k}^{sgs} = -\overline{\rho}(\widetilde{u_i Y_k} - \tilde{u}_i \widetilde{Y}_k) = -\overline{\rho}\left(D_k^t \frac{W_k}{\bar{W}}\frac{\partial \widetilde{X}_k}{\partial x_i} - \widetilde{Y}_k \widetilde{V}_i^{c,t}\right) \text{ with } \widetilde{V}_i^{c,t} = \sum_{k=1}^{N} \bar{D}_k^t \frac{\partial \widetilde{X}_k}{\partial x_i}$$

  and $D_k^t = \mu_t / \rho Sc_k^t$ where Sc_k^t is the turbulent Schmidt number of species k, equal for all species $Sc_k^t = Sc^t = 0.6$

- **Unresolved energy flux**

  $$\overline{q}_i^{sgs} = -\overline{\rho}(\widetilde{u_i E} - \tilde{u}_i \widetilde{E}) = -\lambda_t \partial \widetilde{T}\frac{}{\partial x_i} + \sum_{k=1}^{N} \bar{J}_{i,k} \tilde{h}_{s,k} \text{ with } \lambda_t = \mu_t \bar{C}_p / Pr^t$$

LES models for the subgrid stress tensor only differ through the estimation of ν_t. They are derived on the theoretical ground that the LES filter is spatially and temporally invariant. Three example of models are presented below:

- **The Smagorinsky model** (**Smagorinsky, 1963**) The SGS turbulent viscosity is written as:

  $$\nu_t = (C_s \Delta_x)^2 \sqrt{2\widetilde{S}_{ij}\widetilde{S}_{ij}} \tag{105}$$

  where Δ_x is the filter characteristic length and C_s is the model constant usually equal to 0.18 but which can vary between 0.1 and 0.8 depending on the flow configuration. The Smagorinsky model (1963) was developed in the 1960s and heavily tested for multiple flow configurations. This closure has the particularity of supplying the right amount of dissipation of kinetic energy in homogeneous isotropic turbulent flows. Locality is however lost and only global quantities are maintained. It is known to be too dissipative for transitioning flows (Sagaut, 2002). Moreover, it does not vanish fast enough when approaching wall regions and therefore cannot be used when boundary layers are resolved, i.e., without wall models. To overcome part of these limitations, the dynamic Smagorinsky (or Germano) model has been introduced where the model constant is evaluated by explicit filtering of the LES solution (Germano, 1986).

- **The WALE model** (**Ducros et al., 1998**) was developed for wall-bounded flow in an attempt to recover the correct behavior of wall flows. Similar to the Smagorinsky model, locality is lost and only global quantities are to be trusted. The SGS turbulent viscosity writes:

$$\nu_t = (C_w \Delta_x)^2 \frac{\left(s_{ij}^d s_{ij}^d\right)^{3/2}}{\left(\widetilde{S}_{ij} \widetilde{S}_{ij}\right)^{5/2} + \left(s_{ij}^d s_{ij}^d\right)^{5/4}} \tag{106}$$

with

$$s_{ij}^d = \frac{1}{2}\left(\widetilde{g}_{ij}^{\,2} + \widetilde{g}_{ij}^{2}\right) - \frac{1}{3}\widetilde{g}_{kk}^{2}\delta_{ij} \tag{107}$$

where $C_w = 0.4929$ is the model constant and $\widetilde{g}_{ij} = \partial \tilde{u}_i / \partial x_j$ is the resolved velocity gradient.

- **The SIGMA model** (**Nicoud et al., 2011**) uses an operator based on singular values (σ_1, σ_2, and σ_3) of a tensor built with resolved velocity gradients:

$$\nu_t = (C_\sigma \Delta_x)^2 \frac{\sigma_3(\sigma_1 - \sigma_2)(\sigma_2 - \sigma_3)}{\sigma_1^2} \tag{108}$$

where $C_\sigma = 1.35$ is the model constant. Note that the singular values σ_i have the following near wall behaviors:

$$\begin{cases} \sigma_1 = O(1) \\ \sigma_2 = O(y) \\ \sigma_3 = O(y^2) \end{cases} \tag{109}$$

where y is the normal coordinate to the wall surface and $O(y^p)$ denotes a term of the order of p, i.e., behaving like y^p when the distance to the solid boundary tends to 0. Thus, the SIGMA model behaves correctly in the near-wall region as $\nu_t = O(y^3)$.

4.2 Turbulent Combustion Modeling in LES

The two words of "turbulent combustion" summarize the two major issues of its modeling, also reflected in the two scientific communities interested in the field: chemistry and fluid mechanics.

Combustion chemistry and molecular transport properties are complex and involve a very large number of chemical species: if it stays limited to about 10 species for hydrogen combustion in air, it is higher for hydrocarbons from ≈50 for methane, and up to several hundreds for octane of heavier hydrocarbons. In addition, chemical reactions, and in particular those involving highly reactive radicals, may be very stiff. Note that chemistry must be associated to multicomponent transport properties, which can be simplified as explained in a previous section. Including all details of thermochemistry and transport in the simulation of turbulent combustion only makes sense in the case of DNS, where the turbulence is also calculated with the highest accuracy, as in Kim et al. (2015). In the case of LES, the description of combustion chemistry must be simplified to keep the computational cost at a reasonable level and avoid stiffness issues.

Turbulence, and its interaction with a flame, is another issue. In the context of LES, in addition to the fact that the SGS turbulence is not resolved, combustion introduces other strongly nonlinear phenomena with space and time scales that can be also largely underresolved. This raises new modeling issues for both the heat and mass transport and the chemical source terms in the reaction zone.

4.2.1 Description of Chemical Kinetics in LES

The cost of direct resolution of combustion chemistry in LES is linked to the necessity of adding to the system of flow equations as many transport equations as chemical species. Logically, there are two options to overcome this problem: either not resolving the chemistry directly or reducing the number of species.

The existence of low-dimensional manifolds in the composition space allows to describe the chemical structure of a flame with a reduced number of variables, so-called reduced coordinates or progress variables. The species transport equations can therefore be replaced by a limited set of transport equations for these quantities. The identification of these low-dimensional manifolds is however a complex task and is based on 0D reactor or 1D flame calculations. Although this approach is very appealing and has given excellent results in various configurations, its main drawback is that it becomes very complex if many physical effects must be included, such as heat losses, transient phenomena (ignition and extinction), two-phase combustion, or strong interaction with turbulence.

An alternative is to reduce the number of species, while keeping the maximum accuracy. Starting from the most detailed chemical schemes, three levels of reduction can be identified:

- Skeleton mechanisms describe combustion with some hundred or less species and reactions, at the cost of a slightly reduced range of applicability compared to the original mechanisms. Such mechanisms are obtained by applying grouping or lumping of chemical species, or removing submechanisms that are not useful. Reaction path analysis (Directed Relation Graph Analysis (Lu and Law, 2006)) and sensitivity analysis (Tomlin et al., 1992) are used to identify the species that can be dropped. Such reduction stays moderate, which gives to skeleton mechanisms the advantage of being valid in a large parameter range, but still involves a relatively large number of chemical species.
- Analytically reduced mechanisms result from a further reduction of skeleton mechanisms down to 20 species, based on the quasi-steady state (QSSA) or partial equilibrium (PEA) assumptions for the fastest species (Goussis and Maas, 2011). Sophisticated methods based on CSP (computational singular perturbation) (Lam and Goussis, 1994), instantaneous QSSA error (Turanyi et al., 1993), or level of importance (Loevas, 2009) are used to build such schemes. They lead to the identification of a subset of nontransported species, which are still part of the chemical scheme, i.e., appear in the chemical source terms, and are known from the transported species via the QSSA and PEA algebraic relations. Logically the validity is further reduced compared to skeleton mechanisms, but still covers a range of parameters that is sufficiently large for combustion applications.
- Global mechanisms refer to one- to four-step chemistry, including a maximum of 10 species. They are built with a data-fitting procedure to reproduce macroscopic flame quantities such as the premixed flame speed or burnt gas state (see, e.g., Franzelli et al., 2010; Gicquel et al., 2000). These mechanisms are easy to build and implement in LES codes, and computationally inexpensive. However, their range of validity is very limited and the chemical constants have to be reevaluated for each new application. In addition, all information about intermediate species and the intricate flame structure is lost.

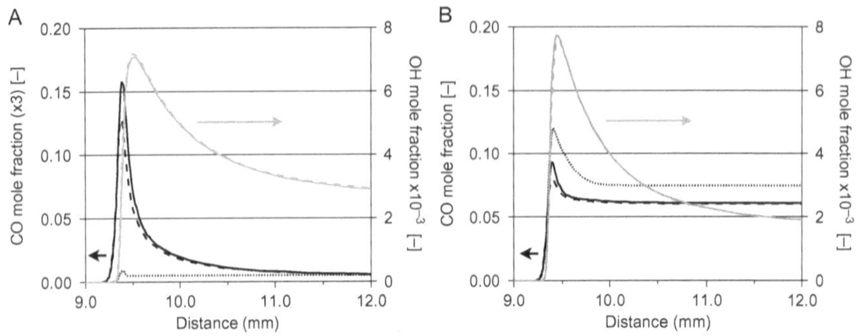

Fig. 49 Structure of a 1D laminar premixed flame for a lean ($\Phi = 0.8$, A) and a rich ($\Phi = 1.2$, B) case. Profiles of CO (*left scale*) and OH (*right scale*) mole fractions are displayed. Detailed mechanism (*solid line*), reduced (*dashed line*), and two steps (*circle*) (for CO only) (Cuenot et al., 2016).

It is interesting to observe that the difference in terms of number of transported species is not much between analytically reduced schemes and global schemes, and can be easily compensated by the use of more processors on parallel computers, as chemistry calculations, being only local, have a very good parallel scaling. With only a slight increase of computational cost, analytically reduced schemes give access to a relatively detailed chemical structure of the flame over a large range of conditions and to any pollutant provided that it has been included in the scheme. Fig. 49 shows, for example, the OH and CO evolutions in an ethylene–air flame obtained with the detailed scheme of Narayanaswamy et al. (2010), a global scheme (for CO only), and an analytically reduced scheme of 14 species, derived from the same detailed mechanism, originally composed of 158 species and 1804 irreversible reactions (Cuenot et al., 2016). The reduced scheme predicts almost perfectly the OH evolution for both cases. In particular, the peak is well captured. The CO profile is also predicted with great accuracy, with equilibrium levels recovered in both cases. Predictions of the two-step mechanism, however, are only qualitative, with errors as far as several orders of magnitude in the lean case.

Reduced chemistry has also proven very efficient for pollutant prediction. It was used in Jaravel et al. (2016) to predict NO_x and CO in aeronautical combustors. The accuracy of the approach was demonstrated on the SANDIA flame D by Jaravel (2016), where all flame variables and species are perfectly recovered with an analytically reduced scheme. Fig. 50 shows an illustration of the flame obtained with DNS and the comparison of experimental and numerical axial profiles of CO and NO_x.

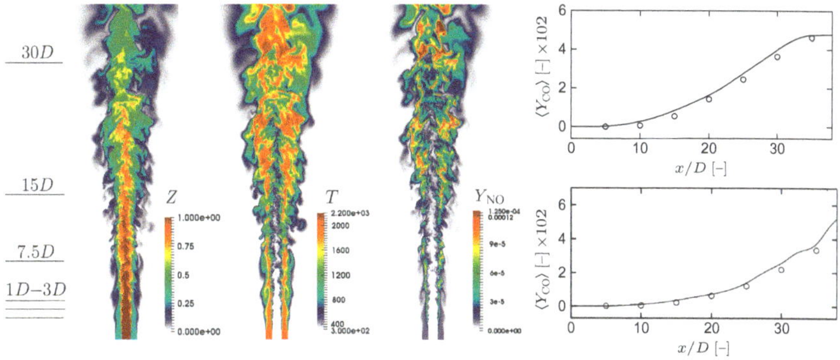

Fig. 50 DNS of the Sandia D flame. *Left*: View of the mixture fraction, temperature, and NO mass fraction. *Right*: Axial profiles of CO and NO_x (*symbols*: exp., *line*: LES) (Jaravel, 2016).

4.2.2 Turbulence–Flame Interaction in LES

Except for the well-stirred reactor burning regime, i.e., high intensity turbulence and slow chemistry, combustion occurs in thin reaction zones below the subgrid, unresolved scale. For example, the flame thickness at atmospheric pressure for kerosene/air combustion is about 0.1–0.5 mm and cannot be resolved on an LES grid. This highly nonlinear process, expressed in the Arrhenius form of the reaction rate, must be therefore modeled. Similar to the modeling of the SGS motion, the objective of a turbulent combustion model in LES is to describe the effect of the SGS mechanisms on the resolved scale filtered turbulent flame. As seen in the previous section, the subgrid turbulence generates flame wrinkling and local high strain and curvature which may significantly alter the inner flame structure, with even a possibility of local flame quenching. Various turbulent combustion models have been developed to evaluate the filtered species reaction rates $\bar{\dot{\omega}}_k$ (see Eq. 103), including the subgrid contribution, from the resolved quantities. Reviews proposed by Veynante and Vervisch (2002), Pitsch (2006), or the textbook of Poinsot and Veynante (2005) give a comprehensive overview of existing combustion models.

Models can be classified into three categories:

- **Statistical approach.** The turbulent flame is here described with the concept of filtered probability density function *pdf* of the species mass fractions and temperature (Colucci et al., 1998; Dopazo, 1994; Jaberi et al., 1999). This allows to write:

$$\langle \dot{\omega}_k \rangle = \int \dot{\omega}_k(Y_k, T) pdf(Y_k, T) dY_k dT \tag{110}$$

The modeling problem is then to determine the joint *pdf* of mass frac tions and temperature. Attempts to find generic forms of this joint *pdf* did not lead to sufficiently accurate results (Landenfeld et al., 2002), and a common simplification is to assume independent statistics of the variables, so that:

$$pdf(Y_k, T) \approx \prod_k pdf(Y_k) pdf(T) \tag{111}$$

This is a very strong hypothesis as in the flame, it is obvious that mass fractions and temperature are not independent. It is then more appropriate to express the chemical source term as a function of the progress variable c and the mixture fraction Z, introduced in Section 3, for which statistical independence is a more reasonable assumption:

$$\langle \dot{\omega}_k \rangle = \int \dot{\omega}_k(c, Z) pdf(c) pdf(Z) dc dZ \tag{112}$$

The source terms $\dot{\omega}_k(c, Z)$ are calculated a priori from representative one-dimensional laminar flames and tabulated for use in the turbulent flame simulation. In many cases the *pdf* of c and Z can have an presumed shape, parameterized with their first-order moments (mean and variance) for which transport equations are written. For example, one very popular function for these *pdfs* is the β-distribution:

$$P(Z) = \frac{Z^{\alpha-1}(1-Z)^{\beta-1}\Gamma(\alpha+\beta)}{\Gamma(\alpha)\Gamma(\beta)} \tag{113}$$

where Γ is the gamma function. The two parameters α and β are related to the mean and variance of the mixture fraction as:

$$\alpha = \widetilde{Z}\gamma \quad \beta = (1-\widetilde{Z})\gamma \tag{114}$$

and

$$\gamma = \frac{\widetilde{Z}(1-\widetilde{Z})}{\widetilde{Z'^2}} - 1 \tag{115}$$

This distribution is well adapted to flames as it reproduces a bimodal distribution for low γ (high variance), characteristic of high segregation between reactants and products, or between fuel and oxidizer, and tends to a Gaussian distribution when γ becomes large (low variance) which describes well-mixed reactants and products.

Another possibility is to solve transport equations for the *pdf* itself (Pope, 1985). Such approach has the great advantage to give direct access to the turbulent reacting term, but requires a closure for mixing. In addition, the high dimensionality of *pdf* transport equations makes them difficult to solve with standard finite-volume or finite-difference techniques, and usually costly Monte-Carlo simulation techniques are employed.

- **Geometrical approach.** The flame is seen as a moving interface between reactants and products, or between fuel and oxidizer, and marking the location of heat release. If the Damköhler number is high enough, i.e., the turbulence does not significantly alter the inner flame structure, a laminar-like flame structure may be attached to the flame surface. The flame front displacement is due to both the turbulent transport and propagation, whereas the laminar flame structure can be described with the generic one-dimensional flames described in Section 3. The flame front is tracked with the flame surface density Σ, defined as the area of flame surface in a given volume. It is possible to write a transport equation for Σ (Boger et al., 1998; Hawkes and Cant, 2000), which again introduces unclosed terms for flame surface creation (by stretch) or destruction (by self-annihilation).

  Then, it is easy to recover the flame heat release or chemical source terms with:

$$\langle \dot{\Omega}_k \rangle = \dot{\Omega}_k \Sigma \tag{116}$$

- **Thickened flame approach.** All the above approaches do not solve chemistry directly in the flow solver, but use one-dimensional laminar flames to describe the local flame structure or global quantities (Gicquel et al., 2000; Maas and Pope, 1992). These flames are usually computed prior to the turbulent flow simulation, and stored in look-up tables as functions of c and Z. However this implies a frozen flame structure, which may not represent transient situations such as ignition or extinction, and cannot adapt to local and changing conditions. Taking into account nonadiabaticity or two-phase flow, for example, is difficult and requires a number of additional parameters which increase the dimension of the look-up tables (Fiorina et al., 2009; Veynante et al., 2008). Finally, it has been shown that the choice of the prototype flame used to create the table is not obvious in complex configurations where the combustion regime is unknown.In the F-TACLES approach

(Fiorina et al., 2010), the look-up table describes an already filtered laminar flame solution and is more consistent with the LES formulation. To avoid flame pretabulation, the artificially thickened flame (TF) model (Colin et al., 2000; Légier et al., 2000) introduces a radically different concept and allows direct integration of chemistry. The principle is to take advantage of premixed flame properties, which scale as $\sqrt{AD}$ for the flame speed and $\sqrt{D/A}$ for the flame thickness. This means that if the diffusivity D and the flame chemical time scale $1/A$ are both multiplied by the same factor $\mathcal{F}$, the flame speed is unchanged, while the flame thickness is increased by the factor $\mathcal{F}$. In that way, the flame may be thickened enough to be resolved on the LES grid, keeping the correct value for the flame consumption rate. Following Butler and O'Rourke (1977), since the reaction rate is still expressed using Arrhenius law, various phenomena can be accounted for without requiring any additional submodels (ignition, flame/wall interactions, heat losses, etc.). The thickening process may be viewed as a simple change of the spatial and temporal variables:

$$x \mapsto \mathcal{F}x \text{ and } t \mapsto \mathcal{F}t \tag{117}$$

The thickened species and temperature source terms are then:

$$\tilde{\dot{\omega}}_k^* = \frac{\widetilde{\dot{\omega}_k}}{\mathcal{F}} \quad \text{and} \quad \tilde{\dot{\omega}}_T^* = \frac{\widetilde{\dot{\omega}_T}}{\mathcal{F}} \tag{118}$$

while the species and temperatures diffusivities become:

$$D^* = \frac{D}{\mathcal{F}} \tag{119}$$

This flame transformation has however an impact on the interaction between the flame and the turbulence, reflected in the change of the Damköhler number Da (see Eq. 92):

$$Da = \frac{\tau_t}{\tau_c} = \frac{l_t}{u'}\frac{S_L^0}{\delta_L^0} \mapsto \frac{l_t}{u'}\frac{S_L^0}{\mathcal{F}\delta_L^0} = \frac{\tau_t}{\mathcal{F}\tau_c} \tag{120}$$

As the Da number is decreased, the flame gets closer to the thick flame regime (see Fig. 51), and vortices may affect the reaction zone. This means in particular that the flame wrinkling is decreased, which directly reflects on the turbulent flame consumption rate. This point has been investigated using DNS by Angelberger et al. (1998) and Colin et al. (2000). To properly recover the unresolved wrinkling effect, the

Fig. 51 The TFLES model: effect of thickening on the flame wrinkling (Poinsot and Veynante, 2005). *From Poinsot T, Veynante D:* Theoretical and numerical combustion. *ISBN 978274663990, elearning.cerfacs.fr.*

so-called efficiency function $\mathcal{E}$ has been proposed to be applied to the filtered reaction rates and the diffusivities as:

$$\tilde{\dot{\omega}}_k^* = \mathcal{E}\frac{\widetilde{\dot{\omega}_k}}{\mathcal{F}} \text{ and } \tilde{\dot{\omega}}_T^* = \mathcal{E}\frac{\widetilde{\dot{\omega}_T}}{\mathcal{F}} \tag{121}$$

$$D^* = \mathcal{E}\mathcal{F}D \tag{122}$$

so that:

$$S_L^* = \mathcal{E}S_L = S_T \tag{123}$$

where S_T is the turbulent flame speed.

The evaluation of the efficiency function is a delicate part of the TFLES model and the topic of ongoing research. It is based on the hypothesis that equilibrium is reached between the turbulent motion and the flame surface at the SGS level, i.e., the turbulence-induced strain and curvature applied to the flame front compensate each other so that there is no creation or destruction of flame surface at the SGS level.

- In the *Colin model*, Colin et al. (2000) proposed an efficiency function $\mathcal{E}_{Colin}$ which compares the flame wrinkling with and without thickening, using the wrinkling factor Ξ:

$$\Xi(\delta_L^0) = 1 + \alpha\Gamma_{Co}\left(\frac{\Delta_e}{\delta_L^0}, \frac{u'_{\Delta_e}}{S_L^0}\right)\frac{u'_{\Delta_e}}{S_L^0} \tag{124}$$

  where α is a constant and Γ_{Co} is a function taking into account the subgrid strain rate depending on the effective filter size Δ_e and the fluctuating subgrid velocity u'_{Δ_e}. Note that α requires an estimation of the integral length scale l_t, which is often unknown and varies in space and time. The efficiency function is then given by:

$$\mathcal{E}_{Colin} = \frac{\Xi(\delta_L^0)}{\Xi(\mathcal{F}\delta_L^0)} \tag{125}$$

Finally, this combustion model is closed by approximating the SGS turbulent velocity u'_{Δ_e} with an operator based on the rotational of the velocity field to remove the dilatation part of the velocity and by estimating the constant α using HIT simulations (Angelberger et al., 1998).

- In the *Charlette model*, Charlette et al. (2002) proposed a power-law model for the wrinkling factor which extends the Colin model:

$$\Xi(\delta_L^0) = \left(1 + \min\left[\frac{\delta_{\Delta_e}}{\delta_L^0}, \Gamma_{Ch}\left(\frac{\delta_{\Delta_e}}{\delta_L^0}, \frac{u'_{\Delta_e}}{S_L^0}, Re_{\Delta_e}\right)\frac{\delta_{\Delta_e}}{\delta_L^0}\right]\right)^{\beta} \tag{126}$$

where β is the model parameter to be specified and Re_{Δ_e} is the turbulent Reynolds at the effective filter size: $Re_{\Delta_e} = S_L^0\delta_L^0/\nu$. It has been recently shown that the parameter β is not constant and that a dynamic procedure could be used to evaluate it (Wang et al., 2011).

In order to tackle nonpremixed configuration where fuel and oxidizer can partially mix before burning, a dynamic thickening method (called DTFLES), where thickening is applied only at the flame location depending on the local mesh resolution, has been developed by Légier et al. (2000). This method is called DTFLES and uses a thickening factor $\mathcal{F}$ that goes from unity in nonreacting zones to $\mathcal{F}_{max}$ in the flame. This is obtained by writing:

$$\mathcal{F} = 1 + (\mathcal{F}_{max} - 1)\mathcal{S} \tag{127}$$

where $\mathcal{S}$ is a sensor depending on the local reaction rate.

5. NUMERICAL METHODS

A numerical setup is the combination of a discretization scheme and a mesh, selected according to the simulation approach (RANS, DNS, or LES). The requirements in terms of accuracy are lowest for RANS and highest for DNS, but the sensitivity of the results to the numerical setup is particularly important for LES. Indeed, the low-order schemes applied on coarse grids in RANS all lead to similar results, whereas DNS minimizes the sensitivity to numerics by using the finest grids and high-order discretization schemes. In the case of LES, the best compromise must be found between numerical scheme accuracy, mesh refinement, and CPU cost.

5.1 Discretization Schemes for LES

LES most commonly uses second-/third-order finite-difference, finite-volume, or finite-element methods in order to discretize the derivative terms in the Navier–Stokes equations (Hesthaven and Warburton, 2008). Finite-volume methods are the most popular, as they are well suited to the unstructured meshes required by complex geometries of the practical applications targeted by LES (Hirsch, 2007). Second-/third-order schemes have the great advantage of having a compact stencil, i.e., using values at two neighboring cells at most to compute fluxes. They also generate significant numerical dissipation, which tends to damp out spurious oscillations, thereby enhancing robustness, but at the cost of lower accuracy. The discretization error has a direct impact on the flow solution, as illustrated in Fig. 52 showing the convection of a scalar by a second-order (left) and a third-order scheme (right).

Using finer grids allows to decrease the dissipation and dispersion errors, but it also becomes rapidly computationally too expensive, and at some point it may be more efficient to use high-order methods (greater than second/third order) on a coarser grid (Kravchenko and Moin, 1997).

A current topic of research is the use in LES of discontinuous spectral element methods (DSEMs), such as discontinuous Galerkin method

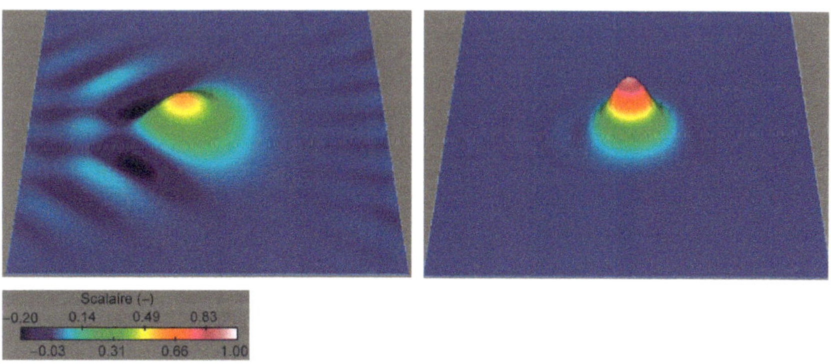

Fig. 52 Convection of a scalar bump: solution after two turnover times. *Left*: Second-order Lax–Wendroff scheme. *Right*: Second-order TTGC scheme (Colin and Rudgyard, 2000).

(Cockburn et al., 1990; Hesthaven and Warburton, 2008; Karniadakis and Sherwin, 1999), or more recently, high-order spectral difference (Jameson, 2010; Kopriva, 1996; Sun et al., 2007) (SD) and flux reconstruction (Huynh, 2007, 2009; Vincent et al., 2010) schemes. Thanks to a polynomial interpolation used to approximate the solution within the element, these last methods are able to reach an arbitrary order of accuracy, still using a compact stencil on unstructured or hybrid grids. In addition, they are particularly efficient in parallel environments. Compared to standard high-order methods, DSEM and SD methods are less sensitive to mesh quality and can handle large cell aspect ratio (typically of 2) and skewed elements. These methods allow both spatial refinement (h-refinement) and localized increase in accuracy (p-refinement, i.e., increase the order of the polynomial interpolation). The h-refinement is useful in a region where a discontinuity appears, while the p-refinement is generally considered in regions where the flow is smooth and the accuracy must be increased. Many results were presented in the literature regarding hp-adaptation for steady solutions of the Navier–Stokes equations (Hartmann and Houston, 2010).

Time discretization must be also at least second or third order and can be explicit or implicit. As combustion problems are usually associated to low-Mach number flows, implicit time integration is appealing to allow for large time steps. However, combustion chemistry introduces very small characteristic times, which impose small time steps. In addition, thermo-acoustic instabilities require the computation of acoustics, i.e., waves traveling at the speed of sound, which again induces very small time steps. As a consequence, explicit time integration keeps significant advantages and is often used in LES of combustion.

5.2 Mesh for LES

The mesh has a particular importance in LES as it is in practice the low-pass filter that defines the resolved equations. From the theory and concept of LES, the mesh size should ideally correspond to the cutoff scale chosen anywhere in the dissipation zone of the turbulent energy spectrum. The problem is that in real geometries, turbulence is far from being homogeneous and the cutoff scale is local and usually unknown. There is therefore no other choice than performing mesh optimization and convergence, i.e., improving and testing the mesh until the mean solution becomes sufficiently independent of it. This requires a strong expertise and know-how in turbulent flows and mesh generation techniques and can be a very lengthy process. In addition, wall flows require specific effort, and different cell sizes will be

Fig. 53 Capture of miscible fluids' interface: solution (*left*) and adapted mesh (*right*) (Alauzet et al., 2007).

used depending if the boundary layer is resolved or not. Typically about five different meshes will be produced before obtaining a satisfactory result.

This however does not guarantee that the final mesh is the best possible one, and an important current topic of research is the development of algorithms for the automatic mesh optimization, based on predefined criteria based on either geometrical properties of the mesh elements or numerical solution of the problem. The techniques consist in node insertion, deletion, or relocation, as well as edge swapping, i.e., reconnection of grid points to modify element faces (Falese, 2013). An example is given in Fig. 53, from Alauzet et al. (2007), showing how the mesh can be refined to capture a sharp interface.

5.3 High-Performance Computing

HPC is essential for LES and has significantly contributed to its success. The capability to efficiently use a large number of processors on parallel computers allows to compute large size problems, which increases accuracy in two ways: the more refined meshes better represent small-scale phenomena and reduces the impact of modeling, and the inclusion of more physics better describes the complexity of turbulent combustion. It is therefore crucial to develop codes with good scalability, as examplified in Fig. 54. The target is to obtain good performance in terms of strong scaling, i.e., a good speedup when increasing the number of processors but keeping a fixed grid size, compared to weak scaling where the speedup is maintained only if the grid size is increased.

Parallel algorithms are usually based on the domain decomposition and use MPI instructions to distribute the tasks over the processors. To be efficient, the domain decomposition must maximize the volume-to-surface ratio of each subdomain and minimize the communication cost between processors. It becomes delicate when constrained by more than one criterion, for example, in two-phase flows where it must take into account the number of nodes and the number of droplets for load balancing.

Recent architectures of parallel computers have allowed to introduce multithreading (using OpenMP) as a second level of parallelism, increasing

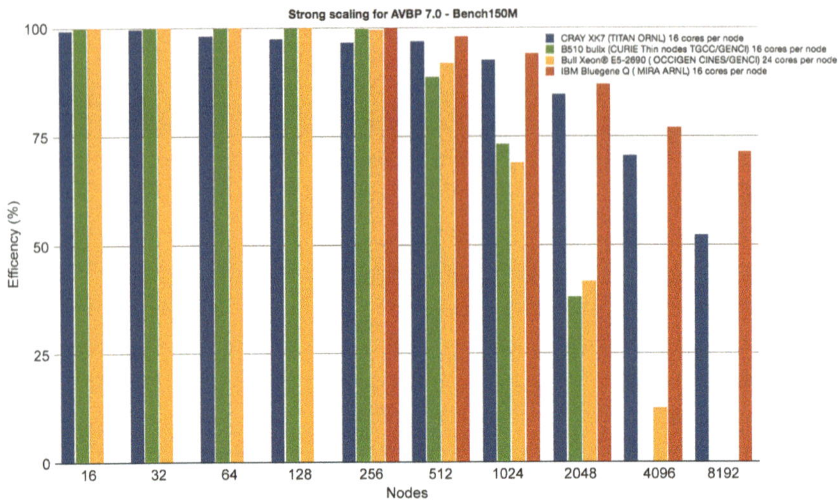

Fig. 54 Efficiency (100% means perfect scaling) of AVBP on various parallel computers and a 150-million point grid. Nodes contain different numbers of processors as indicated in the legend.

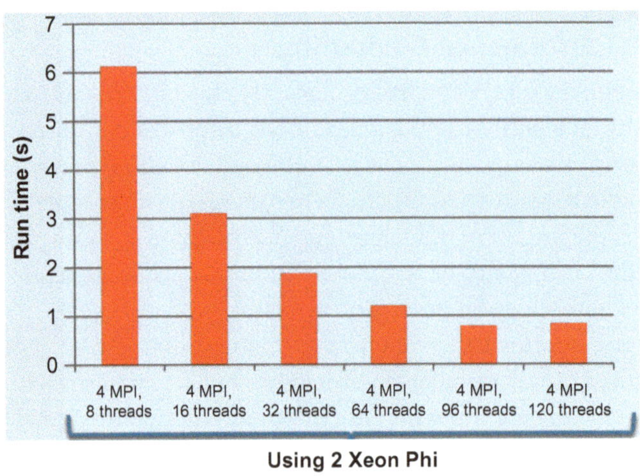

Fig. 55 Impact of multithreading on code performance, using two Intel Xeon Phi processors. *Source: Courtesy of G. Staffelbach, CERFACS.*

significantly the parallel efficiency as shown in Fig. 55. Graphic processing units also offer an alternative model for parallel computing, but require a deep modification of the programming and proved to be quite difficult in flow solvers. Another current challenge is the efficient parallel computing of coupled solvers, for the simulation of multiphysics problems.

6. EXAMPLES OF APPLICATION OF LES TO ENGINES

6.1 Validation and Performance of LES: The PRECCINSTA Burner

The PRECCINSTA experiment was built at DLR (Meier et al., 2007; Weigand et al., 2006) in the framework of a European Project and was computed by many groups for validation purposes (Albouze et al., 2009; Fiorina et al., 2010; Franzelli et al., 2012; Galpin et al., 2008; Moureau et al., 2011; Roux et al., 2005). This methane/air swirled combustor was derived from an industrial engine by Turbomeca and was designed initially to study combustion instabilities. The availability of extensive measurements of velocity, temperature, and chemical species fields has made the PRECCINSTA burner a reference experiment in the scientific community and is shown here to illustrate the standard state of the art of LES of such semiindustrial configurations.

The geometry shown in Fig. 56 includes a plenum, where dry air at ambient temperature is injected through one large hole, an injector, where the air flow is swirled by 12 radial veins, a combustion chamber with square cross subsection ($85 \times 85\ mm^2$) equipped with 1.5-mm-thick quartz walls to enable optical measurements and a converging duct which connects the combustor to the atmosphere. Methane is injected into the air flow through 12 small holes (one for each vane) of 1 mm diameter within the radial swirler. Laser Doppler velocimetry and laser Raman scattering were used

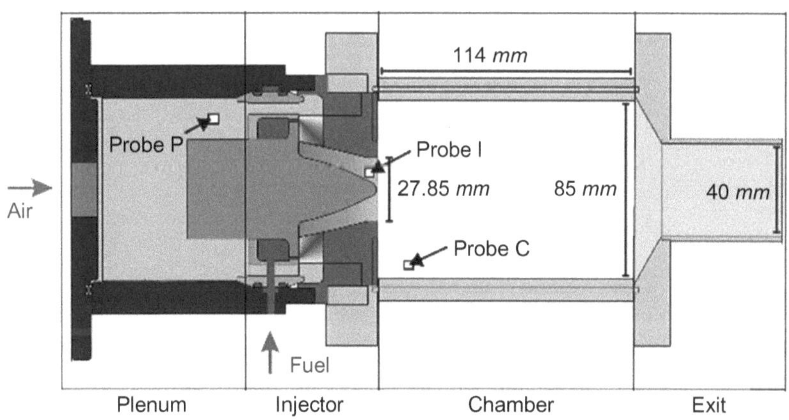

Fig. 56 Schematic of the experimental burner design (Meier et al., 2007; Weigand et al., 2006).

to measure velocity and major species. The burner operating conditions correspond to a quiet flame stabilized at the nozzle exit. Results shown here were obtained by Roux et al. (2005) and Franzelli et al. (2012) with the compressible LES code AVBP (Gourdain et al., 2009a,b), which solves the Navier–Stokes equations on hybrid (structured and unstructured) grids with real thermochemistry. Numerical integration was performed with a third-order Taylor–Galerkin weighted residual central distribution scheme (Colin, 2000; Colin and Rudgyard, 2000; Moureau et al., 2005). The turbulence–combustion interaction was modeled by the TF model introduced in Section 3 (Colin et al., 2000). A two-step reduced scheme for laminar premixed methane/air flames was applied (Franzelli et al., 2010). The inlets and the outlet are described by Navier–Stokes characteristic boundary conditions (Poinsot and Lele, 1992).

The full geometry is computed including the 12 holes located upstream of the swirler, using an unstructured mesh containing 5 millions tetrahedra (Fig. 57) and refined inside the swirler veins to capture mixing, down to a cell size of 0.2 mm (Fig. 58).

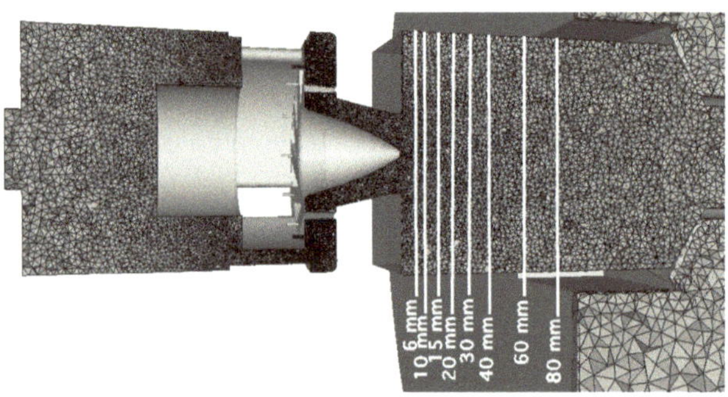

Fig. 57 Computational half-domain mesh (Franzelli et al., 2012; Roux et al., 2005).

Fig. 58 Detail of the 12 computational holes upstream of the swirler for the methane injection: instantaneous iso-surface of methane mass fraction equal to 0.5 (Franzelli et al., 2012; Roux et al., 2005).

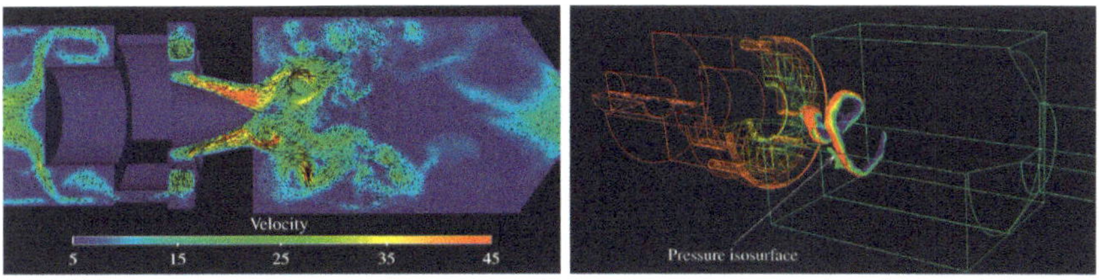

Fig. 59 LES of the PRECCINSTA burner (Franzelli et al., 2012; Roux et al., 2005). *Left*: Cut of the velocity magnitude and velocity vectors. *Right*: The precessing vortex core (PVC) identified with a pressure iso-surface.

Fig. 59 shows an instantaneous view of the flow obtained with LES, illustrating the typical swirled flow of aeronautical burners. Inside the combustion chamber, the jet opens and rapidly transitions to turbulence. The confined geometry generates CRZs, while the opening jet creates an inner recirculation zone (IRZ) where the flame can stabilize. A particular feature is the precessing vortex core (PVC) resulting from a vortex breakdown process well defined by Hall (1972). This unsteady flow structure plays an important role in burner stability and is ideally captured with LES. Figs. 60–62 show mean and fluctuation velocity profiles at five axial locations close to the injector. The CRZ and IRZ are again clearly visible with the zones of negative axial velocity, while the jet opening is marked with the location of velocity peaks. Measurements are also reported for comparison, showing a very good agreement, not only for mean quantities but also for fluctuations, which is a major contribution of LES.

Fig. 63 now compares the numerical and experimental mean and fluctuation temperature profiles at five different axial locations. The solid lines and dashed lines correspond to simulations run with and without a perfect premixing assumption, respectively. The agreement is again very good, with only some overestimation of temperature near the walls since wall heat losses and radiation effects were not taken into account. When air and methane are injected separately, the temperature fluctuations are better described in the reaction zone. The mean and fluctuation profiles of CO_2 provide similar levels of agreement with the experiment (Fig. 64), with again a slight improvement of the CO_2 fluctuations when injecting methane and air separately. All other species are correctly described with a similar quality.

The above simulations were run few years ago and were limited by the computing power available at that time. Despite the relatively coarse mesh,

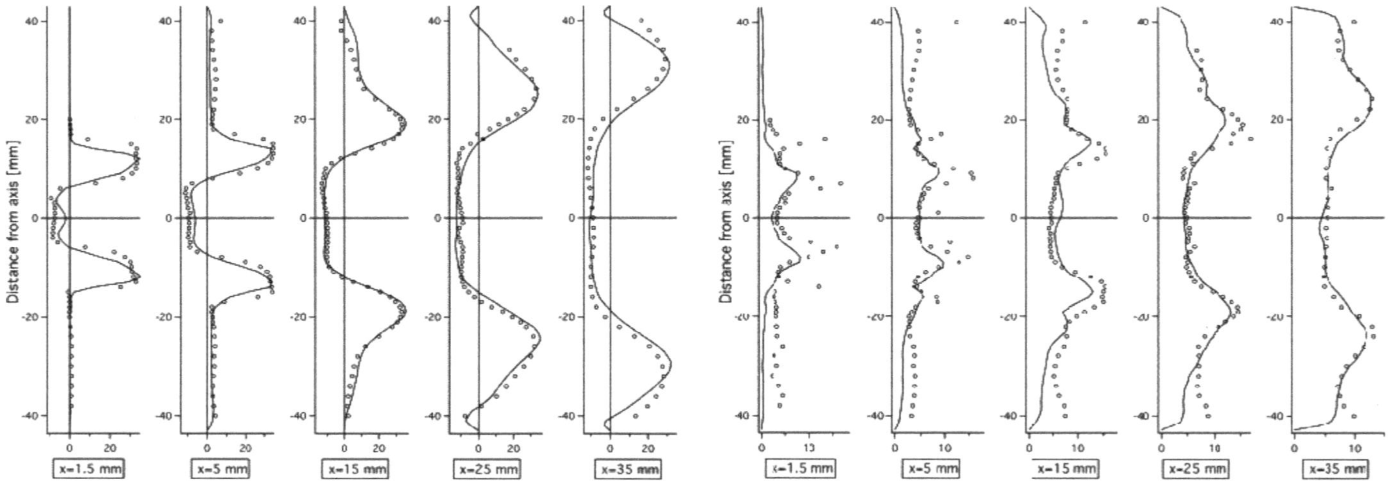

Fig. 60 Comparison of LES (*lines*) and experiment (*symbols*) in the PRECCINSTA burner (Franzelli et al., 2012; Roux et al., 2005). *Left*: Mean axial velocity. *Right*: rms axial velocity.

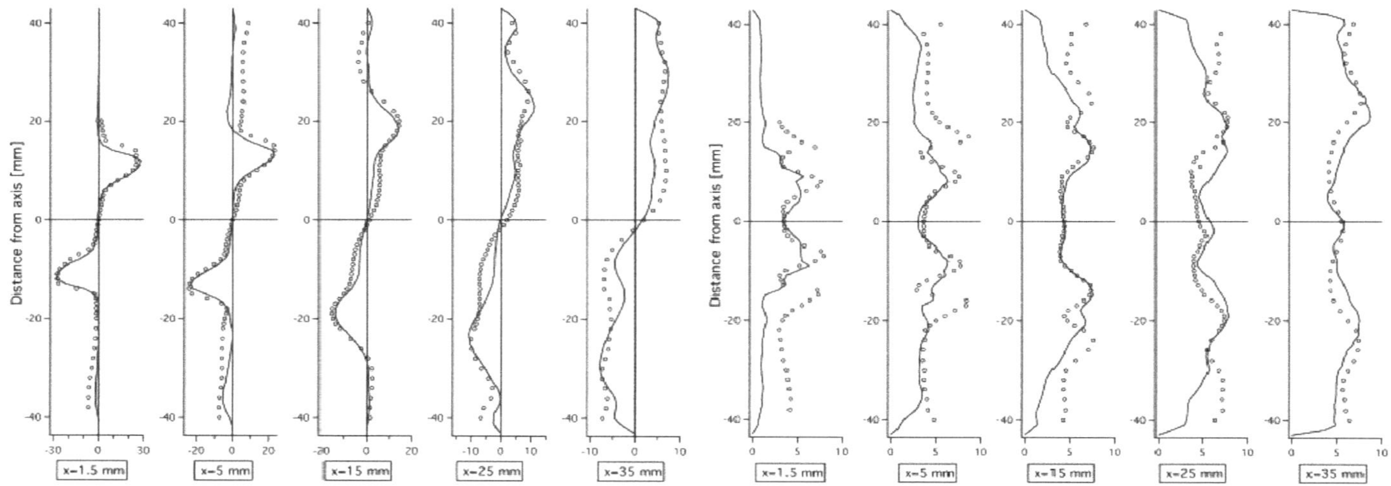

Fig. 61 Comparison of LES (*lines*) and experiment (*symbols*) in the PRECCINSTA burner (Franzelli et al., 2012; Roux et al., 2005). *Left*: Mean tangential velocity. *Right*: rms tangential velocity.

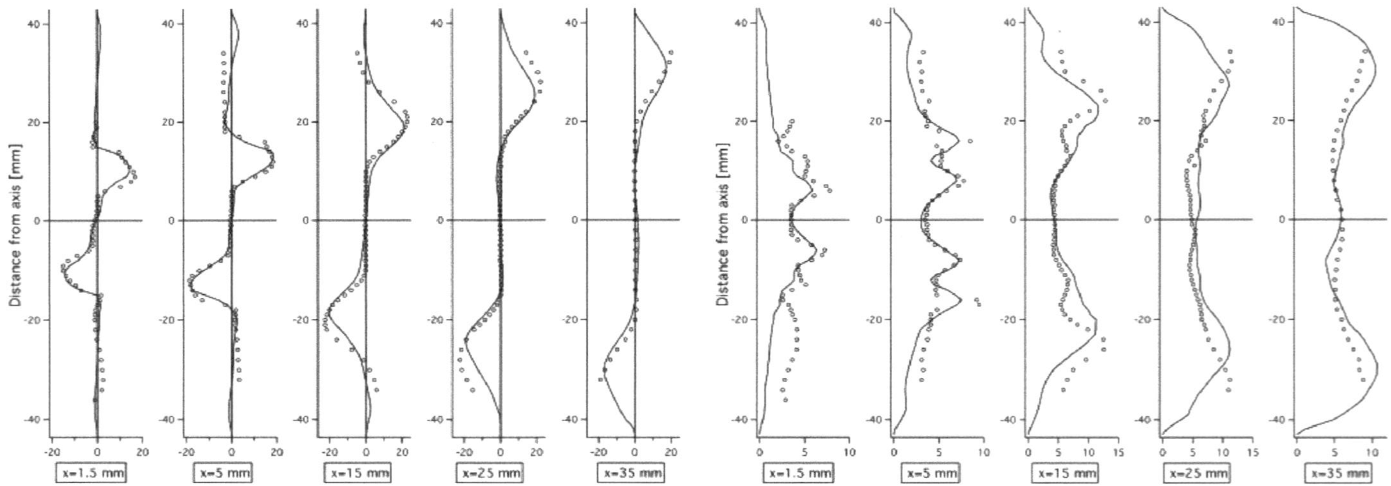

Fig. 62 Comparison of LES (*lines*) and experiment (*symbols*) in the PRECCINSTA burner (Franzelli et al., 2012; Roux et al., 2005). *Left*: Mean radial velocity. *Right*: rms radial velocity.

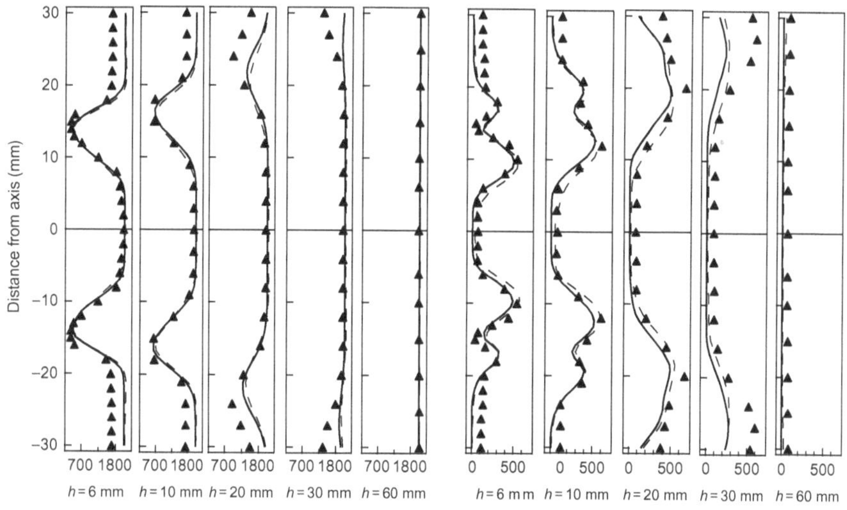

Fig. 63 Comparison of LES (*lines*) and experiment (*symbols*) in the PRECCINSTA burner (Franzelli et al., 2012; Roux et al., 2005). *Left*: Mean temperature. *Right*: rms temperature. Two sets of numerical data are plotted: perfect premixed (*solid line*) and nonperfect premixed simulation (*dashed line*).

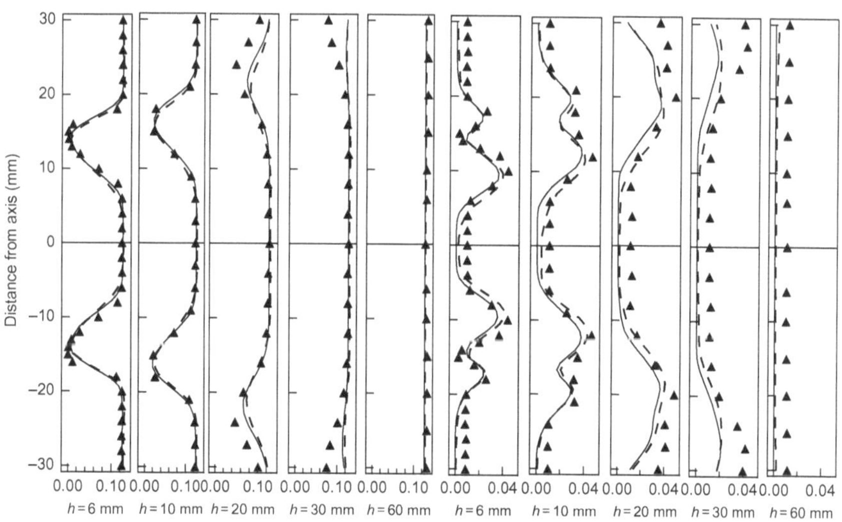

Fig. 64 Comparison of LES (*lines*) and experiment (*symbols*) in the PRECCINSTA burner (Franzelli et al., 2012; Roux et al., 2005). *Left*: Mean CO_2 mass fraction. *Right*: rms CO_2 mass fraction. Two sets of numerical data are plotted: perfect premixed (*solid line*) and nonperfect premixed simulation (*dashed line*).

the results are very good and are shown here as a standard for LES, which should be expected when computing similar flow on small meshes. Although current LES now run on much finer grids, it would not be very useful on this particular case. This was confirmed by the DNS of Moureau et al. (2011) who was able to obtain higher order statistics but did not significantly improve the first-order ones.

6.2 Simulation of Cyclic Variation in ICE Engine

The capacity of LES to predict piston engine behaviors, including cycle-to-cycle variations (CCVs), has been shown in numerous studies (Celik et al., 2001; Goryntsev et al., 2007; Haworth, 1999; Richard et al., 2007; Thobois et al., 2007; Toledo et al., 2007; Vermorel et al., 2007). A recent example is shown here, where the aim was to demonstrate that LES can be used to evaluate the stability (in terms of CCVs) of a given design and/or operating point (Granet et al., 2012). The simulated configuration is the experiment of Lacour et al. (2009), where many operating points were acquired with or without combustion, and with low or high CCV levels. It is a single cylinder, four valves SI engine with a flat piston, and a compression ratio of 9.9. The experimental setup is shown in Fig. 65. Air is introduced and mixed with gaseous propane in a mixing plenum sufficiently large to obtain an homogeneous mixture. At the engine exhaust, gases are tranquilized in another plenum. Flame arrestors are placed for safety reasons between the engine and the plenums.

CCVs are best described with the evolution of in-cylinder pressure, which is very sensitive to boundary conditions. To avoid errors linked to uncertain or not well-controlled inlet and outlet boundaries, the LES

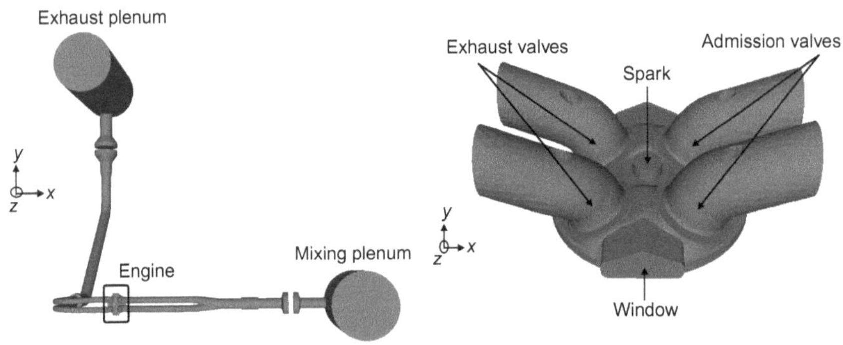

Fig. 65 *Left*: Configuration of the ICE engine test bench. *Right*: Zoom on the single-cylinder engine (Granet, 2011).

methodology includes the whole test bench, including the plenum and pipes of the experiment. The engine itself is also exactly reproduced.

This complex geometry is meshed with tetrahedra, which gives the best mesh quality. The resolution is of the order of 0.8 mm in the cylinder and is locally refined around the valves. One particularity of ICE engines is the moving geometry, with the valve opening and closing. This is mostly handled with mesh deformation following the moving surfaces, using the adaptative mesh refinement technique. However, this may lead locally to cells having a very small volume or bad aspect ratio, which deteriorates the simulation accuracy and performance. This happens in particular around the valves when they are closing. To avoid this adverse effect, the solution is interpolated on a new mesh as soon as the mesh quality has been too much decreased by deformation. For the simulations presented in this section, typically 40 meshes were necessary to describe a cycle. An example of mesh is shown in Fig. 66. Note that about 80% of the cells are inside the cylinder and around the valves, which means that the added cost of including the plenums and pipes is of the order of 20%.

To reach sufficient statistical convergence, a minimum of 25 cycles must be simulated (Goryntsev et al., 2007; Sagaut and Deck, 2009; Vermorel et al., 2009). Simulations were run with the Smagorinsky model (1963) for SGS stresses, and a two-step combustion chemistry.

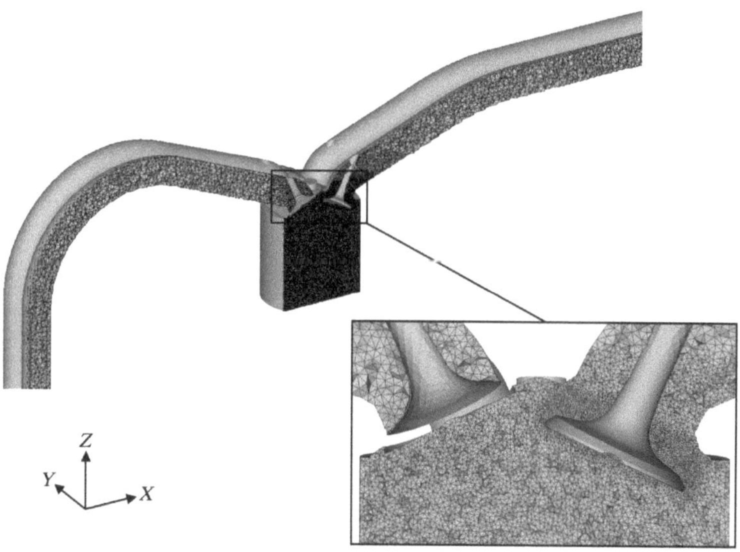

Fig. 66 Mesh of the engine bench (Enaux et al., 2011).

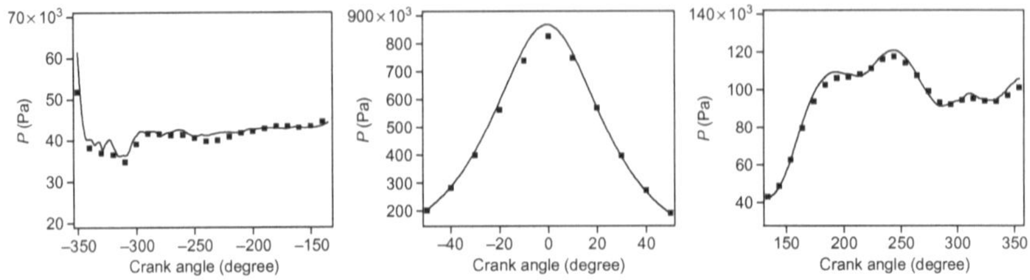

Fig. 67 Evolution of the mean in-cylinder pressure during a cycle. Experiment (*symbols*) and LES (*solid line*) (Enaux et al., 2011). (A) From − 355 CA to closing of admission valves. (B) Compression and expansion. (C) From exhaust valves opening to 365 CA.

The evolution of the mean in-cylinder pressure during one cycle is shown in Fig. 67 for a nonreacting case. The comparison with experiment results shows an excellent agreement and demonstrates that LES is able to capture the complex flow dynamics in the engine cylinder.

The flow motion inside the cylinder is essentially driven by the tumble. Fig. 68 compares the mean velocity fields of LES and experimental PIV for the axial component at 4 crank angles (CA). The agreement is again excellent and shows how well the LES captures the unsteady flow dynamics. The mean pressure evolution is shown in Fig. 69 for two different cases: one reference stable case, and one N_2-diluted unstable case. The statistical envelope corresponding to 95% of probability is also shown. It can be seen that if the mean pressure is very similar for both cases, the statistical deviation is much higher in the unstable case. LES recovers remarkably well the pressure statistics, even with a smaller number of cycles. This cycle-to-cycle variation is illustrated in Fig. 70 for both cases. The stable case is very different from the unstable case, with a much weaker dispersion in both experiment and LES and a maximum pressure in the range of 17–22 bars, while the unstable case varies between 12 and 22 bars. This strong cyclic variation is mostly due to turbulence and may even lead to incomplete combustion, which is the source of severe pollutant emission.

The CCV of the cylinder pressure is linked to the turbulent combustion, as illustrated in Figs 71 and 72 with an iso-surface of heat release rate colored with the velocity magnitude for nine consecutive cycles at a fixed CA. In the stable case, combustion does not much vary from one cycle to the other, the flame always starting from the exhaust side (left side in the image) due to the tumble motion. For the unstable diluted case (Fig. 72), combustion effectively differs between cycles. Cycles 12 (fast burning, $P_{max} =$ 20 bars) and 16 (slow burning, $P_{max} = 14$ bars) develop flames of different sizes and at different locations in the chamber.

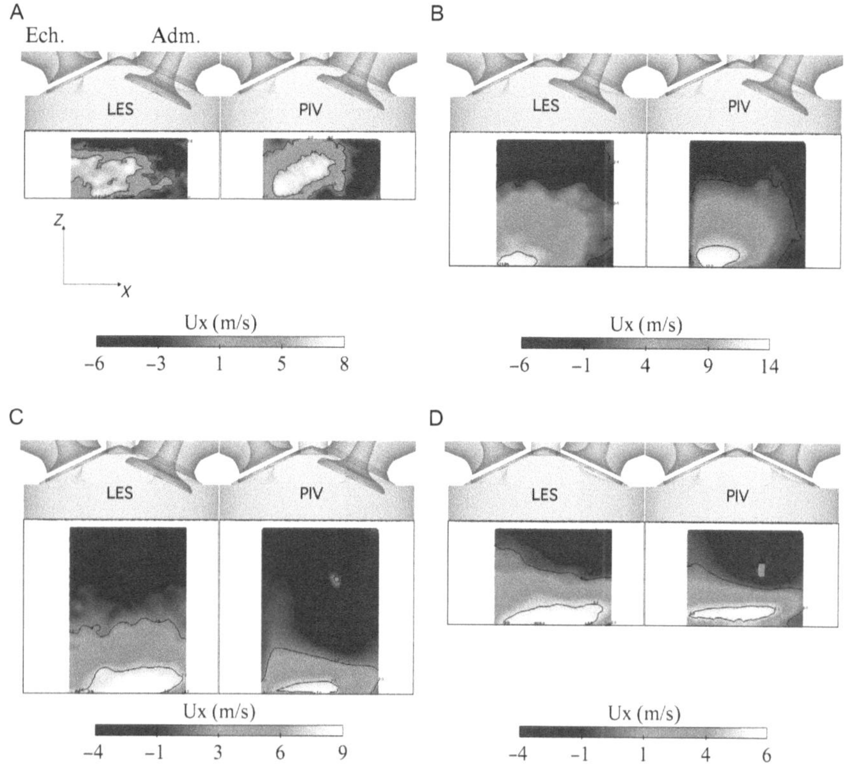

Fig. 68 Comparison of mean axial velocity fields from LES (*left*) and PIV (*right*) at 4 CA (Enaux et al., 2011). (A) 280 DV avant PMH. (B) 240 DV avant PMH. (C) 180 DV avant PMH. (D) 100 DV avant PMH.

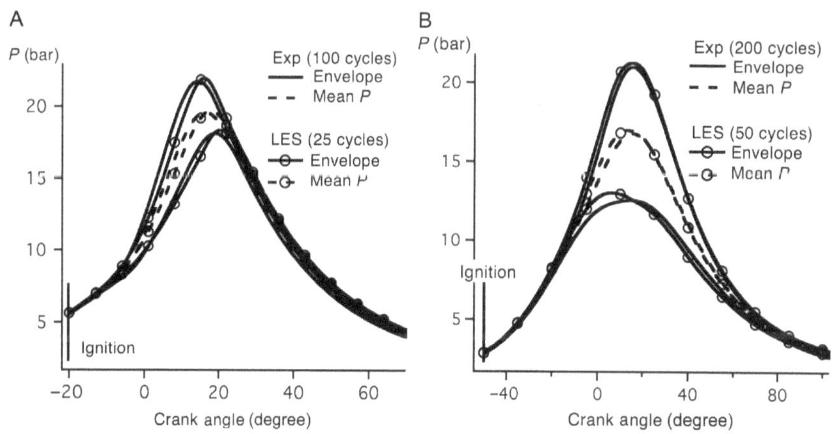

Fig. 69 Comparison of the mean and statistical envelope of the cylinder pressure, experiment, and LES (Granet, 2011; Granet et al., 2012). (A) Stable case. (B) Diluted unstable case.

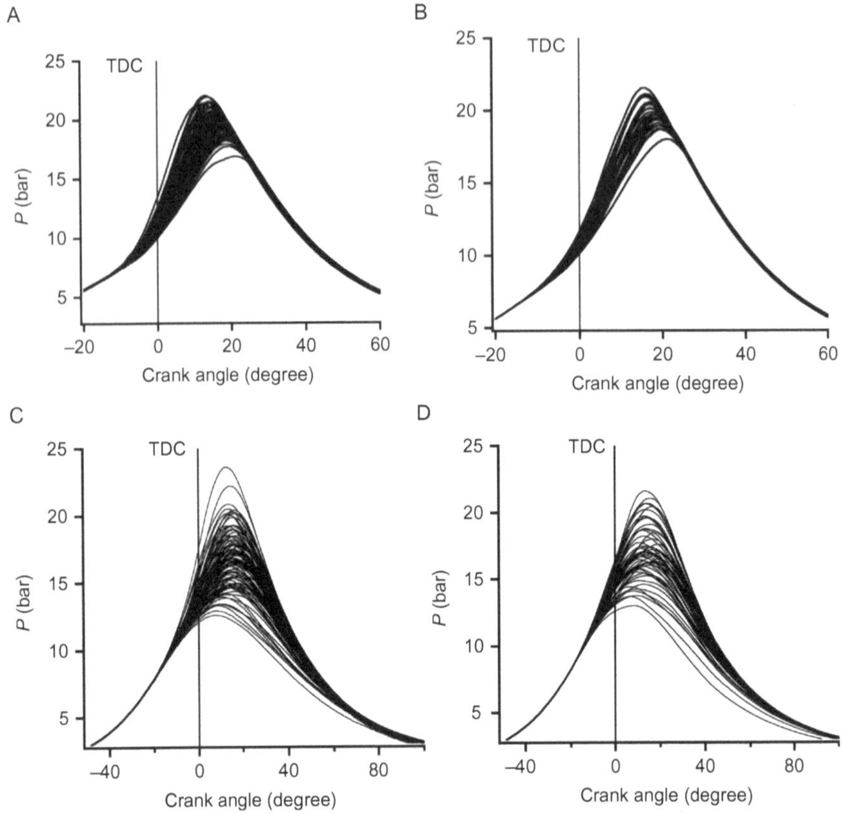

Fig. 70 Comparison of cycle-to-cycle cylinder pressure evolution, experiment and LES (Granet, 2011; Granet et al., 2012). (A) 100 cycles of the stable case, experiment. (B) 25 cycles of the stable case, LES. (C) 100 cycles of the diluted unstable case, experiment. (D) 50 cycles of the diluted unstable case, LES.

6.3 Ignition of an Aeronautical Engine

Ignition is a major safety issue for aeronautical burners. In particular, high-altitude relight capability is required for engine certification and is difficult to guarantee due to low pressure and temperature conditions. Ignition also plays an important role for engine design, as it drives the size of the combustor primary zone as well as the number of injection systems. Indeed, as described in Lefebvre (1999), ignition is triggered by one or two spark plugs, which locally create flame kernels. If local conditions are favorable (Kelley et al., 2009), the kernel transitions to a propagating flame that ignites the first injector, and subsequently propagates to its neighbors. This last

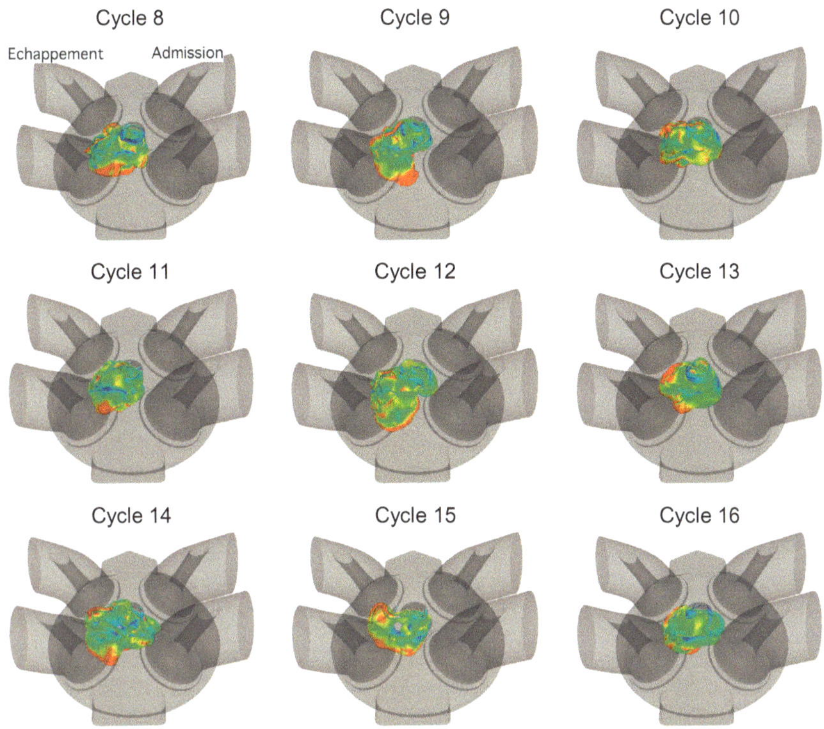

Fig. 71 Flame visualization by an iso-surface of heat release rate *colored* with the velocity magnitude (*blue* 0 m/s and *red* > 10 m/s) at 15 CA after ignition for 9 consecutive cycles of the stable case (Granet, 2011; Granet et al., 2012).

phase is critical and is favored by short distance between consecutive injectors, which implies more injectors, i.e., increase of manufacturing and maintenance cost as well as increased weight of the combustor. A key aspect of engine design is therefore to find the maximum distance between injectors that still guarantee a reliable and fast ignition.

If LES has proven to capture the stochastic nature of ignition in the first instants (Ahmed and Mastorakos, 2006; Ahmed et al., 2007; Mastorakos, 2009), the construction of ignition probability requires long series of runs which are much too demanding in terms of computational cost (Esclapez et al., 2015). LES of this phenomenon is used instead to derive faster predictive models based on one nonreacting LES solution (Esclapez, 2015; Eyssartier et al., 2013; Neophytou et al., 2012).

The subsequent light-around phase requires the simulation of the full ignition sequence, as evidenced by the pioneering work of Boileau et al.

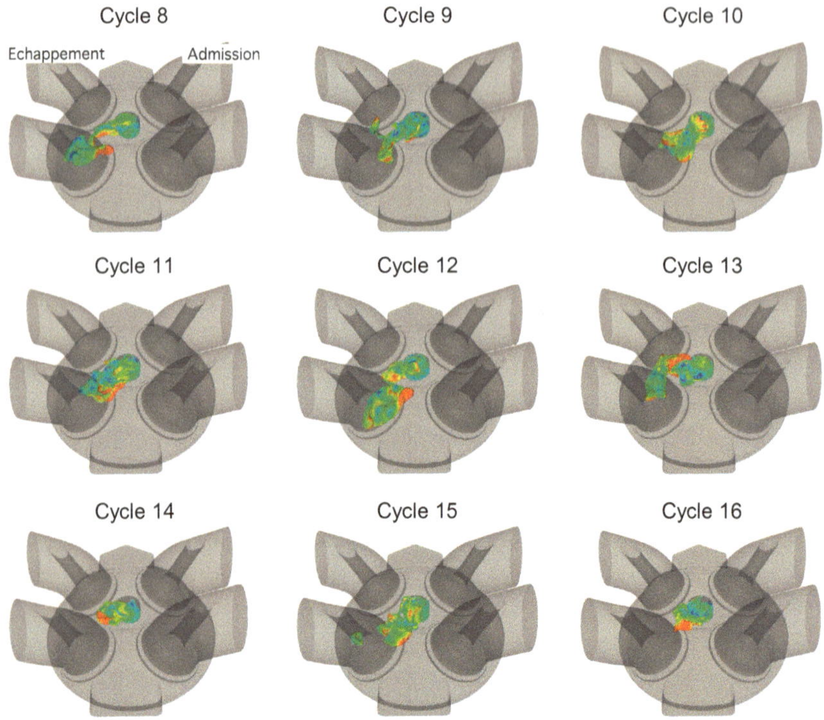

Fig. 72 Flame visualization by an iso-surface of heat release rate *colored* with the velocity magnitude (*blue* 0 m/s and *red* > 10 m/s) at 35 CA after ignition for 9 consecutive cycles of the unstable diluted case (Granet, 2011; Granet et al., 2012).

(2008). To observe this phenomenon, Bourgouin et al. (2013) proposed a propane/air annular burner fitted with 16 swirled injectors and equipped with transparent chamber walls for a full optical access to the flame (Fig. 73, left). Philip et al. (2015) performed the LES of the ignition of this burner with two conceptually different turbulent combustion models, namely the filtered tabulated chemistry (F-TACLES) approach (Fiorina et al., 2010; Vicquelin et al., 2009) and the flame thickening model (TFLES) (Colin et al., 2000), previously presented in Section 3. Their work was the first to present a direct comparison between numerical and experimental results of a full burner ignition. The simulation required a very large mesh of 310 million tetrahedral cells, refined in the regions of high velocity gradients and intense turbulence (Fig. 73, right).

Fig. 74 shows a direct comparison between the TFLES calculation, represented by an iso-surface of temperature colored by axial velocity, and the experimental images showing light intensity emitted by the flame. The

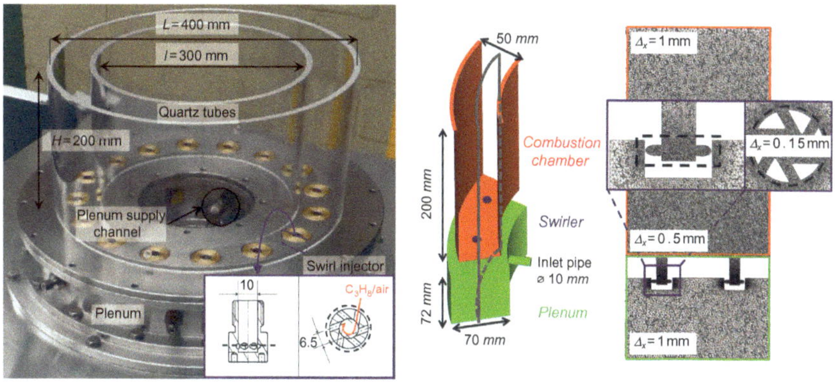

Fig. 73 *Left*: Direct view of the MICCA combustor. A sketch of the swirler appears in the bottom right-hand corner. *Right*: Zoom on a two-sector domain and its matching cylindrical mesh. Δ_x corresponds to the size of the cell (Philip et al., 2015).

swirled jets generated by the injectors are also represented by an iso-surface of the velocity field. After a slight delay occurring at the first instants in the simulation, the flame shapes at the largest scales are very well reproduced by LES. Five phases can be identified. First, the deposit of energy by the spark plug produces a small flame kernel that expands rapidly in the fresh propane/air mixture (time 2 ms); then the flame front propagates to the surrounding burners in the form of an arch, which opens when reaching the upper exit of the chamber (time 10 ms); at this stage a two-front propagation takes over, leading to a burner-to-burner propagation (times 15.5 and 32.5 ms), until they join and merge at the opposite of the ignitor location (time 47.5 ms). Finally the burnt gases evacuate the chamber and a steady burning flame is obtained.

The five phases of the ignition sequence also appear in Fig. 75, where the integrated experimental signal is compared to the numerical reaction rates integrated over the chamber volume for the two combustion models. Note that some fresh gas had time to exit the chamber before ignition in the experiment, and reacts outside the chamber producing some error on the integrated signal, particularly in the last phases of ignition. In the initial and arch-like phases, both the simulations and the experiment feature a sharp increase of the reaction rate due to the hot gas expansion which is maximal when the flame front closes. In the following phase, the reaction rates continue to increase with lower slopes. The quantitative agreement between experiment and LES is overall very good and confirms the capacity of LES to predict such phenomena.

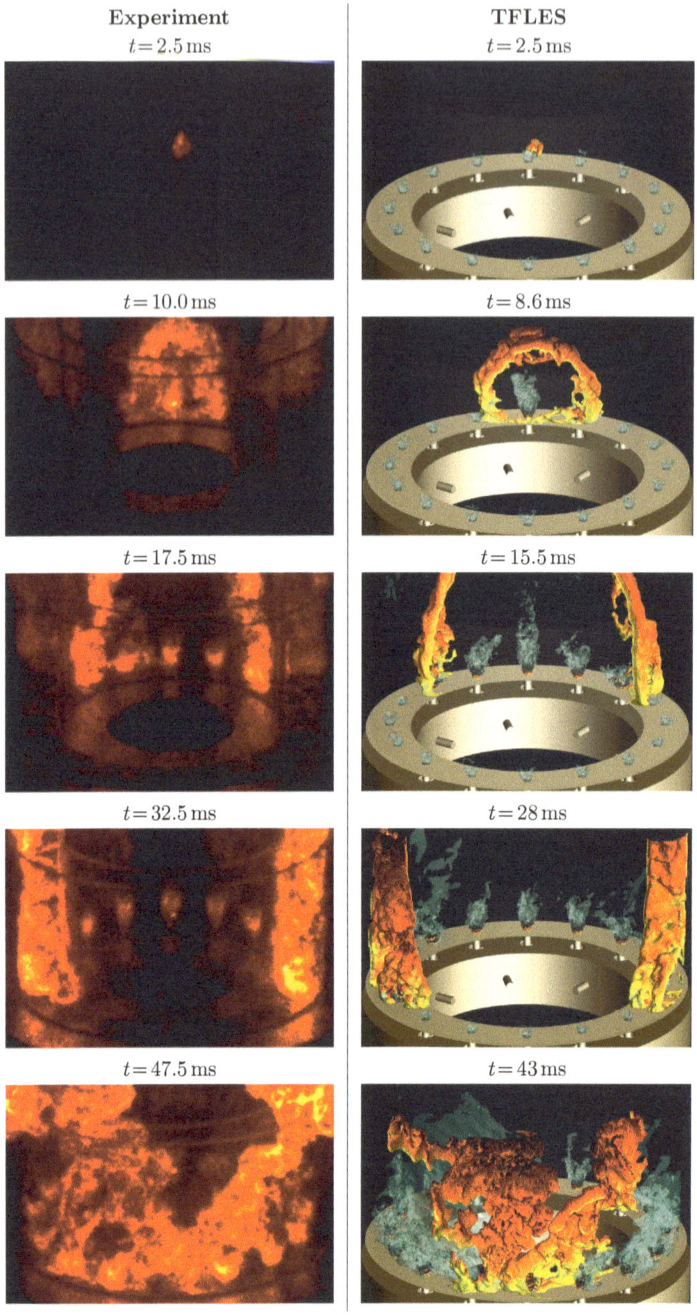

Fig. 74 Five instants of an ignition sequence. *Left*: Experimental images showing light intensity emitted by the flame during the process of light round and represented in *false colors* to improve visualization. *Right*: TFLES simulation. The flame fronts are represented by an iso-surface of temperature $\widetilde{T} = 1781$K for TFLES, *colored* by axial velocity (*light yellow*: -30 m/s; *black*: $+15$ m/s). *Blue iso-surfaces* correspond to the velocity field $U = 25$ m/s (Philip et al., 2015).

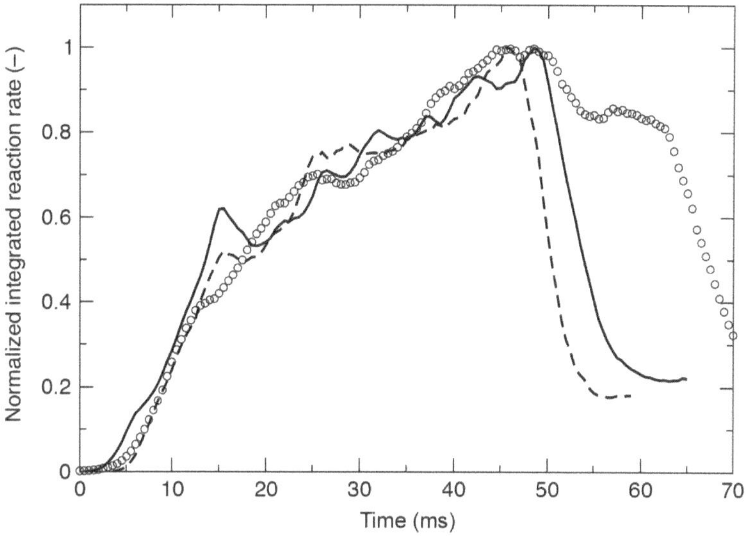

Fig. 75 Time evolution of the integrated direct light emission (experiment) and reaction rate (simulations): comparison of the three signals normalized by their respective maximal value. *Symbols*, experiment; *solid line*, F-TACLES; *dashed line*: TFLES (Philip et al., 2015).

The methodology has been then applied to a real combustor by Esclapez (2015) where it was important to characterize relight at high-altitude conditions with a low-NO_x injection system. Fig. 76 illustrates the obtained ignition dynamics, with the flame propagating from one burner to the other. In addition to the increased complexity of the geometry of the industrial application, this simulation also includes two important key features:

- the burner is fed with liquid fuel, highlighting the effect of two-phase flow on ignition.
- the presence of a secondary air flow significantly modifies the features of the turbulent flow and its interactions with the flame kernel and the propagating flame.

6.4 Combustion Instabilities in a Rocket Engine

Thermoacoustic instabilities in rocket engines may lead to severe damage, system failure, and in extreme cases spectacular explosions of the propulsion system (Culick, 1987; Oefelein and Yang, 1993; Young, 1995). The understanding of the underlying mechanism is attributed to Rayleigh (1878) who related it to the synchronization of pressure and heat release rate fluctuations,

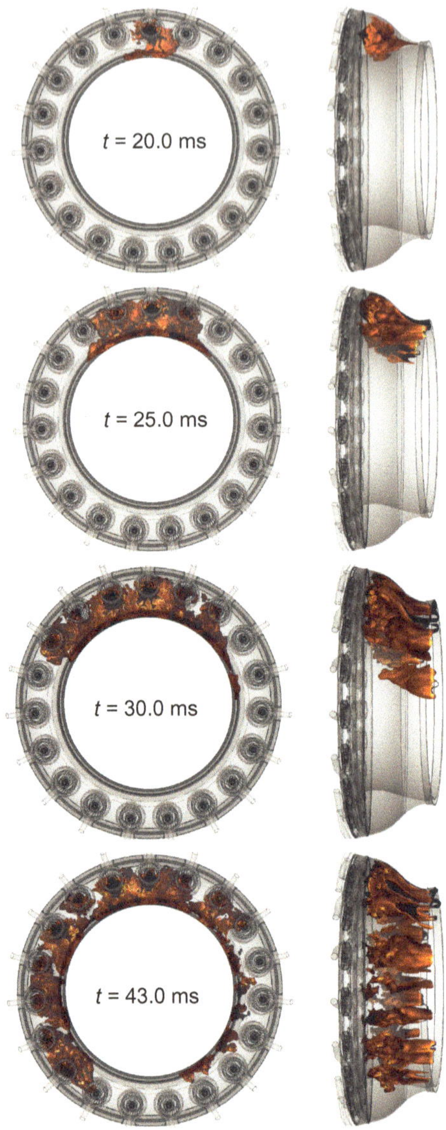

Fig. 76 Snapshots of the ignition sequence of the annular low-NO_x combustor LEMCOTEC (Esclapez, 2015).

itself linked with the combustion delay. This led to the sensitive time lag (STL) theory (Crocco, 1951, 1952, 1965; Marble, 1953; Summerfield, 1951; Tsien, 1952) where the heat release rate evolution with respect to pressure is expressed in terms of an interaction index n and a time lag τ.

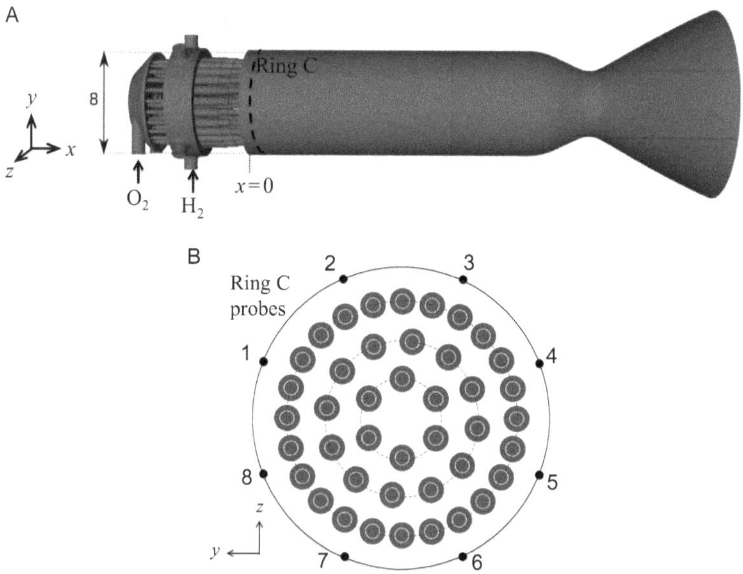

Fig. 77 BKD experiment operated at DLR Lampoldshausen (Gröning et al., 2013, 2014). (A) Full geometry. (B) Injector pattern.

Although a significant amount of work has been devoted to this topic (Culick and Kuentzmann, 2006; Hardi et al., 2014; Harrje and Reardon, 1972), providing useful data for engineering design, the prediction of thermoacoustic instabilities is still an open issue.

An important step was made by Urbano et al. (2016) who applied LES to the BKD experiment, a H_2–O_2 thrust chamber with 42 coaxial injectors (Gröning et al., 2016), operating at supercritical conditions and exhibiting both stable and unstable modes. Geometrical details are given in Fig. 77, which also shows the injector pattern and the location of the experimental pressure transducers, C_1 to C_8 (Fig. 77B). Two operating points were computed: LP1 was a stable case, while LP4 exhibited self-excited oscillations. The major differences between LP1 and LP4 are the chamber pressures (respectively 70 and 80 bars), the oxidizer to fuel ratios (ROF) (respectively 4 and 6), and the power (respectively 66 and 86.2 MW). Both cases are supercritical, requiring real-gas thermodynamics based on a cubic equation of state (Schmitt et al., 2010).

LES was performed on a 70-million element mesh (Fig. 78), using the WALE model for the SGS stress tensor (Ducros et al., 1998) and equilibrium chemistry. Multicomponent real-gas thermodynamics and transport were

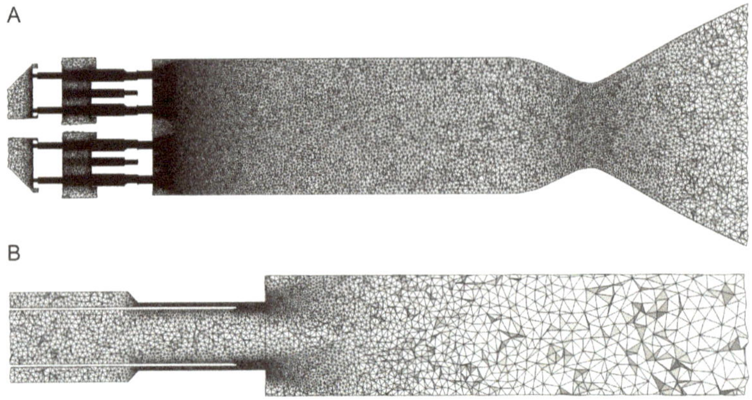

Fig. 78 Unstructured mesh for the LES of the BKD experiment (Urbano et al., 2016). (A) Mesh overview. (B) Closeup on the injection region.

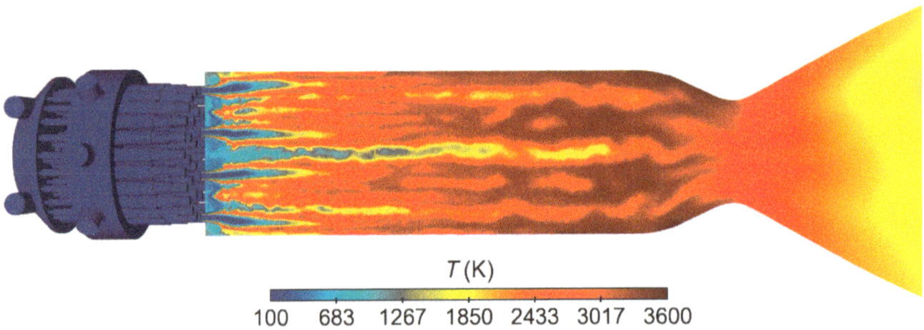

Fig. 79 Longitudinal cut of instantaneous temperature for LP4 (Urbano et al., 2016).

described with the Soave–Redlich–Kwong equation of state (Soave, 1972) and the corresponding-state model of Chung et al. (1988).

A longitudinal slice of the instantaneous temperature field is shown in Fig. 79. Because of the high reactivity of hydrogen, a diffusion flame is stabilized right at the injector lip and expands rapidly downstream. The flames are relatively long because of the delay of mass transfer from the dense oxygen to its lighter surroundings. Note that some cold pockets of unburnt gases reach the nozzle, indicating that combustion is not complete.

To trigger possible instabilities, a pressure perturbation of amplitude Δp is applied initially. In agreement with the experimental data, in the LP1 case the imposed perturbation decays after a short period regardless of the initial amplitude, indicating that case is stable. For LP4 however oscillations increase and eventually reach a limit cycle for sufficient initial amplitude ($\Delta p > 11\%$ of the chamber pressure). A frequency analysis of the pressure signal shows several peaks in the spectral density (Fig. 80). A strong peak

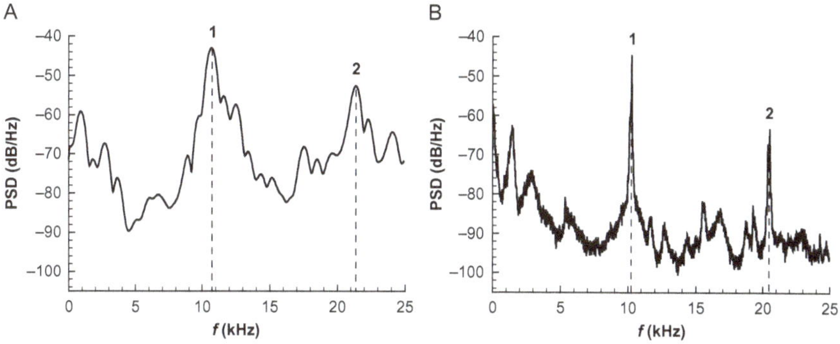

Fig. 80 PSD of the pressure perturbation for LP4 (Δp = 10 bars). Comparison with experimental spectra (raw experimental data courtesy of DLR, processed with the same tools as the LES results (Urbano et al., 2016)). (A) LP4-LES: f_1 = 10, 700 Hz, f_2 = 21, 400 Hz. (B) LP4 experiment: f_1 = 10, 260 Hz, f_2 = 20, 500 Hz.

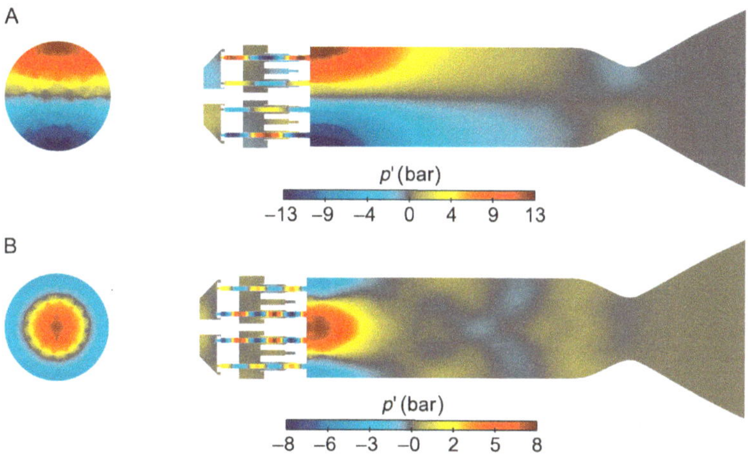

Fig. 81 Fields of power spectral density (PSD) for the two dominant frequencies of LP4 triggered to a limit cycle (Urbano et al., 2016). (A) f_1 = 10, 700 Hz. (B) f_2 = 21, 400 Hz.

at the frequency 10, 700 Hz corresponds to the first transverse mode and matches very well the experiment frequency of 10, 260 Hz. A second peak is also clearly visible at 21, 400 Hz in the LES and 20, 500 Hz in the experiment.

The fields of pressure oscillation associated to the peak frequencies are plotted in Fig. 81. As expected the first frequency corresponds to the first transverse mode of the chamber, but it appears to be coupled with a longitudinal mode of the oxygen injectors, which is also supported by experimental findings (Gröning et al., 2016). The nodal line, initially aligned with the y-axis, is marginally shifted, which indicates a well-defined standing nature. At the second peak frequency, an intense longitudinal acoustic activity in the injectors appears. Because the inner and outer rings of injectors are out of phase, the mode in the chamber has a radial shape in the first part of the chamber. Its amplitude is attenuated rather rapidly in the axial direction.

7. INCLUDING MORE PHYSICS

The examples above only consider gaseous combustion in adiabatic configurations. However most engines burn liquid fuel, which is easier to store and more stable. This introduces additional physical phenomena, such as liquid atomization, evaporation, liquid films on walls, and complex flame structures and therefore has a strong effect on the engine behavior. Heat transfer is another critical process for the engine operation and the thermal fatigue and is difficult to evaluate as the exchanges with the engine environment are usually unknown. Strategies to address these issues are briefly discussed and illustrated below.

7.1 Two-Phase Combustion

The liquid phase is injected in the combustion chamber in the form of a diluted spray, usually through pressurized injectors in ICE or airblast injectors in GTs. The atomization of a liquid jet is a very complex problem which is not yet fully understood and very difficult to model (Fuster et al., 2009; Menard et al., 2007; Tomar et al., 2010). Simulating the atomization process in LES of real engine is very demanding in computing time and still contains high uncertainty. For this reason, it is often preferred to skip this mechanism and directly introduce a spray, in the form of a prescribed droplet velocity and size distribution, taken, for example, from measurements, or guessed from the injector characteristics (Sanjosé et al., 2011).

The spray then interacts with the ambient turbulence in a two-way coupling mode; i.e., droplets are influenced by the turbulent motion, but droplets also impact turbulence through the drag force. Droplet/turbulence interaction is characterized by the Stokes number $St = \tau_d/\tau_f$, where τ_f is a characteristic flow time and $\tau_d = \rho_d d^2/18\mu$ is the droplet relaxation time, with ρ_d the droplet density, d its diameter, and μ the gaseous dynamic viscosity. High Stokes numbers correspond to very inertial droplets, which are little deviated from their initial trajectory by the gaseous flow. On the contrary, small Stokes numbers characterize a tracer-like behavior, where the droplet very rapidly takes the gaseous velocity. A particular behavior

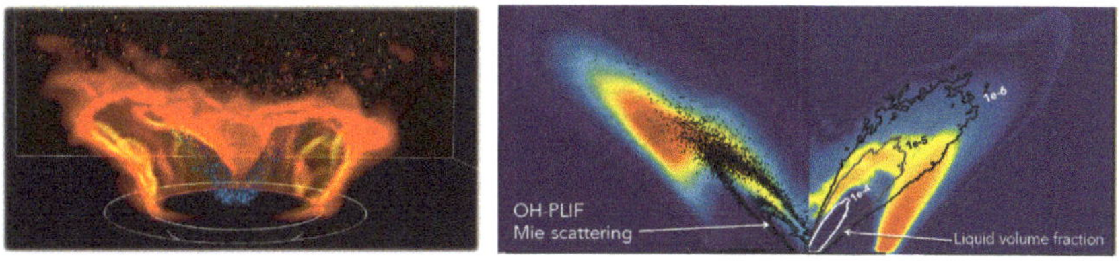

Fig. 82 Two-phase turbulent swirled flame, configuration of Kariuki et al. (2012). LES result, from Paulhiac (2015) (*left*) and comparison with experiment (*right*): field of heat release compared to OH PLIF (*blue* = 10^6; *red* = 10^8, log color scale) and droplets.

called preferential segregation is observed near unity Stokes number, where droplets tend to concentrate at the periphery of vortices (Masi et al., 2014; Réveillon and Demoulin, 2007).

Droplet evaporation and burning finally occur mostly inside the flame front, where the hot gas heats up the liquid almost instantaneously to its boiling temperature. Spray flames exhibit various regimes, mainly depending on the droplet number density (Réveillon and Vervisch, 2005). High number densities favor a flame stabilization at the border of the spray, while isolated droplets may burn individually. The droplet evaporation time also plays a key role, in particular if droplets are not fully preevaporated prior to reaching the flame front. Fig. 82 gives an illustration of typical flames found in swirled burners, obtained by Paulhiac (2015), with an M-shape, premixed flame stabilizing around the hollow-cone evaporating spray, and droplets crossing the flame front and burning in the hot products, either individually or in a partially premixed mode.

When droplets evaporate inside the reacting zone, they alter the flame structure as they modify the balance between fuel vapor release by evaporation and fuel consumption by combustion. This balance is also sensitive to the slip velocity between the liquid and the gas phase. Fig. 83 gives an example of such flame propagating through a mixture of air and fuel droplets, where the varying equivalence ratio through the flame strongly modifies the species and temperature profiles compared to a classical premixed flame. Note that the case of simultaneous evaporation and combustion leads to both a higher temperature and a higher fuel vapor in the burnt gas compared to the prevaporized case, which could never happen in a purely gaseous flame.

Two formalisms exist for the numerical description of a two-phase disperse phase. In the Lagrangian formalism, particle trajectories are tracked individually and a set of equations for the position, momentum, mass,

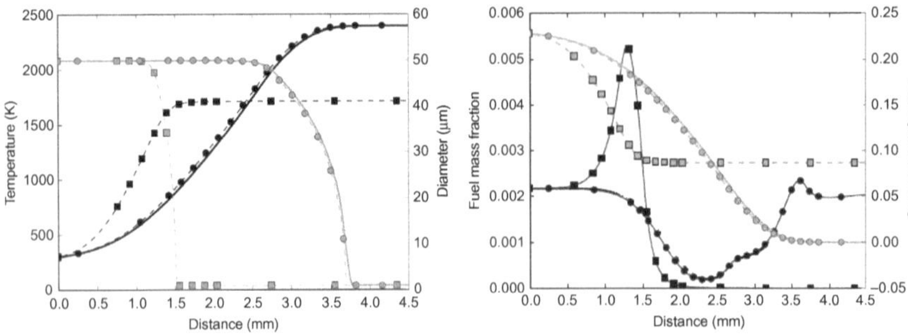

Fig. 83 Temperature and species profiles in a 1D flame propagating through a mixture of air and fuel droplets with nonzero relative velocity. *Squares*, preevaporated case; *circles*, simultaneous evaporation and combustion (Paulhiac, 2015).

and enthalpy are solved for each droplet. Exchanges of mass, momentum, and energy with the gaseous phase, which is discretized on a grid, require to interpolate between grid nodes and droplet position. With this formulation, droplets may be viewed either as physical droplets, which does not need any additional model but implies to track all droplets, or as statistical particles in a Monte-Carlo-like approach, which may allow to reduce the number of calculated droplets but requires stochastic closures. The other formalism is based on an Eulerian description of the liquid phase, which arises from the statistical averaging of the spray. In this approach, statistical moments are calculated as solutions of continuous transport equations which are very similar to the gaseous phase equations. The cost is therefore independent of the number of droplets, but the averaging operation leads to unclosed terms which require modeling. Detailed description and comparison of both formalisms may be found in Senoner et al. (2009) and Jaegle et al. (2011). A general conclusion is that Lagrangian formalisms are more difficult in terms of numerics and parallelism, while Eulerian formalisms are less accurate due to the averaging and closure models. There is no absolute advantage of one formalism over the other, and the choice depends on the application and the objectives of the study.

Fig. 84 shows a typical result of a two-phase burner LES. The simulation used the Lagrangian formulation, which allows a polydisperse spray to form, evaporate and burn. The turbulent flame is computed with the TFLES model (see Section 3) which is able to take into account the liquid phase without further modeling. It is usually observed that the presence of the spray favors flame lift-off and complex flame structures, combining premixed and nonpremixed combustion regimes.

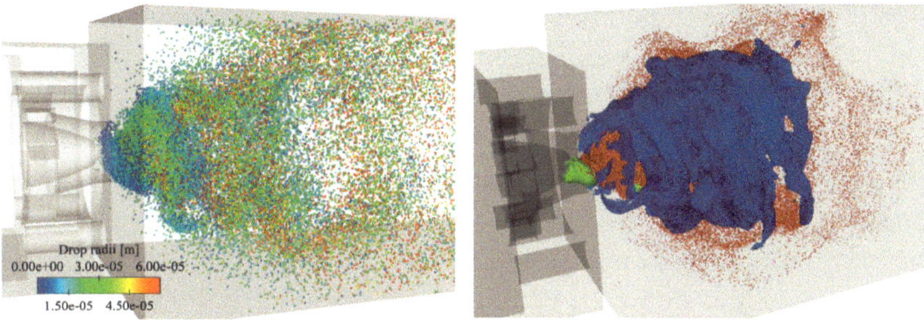

Fig. 84 LES of two-phase combustion in the MERCATO burner of ONERA (Hannebique et al., 2013). *Left*: View of the Lagrangian particles. *Right*: Iso-surface of pressure highlighting the PVC (*green*), iso-surface of heat release (*blue*), and spray (*red*).

7.2 The Thermal Problem: Adding Solid Conduction and Thermal Radiation

A major technical issue in combustion systems is the thermal fatigue. High heat loads on walls may deteriorate the mechanical properties of the solid material and even create cracks which may endanger the integrity of the system. This issue is even more critical when a flame gets into contact with the wall, generating not only high heat fluxes but also chemical reactions with the wall material. To avoid such catastrophic events, efficient cooling systems must be designed using air or other available cooling fluids (as, for example, the cryogenic fuel in rocket engines). One consequence is that the exact thermal environment of the combustion chamber is often unknown, and must be guessed in the form of wall temperature or wall heat flux. This at the end has a strong impact on the combustor behavior and limits the prediction capabilities of the simulation. To alleviate this source of uncertainty, an efficient strategy is to include all thermal phenomena in the burner simulation, by simultaneously solving the coupled problems of combustion, heat conduction in its solid parts, and thermal radiation, taking into account the external environment of the engine (Amaya et al., 2010b; Errera and Chemin, 2004; Poitou et al., 2012).

The heat exchange between a fluid and a solid takes place at the solid surface, where energy is brought by convection, conduction, and radiation in the gas phase. Radiation is known to have little influence on the flame in combustion chambers, where the small volume of hot gas limits the radiative source term to only a few percents of the flame power (contrary, for example, to fires where radiation is the main controlling mechanism due to the large volume of hot gas). However, thermal radiation has been shown to play a significant role in pollutant emission such as soot and NO_x, and on the wall heat flux (Amaya et al., 2010a; Duchaine et al., 2008).

The energy exchange with the solid wall occurs in a very thin layer near the wall, which implies a fine resolution to capture the correct velocity and temperature gradients. An alternative is to construct a model which predicts the wall fluxes from quantities available in the coarse mesh. DNS is often used to study and derives such models for LES and RANS applications, and numerous versions of so-called wall models or wall laws are available in the literature. The impact of radiation on the wall heat flux has been addressed only very recently in combustion systems (Zhang et al., 2013). Models for thermal radiation are usually classified depending on their representation of the gas absorption spectra, from the 1D gray gas approximation (Viskanta, 1966) to the more complex *SNBcK* model studied (Soufiani and Djavdan, 1994), with intermediate solutions such as *FS-SNBcK* or tabulation techniques (Poitou et al., 2009).

The full thermal problem is then the combination of three different physical phenomena, which have very different time and scale characteristics. Heat conduction in solids is much slower than convection and heat diffusion in turbulent flows, itself much slower again compared to thermal radiation which can be considered instantaneous. As a consequence, the most efficient strategy to include these physics in a single simulation is to couple different codes, developed independently and optimized for each physical phenomenon. Coupling different solvers is however a difficult numerical and computational challenge. It involves interpolation techniques and must guarantee stability and well-posedness of the coupled system, while preserving the accuracy of each solver. A special effort is also required for the efficient use of massively parallel machines, which implies to transfer data from a distributed interface to another distributed interface in parallel and on a very large number of processors while maintaining the scalability. Dedicated softwares exist, as, for example, the open-source software Open-Palm, co-developed by CERFACS and ONERA (http://www.cerfacs.fr/globc/PALM_WEB). Using MPI technology, the software is able to define complex coupling algorithms and to provide efficient and flexible data transfer between the codes.

Fig. 85 shows an example of result obtained with LES of combustion coupled to heat transfer considering solid conduction and thermal radiation. Such calculation leads to an accurate prediction of the temperature on the blade surfaces, which is a critical parameter for the design of the turbine.

Fig. 85 Determination of heat loads in an industrial combustion chamber obtained by coupling simulations of reactive fluid dynamics, radiation, and conduction (Amaya et al., 2010a).

8. CONCLUSIONS

As it is able to predict turbulent combustion with unrivalled accuracy and without prior parameter calibration, LES has deeply changed the landscape in this field. As was shown in this chapter, it is today widely used as a research tool because it provides detailed information on the physical processes and allows a better understanding of the complexity of turbulent reacting flows. Its prediction accuracy has also modified the relation between experiment and simulation, which now goes beyond the simple validation of calculations, and combines the two complementary approaches for an improved analysis of the studied system.

The LES methodology developed for turbulent combustion in the last decade is now sufficiently mature to be extended to any kind of turbulent reacting flows, and in particular those encountered in the field of chemical processing. An example is given in Fig. 86, showing an LES result of chemical conversion in a turbulent ribbed pipe (Zhu et al., 2014).

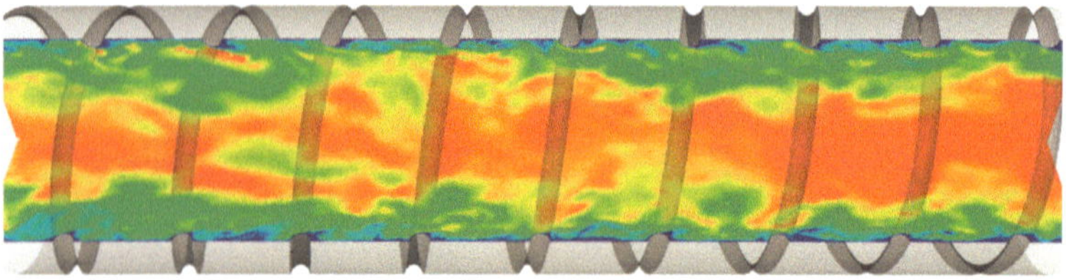

Fig. 86 LES of chemical conversion in a ribbed pipe (Zhu et al., 2014).

REFERENCES

Abdel-Gayed RG, Bradley D: Criteria for turbulent propagation limits of premixed flames, *Comb Flame* 62:61–68, 1985.

Ahmed SF, Mastorakos E: Spark ignition of lifted turbulent jet flames, *Comb Flame* 146:215–231, 2006.

Ahmed SF, Balachandran R, Marchione T, Mastorakos E: Spark ignition of turbulent nonpremixed bluff-body flames, *Comb Flame* 151:366–385, 2007.

Alauzet F, Frey PJ, George PL, Mohammadi B: 3D transient fixed point mesh adaptation for time-dependent problems: application to CFD simulations, *J Comp Phys* 222(2):592–623, 2007.

Albouze G, Poinsot T, Gicquel LYM: Chemical kinetics modelisation and LES combustion model effects on a perfectly premixed burner, *C R Acad Sci Mécanique* 337(6–7):318–328, 2009.

Amaya J, Cabrit O, Poitou D, Cuenot B, El Hafi M: Unsteady coupling of Navier-Stokes and radiative heat transfer solvers applied to an anisothermal multicomponent turbulent channel flow, *J Quant Spec Rad Transf* 111(2):295–301, 2010a. http://dx.doi.org/10.1016/j.jqsrt.2009.06.014.

Amaya J, Collado E, Cuenot B, Poinsot T: Coupling LES, radiation and structure in gas turbine simulations. In *Proceedings of summer program*, 2010b, Center for Turbulence Research, pp 239–249.

Angelberger C, Veynante D, Egolfopoulos F, Poinsot T: Large eddy simulations of combustion instabilities in premixed flames. In *Proceedings of the summer program*, 1998, Center for Turbulence Research, NASA Ames/Stanford University, pp 61–82.

Arcoumanis C, Godwin SN, Kim JW: Effect of tumble strength on combustion and exhaust emissions in a single-cylinder, four-valve, spark-ignition engine, 1998.

Baum E, Peterson B, Böhm B, Dreizler A: On the validation of LES applied to internal combustion flows: Part 1: Comprehensive experimental database, *Flow Turb Comb* 92:269–297, 2014.

Blint RJ: The relationship of the laminar flame width to flame speed, *Combust Sci Technol* 49:79–92, 1986.

Boersma BJ: Direct numerical simulation of turbulent pipe flow up to a Reynolds number of 61,000, *J Phys Confer Ser* 318(4):042045, 2011.

Boger M, Veynante D, Boughanem H, Trouvé A: Direct numerical simulation analysis of flame surface density concept for large eddy simulation of turbulent premixed combustion. In *27th symposium (international) on combustion, Boulder*, 1998, The Combustion Institute, Pittsburgh, pp 917–927.

Boileau M, Staffelbach G, Cuenot B, Poinsot T, Bérat C: LES of an ignition sequence in a gas turbine engine, *Comb Flame* 154(1–2):2–22, 2008. http://www.cerfacs.fr/≃cfdbib/repository/TR_CFD_07_112.pdf.

Boudier G, Gicquel LYM, Poinsot T, Bissières D, Bérat C: Comparison of LES, RANS and experiments in an aeronautical gas turbine combustion chamber, *Proc Combust Inst* 31:3075–3082, 2007.

Bourgouin J-F, Durox D, Schuller T, Beaunier J, Candel S: Ignition dynamics of an annular combustor equipped with multiple swirling injectors, *Comb Flame* 160(8):1398–1413, 2013.

Boussinesq J: Théorie de l'écoulement tourbillant, *Mém Présentés par Divers Savants Acad Sci Inst Fr* 23:46–50, 1877.

Boyce MP: *Gas turbine engineering handbook,* Oxford, UK, 2011, Elsevier.

Butler TD, O'Rourke PJ: A numerical method for two-dimensional unsteady reacting flows. In *16th symposium (international) on combustion*, 1977, The Combustion Institute, pp 1503–1515.

Celik I, Yavuz I, Smirnov A: Large eddy simulations of in-cylinder turbulence for internal combustion engines: a review, *Int J Engine Res* 2(2):119–148, 2001.

Charlette F, Veynante D, Meneveau C: A power-law wrinkling model for LES of premixed turbulent combustion: part I—non-dynamic formulation and initial tests, *Comb Flame* 131:159–180, 2002.

Chung TH, Ajlan M, Lee LL, Starling KE: Generalized multiparameter correlation for nonpolar and polar fluid transport properties, *Ind Eng Chem* 27(4):671–679, 1988.

Cockburn B, Hou S, Shu C: The Runge-Kutta local projection discontinuous Galerkin finite element method for conservation laws. IV. The multidimensional case, *Math Comput* 54(190):545–581, 1990.

Colin, O., 2000. Simulations aux grandes échelles de la combustion turbulente prémélangée dans les statoréacteurs. PhD Thesis, INP Toulouse.

Colin O, Rudgyard M: Development of high-order Taylor-Galerkin schemes for unsteady calculations, *J Comput Phys* 162(2):338–371, 2000.

Colin O, Ducros F, Veynante D, Poinsot T: A thickened flame model for large eddy simulations of turbulent premixed combustion, *Phys Fluids* 12(7):1843–1863, 2000.

Colucci PJ, Jaberi FA, Givi P, Pope SB: Filtered density function for large eddy simulation of turbulent reacting flows, *Phys Fluids* 10:499–515, 1998.

Crocco L: Aspects of combustion instability in liquid propellant rocket motors. Part I, *J Am Rocket Soc* 21:163–178, 1951.

Crocco L: Aspects of combustion instability in liquid propellant rocket motors. Part II, *J Am Rocket Soc* 22:7–16, 1952.

Crocco L: Theoretical studies on liquid propellant rocket instability. In The Combustion Institute, editors: *10th symposium (international) on combustion*, 1965, The Combustion Institute, Pittsburgh, pp 1101–1128.

Cuenot B, Poinsot T: Effects of curvature and unsteadiness in diffusion flames. Implications for turbulent diffusion flames, *Proc Combust Inst* 25:1383–1390, 1994.

Cuenot, B, Felden, A. Riber, E: Effect of the chemistry description on LES of a realistic swirled non-premixed combustor, *Proc Combust Inst* **36**, 2016 (in press).

Culick FEC: Combustion instabilities in liquid-fueled propulsion systems—an overview. In *AGARD 72B PEP meeting*, 1987.

Culick FE, Kuentzmann P: *Unsteady motions in combustion chambers for propulsion systems,* 2006, Technical report, DTIC Document.

Doerr T: Introduction to aero-engine gas turbine combustion. In Von Karman Institute for Fluid Dynamics, editor: *Lecture series 2012-04*, Rhode-Saint-Genèse, 2012, Von Karman Institute for Fluid Dynamics.

Dopazo C: Recent developments in pdf methods. In Libby PA, Williams FA, editors: *Turbulent reacting flows*, London, 1994, Academic Press, pp 375–474.

Duchaine F, Mendez S, Nicoud F, Corpron A, Moureau V, Poinsot T: Coupling heat transfer solvers and large eddy simulations. In *Proceedings of the summer program*, 2008, Center for Turbulence Research, Stanford University.

Ducros F, Nicoud F, Poinsot T: Wall-adapting local eddy-viscosity models for simulations in complex geometries. In Baines MJ, editor: *Numerical Methods for Fluid Dynamics VI*, Oxford, 1998, ICFD, pp 293–300.

Echeckki T, Mastorakos E: Advances, new trends and perspectives. In Echeckki T, Mastorakos E, editors: *Turbulent combustion modeling, Fluid mechanics and its applications*, vol. 95, 2011, Springer.

Enaux B, Granet V, Vermorel O, et al: Large eddy simulation of a motored single-cylinder piston engine: numerical strategies and validation, *Flow Turb Comb* 86(2):153–177, 2011.

Errera M, Chemin S: A fluid-solid thermal coupling applied to an effusion cooling system. In *34th AIAA fluid dynamics conference and exhibit, Portland, Oregon*, , 2004.

Esclapez, L., 2015. Numerical study of ignition and inter-sector flame propagation in gas turbine. PhD Thesis, Toulouse, INPT.

Esclapez L, Riber E, Cuenot B: Ignition probability of a partially premixed burner using LES, *Proc Combust Inst* 35(3):3133–3141, 2015.

Eyssartier A, Cuenot B, Gicquel L, Poinsot T: Using LES to predict ignition sequences and ignition probability of turbulent two-phase flames, *Comb Flame* 160(7):1191–1207, 2013.

Falese, M., 2013. A study of the effects of bifurcations in swirling flows using large eddy simulation and mesh adaptation. PhD Thesis, Toulouse, INPT.

Fiorina B, Gicquel O, Veynante D: Turbulent flame simulation taking advantage of tabulated chemistry self-similar properties, *Proc Combust Inst* 32:1687–1694, 2009.

Fiorina B, Vicquelin R, Auzillon P, Darabiha N, Gicquel O, Veynante D: A filtered tabulated chemistry model for LES of premixed combustion, *Comb Flame* 157(3):465–475, 2010.

Franzelli B, Riber E, Sanjose M, Poinsot T: A two-step chemical scheme for kerosene-air premixed flames, *Comb Flame* 157(7):1364–1373, 2010.

Franzelli B, Riber E, Gicquel LYM, Poinsot T: Large-eddy simulation of combustion instabilities in a lean partially premixed swirled flame, *Comb Flame* 159(2):621–637, 2012.

Fuster D, Bagué A, Boeck T, et al: Simulation of primary atomization with an octree adaptive mesh refinement and VOF method, *Int J Multiphase Flow* 35(6):550–565, 2009.

Galpin J, Naudin A, Vervisch L, Angelberger C, Colin O, Domingo P: Large-eddy simulation of a fuel-lean premixed turbulent swirl-burner, *Comb Flame* 155(1–2):247–266, 2008.

Germano M: A proposal for a redefinition of the turbulent stresses in the filtered Navier-Stokes equations, *Phys Fluids* 29(7):2323–2324, 1986. http://dx.doi.org/10.1063/1.865568.

Gicquel O, Darabiha N, Thévenin D: Laminar premixed hydrogen/air counterflow flame simulations using flame prolongation of ILDM with differential diffusion, *Proc Combust Inst* 28:1901–1908, 2000.

Goryntsev D, Klein M, Sadiki A, Janicka J: Large eddy simulation of fuel-air mixing in a direct injection SI engine. In 5th symposium on turb. and shear flow phenomena, *Munich, Germany*, , 2007.

Gourdain N, Gicquel LYM, Montagnac M, et al: High performance parallel computing of flows in complex geometries—part 1: methods, *Comput Sci Discov* 2, 2009. 015003 (26 pp).

Gourdain N, Gicquel LYM, Staffelbach G, et al: High performance parallel computing of flows in complex geometries—part 2: applications, *Comput Sci Discov* 2, 2009. 015004 (28 pp).

Goussis DA, Maas U: Model reduction for combustion chemistry. In *Turb Comb Model* 2011, Springer, pp 193–220.

Granet, V., 2011. La Simulation aux Grandes Echelles: un outil pour la prédiction des variabilités cycliques dans les moteurs à allumage commandé. PhD Thesis, Toulouse, INPT.

Granet V, Vermorel O, Lacour C, Enaux B, Dugué V, Poinsot T: Large-eddy simulation and experimental study of cycle-to-cycle variations of stable and unstable operating points in a spark ignition engine, *Comb Flame* 159:1562–1575, 2012.

Gröning S, Suslov D, Oschwald M, Sattelmayer T: Stability behaviour of a cylindrical rocket engine combustion chamber operated with liquid hydrogen and liquid oxygen. In *5th European conference for aerospace sciences (EUCASS)*, 2013.
Gröning S, Suslov D, Hardi J, Oschwald M: Influence of hydrogen temperature on the acoustics of a rocket engine combustion chamber operated with LOX/H2 at representative conditions. In *Proceedings of space propulsion*, 2014.
Gröning S, Hardi J, Suslov D, Oschwald M: Injector-driven combustion instabilities in a hydrogen/oxygen rocket combustor, *J Propuls Power* 32:560–573, 2016.
Hall MG: Vortex breakdown, *Annu Rev Fluid Mech* 4:195–217, 1972.
Hannebique G, Sierra P, Riber E, Cuenot B: Large eddy simulation of reactive two-phase flow in an aeronautical multipoint burner, *Flow Turbul Combust* 90(2):449–469, 2013.
Hardi J, Beinke S, Oschwald M, Dally B: Coupling of cryogenic oxygen-hydrogen flames to longitudinal and transverse acoustic instabilities, *J Propuls Power* 30(4):991–1004, 2014.
Harrje DJ, Reardon FH: *Liquid propellant rocket instability,* 1972, NASA. Technical report SP-194.
Hartmann R, Houston P: Error estimation and adaptive mesh refinement for aerodynamic flows. In Deconinck H, editor: *VKI LS 2010-01: 36th CFD/ADIGMA course on HP-adaptive and HP-multigrid methods, October 26–30, 2009*, Rhode-Saint-Genèse, 2010, Von Karman Institute for Fluid Dynamics.
Hawkes ER, Cant SR: A flame surface density approach to large eddy simulation of premixed turbulent combustion. In The Combustion Institute, editors: *28th symposium (international) on combustion*, 2000, The Combustion Institute, Pittsburgh, pp 51–58.
Haworth D: Large-eddy simulation of in-cylinder flows, *Oil Gas Sci Tech Rev l'IFP* 54(2):175–185, 1999.
Hesthaven JS, Warburton T: *Nodal discontinuous Galerkin methods: algorithms, analysis, and applications,* 2008, Springer Science + Business Media, LLC.
Heywood JB: Internal combustion engine fundamentals, McGraw and Hill series in mechanical engineering (New York, 1988, McGraw-Hill.
Hirsch C: Finite volume method and conservative discretization with an introduction to finite element method. In *Numerical computation of internal & external flows: fundamentals of computational fluid dynamics*, ed 2, New York, 2007, John Wiley & Sons, pp 203–248.
Hirschfelder JO, Curtiss CF, Bird RB: *Molecular theory of gases and liquids,* New York, 1969, John Wiley & Sons.
Huynh H: A flux reconstruction approach to high-order schemes including discontinuous Galerkin methods. In *18th AIAA computational fluid dynamics conference, Miami, FL, June 25–28*2007, AIAA, pp 1–42.
Huynh H: A reconstruction approach to high-order schemes including discontinuous Galerkin for diffusion. In *7th AIAA aerospace sciences meeting, Orlando, FL, January 5–8*, vol. 2009–2403, 2009, AIAA, pp 1–34.
Inagaki K, Fuyuto T, Nishikawa K, Nakakita K, Sakata I: Dual-fuel PCI combustion controlled by in-cylinder stratification of ignitability, 2006.
Jaberi F, Colucci PJ, James S, Givi P, Pope SB: Filtered mass density function for large-eddy simulation of turbulent reacting flows, *J Fluid Mech* 401:85–121, 1999.
Jaegle J, Senoner J-M, Garcia M, et al: Eulerian and Lagrangian spray simulations of an aeronautical multipoint injector, *Proc Combust Inst* 33(2):2099–2107, 2011.
James S, Zhu J, Anand M: Large eddy simulation as a design tool for gas turbine combustion systems, *Am Inst Aeronaut Astronaut J* 44:674–686, 2006.
Jameson A: A proof of the stability of the spectral difference method for all orders of accuracy, *J Sci Comput* 45(1):348–358, 2010.
Jaravel, T., 2016. Prediction of pollutants in gas turbines using large eddy simulation. PhD Thesis, Toulouse, INPT.

Jaravel T, Riber E, Cuenot B, Bulat G: Large eddy simulation of an industrial gas turbine combustor using reduced chemistry with accurate pollutant prediction, *Proc Combust Inst* **36**, 2016 (in press).

Kariuki J, Dawson J, Mastorakos E: Measurements in turbulent premixed bluff body flames close to blow-off, *Comb Flame* 159(8):2589–2607, 2012.

Karniadakis G, Sherwin S: *Spectral/hp element methods for CFD,* USA, 1999, Oxford University Press.

Kelley A, Jomaas G, Law CK: Critical radius for sustained propagation of spark-ignited spherical flames, *Comb Flame* 156(5):1006–1013, 2009.

Kim SO, Luong MB, Chen JH, Yoo CS: A DNS study of the ignition of lean PRF/air mixtures with temperature inhomogeneities under high pressure and intermediate temperature, *Comb Flame* 162(3):717–726, 2015.

Kokjohn SL, Hanson RM, Splitter DA, Reitz RD: Experiments and modeling of dual-fuel HCCI and PCCI combustion using in-cylinder fuel blending, 2009.

Kolmogorov AN: The local structure of turbulence in incompressible viscous fluid for very large Reynolds numbers, *C R Acad Sci USSR* 30:301, 1941.

Kopriva D: A conservative staggered-grid Chebyshev multidomain method for compressible flows. II. A semi-structured method, *J Comput Phys* 128(2):475–488, 1996.

Kravchenko AG, Moin P: On the effect of numerical errors in large eddy simulations of turbulent flows, *J Comput Phys* 131(2):310–322, 1997. http://dx.doi.org/10.1006/jcph.1996.5597.

Kuo KK: *Principles of combustion,* ed 2, Hoboken, NJ, 2005, John Wiley & Sons, Inc.

Labarrere, L., 2016. Etude théorique et numérique de la combustion isochore appliquée au cas du thermoréacteur. PhD Thesis, Toulouse, INPT.

Lacour C, Pera C, Enaux B, Vermorel O, Angelberger C, Poinsot T: Exploring cyclic variability in a spark-ignition engine using experimental techniques, system simulation and large-eddy simulation. In *Proceedings of the 4th European combustion meeting,* , 2009.

Lam SH, Goussis DA: The CSP method for simplifying kinetics, *Int J Chem Kinet* 26(4):461–486, 1994.

Landenfeld T, Sadiki A, Janicka J: A turbulence-chemistry interaction model based on a multivariate presumed beta-pdf method for turbulent flames, *Flow Turbul Combust* 68(2):111–135, 2002.

Lefebvre AH: *Gas turbines combustion,* Abigdon, 1999, Taylor & Francis.

Légier J-Ph, Poinsot T, Veynante D: Dynamically thickened flame LES model for premixed and non-premixed turbulent combustion. In *Proceeding of the summer program,* 2000, Center for Turbulence Research, NASA Ames/Stanford University, pp 157–168.

Loevas T: Automatic generation of skeletal mechanisms for ignition combustion based on level of importance analysis, *Comb Flame* 156(7):1348–1358, 2009.

Lu T, Law CK: Linear time reduction of large kinetic mechanisms with directed relation graph: n-heptane and iso-octane, *Comb Flame* 144(1):24–36, 2006.

Lu T, Law CK: A criterion based on computational singular perturbation for the identification of quasi steady state species: a reduced mechanism for methane oxidation with NO chemistry, *Comb Flame* 154(4):761–774, 2008.

Lumley JL: *Engines: an introduction,* Cambridge, 1999, Cambridge University Press.

Maas U, Pope SB: Implementation of simplified chemical kinetics based on low-dimensional manifolds, *Proc Combust Inst* 24:719–729, 1992.

MacIsaac B, Langton R: *Gas turbine propulsion systems,* vol. 49, Hoboken, 2011, John Wiley & Sons.

Marble F: Servo-stabilization of low-frequency oscillations in a liquid bipropellant rocket motor, *J Am Rocket Soc* 23(2):63–74, 1953.

Masi E, Simonin O, Riber E, Sierra P, Gicquel L: Development of an algebraic-closure-based moment method for unsteady Eulerian simulations of particle-laden turbulent flows in very dilute regime, *Int J Multiphase Flow* 58:257–278, 2014.

Mastorakos E: Ignition of turbulent non-premixed flames, *Prog Energy Comb Sci* 35(1):57–97, 2009.
Meier W, Weigand P, Duan XR, Giezendanner-Thoben R: Detailed characterization of the dynamics of thermoacoustic pulsations in a lean premixed swirl flame, *Comb Flame* 150(1–2):2–26, 2007.
Menard T, Tanguy S, Berlemont A: Coupling level set/VOF/ghost fluid methods: Validation and application to 3D simulation of the primary break-up of a liquid jet, *Int J Multiphase Flow* 33:510–524, 2007.
Moureau V, Lartigue G, Sommerer Y, Angelberger C, Colin O, Poinsot T: Numerical methods for unsteady compressible multi-component reacting flows on fixed and moving grids, *J Comput Phys* 202(2):710–736, 2005. http://dx.doi.org/10.1016/j.jcp.2004.08.003.
Moureau V, Domingo P, Vervisch L: From large-eddy simulation to direct simulation of a lean premixed swirl flame: filtered laminar flame-PDF modeling, *Comb Flame* 158(7):1340–1357, 2011. http://dx.doi.org/10.1016/j.combustflame.2010.12.004.
Narayanaswamy K, Blanquart G, Pitsch H: A consistent chemical mechanism for oxidation of substituted aromatic species, *Comb Flame* 157(10):1879–1898, 2010.
Neophytou A, Richardson ES, Mastorakos E: Spark ignition of turbulent recirculating non-premixed gas and spray flames: a model for predicting ignition probability, *Comb Flame* 159(4):1503–1522, 2012.
Nicholas J: *The jet engine,* ed 6, Hoboken, 2005, John Wiley & Sons.
Nicoud F, Hubert Baya Toda H, Cabrit O, Bose S, Lee J: Using singular values to build a subgrid-scale model for large eddy simulations, *Phys Fluids* 23, 2011(paper no. 085106).
Novick AS, Troth DL: Low NO_x heavy fuel combustor concept program. In *AAS/Division of dynamical astronomy meeting*, vol. 1, 1981.
Oefelein J, Yang V: Comprehensive review of liquid-propellant combustion instabilities in F-1 engines, *J Propuls Power* 9(5):657–677, 1993.
Paulhiac, D., 2015. Modélisation de la combustion d'un spray dans un bruleur aéronautique. PhD Thesis, Toulouse, INPT.
Philip M, Boileau M, Vicquelin R, et al: Large eddy simulations of the ignition sequence of an annular multiple-injector combustor, *Proc Combust Inst* 35(3):3159–3166, 2015.
Pilling MJ, Turanyi T, Tomlin AS: On the error of the quasi-steady-state approximation, *J Phys Chem* 97:163–172, 1993.
Pitsch H: Large eddy simulation of turbulent combustion, *Annu Rev Fluid Mech* 38:453–482, 2006.
Poinsot T, Lele S: Boundary conditions for direct simulations of compressible viscous flows, *J Comput Phys* 101(1):104–129, 1992. http://dx.doi.org/10.1016/0021-9991(92)90046-2.
Poinsot T, Veynante D: *Theoretical and numerical combustion,* 2005, RT Edwards, Inc.
Poinsot T, Misdariis A, Vermorel O: LES of knocking in engines using dual heat transfer and two-step reduced schemes, *Comb Flame* 162:4304–4312, 2015.
Poitou D, Amaya J, Singh CB, Joseph D, Hafi ME, Cuenot B: Validity limits for the global model FS-SNBcK for combustion applications, *Proceedings of Eurotherm83—computational thermal radiation in participating media III,* 2009.
Poitou D, Amaya J, El Hafi M, Cuenot B: Analysis of the interaction between turbulent combustion and thermal radiation using unsteady coupled LES/DOM simulations, *Comb Flame* 159(4):1605–1618, 2012.
Pope SB: Pdf methods for turbulent reactive flows, *Prog Energy Comb Sci* 19(11):119–192, 1985.
Reitz RD, Duraisamy G: Review of high efficiency and clean reactivity controlled compression ignition (RCCI) combustion in internal combustion engines, *Prog Energy Combust Sci* 46:12–71, 2015.
Rayleigh L: The explanation of certain acoustical phenomena, *Nature* 18:319–321, 1878.
Réveillon J, Demoulin F-X: Effects of the preferential segregation of droplets on evaporation and turbulent mixing, *J Fluid Mech* 583:273–302, 2007.

Réveillon J, Vervisch L: Analysis of weakly turbulent diluted-spray flames and spray combustion regimes, *J Fluid Mech* 537:317–347, 2005.
Richard S, Colin O, Vermorel O, Benkenida A, Angelberger C, Veynante D: Towards large eddy simulation of combustion in spark ignition engines, *Proc Combust Inst* 31:3059–3066, 2007.
Roux S, Lartigue G, Poinsot T, Meier U, Bérat C: Studies of mean and unsteady flow in a swirled combustor using experiments, acoustic analysis and large eddy simulations, *Comb Flame* 141:40–54, 2005.
Sagaut P: *Large eddy simulation for incompressible flows,* 2002, Springer.
Sagaut P, Deck S: Large eddy simulation for aerodynamics: status and perspectives, *Philos Trans R Soc Lond A Math Phys Eng Sci* 367(1899):2849–2860, 2009.
Sanjosé M, Senoner J, Jaegle F, Cuenot B, Moreau S, Poinsot T: Fuel injection model for Euler-Euler and Euler-Lagrange large-eddy simulations of an evaporating spray inside an aeronautical combustor, *Int J Multiphase Flow* 37(5):514–529, 2011.
Schmitt T, Selle L, Ruiz A, Cuenot B: Large-eddy simulation of supercritical-pressure round jets, *AIAA J* 48(9):2133–2144, 2010.
Senoner J, Sanjosé M, Lederlin T, et al: Eulerian and Lagrangian large-eddy simulations of an evaporating two-phase flow, *C R Mécanique* 337(6):458–468, 2009.
Smagorinsky J: General circulation experiments with the primitive equations: 1. The basic experiment, *Mon Weather Rev* 91:99–164, 1963.
Soave G: Equilibrium constants from a modified Redlich-Kwong equation of state, *Chem Eng Sci* 27:1197–1203, 1972.
Soufiani A, Djavdan E: A comparison between weighted sum of gray gases and statistical narrow-band radiation models for combustion applications, *Comb Flame* 97(2): 240–250, 1994.
Summerfield M: A theory of unstable combustion in liquid propellant rocket systems, *J Am Rocket Soc* 21(5):108–114, 1951.
Sun Y, Wang Z, Liu Y: High-order multidomain spectral difference method for the Navier-Stokes equations on unstructured hexahedral grids, *Commun Comput Phys* 2(2):310–333, 2007.
Sutton GP, Biblarz O: *Rocket propulsion elements,* Hoboken, 2010, John Wiley & Sons.
Syred N, Beer JM: Combustion in swirling flows: a review, *Comb Flame* 23(2):143–201, 1974.
Tacina RR: *Combustor technology for future aircraft*, 1990. NASA technical memorandum 103268, AIAA-90-2400.
Thobois L, Lauvergne R, Poinsot T: Using LES to investigate reacting flow physics in engine design process, *SAE*, 2007 (paper no. 2007-01-0166).
Toledo MS, Le Penven L, Buffat M, Cadiou A, Padilla J: Large eddy simulation of the generation and breakdown of a tumbling flow, *Int J Heat Fluid Flow* 28(1):113–126, 2007.
Tomar G, Fuster D, Zaleski S, Popinet S: Multiscale simulations of primary atomization, *Comput Fluids* 39(10):1864–1874, 2010.
Tomlin AS, Pilling MJ, Turányi T, Merkin JH, Brindley J: Mechanism reduction for the oscillatory oxidation of hydrogen: sensitivity and quasi-steady-state analyses, *Comb Flame* 91(2):107–130, 1992.
Tsien HS: Servo stabilization of combustion in rocket motors, *J Am Rocket Soc* 22(5):256–262, 1952.
Urbano A, Selle L, Staffelbach G, et al: Exploration of combustion instability triggering using large eddy simulation of a multiple injector liquid rocket engine, *Comb Flame* 169:129–140, 2016.
Vermorel O, Richard S, Colin O, Angelberger C, Benkenida A, Veynante D: Multi-cycle LES simulations of flow and combustion in a PFI SI 4-valve production engine, *SAE* (2007-01-0151), 2007.

Vermorel O, Richard S, Colin O, Angelberger C, Benkenida A, Veynante D: Towards the understanding of cyclic variability in a spark ignited engine using multi-cycle LES, *Comb Flame* 156(8):1525–1541, 2009.
Veynante D, Vervisch L: Turbulent combustion modeling, *Prog Energy Comb Sci* 28:193–266, 2002.
Veynante D, Fiorina B, Domingo P, Vervisch L: Using self-similar properties of turbulent premixed flames to downsize chemical tables in high-performance numerical simulations, *Combust Theory Model* 12(6):1055–1088, 2008.
Vicquelin R, Fiorina B, Darabiha N, Gicquel O, Veynante D: Coupling tabulated chemistry with large eddy simulation of turbulent reactive flows, *C R Mécanique* 337(6):329–339, 2009.
Vincent P, Castonguay P, Jameson A: A new class of high-order energy stable flux reconstruction schemes, *J Sci Comput* 47(1):1–23, 2010.
Viskanta R: Radiation transfer and interaction of convection with radiation heat transfer, *Adv Heat Transf* 3:175–251, 1966.
Wang G, Boileau M, Veynante D: Implementation of a dynamic thickened flame model for large eddy simulations of turbulent premixed combustion, *Comb Flame* 158(11):2199–2213, 2011.
Weigand P, Meier W, Duan XR, Stricker W, Aigner M: Investigations of swirl flames in a gas turbine model combustor: I. flow field, structures, temperature, and species distributions, *Comb Flame* 144(1–2):205–224, 2006.
Young V: *Liquid rocket engine combustion instability,* vol. 169, Washington, DC, 1995, AIAA.
Zhang Y, Vicquelin R, Gicquel O, Taine J: Physical study of radiation effects on the boundary layer structure in a turbulent channel flow, *Int J Heat Mass Transf* 61:654–666, 2013.
Zhu M, Riber E, Cuenot B, Bodart J, Poinsot T: Wall-resolved large eddy simulation in refinery ribbed pipes. In *Proceedings of the summer program*, 2014, p 457.

INDEX

Note: Page numbers followed by "*f*" indicate figures, and "*t*" indicate tables.

D

E

F

G

H

I

Q

R

S

CONTENTS OF VOLUMES IN THIS SERIAL

Volume 1 (1956)
J. W. Westwater, *Boiling of Liquids*
A. B. Metzner, *Non-Newtonian Technology: Fluid Mechanics, Mixing, and Heat Transfer*
R. Byron Bird, *Theory of Diffusion*
J. B. Opfell and B. H. Sage, *Turbulence in Thermal and Material Transport*
Robert E. Treybal, *Mechanically Aided Liquid Extraction*
Robert W. Schrage, *The Automatic Computer in the Control and Planning of Manufacturing Operations*
Ernest J. Henley and Nathaniel F. Barr, *Ionizing Radiation Applied to Chemical Processes and to Food and Drug Processing*

Volume 2 (1958)
J. W. Westwater, *Boiling of Liquids*
Ernest F. Johnson, *Automatic Process Control*
Bernard Manowitz, *Treatment and Disposal of Wastes in Nuclear Chemical Technology*
George A. Sofer and Harold C. Weingartner, *High Vacuum Technology*
Theodore Vermeulen, *Separation by Adsorption Methods*
Sherman S. Weidenbaum, *Mixing of Solids*

Volume 3 (1962)
C. S. Grove, Jr., Robert V. Jelinek, and Herbert M. Schoen, *Crystallization from Solution*
F. Alan Ferguson and Russell C. Phillips, *High Temperature Technology*
Daniel Hyman, *Mixing and Agitation*
John Beck, *Design of Packed Catalytic Reactors*
Douglass J. Wilde, *Optimization Methods*

Volume 4 (1964)
J. T. Davies, *Mass-Transfer and Inierfacial Phenomena*
R. C. Kintner, *Drop Phenomena Affecting Liquid Extraction*
Octave Levenspiel and Kenneth B. Bischoff, *Patterns of Flow in Chemical Process Vessels*
Donald S. Scott, *Properties of Concurrent Gas–Liquid Flow*
D. N. Hanson and G. F. Somerville, *A General Program for Computing Multistage Vapor–Liquid Processes*

Volume 5 (1964)
J. F. Wehner, *Flame Processes—Theoretical and Experimental*
J. H. Sinfelt, *Bifunctional Catalysts*
S. G. Bankoff, *Heat Conduction or Diffusion with Change of Phase*
George D. Fulford, *The Flow of Lktuids in Thin Films*
K. Rietema, *Segregation in Liquid–Liquid Dispersions and its Effects on Chemical Reactions*

Volume 6 (1966)
S. G. Bankoff, *Diffusion-Controlled Bubble Growth*
John C. Berg, Andreas Acrivos, and Michel Boudart, *Evaporation Convection*
H. M. Tsuchiya, A. G. Fredrickson, and R. Aris, *Dynamics of Microbial Cell Populations*
Samuel Sideman, *Direct Contact Heat Transfer between Immiscible Liquids*
Howard Brenner, *Hydrodynamic Resistance of Particles at Small Reynolds Numbers*

Volume 7 (1968)

Robert S. Brown, Ralph Anderson, and Larry J. Shannon, *Ignition and Combustion of Solid Rocket Propellants*

Knud Østergaard, *Gas–Liquid–Particle Operations in Chemical Reaction Engineering*

J. M. Prausnilz, *Thermodynamics of Fluid–Phase Equilibria at High Pressures*

Robert V. Macbeth, *The Burn-Out Phenomenon in Forced-Convection Boiling*

William Resnick and Benjamin Gal-Or, *Gas–Liquid Dispersions*

Volume 8 (1970)

C. E. Lapple, *Electrostatic Phenomena with Particulates*

J. R. Kittrell, *Mathematical Modeling of Chemical Reactions*

W. P. Ledet and D. M. Himmelblau, *Decomposition Procedures foe the Solving of Large Scale Systems*

R. Kumar and N. R. Kuloor, *The Formation of Bubbles and Drops*

Volume 9 (1974)

Renato G. Bautista, *Hydrometallurgy*

Kishan B. Mathur and Norman Epstein, *Dynamics of Spouted Beds*

W. C. Reynolds, *Recent Advances in the Computation of Turbulent Flows*

R. E. Peck and D. T. Wasan, *Drying of Solid Particles and Sheets*

Volume 10 (1978)

G. E. O'Connor and T. W. F. Russell, *Heat Transfer in Tubular Fluid–Fluid Systems*

P. C. Kapur, *Balling and Granulation*

Richard S. H. Mah and Mordechai Shacham, *Pipeline Network Design and Synthesis*

J. Robert Selman and Charles W. Tobias, *Mass-Transfer Measurements by the Limiting-Current Technique*

Volume 11 (1981)

Jean-Claude Charpentier, *Mass-Transfer Rates in Gas–Liquid Absorbers and Reactors*

Dee H. Barker and C. R. Mitra, *The Indian Chemical Industry—Its Development and Needs*

Lawrence L. Tavlarides and Michael Stamatoudis, *The Analysis of Interphase Reactions and Mass Transfer in Liquid–Liquid Dispersions*

Terukatsu Miyauchi, Shintaro Furusaki, Shigeharu Morooka, and Yoneichi Ikeda, *Transport Phenomena and Reaction in Fluidized Catalyst Beds*

Volume 12 (1983)

C. D. Prater, J, Wei, V. W. Weekman, Jr., and B. Gross, *A Reaction Engineering Case History: Coke Burning in Thermofor Catalytic Cracking Regenerators*

Costel D. Denson, *Stripping Operations in Polymer Processing*

Robert C. Reid, *Rapid Phase Transitions from Liquid to Vapor*

John H. Seinfeld, *Atmospheric Diffusion Theory*

Volume 13 (1987)

Edward G. Jefferson, *Future Opportunities in Chemical Engineering*

Eli Ruckenstein, *Analysis of Transport Phenomena Using Scaling and Physical Models*

Rohit Khanna and John H. Seinfeld, *Mathematical Modeling of Packed Bed Reactors: Numerical Solutions and Control Model Development*

Michael P. Ramage, Kenneth R. Graziano, Paul H. Schipper, Frederick J. Krambeck, and Byung C. Choi, *KINPTR (Mobil's Kinetic Reforming Model): A Review of Mobil's Industrial Process Modeling Philosophy*

Volume 14 (1988)

Richard D. Colberg and Manfred Morari, *Analysis and Synthesis of Resilient Heat Exchange Networks*

Richard J. Quann, Robert A. Ware, Chi-Wen Hung, and James Wei, *Catalytic Hydrometallation of Petroleum*

Kent David, *The Safety Matrix: People Applying Technology to Yield Safe Chemical Plants and Products*

Volume 15 (1990)

Pierre M. Adler, Ali Nadim, and Howard Brenner, *Rheological Models of Suspenions*

Stanley M. Englund, *Opportunities in the Design of Inherently Safer Chemical Plants*

H. J. Ploehn and W. B. Russel, *Interations between Colloidal Particles and Soluble Polymers*

Volume 16 (1991)

Perspectives in Chemical Engineering: Research and Education

Clark K. Colton, *Editor*

Historical Perspective and Overview

L. E. Scriven, *On the Emergence and Evolution of Chemical Engineering*

Ralph Landau, *Academic—industrial Interaction in the Early Development of Chemical Engineering*

James Wei, *Future Directions of Chemical Engineering*

Fluid Mechanics and Transport

L. G. Leal, *Challenges and Opportunities in Fluid Mechanics and Transport Phenomena*

William B. Russel, *Fluid Mechanics and Transport Research in Chemical Engineering*

J. R. A. Pearson, *Fluid Mechanics and Transport Phenomena*

Thermodynamics

Keith E. Gubbins, *Thermodynamics*

J. M. Prausnitz, *Chemical Engineering Thermodynamics: Continuity and Expanding Frontiers*

H. Ted Davis, *Future Opportunities in Thermodynamics*

Kinetics, Catalysis, and Reactor Engineering

Alexis T. Bell, *Reflections on the Current Status and Future Directions of Chemical Reaction Engineering*

James R. Katzer and S. S. Wong, *Frontiers in Chemical Reaction Engineering*

L. Louis Hegedus, *Catalyst Design*

Environmental Protection and Energy

John H. Seinfeld, *Environmental Chemical Engineering*

T. W. F. Russell, *Energy and Environmental Concerns*

Janos M. Beer, Jack B. Howard, John P. Longwell, and Adel F. Sarofim, *The Role of Chemical Engineering in Fuel Manufacture and Use of Fuels*

Polymers

Matthew Tirrell, *Polymer Science in Chemical Engineering*

Richard A. Register and Stuart L. Cooper, *Chemical Engineers in Polymer Science: The Need for an Interdisciplinary Approach*

Microelectronic and Optical Material

Larry F. Thompson, *Chemical Engineering Research Opportunities in Electronic and Optical Materials Research*

Klavs F. Jensen, *Chemical Engineering in the Processing of Electronic and Optical Materials: A Discussion*

Bioengineering

James E. Bailey, *Bioprocess Engineering*

Arthur E. Humphrey, *Some Unsolved Problems of Biotechnology*

Channing Robertson, *Chemical Engineering: Its Role in the Medical and Health Sciences*

Process Engineering

Arthur W. Westerberg, *Process Engineering*

Manfred Morari, *Process Control Theory: Reflections on the Past Decade and Goals for the Next*

James M. Douglas, *The Paradigm After Next*

George Stephanopoulos, *Symbolic Computing and Artificial Intelligence in Chemical Engineering: A New Challenge*
The Identity of Our Profession
Morton M. Denn, *The Identity of Our Profession*

Volume 17 (1991)
Y. T. Shah, *Design Parameters for Mechanically Agitated Reactors*
Mooson Kwauk, *Particulate Fluidization: An Overview*

Volume 18 (1992)
E. James Davis, *Microchemical Engineering: The Physics and Chemistry of the Microparticle*
Selim M. Senkan, *Detailed Chemical Kinetic Modeling: Chemical Reaction Engineering of the Future*
Lorenz T. Biegler, *Optimization Strategies for Complex Process Models*

Volume 19 (1994)
Robert Langer, *Polymer Systems for Controlled Release of Macromolecules, Immobilized Enzyme Medical Bioreactors, and Tissue Engineering*
J. J. Linderman, P. A. Mahama, K. E. Forsten, and D. A. Lauffenburger, *Diffusion and Probability in Receptor Binding and Signaling*
Rakesh K. Jain, *Transport Phenomena in Tumors*
R. Krishna, *A Systems Approach to Multiphase Reactor Selection*
David T. Allen, *Pollution Prevention: Engineering Design at Macro-, Meso-, and Microscales*
John H. Seinfeld, Jean M. Andino, Frank M. Bowman, Hali J. L. Forstner, and Spyros Pandis, *Tropospheric Chemistry*

Volume 20 (1994)
Arthur M. Squires, *Origins of the Fast Fluid Bed*
Yu Zhiqing, *Application Collocation*
Youchu Li, *Hydrodynamics*
Li Jinghai, *Modeling*
Yu Zhiqing and Jin Yong, *Heat and Mass Transfer*
Mooson Kwauk, *Powder Assessment*
Li Hongzhong, *Hardware Development*
Youchu Li and Xuyi Zhang, *Circulating Fluidized Bed Combustion*
Chen Junwu, Cao Hanchang, and Liu Taiji, *Catalyst Regeneration in Fluid Catalytic Cracking*

Volume 21 (1995)
Christopher J. Nagel, Chonghum Han, and George Stephanopoulos, *Modeling Languages: Declarative and Imperative Descriptions of Chemical Reactions and Processing Systems*
Chonghun Han, George Stephanopoulos, and James M. Douglas, *Automation in Design: The Conceptual Synthesis of Chemical Processing Schemes*
Michael L. Mavrovouniotis, *Symbolic and Quantitative Reasoning: Design of Reaction Pathways through Recursive Satisfaction of Constraints*
Christopher Nagel and George Stephanopoulos, *Inductive and Deductive Reasoning: The Case of Identifying Potential Hazards in Chemical Processes*
Keven G. Joback and George Stephanopoulos, *Searching Spaces of Discrete Soloutions: The Design of Molecules Processing Desired Physical Properties*

Volume 22 (1995)
Chonghun Han, Ramachandran Lakshmanan, Bhavik Bakshi, and George Stephanopoulos, *Nonmonotonic Reasoning: The Synthesis of Operating Procedures in Chemical Plants*
Pedro M. Saraiva, *Inductive and Analogical Learning: Data-Driven Improvement of Process Operations*

Alexandros Koulouris, Bhavik R. Bakshi and George Stephanopoulos, *Empirical Learning through Neural Networks: The Wave-Net Solution*
Bhavik R. Bakshi and George Stephanopoulos, *Reasoning in Time: Modeling, Analysis, and Pattern Recognition of Temporal Process Trends*
Matthew J. Realff, *Intelligence in Numerical Computing: Improving Batch Scheduling Algorithms through Explanation-Based Learning*

Volume 23 (1996)
Jeffrey J. Siirola, *Industrial Applications of Chemical Process Synthesis*
Arthur W. Westerberg and Oliver Wahnschafft, *The Synthesis of Distillation-Based Separation Systems*
Ignacio E. Grossmann, *Mixed-Integer Optimization Techniques for Algorithmic Process Synthesis*
Subash Balakrishna and Lorenz T. Biegler, *Chemical Reactor Network Targeting and Integration: An Optimization Approach*
Steve Walsh and John Perkins, *Operability and Control inn Process Synthesis and Design*

Volume 24 (1998)
Raffaella Ocone and Gianni Astarita, *Kinetics and Thermodynamics in Multicomponent Mixtures*
Arvind Varma, Alexander S. Rogachev, Alexandra S. Mukasyan, and Stephen Hwang, *Combustion Synthesis of Advanced Materials: Principles and Applications*
J. A. M. Kuipers and W. P. Mo, van Swaaij, *Computional Fluid Dynamics Applied to Chemical Reaction Engineering*
Ronald E. Schmitt, Howard Klee, Debora M. Sparks, and Mahesh K. Podar, *Using Relative Risk Analysis to Set Priorities for Pollution Prevention at a Petroleum Refinery*

Volume 25 (1999)
J. F. Davis, M. J. Piovoso, K. A. Hoo, and B. R. Bakshi, *Process Data Analysis and Interpretation*
J. M. Ottino, P. DeRoussel, S., Hansen, and D. V. Khakhar, *Mixing and Dispersion of Viscous Liquids and Powdered Solids*
Peter L. Silverston, Li Chengyue, Yuan Wei-Kang, *Application of Periodic Operation to Sulfur Dioxide Oxidation*

Volume 26 (2001)
J. B. Joshi, N. S. Deshpande, M. Dinkar, and D. V. Phanikumar, *Hydrodynamic Stability of Multiphase Reactors*
Michael Nikolaou, *Model Predictive Controllers: A Critical Synthesis of Theory and Industrial Needs*

Volume 27 (2001)
William R. Moser, Josef Find, Sean C. Emerson, and Ivo M, Krausz, *Engineered Synthesis of Nanostructure Materials and Catalysts*
Bruce C. Gates, *Supported Nanostructured Catalysts: Metal Complexes and Metal Clusters*
Ralph T. Yang, *Nanostructured Absorbents*
Thomas J. Webster, *Nanophase Ceramics: The Future Orthopedic and Dental Implant Material*
Yu-Ming Lin, Mildred S. Dresselhaus, and Jackie Y. Ying, *Fabrication, Structure, and Transport Properties of Nanowires*

Volume 28 (2001)
Qiliang Yan and Juan J. DePablo, *Hyper-Parallel Tempering Monte Carlo and Its Applications*
Pablo G. Debenedetti, Frank H. Stillinger, Thomas M. Truskett, and Catherine P. Lewis, *Theory of Supercooled Liquids and Glasses: Energy Landscape and Statistical Geometry Perspectives*
Michael W. Deem, *A Statistical Mechanical Approach to Combinatorial Chemistry*

Venkat Ganesan and Glenn H. Fredrickson, *Fluctuation Effects in Microemulsion Reaction Media*
David B. Graves and Cameron F. Abrams, *Molecular Dynamics Simulations of Ion–Surface Interactions with Applications to Plasma Processing*
Christian M. Lastoskie and Keith E, Gubbins, *Characterization of Porous Materials Using Molecular Theory and Simulation*
Dimitrios Maroudas, *Modeling of Radical-Surface Interactions in the Plasma-Enhanced Chemical Vapor Deposition of Silicon Thin Films*
Sanat Kumar, M. Antonio Floriano, and Athanassiors Z. Panagiotopoulos, *Nanostructured Formation and Phase Separation in Surfactant Solutions*
Stanley I. Sandler, Amadeu K. Sum, and Shiang-Tai Lin, *Some Chemical Engineering Applications of Quantum Chemical Calculations*
Bernhardt L. Trout, *Car-Parrinello Methods in Chemical Engineering: Their Scope and potential*
R. A. van Santen and X. Rozanska, *Theory of Zeolite Catalysis*
Zhen-Gang Wang, *Morphology, Fluctuation, Metastability and Kinetics in Ordered Block*
Copolymers

Volume 29 (2004)
Michael V. Sefton, *The New Biomaterials*
Kristi S. Anseth and Kristyn S. Masters, *Cell–Material Interactions*
Surya K. Mallapragada and Jennifer B. Recknor, *Polymeric Biomaterias for Nerve Regeneration*
Anthony M. Lowman, Thomas D. Dziubla, Petr Bures, and Nicholas A. Peppas, *Structural and Dynamic Response of Neutral and Intelligent Networks in Biomedical Environments*
F. Kurtis Kasper and Antonios G. Mikos, *Biomaterials and Gene Therapy*
Balaji Narasimhan and Matt J. Kipper, *Surface-Erodible Biomaterials for Drug Delivery*

Volume 30 (2005)
Dionisio Vlachos, *A Review of Multiscale Analysis: Examples from System Biology, Materials Engineering, and Other Fluids-Surface Interacting Systems*
Lynn F. Gladden, M.D. Mantle and A.J. Sederman, *Quantifying Physics and Chemistry at Multiple Length-Scales using Magnetic Resonance Techniques*
Juraj Kosek, Frantisek Steěpánek, and Miloš Marek, Modelling of Transport and Transformation
Processes in Porous and Multiphase Bodies
Vemuri Balakotaiah and Saikat Chakraborty, *Spatially Averaged Multiscale Models for Chemical Reactors*

Volume 31 (2006)
Yang Ge and Liang-Shih Fan, *3-D Direct Numerical Simulation of Gas–Liquid and Gas–Liquid–Solid Flow Systems Using the Level-Set and Immersed-Boundary Methods*
M.A. van der Hoef, M. Ye, M. van Sint Annaland, A.T. Andrews IV, S. Sundaresan, and J.A.M. Kuipers, *Multiscale Modeling of Gas-Fluidized Beds*
Harry E.A. Van den Akker, *The Details of Turbulent Mixing Process and their Simulation*
Rodney O. Fox, *CFD Models for Analysis and Design of Chemical Reactors*
Anthony G. Dixon, Michiel Nijemeisland, and E. Hugh Stitt, *Packed Tubular Reactor Modeling and Catalyst Design Using Computational Fluid Dynamics*

Volume 32 (2007)
William H. Green, Jr., *Predictive Kinetics: A New Approach for the 21st Century*
Mario Dente, Giulia Bozzano, Tiziano Faravelli, Alessandro Marongiu, Sauro Pierucci and Eliseo Ranzi, *Kinetic Modelling of Pyrolysis Processes in Gas and Condensed Phase*
Mikhail Sinev, Vladimir Arutyunov and Andrey Romanets, *Kinetic Models of C1–C4 Alkane Oxidation as Applied to Processing of Hydrocarbon Gases: Principles, Approaches and Developments*
Pierre Galtier, *Kinetic Methods in Petroleum Process Engineering*

Volume 33 (2007)

Shinichi Matsumoto and Hirofumi Shinjoh, *Dynamic Behavior and Characterization of Automobile Catalysts*

Mehrdad Ahmadinejad, Maya R. Desai, Timothy C. Watling and Andrew P.E. York, *Simulation of Automotive Emission Control Systems*

Anke Güthenke, Daniel Chatterjee, Michel Weibel, Bernd Krutzsch, Petr Kočí, Miloš Marek, Isabella Nova and Enrico Tronconi, *Current Status of Modeling Lean Exhaust Gas Aftertreatment Catalysts*

Athanasios G. Konstandopoulos, Margaritis Kostoglou, Nickolas Vlachos and Evdoxia Kladopoulou, *Advances in the Science and Technology of Diesel Particulate Filter Simulation*

Volume 34 (2008)

C.J. van Duijn, Andro Mikelić, I.S. Pop, and Carole Rosier, *Effective Dispersion Equations for Reactive Flows with Dominant Peclet and Damkohler Numbers*

Mark Z. Lazman and Gregory S. Yablonsky, *Overall Reaction Rate Equation of Single-Route Complex Catalytic Reaction in Terms of Hypergeometric Series*

A.N. Gorban and O. Radulescu, *Dynamic and Static Limitation in Multiscale Reaction Networks, Revisited*

Liqiu Wang, Mingtian Xu, and Xiaohao Wei, *Multiscale Theorems*

Volume 35 (2009)

Rudy J. Koopmans and Anton P.J. Middelberg, *Engineering Materials from the Bottom Up – Overview*

Robert P.W. Davies, Amalia Aggeli, Neville Boden, Tom C.B. McLeish, Irena A. Nyrkova, and Alexander N. Semenov, *Mechanisms and Principles of 1 D Self-Assembly of Peptides into β*-Sheet Tapes

Paul van der Schoot, *Nucleation and Co-Operativity in Supramolecular Polymers*

Michael J. McPherson, Kier James, Stuart Kyle, Stephen Parsons, and Jessica Riley, *Recombinant Production of Self-Assembling Peptides*

Boxun Leng, Lei Huang, and Zhengzhong Shao, *Inspiration from Natural Silks and Their Proteins*

Sally L. Gras, *Surface- and Solution-Based Assembly of Amyloid Fibrils for Biomedical and Nanotechnology Applications*

Conan J. Fee, *Hybrid Systems Engineering: Polymer-Peptide Conjugates*

Volume 36 (2009)

Vincenzo Augugliaro, Sedat Yurdakal, Vittorio Loddo, Giovanni Palmisano, and Leonardo Palmisano, *Determination of Photoadsorption Capacity of Polychrystalline TiO2 Catalyst in Irradiated Slurry*

Marta I. Litter, *Treatment of Chromium, Mercury, Lead, Uranium, and Arsenic in Water by Heterogeneous Photocatalysis*

Aaron Ortiz-Gomez, Benito Serrano-Rosales, Jesus Moreira-del-Rio, and Hugo de-Lasa, Mineralization of Phenol in an Improved Photocatalytic Process Assisted with Ferric Ions: *Reaction Network and Kinetic Modeling*

R.M. Navarro, F. del Valle, J.A. Villoria de la Mano, M.C. Alvarez-Galván, and J.L.G. Fierro, *Photocatalytic Water Splitting Under Visible Light: Concept and Catalysts Development*

Ajay K. Ray, *Photocatalytic Reactor Configurations for Water Purification: Experimentation and Modeling*

Camilo A. Arancibia-Bulnes, Antonio E. Jiménez, and Claudio A. Estrada, *Development and Modeling of Solar Photocatalytic Reactors*

Orlando M. Alfano and Alberto E. Cassano, *Scaling-Up of Photoreactors: Applications to Advanced Oxidation Processes*

Yaron Paz, *Photocatalytic Treatment of Air: From Basic Aspects to Reactors*

Volume 37 (2009)

S. Roberto Gonzalez A., Yuichi Murai, and Yasushi Takeda, *Ultrasound-Based Gas–Liquid Interface Detection in Gas–Liquid Two-Phase Flows*

Z. Zhang, J. D. Stenson, and C. R. Thomas, *Micromanipulation in Mechanical Characterisation of Single Particles*

Feng-Chen Li and Koichi Hishida, *Particle Image Velocimetry Techniques and Its Applications in Multiphase Systems*
J. P. K. Seville, A. Ingram, X. Fan, and D. J. Parker, *Positron Emission Imaging in Chemical Engineering*
Fei Wang, Qussai Marashdeh, Liang-Shih Fan, and Richard A. Williams, *Electrical Capacitance, Electrical Resistance, and Positron Emission Tomography Techniques and Their Applications in Multi-Phase Flow Systems*
Alfred Leipertz and Roland Sommer, *Time-Resolved Laser-Induced Incandescence*

Volume 38 (2009)
Arata Aota and Takehiko Kitamori, *Microunit Operations and Continuous Flow Chemical Processing*
Anıl Ağıral and Han J.G.E. Gardeniers, *Microreactors with Electrical Fields*
Charlotte Wiles and Paul Watts, *High-Throughput Organic Synthesis in Microreactors*
S. Krishnadasan, A. Yashina, A.J. deMello and J.C. deMello, *Microfluidic Reactors for Nanomaterial Synthesis*

Volume 39 (2010)
B.M. Kaganovich, A.V. Keiko and V.A. Shamansky, *Equilibrium Thermodynamic Modeling of Dissipative Macroscopic Systems*
Miroslav Grmela, *Multiscale Equilibrium and Nonequilibrium Thermodynamics in Chemical Engineering*
Prasanna K. Jog, Valeriy V. Ginzburg, Rakesh Srivastava, Jeffrey D. Weinhold, Shekhar Jain, and Walter G. Chapman, *Application of Mesoscale Field-Based Models to Predict Stability of Particle Dispersions in Polymer Melts*
Semion Kuchanov, *Principles of Statistical Chemistry as Applied to Kinetic Modeling of Polymer-Obtaining Processes*

Volume 40 (2011)
Wei Wang, Wei Ge, Ning Yang and Jinghai Li, *Meso-Scale Modeling—The Key to Multi-Scale CFD Simulation*
Pil Seung Chung, Myung S. Jhon and Lorenz T. Biegler, *The Holistic Strategy in Multi-Scale Modeling*
Milo D. Meixell Jr., Boyd Gochenour and Chau-Chyun Chen, *Industrial Applications of Plant-Wide Equation-Oriented Process Modeling—2010*
Honglai Liu, Ying Hu, Xueqian Chen, Xingqing Xiao and Yongmin Huang, *Molecular Thermodynamic Models for Fluids of Chain-Like Molecules, Applications in Phase Equilibria and Micro-Phase Separation in Bulk and at Interface*

Volume 41 (2012)
Torsten Kaltschmitt and Olaf Deutschmann, *Fuel Processing for Fuel Cells*
Adam Z.Weber, Sivagaminathan Balasubramanian, and Prodip K. Das, *Proton Exchange Membrane Fuel Cells*
Keith Scott and Lei Xing, *Direct Methanol Fuel Cells*
Su Zhou and Fengxiang Chen, *PEMFC System Modeling and Control*
François Lapicque, Caroline Bonnet, Bo Tao Huang, and Yohann Chatillon, *Analysis and Evaluation of Aging Phenomena in PEMFCs*
Robert J. Kee, Huayang Zhu, Robert J. Braun, and Tyrone L. Vincent, *Modeling the Steady-State and Dynamic Characteristics of Solid-Oxide Fuel Cells*
Robert J. Braun, Tyrone L. Vincent, Huayang Zhu, and Robert J. Kee, *Analysis, Optimization, and Control of Solid-Oxide Fuel Cell Systems*

Volume 42 (2013)
T. Riitonen, V. Eta, S. Hyvärinen, L.J. Jönsson, and J.P. Mikkola, *Engineering Aspects of Bioethanol Synthesis*
R.W. Nachenius, F. Ronsse, R.H. Venderbosch, and W. Prins, *Biomass Pyrolysis*
David Kubička and Vratislav Tukač, *Hydrotreating of Triglyceride-Based Feedstocks in Refineries*

Tapio Salmi, *Chemical Reaction Engineering of Biomass Conversion*
Jari Heinonen and Tuomo Sainio, *Chromatographic Fractionation of Lignocellulosic Hydrolysates*

Volume 43 (2013)
Grégory Francois and Dominique Bonvin, *Measurement-Based Real-Time Optimization of Chemical Processes*
Adel Mhamdi and Wolfgang Marquardt, *Incremental Identification of Distributed Parameter Systems*
Arun K. Tangirala, Siddhartha Mukhopadhyay, and Akhilananand P. Tiwari, *Wavelets Applications in Modeling and Control*
Santosh K. Gupta and Sanjeev Garg, *Multiobjective Optimization Using Genetic Algorithm*

Volume 44 (2014)
Xue-Qing Gong, Li-Li Yin, Jie Zhang, Hai-Feng Wang, Xiao-Ming Cao, Guanzhong Lu, and Peijun Hu, *Computational Simulation of Rare Earth Catalysis*
Zhi-Jun Sui, Yi-An Zhu, Ping Li, Xing-Gui Zhou, and De Chen, *Kinetics of Catalytic Dehydrogenation of Propane over Pt-Based Catalysts*
Zhen Liu, Xuelian He, Ruihua Cheng, Moris S. Eisen, Minoru Terano, Susannah L. Scott, and Boping Liu, *Chromium Catalysts for Ethylene Polymerization and Oligomerization*
Ayyaz Ahmad, Xiaochi Liu, Li Li, and Xuhong Guo, *Progress in Polymer Nanoreactors: Spherical Polyelectrolyte Brushes*

Volume 45 (2014)
M.P. Dudukovic and P.L. Mills, *Challenges in Reaction Engineering Practice of Heterogeneous Catalytic Systems*
Claudia Diehm, Hüsyein Karadeniz, Canan Karakaya, Matthias Hettel, and Olaf Deutschmann, *Spatial Resolution of Species and Temperature Profiles in Catalytic Reactors:* In Situ *Sampling Techniques and CFD Modeling*
John Mantzaras, *Catalytic Combustion of Hydrogen, Challenges, and Opportunities*
Ivo Roghair, Fausto Gallucci, and Martin van Sint Annaland, *Novel Developments in Fluidized Bed Membrane Reactor Technology*

Volume 46 (2015)
Wolfgang Peukert, Doris Segets, Lukas Pflug, and Günter Leugering, *Unified Design Strategies for Particulate Products*
Stefan Heinrich, Maksym Dosta, and Sergiy Antonyuk, *Multiscale Analysis of a Coating Process in a Wurster Fluidized Bed Apparatus*
Johan T. Padding, Niels G. Deen, E.A.J.F. (Frank) Peters, and J.A.M. (Hans) Kuipers, *Euler–Lagrange Modeling of the Hydrodynamics of Dense Multiphase Flows*
Qinfu Hou, Jieqing Gan, Zongyan Zhou, and Aibing Yu, *Particle Scale Study of Heat Transfer in Packed and Fluidized Beds*
Ning Yang, *Mesoscale Transport Phenomena and Mechanisms in Gas–Liquid Reaction Systems*
Harry E. A. Van den Akker, *Mesoscale Flow Structures and Fluid–Particle Interactions*

Volume 47 (2015)
Shuangliang Zhao, Yu Liu, Xueqian Chen, Yuxiang Lu, Honglai Liu, and Ying Hu, *Unified Framework of Multiscale Density Functional Theories and Its Recent Applications*
Linghong Lu, Xuebo Quan, Yihui Dong, Gaobo Yu, Wenlong Xie, Jian Zhou, Licheng Li, Xiaohua Lu, and Yudan Zhu, *Surface Structure and Interaction of Surface/Interface Probed by Mesoscale Simulations and Experiments*

Kai Wang, Jianhong Xu, Guotao Liu, and Guangsheng Luo, *Role of Interfacial Force on Multiphase Microflow—An Important Meso-Scientific Issue*
Wei Wang and Yanpei Chen, *Mesoscale Modeling: Beyond Local Equilibrium Assumption for Multiphase Flow*
Mao Ye, Hua Li, Yinfeng Zhao, Tao Zhang, and Zhongmin Liu, *MTO Processes Development: The Key of Mesoscale Studies*
Mingquan Shao, Youwei Li, Jianfeng Chen, and Yi Zhang, *Mesoscale Effects on Product Distribution of Fischer–Tropsch Synthesis*

Volume 48 (2016)

Jérémi Dauchet, Jean-François Cornet, Fabrice Gros, Matthieu Roudet, and C.-Gilles Dussap, *Photobioreactor Modeling and Radiative Transfer Analysis for Engineering Purposes*
Laurent Pilon and Razmig Kandilian, *Interaction Between Light and Photosynthetic Microorganisms*
Matthias Schirmer and Clemens Posten, *Modeling of Microalgae Bioprocesses*
Marcel Janssen, *Microalgal Photosynthesis and Growth in Mass Culture*
Jeremy Pruvost, Francois Le Borgne, Arnaud Artu, Jean-François Cornet, and Jack Legrand, *Industrial Photobioreactors and Scale-Up Concepts*

Volume 49 (2016)

E. Ranzi, T. Faravelli, and F. Manenti, *Pyrolysis, Gasification, and Combustion of Solid Fuels*
X. Zhou, L.J. Broadbelt, and R. Vinu, *Polymer Pyrolysis Modeling*
Y. Zhang, G. Hu, W. Du, and F. Qian, *Steam Cracking and EDC Furnace Simulation*
B. Cuenot, *Gas Turbines and Engine Simulations*

FSC
www.fsc.org
MIX
Papier | Fördert gute Waldnutzung
FSC® C083411

Kolibri 360 GmbH
Ferdinand-Jühlke-Straße 7
99095 Erfurt, Deutschland
produktsicherheit@kolibri360.de